The Business of Farming

A Guide to Farm Business Management in the Tropics

SECOND EDITION

David T. Johnson

BSc.(Agric.), NDA, MSc.(Trop. Agric. Dev.), FBIM

**MACMILLAN
PUBLISHERS**

First edition 1982
Reprinted 1983
Second Edition 1990

Published by *Macmillan Publishers Ltd*
London and Basingstoke
Associated companies and representatives in Accra,
Auckland, Delhi, Dublin, Gaborone, Hamburg, Harare,
Hong Kong, Kuala Lumpur, Lagos, Manzini, Melbourne,
Mexico City, Nairobi, New York, Singapore, Tokyo

ISBN 0-333-49921-2

Printed in Hong Kong

Contents

Preface

This book has been written to give a broad survey of the whole field of business management as it applies to tropical farming. It is meant not only for students but also to be of practical use to farmers and farm managers by including in one volume material from many sources.

Study of business management not only teaches specific techniques. It has wider educational value as it promotes precise, logical thinking. It also helps to develop sound attitudes towards economy of both time and materials. The need for sound planning is stressed. This should increase initiative and organising ability — two essential aspects for successful management.

After working as an agricultural extension officer and agronomist and before becoming a consultant in tropical agriculture, I taught farm business management at Chibero College of Agriculture in Zimbabwe, Mananga Agricultural Management Centre in Swaziland and at the University of Malawi. During this time I often was asked to recommend a single textbook that adequately covered the whole field of farm business management in the tropics. As no such book seemed to exist, I began preparing this text in 1976, as an integral part of course development for degree and diploma students at the University of Malawi.

For those who have ploughed through many 'farm management' books already and been confused by all the graphs, response surfaces, isoquants, production functions and Greek symbols, I say, 'Join the Club. Don't worry.' They often baffle me and, even if clear, I rarely find them to be of more than academic interest. In any case, such books, whatever their titles, are mainly about production economics, not farm business management as a whole, and few of them are based on tropical conditions.

During its development, farm management gave birth to a child: production economics. In the last two or three decades, agricultural economics has become ever more technical and specialised and has moved away from its earlier 'muddy boots' connection with farm management towards 'black coated' academic economics. The time has come to wean this demanding child so that its parent can resume its initial aim of improving overall management at farm level.

In their zeal to show the world and their colleagues how much they know about economics, some agricultural economists have tended to make their subject one of making common sense hard to understand. Farm management research has become increasingly concerned with methodology and theory and there has been a rapid growth of elaborate and sophisticated mathematical and econometric models bearing little or no relation to reality. Through the use of complex words and phrases and countless obscure and confusing graphs some agricultural economists have made their subject as difficult as the most intricate foreign language and just as hard to understand and apply to local farm management problems. Consequently, little of their output seems relevant to the practical problems of farm managment and farmers make little use of economic theory for decision-making. They are being joined by a growing army of economists outside academic employment and even by some academics.

Unfortunately, farm management has grown apart from general management studies and stresses different fields of management. It covers only part of the work of farm managers and pays little attention to short-term problems at the individual farm level. More stress in farm management training should be placed on business management techniques developed for industrial and commercial management. If farms, or agri-businesses, are to stay solvent in conditions of the cost/price squeeze, farm managers must adopt and, if necessary, adapt management techniques from industry and commerce.

In this book, I try to cover the whole field of financial, personnel, production and marketing management as it applies to both large-scale, commercial farmers and small-scale, mainly subsistence, farmers in the tropics. Financial and personnel management have been neglected in the standard textbooks, despite their growing importance in relation to technical production. The returns from wise financial and personnel decisions can greatly exceed those from the more traditional production and marketing decisions.

Few managers seem to know what management techniques are available, how they can be used, which of them might be useful or what their long names mean. They are often frightened of them because they are wrapped in an atmosphere of magic, mystery or mathematics.

A management technique is a recognised method for analysing or solving a recognised type of management problem in a detailed systematic way. Only simple techniques, needing no more than the use of an electronic calculator, are described in this book. There are plenty of specialist books and consultants for managers who really need more advanced techniques, such as linear programming, critical path analysis or PERT.

As English will not be the first language of many readers, it has been kept as simple as possible, with any technical terms defined in a glossary at the end of each chapter. Much of today's jargon is unnecessary and many so-called 'new' ideas in management are really old ones expressed in more complicated terms.

The monetary unit used in this book is a nominal dollar currency. This does not reflect any real currency.

Parts of this book were written while I was in the employment of Hunting Technical Services Ltd. The views expressed here are my own and may not reflect the policies of my employer.

To all who have helped, inspired or prodded me, either knowingly or not, to write this book — my thanks. However, responsibility for the end result is solely my own. I am grateful to a director of a firm of international accountants and auditors, who must remain unnamed for professional reasons, and to the following people who provided me with much useful criticism:

J.A.M. Alcock, National Bank of Malawi; P.R. Baily, Agricultural Credit Officer, Ministry of Agriculture, Malawi; F.S. Bardo, Professor, University of Zimbabwe; B. Egner, Rural Development Officer, Barclays Bank of Botswana Ltd; J.E. Harrison, Managing Director, Lugg, Harrison & Associates (Pty) Ltd, Swaziland; J. de Jong, Senior Work Study Specialist, Ministry of Agriculture Zimbabwe; S.P.A. Kelly, Chief Executive, Zimbabwe Tobacco Association; D.H. Lloyd, Senior Lecturer in Agricultural Management, University of Reading; P.T.W. Murphy, Senior Agricultural Economist, Ministry of Agriculture, Zimbabwe; J.R. Parkinson, Senior Economist, Zimbabwe National Farmers' Union; P.J. Scott, Assistant Manager, Lakeshore Rural Development Project, Malawi; A.J. Spear, Chief Engineer, Institute of Agricultural Engineering, Zimbabwe; S. Steyn, Lecturer, University of Zimbabwe; B.B. Wilson, Principal Farm Management Specialist, Ministry of Agriculture, Zimbabwe; F.N. Youdale, Principal, Mananga Agricultural Management Centre, Swaziland.

I am especially indebted to Dr D.A.G. Green, who, when Professor of Agricultural Economics at the University of Malawi, did much to encourage me to begin and persist with the drafting of this book.

D.T.J.
Zalingei, Sudan
June 1981.

Preface to Second Edition

Since completing the manuscript for the first edition, in Sri Lanka six and a half years ago, there have been few changes in the subject of this book; apart from a few words and phrases becoming fashionable. What has occurred, as a result of the recent recession, is that inflation is being reduced in many parts of the world as governments rediscover the virtues of good housekeeping. One effect of this is that experienced managers at all levels and in every sphere of activity have had to redirect and refocus their attention on improving performance and effectiveness.

The first edition was received as expected. Bemused academics, from their cosy studies in the western world, twitched their eyebrows in astonishment.

To my initial surprise, the book received wider recognition and distribution in developed countries than in those for which it was intended. This arose because of foreign, hard currency rationing in so many less developed, low income countries. In 1985, through the good offices of the UK Overseas Development Administration and the British Council, the highly-subsidised English Language Book Society reprint was made available in the target areas. For this I am most grateful.

It has become clear that the readership includes not only students, farmers and managers but also project managers, extension workers, trainers and others in both public and private sectors.

Since writing the first edition, I have worked in Ethiopia, Oman, Sudan — twice, Swaziland, Uganda and Zimbabwe.

For this revised edition I have made minimal alterations to the text but a major revision to the reading lists.

Both I and the publishers would be grateful for comments and criticisms from users on how to improve the usefulness of this book.

D.T. Johnson
Entebbe, Uganda
December 1987.

1 Introduction

WHAT IS MANAGEMENT?

There must be as many definitions of management as there are people who try to define it. This is unfortunate because one cannot assume that everyone — even managers themselves — has the same idea of what management means. Absence of an agreed, authoritative definition of management or its essential activities means that there is no accepted terminology. If a field of study is not to become confused by misunderstandings, the first need is for it to be defined.

The very word 'management' has powerful, emotional effects. It may refer to a group of executives together responsible for efficient control of a business and calling themselves the management. Alternatively, the word may not refer to people at all but to the practice of managing and the skills used. In other contexts, the word refers in a wider sense to planning and policy, to matters of long-term administration. Even mention of the word 'management' can be seen as a comment on an individual's overall ability to run his or her own affairs or ability to farm. It is a comment on the person himself that he might find irritating in a way that technical comments about some aspects of his farm would not be.

Because language is alive and the ideas we wish to express in words are dynamic, together with the fact that the leading management experts come from both sides of the Atlantic, this lack of standard terms is not surprising. However, without a definition of management, how can we know who the managers are? Are some or all supervisors managers?

Care should be taken to distinguish between tools and content. Thus, mathematics, operations research, accounting, production economics, etc., are *tools* of management but not, in themselves, part of its content.

As an activity, management is not confined to profit-seeking businesses nor even to a capitalist economy but is needed in any situation involving complex activity. It is needed in public services, religious orders, police forces and universities even though the managers may be called administrators, town clerks, bishops, commissioners, registrars, etc.

Having stated both that there is no widely agreed definition of management and that a reasonable definition is needed, I must clearly try to provide one. Examination of the many definitions available shows that they have much in common. It therefore is not too hard to compile a reasonable working definition as follows: *Management is the active process of making decisions so that use of the available human and material resources of the organisation is planned and controlled to achieve its long- and short-term aims most efficiently. It is neither art nor science; it is both.*

Good management is the means by which a leader creates a united team that achieves agreed objectives within the required time and with economy in the use of resources.

THE WORK OF THE BUSINESS MANAGER

As with the term 'management' itself, no general agreement exists on the basic functions of a manager or what these should be called. This is because, although similar ideas are common to the work of many writers, no common terms have been used to describe these ideas. Again, certain terms, common to many writers, are used to describe different ideas.

Many writers give as functions of managers items that, in fact, are not functions but usually techniques or tools of managers. A function is a natural, proper or characteristic action. A technique is a method or the details of procedure essential for skilled performance in any field. A tool is an implement or anything that serves as a means to an end. The management cycle can be regarded conveniently as to:

- Plan what to do.
- Organise who should do it and how it should be done.
- Co-ordinate and motivate subordinates to operate the plan.
- Control the achievement of the plan.

Figure 1.1 conveniently sets out the whole field of management to me.

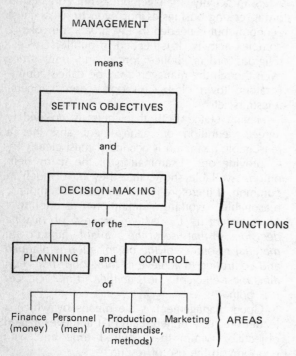

Figure 1.1   The Field of Management

Management begins, or should begin, with setting objectives and the strategic or long-term plans of the business to attain those objectives. For most farming plans a ten-year horizon should be adequate. The objectives of any business need to be much more than a vague philosophy such as 'To maximise profit through commercial farming operations and to minimise risk by diversification'. They need to be clear and

numerical to be used as a basis for planning, organising, motivating and controlling. Useful statements of objectives have four parts:

- An attribute, e.g. return on capital employed.
- A unit of measure, e.g. percentage.
- A quantity, e.g. 20.
- A time unit, e.g. in ten years from now.

A suitable statement might be 'To achieve by December, 1995, profits before tax of $40,000 on a turnover of $250,000 using net assests of $140,000 using constant 1985 prices. Gaining commitment to the objectives is all-important and if they are seen as little more than star-gazing they will soon fall into disrepute.

Decisions must be made later to develop tactical plans for the current and perhaps the next year, that are designed to progress towards the strategic objectives, execute them and exercise control over events. It is mainly responsibility for these functions that sets apart the 'manager', whether he is the owner of the business or not, from the 'non-manager' or operator. The overall duty of all managers is to ensure the best use of resources (people, money and things) to achieve both the short-term and the long-term aims of their organisation.

Many writers list 'organising' as a separate managerial function. By this they mean deciding *who* should implement the plan and, possibly, *how*.

Managers exercise their planning and controlling functions in respect of four identifiable but inter-related business areas — finance, personnel, production and marketing. In farm management, stress has been placed conventionally on production, less on finance and marketing and little attention has been paid to personnel management. Although the average farmer now insists that 'farming is a business', he is still by background, experience, circumstances and inclination, mainly a *producer* to whom the more specialised aspects of business management do not come easily — and often may seem unnecessary. The general tendency today is to lay less stress on the methods of production (husbandry skills), more on financial management and, in particular, more on aspects of personnel management such as organisational structures and staff motivation.

In discussing the areas of management, it is not intended to imply that all managers operate in each of these areas. Managers, like non-managers, may specialise in a specific area or areas, for example, a field manager in production or a company secretary in finance. Most farm managers, however, probably spend about 40, 30, 20 and 10 per cent of their time on production, financial, personnel and marketing management respectively.

A manager is anyone who is responsible for more work than he can achieve with his own hands or brain. He therefore needs the help of others. Managers are those who use formal authority to organise and control other people to co-ordinate their efforts towards the objectives of the organisation. A person is not a manager unless he has this formal authority. All managers, other than sole traders or the ultimate heads of organisations, are themselves managed. The critical difference is between managers and non-managers — not between managers and managed.

THE NEED FOR BUSINESS MANAGEMENT IN FARMING

It is still often thought that to be vague and to adapt to unforeseen or unforeseeable events as they occur is, in some mysterious way, 'practical' while to draw up plans and procedures ahead of anticipated possibilities is somehow 'theoretical'. The theoretical approach is scorned by the 'practical man'. Unfortunately for them, these 'practical' men are finding themselves increasingly out of their depth in modern farming which often involves heavy capital investment and making decisions that will probably affect profit far into the future.

The 'practical man' spends much time repairing his tractors, supervising field labour and generally avoiding paper work.

A good business manager consistently has to make informed decisions over a wide field ranging from economics through agronomy, nutrition and engineering to sociology and psychology. Although he need not be an *expert* in any of these fields, he must be able to obtain and interpret relevant information. An able farm manager must be competent in both the technical activities of production and in business activities such as financial planning, personnel control and marketing. Business ability becomes more important as the size of the business increases — technical specialists can always be obtained for individual enterprises or areas of the business.

An able manager needs not only some inherent ability and training but also 'drive' and a commitment to improve his performance. He is usually both numerate and financially minded, keen, adaptable and has the ability to correctly interpret and apply advice and seek to improve his knowledge without a bias of preconceived ideas. In other words, he brings executive ability to the running of the farm or estate and knows when to do things himself, when to delegate them and when to call in an expert — also when not to follow the 'expert's' advice! What is critical is ability over a wide area. Outstanding ability in a single field, such as stockmanship or labour relations, will rarely compensate for defects in other areas, such as over-capitalisation or poor timing of fieldwork. The best managers are usually above average in all respects but are often outstanding in none. They set aside reasonable time for office work, get reliable advice and make calculations before taking decisions.

Good management will always obtain a better return than poor management, using the same quantities of land, capital and labour; but managerial skill, especially the ability to make good decisions, is one of the hardest resources to measure.

Compared with other managers, farm managers can be professionally, and geographically, somewhat isolated. They may also, because of the seasonality of their work, sometimes be managerially underemployed and be confronted with a higher than normal degree of uncertainty.

Many small farmers in the tropics lack managerial skill when they advance from traditional, subsistence farming to the less-known commercial field. They often fail to distinguish between income and profit and either have an excessive fear of debt or, alternatively, get 'up to their ears' in it. In many countries, the common absence of the able-bodied, male, family head in wage employment means absence of the potential decision-maker or innovator. This retards the rate of adoption of commercial farming practices.

BUSINESS OBJECTIVES

One of the main managerial tasks is to define business objectives and, sometimes, policy. Profit maximisation is usually assumed as the objective of business management. This helps when developing logical analysis and planning techniques. However, in practice few businesses aim simply at profit maximisation. Naturally, profit-making is important. Quite apart from its value as income to the owners of the business, profit is one of the means of keeping a business in being by providing contributions that can be retained in the business to maintain assets in good condition and to add to them for better future performance. If a business makes continuous losses, it will soon cease trading as it will neither produce capital nor attract outside capital into it. Despite this admitted importance of making adequate profits, profit maximisation as an objective is invariably modified by other aspects — the most important being adequate cash income in commercial farming and adequate diet in subsistence farming. Without these neither the business nor the family could survive to aspire to any other objectives.

Profit may in no way reflect the amount of cash available because, although it is generally measured in cash, it may be represented by a rise in the value of assets or a reduction of debt. A highly profitable firm may have insufficient cash available to stay in business.

The individual hopes of top managers will decide the purpose of the business. This will vary widely because objectives and the relative importance placed on them vary greatly between people. Farmers claim or show a wide range of business objectives, many of which limit potential profitability.

The objectives for any organisation should be clear and numerical so that all its managers have a commonly accepted goal. Popular strategic or long-term business and personal objectives in commercial farming include:

- High cash income.
- A high return on capital employed.
- Capital growth. It may, for example, be more important to expand the business to enable a son to join it later than to make maximum profit now. Land may be bought in the hope of rising land prices, thus reducing working capital and current profit.
- Personal satisfaction. Potential profit is often lost in order to maximise satisfaction. For example, a simple, easily controlled system with familiar, preferred enterprises may be chosen to give an agreeable way of life that leaves adequate time for public affairs or leisure. Moslems are unlikely to breed pigs or Apostles to grow tobacco, however profitable they may be! The latest machine may be bought not because it will raise income or reduce cost but because a neighbour has one. A breeding herd of cows may be kept because it is nice to look at and the last litre of milk may be extracted from it, irrespective of cost, because of the prestige of having the highest-

yielding herd in the district.
- Social responsibilities of management. These include sentimental or humanitarian considerations like keeping long-service but old or sick workers. Maintenance of a good public image, raising soil fertility, reducing soil erosion and environmental pollution may all reduce potential profit — especially in the short term.
- Security. Avoidance of risk to ensure survival of the business may be important. Stable profits may be more desirable than maximum average profits.

All economic activity is accompanied by risk and subsistence farmers run the greatest risk of all — they could starve. A subsistence farmer makes his main decisions once or twice a year and there is little he can do to change them. He thus tends to be averse to risk, unwilling to innovate and to adopt strategies that will give him the best results in the worst conditions rather than the best average result. For example, given the information in table 1.1, a commercial farmer would probably choose hybrid maize seed as it gives the greatest expected average return over the years. A subsistence farmer, however, with his shorter time horizon and lack of assets to carry him through a bad year, would probably choose composite seed as it gives the best result in the worst season.

Table 1.1   Farmer Strategies, for example Maize Seed

		Unimproved	*Composite*	*Hybird*
		(Yield of grain expected in quintals (100kg)/hectare)		
State of nature, e.g., season quality	Good	40	64	100
	Average	12	16	64
	Poor	8	10	6
	Mean	20.0	30.0	56.7

Management must consider both the present and the future; both the short run and the long run. It is no use aiming for quick profits if the long-term health, perhaps even the survival, of the business is threatened. The business must be kept going in the present or there will be no business in the future. The main consideration is the production of an adequate cash income or of staple foods.

Management also has to make the business capable of performance, growth and change in the future. Once a business is established, it will be concerned less with making quick profits than with making satisfactory profits on the capital employed — allowing for the risks involved and what could be earned elsewhere. The word 'satisficing' has been used by economists to describe this type of profit motive.

Objectives need deep thought especially with regard to commitments and cash requirements. An effort should be made to foresee needs resulting from changes in family size or expected living standards. Objectives, besides earning an adequate cash income, should be put in some rational order of priority and the picture that emerges should form the background to all business decisions.

Management advisers need to keep in mind constantly the often perfectly valid, but nevertheless non-financial, motives of their clients if they want their advice to be adopted. Profit maximisation can still be the aim in planning farming systems but *within the limits set by the other objectives*.

DECISION-MAKING

This is the activity of selecting, from among possible alternatives, a future course of action. Decision-making is essentially, therefore, choosing between alternatives, even if the alternatives are to accept a proposed change or to continue as at present.

Managers are making decisions almost constantly for if no decisions had to be made, there would be no need for managers. It has been said that decision-making *is* management and managers are often judged on their ability to make decisions. Decision-making is an essential component of management in all the stages of planning and control. However, managing involves more than decision-making alone. Decisions clearly must be both agreed with others and implemented. All managers, whatever their precise roles, are constantly involved in such decisions as how to use resources, determining production plans and investment priorities. On subsistence holdings, the wife or wives of the household head are often completely independent of him in making decisions regarding their own plots. Whatever the nature of the decision, the problem is basically the same. Management needs an answer to the question, 'Which alternative is most worthwhile — A, B, C, etc?' Management needs to measure 'worthwhileness'. No one is a real manager if he avoids this responsibility.

The obvious aspects of decision-making are concerned with making up one's mind and with getting things done. Anyone who can toss a coin can make a choice; anyone with a mouth can give an order; anyone in authority can get some sort of results. The reality of decision-making is concerned with getting the *right* things done in order to solve the *right* problems.

Optimal decision-making needs *rational* selection of a course of action. However, complete rationality can never be achieved as not all alternative courses can be identified let alone analysed. Also decisions are made for the future so the expected results of alternative courses are never certain.

Instead of being based, as far as possible, on reasoning, however, decisions are often influenced in practice by personal feelings, power politics, the influence of others and the decision-maker's own values. Instead of aiming for perfect decisions always, the concept of satisficing will often suffice. This limits the task to finding a course of action that is 'good enough'.

Search for the best, practical decision, with the information and in the time available, is helped if certain systematic, basic steps are followed.

FIND THE PRECISE PROBLEM

It is necessary first to realise that a problem exists and that a solution is needed if one is to be able to think logically about it. If the problem is not clearly understood in the first place, great difficulties may emerge only when it is too late to remove them. A common error in decision-making is stress on finding the right answer rather than the right question. You must be sure you are solving the right problem. For example, raising salaries to solve shortage of a certain kind of technician would probably help if competitors are paying more but not if the real reason for the shortage is a lack of trained personnel or poor working conditions.

DEFINE THE MOST LIKELY SOLUTIONS TO THE PROBLEMS

All possible courses of action should be listed. There are few courses in practice when only one decision is obviously right. There are usually several reasonable alternatives and an alternative that is not obvious often proves best. The more important the decision, the more important it is for managers to spend time searching for and examining possible solutions so that the decision is made with reliable facts, rather than guesses, and within the framework of both available resources, including time, and the policy of the organisation.

When a wide range of choice is available, it is necessary, after considering all possible courses of action, to remove the least likely ones. The fewer promising possibilities can then be examined closely. This needs careful observation, collection and organisation of relevant facts. Whenever possible, decisions should be based on knowledge of facts, though opinions may also be relevant, because decisions based on incomplete information must be partly a gamble. Clearly, the more important the decision, the greater should be the search for relevant facts. In practice, the amount of search tends to vary inversely to the time available for a given decision.

ASSESS THE QUANTITATIVE EFFECTS OF EACH LIKELY SOLUTION

All factors that could possibly influence the effects of the decision, and that can be quantified, should be considered. However, attention can be confined to only those items, such as costs, prices and yields that will vary with the alternative chosen. Various techniques for estimating the quantitative effects of alternative decisions will be described later. These include partial budgets, breakeven budgets, cash-flow budgets and capital investment appraisal methods.

No decision can change what has already happened. The past is history; decisions made now can affect only what will happen in future. It follows, therefore, that the only relevant figures to use are estimates of future costs, prices and yields. These are, by nature, uncertain so there is usually a margin of error in any calculation concerning the future. Consequently, there is no value in taking estimates to several decimal places. In fact, there is a danger of being misled by the false idea of precision that such detail gives. It is a matter of being honest with oneself. Fancy figures should not be produced from rough estimates; nor is a result necessarily precise and valid because it took a long time to calculate!

EVALUATE THE QUALITATIVE EFFECTS OF EACH LIKELY SOLUTION

Alternative solutions will usually have several effects that are hard or impossible to quantify.

Financial calculations make it possible to express the net effect of many factors affecting the decision as a single figure. They reduce therefore those factors to be assessed separately in the final judgement leading to the decision. Put another way, they narrow the area within which judgement must be used. Rarely, if ever, do they remove this crucial need for judgement.

DECIDE THE BEST SOLUTION

Choose which alternative to accept. Despite uncertainties, a decision should be made if as much information is available as can be obtained within a reasonable cost and time limit. In choosing, it is important to concentrate on the main constraint towards achieving the objective, such as, for example, capital shortage or uncertainty. A decision is normally better than none at all! The manager in a dilemma may keep steadily collecting unnecessary detail merely to delay the time when he must make a decision. Some decisions must be made with insufficient facts because time is short or the cost of getting more facts would be too great. Postponing the decision is the same as deciding to continue the existing situation. This may be the worst possible decision! Timing is important. Waiting too long is the same as deciding to do nothing during the time of indecision. In practice, decisions must often be made quickly and the manager may have to make one that is adequate rather than ideal.

TAKE ACTION

It is useless to make a decision if it is not effected. As action must be taken to apply the decision to the problem, it must be communicated to all concerned. A control system is then necessary to check if action has been taken as planned.

REVIEW THE RESULTS OF THE ACTION

The decision will normally have been based on incomplete information and even the 'facts' obtained may change. It is important therefore to observe and measure the results of effecting the decision. This will not only enable remedial action to be taken, if necessary; it should also help to improve future decisions and to avoid repeating mistakes.

PLANNING

Planning is a basic but complex management function combining financial, physical and technical aspects for selecting and developing the best of the alternative ways of achieving stated objectives. Planning itself is concerned with *what* needs to be done. The sub-process of organising deals with *who* should do it and *how* it should be done. Organising is sometimes called administration. The main objective, as stated before, is to maximise profit within the limits of all the effective constraints. Planning therefore is essentially decision-making since it means selecting courses of action, from amongst alternatives, either for the entire business or for any part of it. Scarce resources of land, labour, capital, skill and time, with many alternative uses, are allocated so that objectives are reached most efficiently.

Planning is an essential step in doing all but the very simplest tasks, yet there is a reluctance to give it the importance it merits. The reasons for this are that:

- It needs deep thought.
- It is not visible action and may look to others like indecision.
- It needs the courage to commit oneself.
- It implies imposition of discipline or constraint.
- The planning process may not be understood.

Until recently, most planning in farming concentrated on production and finance. Greater stress is now placed on the other two areas of marketing and personnel. For many products a carefully planned marketing programme is essential for success. This is particularly true for perishable products and it is likely to apply to a growing range of farm products. In personnel management, agriculture has tended to lag behind many other industries. The successful implementation of a farm plan is usually heavily dependent on the effectiveness of the staff in achieving rates of work, yields, quality of product, economical machinery use, etc., as planned. To ensure this effectiveness is mainly a matter of planning. Planning a personnel programme means making careful provision for remuneration, training, health, safety, welfare, leave and many other aspects of 'man management'. Planning of this sort may seem rather vague compared with, say, production planning but it is becoming ever more vital — especially if there is strong competition from other industries for labour.

Planning resource use to achieve objectives is a necessary part of any manager's job whether he is at the top, middle or bottom of the organisational structure. The purpose or objectives of the business form its strategy or long-term plan. Once this has been agreed, tactical plans or programmes of action need to be developed for the shorter-term production cycle. The further ahead planning decisions commit one, the more important it is to check regularly on events and expectations and revise the plans, if necessary, to keep on course towards the desired objectives.

For many, planning is a simple, common-sense job based not on formal planning methods, but rather on experience of what is possible within the limits of available resources and possible activities. Within the growing complexity and need for efficiency in farming during the last 20 years, however, much attention has been paid to developing systematic aids to farm planning. These range from simple, partial budgets to complex computer techniques. Despite vast differences between farms in size, market involvement, ownership and objectives, suitable modern planning or decision-making methods are always useful. They may, however, vary with business size and circumstances.

CONTROL

The managerial function of control is the process by which *action* is adjusted to achieve business objectives. It is not merely collection of information. All managers, but especially those entrusted with effecting plans, exercise control by taking action to return to course when events have not occurred as planned. Put simply, managerial control means knowing where you are going, when you should be there, how to get there, where you are now, changing direction if you are off-course and moving faster or slower if your timing is not as planned — just like driving a car!

Control has been necessary ever since people bothered about the results of their efforts. However, traditional control methods used by managers to answer the question, 'How are things going?' are time-consuming, inefficient and do not give good control. Departures from plan should be found soon enough to allow early effective correction, at reasonable cost. Control is justified only if departures from plan are corrected or the plan is redrawn. An effective control system will tell the manager *what action on his part* is needed to correct the problem.

As the size and complexity of farm businesses have grown, many farmers now have less direct control with what is happening. Consequently, the need for control systems to supply information on the effectiveness of plan execution has risen. In small farm businesses, the owner does many or all the critical tasks himself, has few bought resources, little contact with the outside world and his span of control is small. He is well aware of results and therefore does not need elaborate control techniques. However, in modern, commercial farming with complex, costly production methods, much hired labour, borrowed capital and falling profit margins, even small departures from plan should be detected early. A control system will not stop errors from occurring but many problems can be corrected before they get too serious, if there is constant feedback of information on actual performance.

A question that naturally arises is just what should be controlled. Inability to define this causes most problems of ineffective control of plans so it should be decided when the plan is made. Management without control would be useless. However, a manager who tries to control every detail himself would be overloaded. Intelligent control concentrates on performance in those key areas that will determine the plan's success. Aspects of production in most need of control are usually quantity of both yields and inputs, quality of output, timing and costs. A study of time use involves both the time needed for any given job and the co-ordination of jobs to prevent time wasting.

Strict financial control promotes cost consciousness. Budgetary control is probably the most widely known management control technique. Cash-flow budgeting is one of the most useful forms of financial control. Labour control in the tropics is often thought to mean constant supervision. This is a luxury the manager cannot afford as it reduces the time he can spend managing the farm as distinct from running it. A well-organised, motivated work force should not need continuous, close supervision. The aim should be self-control by the work force and the common system of task work suggests that managers realise the importance of standards in reducing supervision time.

Control is clearly necessary but how should it be practised?

An effective control system requires the following steps:

The key areas to be controlled should be defined when the plan is made

These are also called strategic control points. The earlier in a process that a strategic control point is placed, the more likely it is that departures can be corrected effectively. Examples of strategic control points in farming are pregnancy diagnosis, daily work tasks and the fuel consumption of tractors.

Performance standards should be set for the key areas

Effective control needs objective standards against which results can be compared, for example, 85 per cent grade A seed cotton, 10 absences for each worker a year. These standards should be set *before* the plan is executed but this is seldom done. Farmers are more likely to compare their results either with those of other farms or with their previous results. Neither of these comparisons is adequate. There is a real danger that, by adopting such loose standards, poor results will be repeated. To take a simple example, the average calving percentage in tropical ranch cattle is often below 50 per cent. This is clearly unsatisfactory yet, happy that he is above average at 65 per cent and ignorant that he should get 85 per cent, the farmer happily blunders on.

Average performance levels are not good standards though they may help in fixing them.

Execute the plan and measure the results

Checking actual performance needs both physical records and financial accounts such as labour-use records, production records, cash flows, return on capital, stock levels and labour turnover. Although records are essential control tools, they are not in themselves a component of control. They are a means — not an end. We keep records to help us farm, not farm to keep records. Recording can become a fetish and should be limited to the key performance areas. Despite the failings of many types of farm records, however, it would be foolish to suggest that efficient control is possible on any but the smallest farms without records. What is important is that relevant records should not only be kept, they should be used.

Compare results with the pre-set standards

There could be many reasons for differences between actual and planned performance (deviations or variances). Managers cannot expect to deal with them all as some will be small and some large. In the latter case, some may be controllable and some not. What is important is that significant deviations from plan, where the system is changing its operating characteristics, are brought to their attention. Timely feedback of information on achievement is needed.

Take corrective action

Control is effective only when it incorporates action. If the variance is unacceptably great *and* controllable, corrective action must be taken. If the variance is uncontrollable, such as a rise in fuel cost, then either standards or the plan itself may need to be changed. Clearly, both the cause and the effect of any deviations should be examined. It is all too easy to blame poor yields on the weather when they may be, at least partly, due to, say, late planting, an unsuitable cultivar or low crop stands.

Requirements for a successful control system are that there must be objectives, a plan to achieve these objectives, authority over any junior staff executing the plan or part of it and a way of measuring performance for comparison with predefined standards.

Control methods make use of many devices such as budgets, progress charts, reports and financial ratios. These are all helpful, if not essential, but they are not enough. The experienced manager gets much useful information from personal observation. Two methods often used to reduce the work of control are *control by exception* and *control by sampling*.

A manager's main concern is to know about important departures from plan and not about irrelevant, small details. He should not waste his time on small recording jobs, such as weighing milk and issuing fuel, as these can be delegated to literate and numerate staff. The staff should be taught to report only deviations of a given size. This is control by exception.

Sampling is already widely used in the form of spot checks on labour, bulk fuel supplies, etc. Other possible types of control by sampling include detailed costing of only one enterprise on the farm at a time and occasional checks on individual vehicle fuel consumption or the mass of a bale of hay.

INTER-DEPENDENCE OF PLANNING AND CONTROL

Although planning and control are definable and clearly need different types of thought process, they are not carried out separately within a business; not at different times, by different people or for different occasions. Planning and control are a continuous cycle of management activity. No manager can control without objectives and plans and these need follow-up otherwise they are unlikely to be achieved. Without control, plans become merely dreams!

We can survey the whole process by looking at the basic planning and control cycle in figure 1.2.

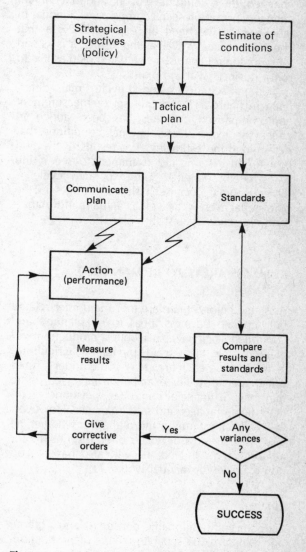

Figure 1.2 The Basic Planning and Control Cycle

From the first definition of objectives and esti-mates of conditions, (or 'environmental analysis and forecast'), form a quantitative view of the environment within which the business must survive and prosper. This has been called assessing the threats and opportunities. You make detailed plans to achieve these objectives and communicate these plans and exact standards of quantity and quality to staff. After action, for example, production, you measure the results, compare them with the standards and give orders for corrective action, if necessary. This basic cycle is a simplification. It reports only variances from plans, assuming that the plans themselves are right. In practice, the plans may need revising; the economic climate may have changed; there could have been sudden technical changes or unforeseen competitive action. In other words we need an expanded cycle that checks the plans as well as the results. It is as bad to try to control against an outdated plan as it is to run a business with no plan!

You can see the expanded cycle's main differ-ence in figure 1.3. Another set of measurement and comparison activities has been added to check the external and internal conditions that determined the estimates. The results of the two monitoring systems are examined to see if the plans need changing. The expanded cycle is more dynamic because it shows changes in the plan's effectiveness as well as actual performance of the business against that plan.

BUSINESS AREAS TO BE MANAGED

As stated before, there are four main inter-related but separate business areas to be planned and controlled — finance, personnel, production and marketing. In large-scale, plantation agriculture, as in industry, each area may come under a func-tional specialist. For example, a large tea estate may have a financial controller, personnel man-ager, field manager and sales manager. However, in small-scale farming the manager, whether he owns the business or not, is often responsible for all four areas. Indeed, he will also have to do non-managerial, manual work.

Finance

Economists traditionally consider land, labour and capital to be separate factors of production. To businessmen, however, capital gives control over any other resources needed — whether that control is made possible with owned or borrowed funds. Without capital, production would be impossible so, basically, efficient farming means effective use of capital. All farmers should obtain sufficient capital to implement effectively their plans.

Financial management includes obtaining funds, ordering inputs to best advantage, keeping stocks at economic level, keeping assets in good working order, ensuring adequate cash flows for current activities and, in the longer term, obtain-ing capital growth.

Personnel

As businesses grow and more hired labour is used, more stress tends to be put on personnel management. It is people who have to be man-aged and who, in turn, organise and use the other resources. Quality of labour is therefore important as well as its quantity. In essence, a business cannot be more efficient than the people who manage and work in it. Labour is also important because it is provided by people with basic physical and psychological needs in which work has a large place.

Important aspects of planning include staff selection (getting the right person into the right job), reward structures and organisation. The latter means fixing the responsibilities by which activities are delegated to staff and the formal relationship set up between staff because of these responsibilities.

Good staff control will keep an adequate, but not excessive, work force and raise its efficiency. Control of junior staff is a continuous, complex job and the methods used vary widely. However, the aim is to communicate with, train and moti-vate staff so that they work efficiently towards the firm's objectives.

Production

Business success in farming depends largely on management organising an efficient production system. All the main management activities are closely related to production. The reason for planning is to relate production systems to pre-set objectives. These can be achieved only by adequate output. Control systems are designed to achieve this and management's concern for finance, personnel and marketing is necessitated mainly by production requirements.

Production management is specially concerned

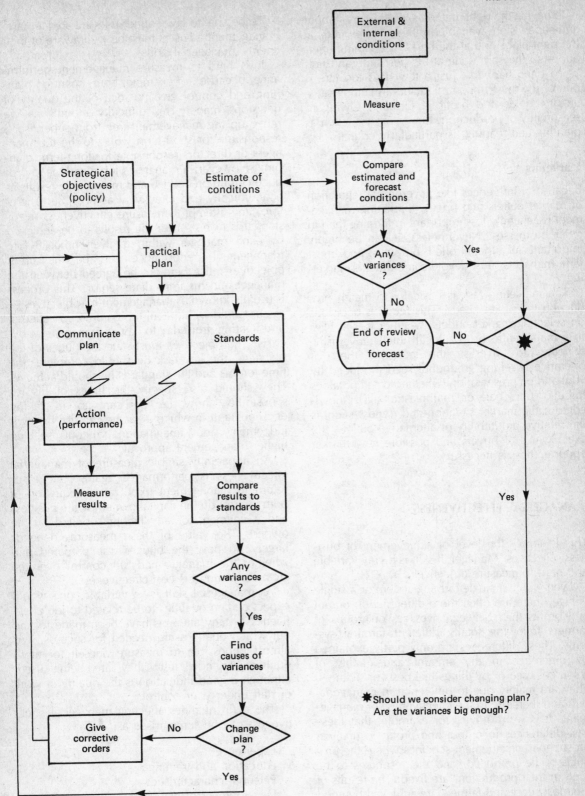

Figure 1.3   The Expanded Planning and Control Cycle

with having the right materials, machinery, labour and supervision available in the right amounts, at the right place and at the best time. Timing operations so they are done at the best time and not, as so often happens, just a few days too late is one of the most important factors in getting satisfactory yields and profit in farming. Owing to seasonality of production, timing of jobs, such as planting and reaping, is particularly critical.

Marketing

Because of influences like government regulation of market outlets and prices, marketing management is generally less important in farming than in other businesses. Nevertheless, it can be highly important for uncontrolled products and most farm managers must devote some time to marketing.

The marketing task is basically to match production to market needs. This requires decisions on selecting market outlets (whether to sell eggs in a main centre or to a local hotel), advertising (whether to advertise the fact that the Co-operative is selling groundnut seed produced by you) and on presentation (whether to sell potatoes in bulk, gunny bags or small plastic bags). Knowledge of the market situation and trends, such as the relative profitability of alternative outlets, the exact wants of buyers and possible new market opportunities, is necessary.

MANAGERIAL EFFECTIVENESS

This is simply the level of achievement of business objectives. Managerial skill is highly variable and hard to measure objectively.

Even within a single farming system or a single village, there are often more difference in output and profits than between averages of groups of farmers following totally different farming systems. These differences are only partly explained by difference in the amount and quality of resources used or by things like pests or sickness. They are mainly due to difference in managerial skill and therefore in the effectiveness of management. It is well known, for example, that these vast differences in output and profit occur even on supervised settlement schemes where the products to be produced, and the resources to be used in their production, are fixed. The results of the most successful farmers are achieved by good management — not luck!

To be able to give sound, relevant advice, farm management advisers must be well aware of their clients' managerial skill.

It is hard to measure management performance, because, for example, hours spent on planning and control give no idea of the quality of decisions made. This difficulty mounts as we move up the management tree from supervisors, responsible for short-term work, to the top managers or directors responsible for long-term planning. However, a manager's job clearly is to produce certain predefined and measurable results as economically and effectively as possible. Therefore, any attempt to measure effectiveness needs standards to be set for the results to be achieved by each manager within a given time. Before management performance can be measured, objectives must therefore be agreed between the manager and his immediate senior. This process is usually known as management-by-objectives — as compared with the all-too-common practice of 'fire-fighting' from day to day or management-by-crisis. The objectives must state the duties of the manager, the key task he should complete, the time for this and the standards he should achieve. They should also specify the limits of his responsibility, how far he can go in taking decisions and in which areas he can use his own judgement and imagination without seeking higher management approval.

An apparently simple measure of managerial effectiveness is performance against budget or return to management (profit after deduction of both interest on the capital owned by the business and a payment for any labour supplied by the owner). The value of these measures depends largely on how the budget was prepared and whether the manager had full control over the factors affecting his cost-effectiveness.

As managerial skill is so variable, one would expect certain qualities to be related to job effectiveness. Many attempts have been made to find out which qualities are needed for success. It is almost impossible to measure oneself for these qualities or how fully they are being used, although a good indication is the quality of work of staff under your control.

The main qualities affecting managerial effectiveness can be summarised as:

Experience
Education and training
Personal characteristics
Development of managerial skills
Age

EXPERIENCE

A person's background is important as it affects his ambitions. Someone who has spent his life in a communal or tribal society may feel obliged to conform to locally accepted norms in farming as well as in everything else. He will tend to have a rigid mental outlook and be slow to adopt new ideas. On the other hand, people who have travelled and experienced a wider world generally have a wider out-look and higher ambitions. In small-scale farming the most successful farmers, therefore, are often those who have been soldiers, policemen, civil servants or those who also run another business such as a store or a butchery. Effective experience in the type of farming in which he is engaged raises a farmer's or farm manager's effectiveness. For example, it would be unwise to introduce tractors or irrigation without experience or training.

EDUCATION AND TRAINING

Several studies indicate that the ideally qualified manager combines adequate general education with both technical training and experience. Flexibility of mind to make the best use of new methods and conditions is probably best encouraged by raising the general level of education of farmers and their families. The best educated farmers tend to make the most use of advisory services and management consultants and to adapt the available advice to suit their own conditions.

PERSONAL CHARACTERISTICS

Certain qualities seem to be common in the most successful managers. These include:

- Willingness to work hard — very hard when necessary.
- Courage — to innovate, invest and expand.
- Toughness — equivalent almost to ruthlessness at times.
- Self-confidence and willpower amounting at times to excessive enthusiasm.
- Honesty. Integrity.
- Ability to inspire, to get on with and control people.
- Resilience to try again when knocked down.
- Good health.

DEVELOPED MANAGERIAL SKILLS

Certain important managerial skills can be developed by experience or training. The most important of these are probably observation, analytical ability, decision-making and implementing, numeracy and communication skills.

AGE

The manager's age often influences his effectiveness. Most people go through a three-phase cycle but the length of each phase varies widely.

1. The learning period.
2. Full maturity and top performance. (The years of discretion between 45 and 54?)
3. Post-maturity/pre-retirement when changed goals and other influences may lower the manager's effectiveness.

Young men, especially those with growing families, have the greatest desire to maximise their incomes. Therefore they tend to be progressive and innovative. However, they often find difficulty in leading older staff who may resent having a younger manager. Older men have more experience but their goals often change to greater stress on leisure. The old manager is often incapable in various ways such as lack of drive and outdated knowledge. The Peter Principle, which states that people tend to rise to their own level of incompetence, plays a part here. This means that if a man is competent at his job, he will probably be promoted. This tends to continue throughout his career until such time as he is no longer thought competent enough in his job for further promotion. Consequently, many older men (and quite a few not so old men!) hold jobs in which they are incompetent.

I have tried to summarise the qualities of the most successful managers. It should be realised, however, that the relative importance of these qualities varies according to the types of managerial job and particularly with the proportion of technical and human relations content in it.

Table 1.2 summarises the results of a survey taken in Britain.

Table 1.2   Ranking of Qualities most Valuable for Top Management

	%People who think it is developed mainly by:	
	Academic work	Professional experience
1. Ability to make decisions	5	90
2. Leadership	0	96
3. Honesty	9	62
4. Enthusiasm	9	66
5. Imagination	14	59
6. Willingness to work hard	35	20
7. Analytical ability	72	21
8. Understanding of others	0	97
9. Ability to see opportunities	6	88
10. Ability to face unpleasant situations	1	98
11. Ability to adapt quickly to change	5	90
12. Willingness to take risks	1	84
13. Readiness to undertake hard jobs	5	84
14. Clear speaking	40	52
15. Practical wisdom (astuteness)	3	91
16. Efficient administration	8	92
17. Unprejudiced to new items (open-minded)	25	60
18. Ability to 'stick to it' (persistence)	35	46
19. Willingness to work long hours	25	32
20. Ambition	21	52
21. Single-mindedness	29	52
22. Clear writing	85	15
23. Curiosity	40	31
24. Skill with numbers	71	21
25. Capacity for abstract thought	74	22

Adapted from *Sunday Times* Survery (London, 26 September 1976).

GLOSSARY

Activity a possible course of action, for example, an enterprise, rotation or production method.

Constraint any limitation or restriction to a course of action.

Deviation a departure from the planned course of action.

Satisficing the concept that businesses are content with a certain minimum (target) level of profit, rate of return, etc., rather than with maximising the level reached.

Span of control the number of people for whose work the manager is accountable.

Variance a departure from the planned course of action.

RECOMMENDED READING
Books

Adair, J., *Training for Decisions* (Macdonald, 1971).

Barnard, C.S. and Nix, J.S., *Farm Planning and Control*, 2nd ed. (Cambridge University Press, 1979).

Brech, E.F.L. (ed.), *The Principles and Practice of Management*, 3rd ed. (Longman, 1975).

Bridge, J. & Dodds, J.C., *Managerial Decision Making* (Croom Helm, 1975).

Buckett, M., *An Introduction to Farm Organisation and Management* (Pergamon, 1981).

Capstick, M., *The Economics of Agriculture* (George Allen & Unwin, 1970).

Cyert, R.N. and Welsch, L.A., *Management Decision Making* (Penguin, 1970).

Deverell, C.S., *Business Administration and Management* (Gee, 1972).

Drucker, P.F., *Management Tasks, Responsibilities, Practices* (Heinemann, 1974).

Drucker, P.F., *Management* (Pan Books, 1979).

Hyre, E.C., *Mastering Basic Management* (Macmillan, 1982).

International Labour Office, *Introduction to Work Study* (Geneva, ILO, 1969).

Giles, A.K. and Stansfield, M., *The Farmer as Manager* (George Allen & Unwin, 1980).

Harrison, E.F., *The Managerial Decision-Making Process* (Houghton-Mifflin, 1975).

Jackson, K.F., *The Art of Solving Problems* (Heinemann, 1980).

Koontz, H. and O'Donnell, C., *Principles of Management*, 5th ed. (McGraw-Hill, 1972).

Leeds, G.A. and Stainton, R.S., *Management and Business Studies* (Macdonald & Evans, 1974).

Rickards, T., *Problems Solving Through Creative Analysis* (Gower Press, 1974).

Sloma, R.S., *How to Measure Managerial Performance* (Macmillan, 1981).

Taylor, W.J. and Watling, T.F. *The Basic Arts of Management* (Business Books, 1972).

Thomas, H., *Decision Theory and the Manager* (Pitman, 1972).

Upton, M., *Farm Management in Africa* (Oxford University Press, 1973).

Periodicals

Boyer, R.S., Giles, A.K. and Salmon, G.D., 'The Overall Role of the Farm Business Manager', *Farm Management* (1973), 2, 6.

Dillon, J.L., 'The Definition of Farm Management,' *J. agric. econ.* (1980), 31, 2.

Giles, A.K., Some Human Aspects of Giving and Taking Farm Management Advice, *J. agric. econ.* (1983) 34,3.

Harwood Long, W., The Definition of Farm Management *J. agric. econ.* (1979) 30, 3.

Nix, J., 'Farm Management: The State of the Art (or Science)', *J. agric. econ.* (1979), 30, 3.

Anon, 'Ranking of Attributes most Valuable for Top Management', *J. agric. econ.* (1979) 30,3.

2 Objectives, land tenure and business organisation in tropical farming

Much was stated in chapter 1 about farming objectives. However, no book on tropical farm management could leave this as it stands because the wide range between subsistence and commercial objectives in the tropics has so much effect on the farming systems followed. For the same reason, the various types of land tenure and business organisation will be described, it is hoped, in sufficient detail to help a farm manager or management adviser.

SUBSISTENCE FARMING

The term 'subsistence farming' refers to the main object of a peasant farmer rather than to his standard of living. In true subsistence farming, no farm products would be sold. In this sense, subsistence farming is rare, existing only in a few isolated areas. After all, towns must be fed! In practice, subsistence farmers are those whose *main* objective motive is to feed themselves and their dependants with their own produce. They usually sell surplus produce in a good season and they usually try to trade such a surplus to find money for items like school fees, tax and goods that cannot be produced on the farm such as clothing, sugar and bicycles.

Although subsistence provision is the main objective of subsistence farmers, they have multiple and complex goals that often compete with each other. In managing his farm, the farmer will try to achieve that mix of goal attainment that gives him the best level of overall satisfaction across his multiple goals.

Tropical peasant farmers, like most other people, usually wish to enjoy a range of material comforts, a stable food supply, reasonable social status, a satisfactory family life, participation in customary social activities, and leisure time diversion, and some security against misfortune. The weights placed on these various objectives vary significantly between groups and between individuals within these groups. As resources to fulfil these objectives are limited, pursuit of one may lead to the dilution of others. The possibility exists therefore that adoption of a new technique that will raise farm output will reduce overall satisfaction because the subjective value of the extra output is less than that of the extra inputs needed of labour, managerial skills, social re-organisation and so on.

A farmer who sells some produce in six years out of ten is probably best called a 'peasant farmer'. Nevertheless, any system where the marketed surplus is small, compared with total production, and unreliable is usually called subsistence farming.

The main features of subsistence farming are:

- Diversified production of each farm.
- Dominance of consumption and survival aims over commercial ones.
- A 'closed' farming system with few bought inputs like fertiliser or pesticides.
- Little difference between the farm firm and the household; between producers and consumers.

The decision on what to produce may depend on maximising the amount of food produced from the available resources. Root crops (like cassava and yam) and plantains provide more food, even if of low quality, per hectare than grain crops;

and cassava, in particular, needs little labour.

Security aspects are likely to dominate decision-making in subsistence farming in the seasonally arid tropics where people so often encounter food shortages. Requirements for the staple foods are usually produced on the farm as farmers are reluctant to rely on an uncertain income from other products to enable the staple to be bought. Plans that maximise average profit in the long term, without considering the effect of a poor growing season on staple food supply, are likely to be rejected. If a farmer cannot save or borrow enough to see him and his family through the development period, the uncertainty inherent in starting a new enterprise or adopting a new technique is simply too great.

Whilst each subsistence farmer may not actively assess uncertainties, he relies on the communal strategies expressed in traditional farming practices. These are usually sensible ways of reducing uncertainty. The customs and organisation of traditional societies give each member some security. The high proportion of time spent on funerals, visiting, celebration, etc., enables the family to stay fully paid-up members of society and thus to be able to claim its social security benefits. Both the farming methods, however inadequate, and the social structure, however limiting, have evolved as the best known way of guaranteeing a living to *all* members of society.

It is considered the duty of the more lucky or able farmers to help members of their extended family when required. This sharing of the fruits of success may impede commercialisation. Furthermore, while traditional customs and organisation provide some security for everyone, they tend to retard innovations – innovators, by straying from accepted standards may take excessive risks and thus may have to make excessive demands for support from their more conservative relatives, they become disliked and risk being denied the benefits of society membership.

Most labour in subsistence farming is provided by the family without cash payment and women do much of the farm work. There is little specialisation of labour apart from the common practice of men concentrating on cattle and sale crops and women on crops for use in the home. As little labour is hired, family size influences the farming system.

The largest, tangible capital assets in subsistence farming are usually cattle in Africa and jewels in Asia. These tend to be saved as stores of wealth instead of being used to produce income. African pastoralists often regard their livestock as a walking bank, a measure of tribal status or a social security fund but rarely as enterprises to be managed to maximise profit. There is little investment in farm equipment, most capital investment being of a non-monetary kind amassed by the direct application of labour to land to improve it for future production. Traditional cultivators cannot pledge land which, being communally owned, cannot be sold to pay their debts. Cattle seldom are acceptable as security for loans because of the difficulty in establishing their true ownership and because they are mobile and likely to disappear if a creditor wants to sell them. Most subsistence farmers, therefore, have little access to credit; they earn little more than necessary to survive and can rarely save enough to buy the inputs needed to raise productivity. It is a vicious circle!

Failure to recognise the various constraints to production – social, technical and institutional – in subsistence farming has probably been the main fault of farm management advice in this area. Maintenance of position in society, to get social security benefits, is one constraint; uncertainty is another. Uncertainty is probably a more important constraint in subsistence farming than in other types of farming.

UNCERTAINTY IN SUBSISTENCE FARMING

Uncertainty and the methods used to reduce its effects help to explain many typical aspects of subsistence farming. Yield uncertainty is largely due to variations in weather, especially the length and quality of the growing season, and to pests, diseases and storage losses. Different crops vary in their response to these factors. Consequently, diversification, mixed cropping, successive plantings and land fragmentation all help to reduce the chance of total crop failure and food shortage in poor seasons. Hoarding of cattle and gems, besides their important role in marriage, serves as an insurance policy that can be drawn on in bad years. Most innovations increase yield variability as well as average yield. It is therefore wise for subsistence farmers to be cautious. Unless they can be given non-farm work or new areas to occupy, the almost universal increase in population in the tropics must reduce the area of individual holdings. As this occurs, their safety-first tendencies, and probably the annoyance of those who advise them, are likely to grow.

Although subsistence farmers already practise

methods of reducing uncertainty, there are several other ways of further reducing it that therefore stand a good chance of being accepted. These include reducing the season length needed, extending the effective growing season, using existing rainfall more effectively and using suitable biocides. Reducing the season length needed requires more careful matching of crops, cultivars and agronomic practices to the climate than is often done now. For example, in short-season areas, sorghum could replace maize, quicker-maturing maize cultivars could be brought in and cotton could be planted more thickly to hasten maturity. Obvious ways to lengthen the growing season include irrigation, drainage and flood control. However, other than on a small scale, these are likely to be worthwhile only with large changes in land use. Less impressive methods, nevertheless, are available to farmers. If the community agrees to exclude cattle from arable lands during the dry season, they can be prepared when the soil is still moist at the end of the rains. If weed growth is prevented, stored moisture in the soil will enable the next crop to be established earlier, either by dry planting or by planting with a little applied water *before* the rains break. Much scarce rainfall is now lost as surface run-off. This loss could be much reduced by establishing a protective plant cover as soon as possible to protect the soil surface from raindrop impact and crusting. Another way of holding rainfall in place, to give it time to soak into the soil, is to plant crops across the slope — preferably on tied ridges. One of the most valuable biocides to reduce food shortages is an effective pesticide to reduce grain shortage losses.

CUSTOMARY LAND TENURE

Land tenure means the way land is owned and held and the rights and duties arising from this. In general, it refers to the laws and customs governing relations between people in using land. Customary land tenure is often called traditional or communal tenure. It is common in subsistence farming.

Before considering the advantages and disadvantages of customary land tenure, its general features need describing. The attitude towards land where customary land tenure prevails differs greatly from that in the west. It often has a spiritual value as the home of those ancestors who, more often than is realised, still play an active and important part in daily life. Land is regarded as a free but limited good belonging to the whole community. It has no market price as it cannot be bought or sold by individuals — being regarded as no more open to private ownership than, say, rain. However, *everyone* accepted as a member of the local community has a basic and exclusive *right to cultivate* and use arable land. He has a right to use a fair share of the available land fixed, by traditional custom, according to his need. The right to use land is decided by the local community through their traditional leaders to whom this power is entrusted under clear, accepted rules. This naturally gives these leaders strong social and political control and partly explains their frequent resistance to efforts to change the system. Customary tenure exists in much of the tropics today in various forms.

Communal tenure produces many problems and helps to keep agricultural productivity low. It is often accused of being a great obstacle to rural development. The main charges against the system are that it leads to lack of individual responsibility, fragmentation, little incentive for improvement, no security of tenure, restricted scale of operations and problems in getting credit. Communal use, without individual, collective or governmental responsibility, can be efficient only if there is little population pressure on the land.

With communal tenure, many people have the right to use communal land for their own uses, like grazing and collecting fuel and water, without central control. Each farmer tends to take as much as he wants and has no incentive to conserve or improve the land. He can graze any number of cattle on the communal grazing area at no cost to himself. Grazing pressure, stock breeding and diseases cannot be controlled. In its original form, communal tenure rapidly damages the land as population pressure increases.

Customary land areas, where population density is high, present a patchwork of small plots with much loss of land and soil due to boundary paths that gully easily. There are problems in working fragmented holdings owing to time wasted in travelling and to adopting crop rotations on scattered plots. Consequently, consolidation often is regarded as necessary for lasting improvement. However, sub-division and fragmentation of holdings is not confined to communal tenure. It has, for example, caused problems in West Germany. Experience shows that wherever land is inherited under a system of individual tenure, and land control is ineffective, sub-division may occur at least once in each generation. In fact, fragmentation helps to reduce uncertainty by spreading the danger of crop failure through hail, excessive rain, drought and different soil types.

A common argument against customary tenure is that it discourages investment in improvements. There is something in this as the greatest security of tenure sometimes occurs together with the most conservative and limiting social surroundings. A young man wanting to try out new ideas may need to seek land as a 'stranger' in a more progressive village than his own. However, improvements, like houses and trees, normally belong to the landholder.

Lack of security of tenure is often quoted as a disadvantage of customary land tenure. However, much security exists in communal systems of tenure. A definite obstacle to commercialisation through individual drive and ability however does exist in the idea of equal rights for all members of the community to use of land according to their needs rather than their ability. There is no provision for commercialisation of land or the right to use it and progressive young cultivators find it hard to get land rights beyond their subsistence needs. Another disadvantage of customary land tenure is that the individual landholder has trouble getting credit as he cannot pledge land which he does not own as security for a loan. However, credit often can be obtained from a cooperative society if all members bind themselves to be collectively responsible for repayment.

Though there is no registered title to customary land, there is much security of tenure if the person concerned is farming in his own village and is not disowned by the community. The traditional organisation provides for mutual duties and reciprocal arrangements which provide security for community members. Tradition puts great stress on security which, agriculturally, means that everyone who fulfils his traditional obligations to the community is sure of his rights to a share in the common land. Members of a kinship group forming a village can usually continue to occupy a plot of land as long as they need it and the right to use it may pass to descendants.

More and more subsistence farmers obtain their cash and other needs by working temporarily in towns either at home or abroad. Nevertheless, they intend to return home and settle down again as members of the community. The fact that this is possible is a convenient 'non-contributory pension scheme' for both themselves and their families.

In many parts of the tropics, customary systems of tenure are being modified by social and economic forces and are slowly changing towards individual land rights. Methods of consolidating and registering rights in family heads, to hasten agricultural development, exist in some countries, such as Kenya, where there is now a market in formerly customary land.

It seems that the main fault in subsistence farming is of land use rather than of land law. Governments should either leave customary land tenure alone or integrate land reforms with customary land laws. 'Reform' in land tenure is not enough, by itself, to raise farm productivity. As human and livestock numbers keep rising, however, it seems likely that soon there will be insufficient land to give every citizen what he traditionally expects as a birth right. Farmers must learn to produce more from the smaller areas they will have in future.

Table 2.1   Comparison Between the Extremes of Subsistence and Commercial Objectives in Farming

Subsistence	Commercial
Main objective:	
Security maximisation	Profit maximisation – within constraints
Products:	
Diversified food products	Specialised sale products
Land tenure:	
Usually customary	Usually freehold or leasehold
Labour source:	
Mainly family and communal	Mainly hired
Capital investment:	
Mainly land improvement with hand labour	Investment in buildings, equipment, conservation
Few bought inputs	works, etc. Growing use of bought inputs
Power sources:	
Mainly hand labour	Growing use of animal and mechanical draught
Use of products:	
Mainly for consumption	Mainly for sale
Income:	
Low but relatively stable	Higher but more variable

Table 2.1 summarises the main difference between subsistence and commercial farming. The rest of this chapter will refer to the latter.

LEASEHOLD TENURE

Leasehold and freehold are both types of individual land tenure. As land must be either bought or rented, it is normally necessary to farm commercially to ensure adequate income to meet the financial commitments. Leasehold is a form of tenure where an agreement is made by a landowner allowing someone else to farm it for a given time in return for payment of a stated rent. The landowner or lessor is often called a landlord and the lessee a tenant. Leasing arrangements are greatly affected by custom and tradition and therefore vary considerably between countries.

Rent is a payment made, per unit of time, for use of land, buildings and any other capital asset supplied by the landowner. It may be a fixed cash sum, a fixed share of income or a fixed share of the physical output of the farm. A fair rent should reflect potential farm profit and the current market value of the farm. It gives the owner a fair return on his capital investment and the tenant a chance to make a fair profit. Too high a rent can be harmful for both owner and tenant as the latter will probably leave and rapid change of tenants will give a bad reputation to both the owner and the farm. Rent levels tend to be fixed by custom and it affects the way land is used as only certain farming systems will make it economic.

Cash rent is a fixed cost regardless of yields and product prices so a tenant bears all the risk of changes in these. However, tenants paying rent in cash are usually free to farm with few restrictions and it may be cheaper than other forms of rent payment because the landowner's income is guaranteed to him. A modification of fixed-cash rent is fixed-product rent. Here, cash is paid to the landowner in relation to long-term average yield at current product prices. The owner thus shares the risk of price variation. The tenant bears the risk of yield variation and reaps the full reward if he can raise yields by better management. Neither of these systems of cash payments allow for a series of poor growing seasons for which the tenant bears all the risk.

An adjustment to allow both yield and price uncertainty to be shared would provide for the owner to take a fixed share of gross sales, profit or of physical output. Share-renting is common in the Islamic world where taking anything like interest often is forbidden. It is common also in other countries. As the owner has a direct interest in production with share-renting, he normally sets conditions on the farming system to be followed. As the tenant must share the benefits of success with the landowner, this can lead to inefficient resource use. For example, if maize responds to fertilisation as shown in table 2.2 with fertiliser costing $6 a bag and a maize price of $9 a bag, the best amount of fertiliser for a tenant on a cash-rent system would be about 19 bags — where marginal (extra) cost would equal marginal revenue. However, if the rent is a half share of production, it would pay the tenant to use only about 11 bags when the value of his half share of the extra yield equals the cost of the fertiliser needed to produce it.

Table 2.2   Best Use of Fertiliser under a Share-rent Lease where Half the Output is paid as Rent

Bags of fertilliser (50kg)	Bags of maize (90 kg) produced	Marginal cost of fertiliser $	Marginal revenue $	Tenant's share of marginal revenue $
0	30	—	—	—
5	38	30	72	36.00
10	45	30	63	31.50
15	50	30	45	22.50
20	53	30	27	13.50
25	54	30	9	4.50
30	54	30	0	0

Rent payment can take many other forms. For example, an absentee owner of undeveloped land who wants to farm it himself after retiring from his present job may want no rent but may specify improvements that the tenant must make during his tenancy. Grazing tenancies may charge rent at a certain yearly sum per animal unit of the maximum safe carrying capacity of the holding.

Tenancy has many advantages over land ownership by the farmer himself. The main one is that it is cheaper to farm this way because capital is not tied up in land and a tenant does not start off loaded with debt. Farmers with skill but little capital can get control over more resources than would be possible with land ownership and use all the available capital to equip, stock and run the farm.

The ideal of farm ownership has caused many farming failures. In good times many people will buy farms on mortgages at prices and rates of interest that the farm's potential profitability cannot justify even in boom times. Capital and much of income are then tied up in land and insufficient cash is available for seeds, fertiliser, labour, etc. When hard times come, the effect is disastrous, necessitating a lower living standard and soil 'mining' in efforts to meet mortgage repayments. Farms then tend, by foreclosure, to be taken over by absentee owners and stay undeveloped. Rent levels are more flexible than mortgage repayments during slumps.

If land prices keep rising, tenancy will probably become the main way for most able young men to start farming on their own and by which successful farmers can expand by moving to larger farms.* The argument that it is better to buy a farm by instalments than to pay rent means risking excessive debt loads; furthermore, few young men can afford to buy a farm that will satisfy later ambitions. Finally, tenancy tends to prevent uneconomic sub-division of farms, generation by generation. This often occurs when the land itself is bought by prospective farmers.

There are, of course, disadvantages to leasehold tenure although most of them can be overcome if the lease agreement is drafted carefully. In some countries, the tenant may not have security of tenure or compensation for disturbance or wrongful eviction and his freedom in farming methods may be limited. If he is not going to be compensated when he leaves for improvements made, he is unlikely to make any and the farm may run down. He may be tempted, in fact, to 'mine' the soil. Lastly, if the owner has mortgaged his farm and fails to honour his contract, the mortgagee can sell the farm, free of lease, unless he has already agreed to that lease.

In negotiating a lease, each party should inquire thoroughly into the character, reputation and experience of the other party. There can be no satisfactory agreement with an incompetent or dishonest man. References should be sought. Both parties need to get all the relevant facts, understand each other's interests and make an effort to reconcile those interests to mutual advantage. Verbal leases are binding, if they can be proved, but they are undesirable and can be a menace if either party dies. Professional legal and accounting advice is needed to prepare a mutually acceptable, written lease agreement. Economising on professional services may be costly in the long term. Lease agreements are prepared normally by the owner's lawyers and therefore are not mainly designed to protect the prospective tenant's interests. Consequently, once an agreement is ready and before any commitment is made, the prospective tenant should have it checked by his own lawyer.

A farm lease is an important business contract designed to protect the interests of both parties and give possession of property for a stated period. The owner's interest is to get a fair rent and to get his farm back undamaged and preferably improved. The tenant's interest is to get a fair

*An advantage of leasing with option-to-buy agreements is that prospective buyers can find the problems of a farm before finally committing themselves to buy it.

return on his investment and effort. The terms of the agreement will affect the enterprises selected and the way capital and skill is used. They should encourage the tenant to farm as well as he can without fear of financial loss during the lease.

Most lease agreements contain only a description of the land, rent, matters connected with non-payment of rent and a few provisions for ending the lease. In such an important contract, as little as possible should be undefined.

Most or all of the following points should be included in a lease agreement to forestall possible disputes as far as possible:

NAMES OF PARTIES

ADDRESSES FOR SERVICE OF NOTICES

DESCRIPTION OF THE PROPERTY

DETAILS OF THE CONDITION OF THE FARM AND EXISTING IMPROVEMENTS ON ENTRY

PERIOD OF LEASE

When does it start and finish? (At least a three-year lease is normally desirable.) How much notice of cancellation is needed?

OPTIONS AND FIRST REFUSALS TO BUY THE FARM OR RENEW THE LEASE

RENT

How is it calculated in terms of money or produce? Where, when and to whom is it to be paid?

USE OF THE FARM

With the present rapid changes in farming methods, restrictions other than 'general farming and ranching' are usually undesirable. However, it may be necessary to exclude certain areas from cropping or grazing or to limit sales of produce on or off the farm, or shooting or fishing rights.

GOOD HUSBANDRY

For example, 'The tenant shall carry on all farming operations on the said property in a proper manner according to the tenets of good husbandry as current from time to time and in such manner as to avoid, insofar as is reasonably poss-ible, damage to the land from soil erosion, over-grazing, cultural practices, fire, weed, pest or disease infestation or impoverishment from any cause whatsoever.'

MAINTENANCE AND REPAIRS

Agreement by one or other party to maintain and repair all fixed equipment such as buildings and water supplies.

IMPROVEMENTS AND ALTERATIONS

It should be stated clearly whether or not the tenant needs the owner's approval to make improvements to the property and how compen-sation will be fixed to repay him for the value of any still useful improvements at the end of the lease; that is, how they are to be depreciated. If no arrangement is made to compensate the tenant for improvements, like conservation works made by him, the property is unlikely to be improved much. This is to be expected as he is renting land to make a living and, at the end of the tenancy, he wants to have made a return on his capital.

TIMBER

Any limits on cutting or sale and any planting requirements.

MINING

Any limits on the tenant prospecting or mining, or a requirement for him to notify the owner of pros-pecting by others.

RIGHTS OF WAY

RIGHT OF OWNER TO INSPECT

CESSION, SUB-LETTING AND ASSIGNMENT

INSURANCE, RATES, TAXES AND ELECTRICITY

Who pays? Are buildings to be insured by the owner? Must the tenant protect his ability to pay rent by insuring against theft, fire, hail or death?

PENAL OR BREACH CLAUSE

The lease will be hard to enforce unless quite definite provision is made for eviction if the tenant breaks the agreement terms.

ARBITRATION

For example, 'Any dispute, not involving a legal matter, arising from this lease shall be submitted to the Director of Agriculture who shall appoint a single arbitrator to arbitrate in the dispute and whose decision shall be legal and binding.'

COSTS OF THE LEASE

Who pays. They are often shared. Stamp duty?

SIGNATURES AND WITNESSES

Both parties should initial each page and any changes.

CONDITIONAL LEASES

If the lease is conditional upon something, like grant of a loan or a relative providing working capital, this must be clearly stated. Otherwise, the tenant may have to pay rent for a property that he cannot farm.

Much bias exists against tenancy and in favour of owner-operation. This is an ever more costly bias often costing the state large sums in subsidies to preserve owner-occupation on farms too small for efficiency. It is not suggested that tenancy is a better system of land tenure than owner-occupaton or otherwise. Both systems have their place but I do suggest that owner-occupation needs much cheap land and credit or eventually farms will become too small. The way in which land is held is only a means to an end — not an end itself.

FREEHOLD TENURE

Freehold tenure means that land has an individual owner with the sole right to use or dispose of it as, and to whom, he wishes. If the owner of freehold title farms his land himself, instead of leasing it to a tenant, he is known as an owner-occupier. A compromise between full ownership and tenancy is mortgaged ownership whereby title is obtained by borrowing most of the purchase price, with the farm itself as security, and repaying the loan, with interest, over the years.

It is often loosely said that owner-occupiers make the best farmers because they can take the long view and will maintain their land and equipment better than tenants — especially insecure tenants. They are likely to work harder, make long-term improvements to the land and other farm assets. However, all these advantages could be obtained with a well-designed lease. Nevertheless, a common aim of land reform is to allocate land to small, individual owner-operators. It is widely believed, especially in East Africa, that individual freehold tenure is almost essential for agricultural development.

Land ownership attracts people for both financial and non-financial reasons whether they want to farm the land themselves or not. Social esteem and a pleasant life go with the prestige of ownership but the main attraction is basic confidence in the long-term stability of the real value of land. Investment in land provides a hedge against inflation. Indeed, as less developed (and generally under-fed) countries try to increase their own food production, rather than rely on supplies from elsewhere, real land values may well rise steeply. This is of doubtful value to the owner-occupier who is short of capital, as the benefit of rising land values can be obtained only on selling part or all of the farm. Landowners, however, do have the advantage of almost complete security of tenure, no rent exploitation, freedom to farm as they want, improvements being for their sole benefit and the ability to mortgage the land as security for loans.

Disadvantages of land ownership include all the advantages of tenancy. Owners are less mobile within the industry and they often have so much capital tied up in land and committed to repair and maintenance of fixed equipment that there is insufficient working capital to finance farming operations so as to maximise current profit.

Purchase of a farm is a crucial financial transaction. The buyer is making a large, long-term investment and probably buying a home as well as an income-producing asset. Before committing himself, the intended buyer should ensure, first, that a written contract is made to record the terms of the agreement and, secondly, that those terms are favourable to him.

A prospective buyer should get lawyers to check the title deeds of the farm to find any servitudes, mortgage bonds, forbidding of sub-division or any other conditions or claims registered against the land. The farm boundaries should also be checked. A property is generally sold subject to existing servitudes or encumbrances so it is too late to query these conditions after an agreement has been signed. The Agreement of Sale is usually drawn up by a lawyer or estate agent acting for the seller so it is important for it to be checked with the buyer's interests in mind before signing. Many Agreements of Sale require payment of a deposit upon signing them. The buyer should ensure that anything he pays on account, before registration of transfer, is held in trust by the seller's lawyer until actual registration of the transfer.

The market value of land is what a wise and willing buyer will pay a willing seller. It is affected by many things with little or nothing to do with its profit potential, such as the standard of the house, hunting and fishing rights, other agreeable facilities, nearness to schools and shopping centres and the community in which the farm is placed. Economic considerations affecting the market value of a farm are those factors that determine the supply and demand for that particular type of holding. These are related to its profit potential, namely:

QUANTITY OF LAND

QUALITY OF LAND

Classification of arable and grazing land classes (which determines the safe intensity of use), fertility level, drainage, etc.

CLIMATE

Length and quality of the growing season, temperatures, availability of water rights, etc.

PRODUCT PRICES EXPECTED

(These four factors affect possible enterprises, yields, quality, price and, therefore, the income potential of the farm.)

COMMUNICATIONS

Distance to all-weather roads and bridges, railways, processing and marketing facilities. This is of more importance with low-value, bulky or perishable products, like maize or vegetables, than with light, high-value, non-perishable products like tobacco or chillies.

LABOUR COST AND AVAILABILITY

(These two factors affect production costs.)

STATE OF DEVELOPMENT

House, buildings, boreholes, dams, roads, dip tanks, fencing, land clearing, availability of electric power and telephone, etc. (This factor allows for the cost of bringing the farm into full production.)

There is little point in quoting an overall value per hectare of farmland because of differences in topography, land quality and state of development even on adjoining farms. A hectare of good date land is not worth the same, in terms of profit potential, as a hectare of rock outcrop! It is more sensible to value the fully developed arable area, with all necessary improvements and facilities, like house, buildings and water supplies, value grazings and then deduct from the total any essential developments still to be completed.

Freehold tenure is most successful in relatively advanced countries, when combined with co-operatives to obtain economies of large-scale production and marketing while keeping the incentives of owner-operation.

There is no universally ideal system of land tenure and any land reform should be gradual and evolutionary otherwise at least a temporary fall in production and greater public expenditure is likely.

TYPES OF BUSINESS STRUCTURE

In common with the land tenure system, type of business organisation has a large effect on farming systems.

No business can exist without adequate finance. This need for finance is reflected in the business structure adopted. It will vary with the scale and nature of business operations — from say $50 for a subsistence farmer to perhaps $500 million for a large, public, sugar-growing and milling company. These vastly different capital needs directly influence the business structure, one being wholly self-financed out of savings and the other needing a complex structure to help it obtain funds from people and institutions who share both the risks and the profits to some extent. Between these two extremes of sole-ownership and public company farming, intermediate business structures are partnership and private, limited liability companies.

Apart from the usual structural development within family farming, which is normally development from subsistence farming, through partly commercial farming as a sole trader, to fully commercial farming, maybe as a private company, there are other forms of business structure as shown in figure 2.1. Before describing sole ownership, partnerships and company farming in detail, it is useful to examine co-operative farming, collective farming and plantation agriculture.

Co-operative farming faces problems such as lack of a clear connection between individual effort and gain, lack of qualified leadership and the conflict between the co-operative ideal of all members having an equal say and the practical need for one competent decision-maker. Co-operative farming seems to succeed only where, as seemingly in Israel, members are more interested in working for the common good than for themselves. Collective farming with state direction is common in many communist and socialist countries but is not yet important in the tropics although *ujamaa* in Tanzania seems to be a form of it and it may grow in those countries where communist influence is growing. Collective farming has often led to poor performance through faulty and rigid management policies, lack of full co-operation from the workers and other labour problems due mainly to killing personal initiative and incentive. Some states have recognised this.

Co-operative and collective structures need skilled management and this is often scarce in less developed countries. It also seems unlikely that people will work as hard for co-operative or other authorities as they would for themselves. These business structures are suitable only if skilled management and community spirit are favourable.

Plantation agriculture is run by large private or public companies, usually controlled from a developed country or by the state. They generally use foreign capital and management and send

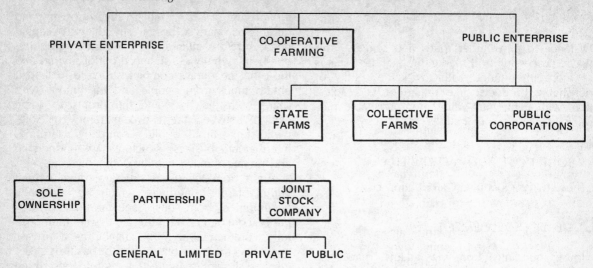

Figure 2.1   Main Forms of Business Structure in Farming

some of their profits home. This structure is still important in countries producing mainly export, but also import substitution, products. Examples include rubber in Malaysia and Liberia, sugar in Sri Lanka and Kenya and tea in Malawi. Plantation agriculture enables tropical countries to attract foreign capital and skills that can be taught to local people. It enables new areas to be developed rapidly and increases both revenue-earning exports and self-sufficiency in basic foods and raw materials.

SINGLE OWNERSHIP (SOLE TRADING)

The oldest, simplest business form is the one-man business with a single owner called a sole trader. This is still the most typical business structure in farming. Sole traders are self-employed and responsible only to themselves. They can make their own decisions and change their policies as often as they wish. They run their business with or without a manager, take full responsibility for business results and share neither profit (except, possibly, with a manager or son) nor loss. Anyone working for a sole trader, even his own son or wife, is an employee entitled to a wage but not to any of the profit. A sole trader owns the *business* but not necessarily the land.

A sole trader provides all or most of the capital himself and the business assets form part of his personal estate on sale or death. Desire for independence and the limited funds needed to start are the usual reasons why people adopt this form

of structure. It has, however, several disadvantages.

The main disadvantage of single ownership is that the owner has to provide most of the business finance either from his own resources or from friends and relatives who will trust him with use of their money in his business. With the risks involved, he may have to pay high interest on their loans. As farming becomes increasingly more capital intensive, sole owners are finding that shortage of capital for farming, and possibly also land purchase, limits their scale of operation. With limits on the size of his business, there is little chance for division of labour or for growth to obtain economies of scale. The latter are less prominent than in factory operations but they are still important in some areas like mechanical cultivations, harvesting and some intensive livestock enterprises. Sole owners must be prepared to live frugally even when the business is profitable as the need to grow and consolidate means using profit mainly for finance within the business.

Another disadvantage of single ownership is lack of limited liability. The farmer is liable for any losses to the full extent of his personal wealth. Creditors can take not only the business, if it fails, but also most of an owner's personal assets.

While many large farms began under single ownership, so much capital is needed for development and efficient operation that many owners found this form of business structure unsuitable. Some of the disadvantages of sole trading can be overcome by forming a partnership.

PARTNERSHIPS

A partnership is a business owned by an association of two or more people who share both risk and profit.

If a business needs more capital for growth than a sole trader can supply himself, he may consider sharing his ownership with others in an association that will enable his co-owners to bring capital and/or skills into the business. In a partnership, the partners, often relatives, agree to introduce new capital either in cash or in other assets and to own and manage the business jointly.

Partnership is a useful business arrangement for friends wanting to start farming together by pooling their limited capital. Alternatively, it can be used to persuade a wealthy non-farmer to invest as a 'sleeping partner', taking no active part in management, or to enable two or more farm businesses to combine to obtain economies of scale. Partnerships often are arranged between father and son so the father can keep some control over the business without frustrating his son's ambitions and at the same time allow the father to fall into a lower tax bracket. They also can be used to retain an experienced, salaried manager by giving him a financial interest in the business or enable a landowner to invest in his tenant's farming activities.

Each partner contributes an agreed quantity of funds, other assets or skill to the business. At any time, a partner may withdraw some or all of his capital at short notice even though this may force the firm out of business. Farming profit is divided between the partners by mutual consent. It is often, but not always, divided equally. Also, partners may receive interest on their individual capital inputs and a management salary before the remaining profit is split. Partnerships are generally most successful when each partner has a similar stake but partners often do not put in equal capital, draw equal shares of profit or share profit in proportion to their capital. A son may work longer hours than his partner-father yet have only a small sum in the business. A fair share is a matter for agreement.

The property of the firm belongs jointly to all partners. If this is not acceptable, a father could retain land ownership and rent the farm to the business partnership he has entered into with, say, his son. Partners are legally equal. Each one is entitled to take part in the daily management, to place orders and to incur debts on behalf of the business. Each partner usually has an equal voice in managing the partnership whatever his financial stake. However, management may be placed in anyone the partners mutually agree upon. Each partner is jointly and separately liable for business debts. If the business fails, the personal assets of any or all partners could be seized to repay business debts. Individual partners cannot be sued, in their capacity as partners, while the partnership exists. Action is taken against all the partners, judgement taken against the partnership and, if necessary, a writ of execution is issued against partnership property. Only if this is insufficient to repay creditors can the assets of individual partners be seized. Clearly, mutual trust is essential in a successful partnership.

Joint management can successfully continue only if the partners can set up an effective way of settling differences of opinion. The need for this should seldom arise if the agreement calls for all partners to share in fixing policies before the start of the production period but for only one to be the active manager and decision-maker from then until the season is over. However, if day-to-day management is shared, it may be necessary to continually adjust attitudes and agree on decisions.

In most countries, partnerships are treated as having the same identity as their members. They are regarded simply as groups of people who have combined to seek common aims as stated in their agreements. Consequently, partnerships as such pay no tax. Each partner is taxed personally on his share of the profit. Any salary drawn by a partner cannot be charged as a business expense but is regarded as merely an advance drawing of profit. Losses can often be offset against any other incomes of the partners.

Partnership usually promotes larger, more efficient business than sole trading. Many costs can be spread over a greater output thus gaining economies of scale. They are flexible as the objects and constitution can be changed by unanimous agreement. For example, children can increase their share in the business as their parents gradually withdraw from farming. When profits are not too high, total tax liability is usually less than it would be with a company and there are less regulations to be observed.

The main disadvantages of a partnership are limited capital availability, unlimited liability for debts and lack of continuity of the business if a partner dies or withdraws. Sources of finance are limited to the partners, their friends and relatives. Most partnerships are general ones with unlimited liability. Each partner's interest and credit is liable to danger from every other partner because a partner who buys, sells or otherwise contracts on behalf of the firm binds the other partners.

In some countries, it is possible to limit the liability of those 'sleeping partners' who take no active part in management. These partners are liable only for their stake in the business but cannot withdraw their capital without the consent of the general partners. In the rather rare case of a limited partnership, one or more general partners must have unlimited liability.

A great snag of partnership is that any change in membership, when a partner dies or withdraws, ends the partnership and confuses several years' tax caluculations. The business may have to be liquidated or the remaining partners may reach a new agreement after finding a new partner or selling assets to repay the outgoing partner or his executors. Adjusting one partnership and starting a new one can work out very unfairly and seriously upset the business. If a partner dies and the others can find neither sufficient cash to buy his share nor a suitable new partner with ready funds, it may be possible to arrange with the dead partner's executors to pay his widow an annuity for the rest of her life. However, she may want to withdraw the capital from the business. Life assurance can help. Let us say Mr Yesufu is in partnership with Mr Edje. Each of them takes out a whole life policy on his partner's life. If Mr Yesufu dies, Mr Edje pays the proceeds of the policy on his life to Mrs Yesufu and she uses this to live on for a few years. Mr Yesufu's share of the partnership capital is reduced and Mr Edje's is raised by the proceeds of the policy. Mr Edje can pay any balance due to Mrs Yesufu gradually out of income and become the sole owner. Alternatively, he has time to find a new partner who will buy in and pay off Mr Yesufu's executors.

As soon as two or more people combine as owners, or in profit sharing, complications arise. It is strongly recommended that a written Deed of Partnership is drafted by a lawyer and signed by all parties. It should clearly set out, in legal terms, the whole arrangement. It might include:

- The way in which profits will be shared. (Unless shared equally, this is essential.)
- The annual drawings permitted to a partner.
- How the partnership can be ended or new partners be introduced.
- The duties of each partner in daily management.
- A clause stating that one partner cannot inject

or withdraw capital without the agreement of the others.
- Arrangements for preparing and auditing proper annual accounts.

The deed of partnership should be prepared professionally as soon as possible. While affairs are smooth and friendly, it looks so simple and unnecessary; when problems occur, the terms may suddenly become almost impossible.

Table 2.3 depicts a typical partnership agreement.

Table 2.3   A Typical Parnership Agreement

	Mr Avila	Mrs Perreira
Capital invested	$12000	$20000
Management salary	2400	—
Proportion of net profit	40%	60%
Farming profit available for sharing	$11000	
Interest on capital at 12%	1440	2400
Salary	2400	
Share of net profit	1904	2856
Totals	$5744	$5256

COMPANY FARMING

Most farmers are in business on their own account either as sole traders or in partnerships. Only a few of them trade as a private, limited liability company although this is the usual business form in industry. Company status has not been used to the same extent in farming as in other businesses owing to a natural reluctance of farmers to load themselves with the extra administrative burdens and disclosure requirements of a company, especially if profits are not high enough for large tax benefits to result. However, company farming is growing in popularity for large farms.

There are two types of company limited by shares — private and public ones. There are more private than public ones as they are cheaper and easier to form. They are often, in practice, partnerships that have become limited companies to obtain benefits such as reduced tax liability. Generally, in farming, private companies are family businesses run by members of the family or close business associates.

A limited company is a separate legal entity in its own right with a distinct identity and separate existence from its owners (called shareholders) or the controllers of its activities (called directors).

There must be at least two but, usually, no more than 50 shareholders in a private company. Generally, there are less than a dozen and they often belong to only one or two families. Capital is obtained by selling shares in the ownership of the business. Once the shares have been sold, they are permanent for as long as the company lasts. A shareholder who wants to withdraw from part-ownership of the business may sell his shares to someone else but he cannot cancel them or demand his money back from the business. Even selling his shares depends on finding someone willing to buy them at a price he is prepared to accept and that person usually has to be approved by the other shareholders. There are restrictions on the transfer of shares in private companies and they may not be offered to the general public.

Share capital takes three forms:

NOMINAL (AUTHORISED/REGISTERED) CAPITAL

This is the money value of the maximum number of shares the company is allowed to issue. The actual number may be less so that the company can later increase its capital up to the full amount without further permission.

ISSUED (SUBSCRIBED) SHARE CAPITAL

This is the value of the number of shares which the company has so far agreed to issue.

PAID-UP SHARE CAPITAL

The value of shares sold.

Control of a company rests in its shareholders in proportion to their shareholding. Management is undertaken by one or more of the directors elected by the shareholders. Payment of directors is by their agreement, subject to approval by the shareholders. It consists of a salary for managing directors and fees for any other directors. These payments are chargeable as business expenses in calculating the profit on which the company is taxed. Furthermore, they are drawn whether the company made a profit or not in any one year.

Most people regard companies as vast organisations but this need not be so. A farmer may be both the main shareholder and the main (or only) director but the company will still keep its separate identity. For example, it is common for a farmer to be sole director and also sole share-holder apart from one share which is transferred to his wife or manager as his nominee. This is called a 'one man' company.

The advantages of company farming are limited liability, greater ease in attracting funds for growth and continuity of ownership and the chance of reducing total tax liability.

The shareholders' (owners') liability for company debts is usually limited to their shareholding if the business fails. Once a share has been issued, the shareholder is liable only for the value of that share. If he has paid the amount due, he cannot be made to meet any outstanding company debts — unless he is a debtor in the normal course of business. Limitation of liability to the value of his shares protects the interests of the investor as his personal assets are safe unless he has either given a personal guarantee or acted dishonestly.

The public is expected to know when a business has limited liability and to deal with it accordingly. In practice, therefore, a bank or some other creditor may insist on the directors/shareholders giving personal guarantees to secure company debts, thus destroying the limited liability. On liquidation, incidentally, the shareholders, as creditors themselves, will get a share in the distribution of the value of assets sold, if the company is solvent.

Formation of a company enables capital to be raised, often from non-farming investors and be brought into the business for expansion without the farmer losing control. People and institutions may be persuaded to invest in the business in return for dividends and possible capital growth. Though the minority shareholders who contributed capital have only limited powers and no voice in day-to-day management, they usually can sell their shares although not necessarily for the same amount as they paid.

Allocation of shares enables ownership of a farm business to be divided among several people without breaking up the farming unit. As a company is a separate entity and cannot die, it is permanent and not destroyed if a large shareholder dies or sells his shares. Only by the vote of a large majority, say 75 per cent, of shareholders is it possible to force closure of a solvent company. When a non-company farmer or a partner dies, his widow and/or childern often face severe problems in keeping the business going until the will is proved; also his personal estate is liable to death duty. If a company farmer dies, there should be no problem in continuing the business as the company, as such, is unaffected and alive.

No death duties are payable and the bank account and other financial arrangements continue.

Control of a family farm can be smoothly passed to a son before or after the original farmer's death. If he wants to split his estate on death without forcing sale of the business, he can simply leave shares. Consequently, although company farming is often seen as an alternative to family farming, it can, in fact, help to maintain family farm units. When retirement or death forces a change in the pattern of farm ownership, a company is far more satisfactory than a partnership. The remaining directors can continue their work uninterrupted while the legal procedures take their course. A further advantage of company farming is that junior directors, managers and other staff can be allowed to obtain shares to give them a direct, personal interest in the success of the business.

The choice between a partnership and a company is decided usually by whichever is most favourable from a tax standpoint. A company is especially useful in farming where profits are notoriously variable with losses and profits often alternating from year to year. Trading profits can either be paid as fees and salaries to directors, according to their personal tax positions in any year (allowable as a company expense), be distributed as a dividend to shareholders or be retained in the business for growth. Although trading losses can be carried forward and offset against future profits, this is not allowed for personal income tax allowances which are thus lost in a year when no profit is made. A company can pay its directors salaries and fees each year at levels that will enable them to make full use of their personal allowances and pay a maximum rate of personal income tax just less than the rate of company tax. This is a complex subject best illustrated by an example and farmers should always get professional advice to minimise their total income/company tax liability. This total consists of the income tax paid by each director on his total personal income and the company tax paid by the business on its profits after payment of directors' fees and salaries. In many countries, limits are set on how much of a director's pay is allowed as a company expense.

Table 2.4 shows how a farmer with a high, average but varying income can benefit from reduced total tax liability by trading as a company. This is because company tax is often less than marginal income tax when personal income is high.

Company farming can be useful for quite small, family farmers as well as big corporations. However, the latter themselves help small farmers in many tropical countries. Large public companies with much capital often set up nucleus estates with extensive, factory processing plants. Use of these plants and provision of finance and other services is then extended to surrounding small farmers or outgrowers. This applies to such crops as tea in Kenya, sugar in Sri Lanka, tobacco in Malawi and Zambia and palm oil in Papua New Guinea. However, operating as a company is often advantageous even to fairly small farmers themselves. There is flexibility in transferring property to heirs and it is much better than partnership for surviving hard times. Let us look at some examples.

- Mr Jere has farmed successfully for years but now his son has grown up and returned, full of new ideas, from agricultural college wanting a job on the farm. He could be brought in as a junior director, gradually be given more authority and have shares transferred to him by stages.
- Mr Dhlamini dies suddenly and his widow and son want to continue farming themselves. Mrs Dhlamini could be the main shareholder but hand over daily control to her son as managing director.
- Mr Rashid is a capable young farmer but direly short of capital. His neighbour does not know the difference between a sheep and a goat but would invest funds in the farm in return for a share in the profit if his liability is limited. The safest way to bring in the neighbour's funds would be through a limited company rather than by making him a 'sleeping partner'.
- Abu Riyadh wants a retirement pension. A company pension scheme is usually better than one for a self-employed person. As contributions are allowed as an expense in calculating company profit, they can be increased when trading profit and potential tax is high and decreased when it is low.

It is simple to start a company. The promoters draw up a statement of purpose for which it is being formed. This is the articles of association and is usually loosely worded to allow the company to undertake almost any legal activity. It contains the name of the company, for example, Baquba Date Estates Ltd, and the nominal amount of its capital. Despite the aura of wealth and importance engendered by forming a company and appointing one or more directors, it is not

Table 2.4   Illustration of Difference in Tax Liability Between Company and Non-company Farming (A Married Farmer with Three Children)

		Year 1	Year 2	Year 3	Year 4	Year 5	Year 6	Year 7	Average annual tax
		$	$	$	$	$	$	$	
FARMING PROFIT		8000	5000	(1600)	10000	(1000)	6000	5000	
A. A SINGLE OWNER OR PARTNER									
Primary allowance		1500	1500	1500	1500	1500	1500	1500	
Allowance for children		1000	1000	1000	1000	1000	1000	1000	
Taxable income (allowing for previous year's losses)		5500	2500	NIL	5900	NIL	2500	2500	
Income tax:									
5% on first	$500	25	25	—	25	—	25	25	
10% on second	$500	50	50	—	50	—	50	50	
15% on third	$500	75	75	—	75	—	75	75	
20% on fourth	$500	100	100	—	100	—	100	100	
25% on fifth	$500	125	125	—	125	—	125	125	
30% on the rest	$500	900	—	—	1020	—	—	—	
Total income tax payable	$500	1275	375	NIL	1395	NIL	375	375	542
B. A COMPANY FARMER		4000	4000	4000	4000	4000	4000	4000	
Income as director's salary and fees		2500	2500	2500	2500	2500	2500	2500	
Allowances as above		1500	1500	1500	1500	1500	1500	1500	
Taxable income		150	150	150	150	150	150	150	150
Total income tax payable		4000	1000	(5600)	6000	(5000)	2000	1000	
Company profit after salary and fees		4000	1000	NIL	400	NIL	NIL	NIL	
Taxable amount		1600	400	NIL	160	NIL	NIL	NIL	309
Company tax at 40%									
		1750	550	150	310	150	150	150	459
Total tax payable									

(N.B. Figures in brackets are negative.)

costly, needing only, say, two one-dollar shares.

Having formed the company, the farmer will subscribe for shares in it. He may initially transfer $300 to the company bank account in return for 300 $1 ordinary shares; if his wife, friend or manager buys another 50, the issued and paid-up capital of the company will be 350 $1 ordinary shares. The final stage is for the shareholders to appoint company officers — the director(s) and a secretary. In most small, private, farming companies, the main shareholders will also be directors and one of them may also be appointed secretary. The directors are elected by the shareholders and are responsible for guiding the company's affairs.

Besides keeping financial accounts, like any other business, companies must also keep a number of registers such as a register of members (shareholders), a register of charges, a register of directors and secretaries, a register of directors' shareholdings and a minute book. Resolutions must be recorded and an annual general meeting of shareholders held each year. Accountants will normally charge more for preparing company than non-company accounts as the accounts must be audited. Many farmers regard these details as annoying enough to deter them from forming a company.

Two further possible disadvantages of company operation are difficulties in getting credit and reluctance of owners to rent land to a farming company. Credit difficulties arise from limited liability although many companies using large amounts of credit have a paid-up capital of only $2 or so. As previously stated, this problem can be overcome, if it arises, by shareholders giving personal guarantees. Reluctance of landowners to rent their farms to a company is understandable because the company never dies and their prospects of obtaining vacant possession in the future may, in some countries, be reduced. If this difficulty arises, it is possible for the farmer to rent the land personally and farm it as a company.

GLOSSARY

Allowances tax-free amounts deducted from total income to give taxable income.

Annuity a constant, annual payment.

Auditor an independent, qualified person who officially examines accounts to ensure that they represent correctly the financial affairs of a business.

Bankruptcy declaration by a court that a person or company is insolvent, that is, cannot pay due debts. A bankrupt company may be **liquidated.** The assets of a bankrupt business are distributed for the benefit of all creditors.

Contract a legally binding agreement that the law will enforce.

Director member of a board of a company.

Dividend the amount of a company's profit that the directors decide to distribute to ordinary shareholders.

Estate the assets of a person at the time of his death.

Estate agent someone who buys, sells or lets property in exchange for commission.

Executor a person or institution appointed by a person in his will to see that his wishes are carried out.

Hedge action taken by a buyer or seller to protect his business or assets against a change in prices; that is, to secure against loss through a fall in real, as against actual, prices.

Lawyer a trained member of the legal profession such as an attorney or solicitor.

Liquidate to wind up or dissolve a company, determine its liabilities and distribute its assets. Liquidation may be initiated voluntarily by the shareholders or directors, or by its creditors or by a court order if the company is insolvent.

Mortgage a legal agreement transferring conditional ownership of assets, such as a farm, as security for a loan and ceasing when the debt is repaid. Loans for initial purchase of a farm are often obtained in this way.

Pledge to hand something over as security for fulfilment of a contract, payment of debt, etc., and liable to be forfeited in the event of failure to keep to the terms of the contract.

Servitude liability of any kind of permanent property to a right of way or similar right by someone other than the owner.

Sue to prosecute in a law court.

Title a legal document providing ownership of property either with or without possession.

RECOMMENDED READING
Books

Castle, E.N., Becker, M.H. and Smith, F.J., *Farm Business Management,* 2nd ed. (Macmillan, 1972).

Dillon, J.I. & Hardaker, J.B., 'Farm Management Research for Small Farmer Development' *F.A.O. Agric. Serv. Bull.* (1980) 41.

Hall, A.G., *Introduction to Modern Accounting,* 2nd ed. (Heinemann, 1974).

Leeds, C.A. and Stainton, R.S., *Management and Business Studies* (MacDonald & Evans, 1974).

Uption, M., *Farm Management in Africa* (Oxford University Press, 1973).

Periodicals

Furlong, F.A.C., 'Farming as a Limited Company,' *Farm Management* (1987) 6, 5.

Gasson, R., 'Goods and Values of Farmers' *J. agric. econ.* (1973) 24, 521.

Harding, T.G., 'Farming Management Advice to Peasant Agriculture: The Transfer of Technology.' *J. agric. econ.* (1982) 33, 1.

Junankar, P.N., 'Tests of the Profit Maximisation Theory: A Study of Indian Agriculture.' *J. Dev. Studies* (1980), 16, 2.

Midland Bank, 'Farm Partnerships', *Midland Bank Review* (1979)

Nicholson, J.A.H., 'MBO Re-considered', *Farm Management* (1983) 5, 2.

Laventure, B., 'Companies and Partnerships', *Big Farm Management* (October 1976).

Panes, C., 'Farm Partnerships', *Farm Management* (1980), 4, 3.

3   Basic management concepts

THE PRODUCTIVE RESOURCES

Resources are often called factors of production. Economics is concerned with the sensible use of scarce resources. Resources traditionally and conveniently are classified by economists into four main groups — natural resources (sometimes simply called land), labour, capital and management. Each of these covers a large set of productive services. Managerial skill, which is simply for farmers and other businessmen combining the use of the other available resources to achieve their objectives, was described in chapter 1. It will not be examined again here. Instead, time, a fifth but often ignored resource, will be described in this chapter. The resources used, or the productive services they provide, are the farm inputs. This distinguishes them from the outputs produced which are either sold, consumed or used as capital inputs in the next production cycle.

As farmers use resources to achieve their business objectives, it follows that without a supply of them, production would be impossible. Businessmen often face a situation where the demand for one or more resources exceeds supply. This relative scarcity necessitates allocating available resources to best advantage in an effort to achieve business objectives as closely as possible. Compared with farmers in western Europe and North America, where most farm management theory was developed, peasant farmers in the tropics usually control few productive resources.

Many resource inputs are indivisible. Such investments therefore need a minimum scale of activity to justify their use in the business. For example, a bull should service about 20 cows; a farmer with only five cows cannot buy a quarter of a bull! A minimum arable area is needed to justify investment in draught animals and the related implements. Regular hired labour is also a lumpy resource. Besides often being indivisible, resources are also, fortunately, partly substitutable for each other. For example, mechanisation is substitution of capital for labour; more intensive land use is substitution of labour, and probably capital, for land. The distinctions between the different types of resource are not always clear. A skilled worker, for example, may be seen as a capital asset rather than as simply a unit of labour. However, there is little practical use in arguing about the basic difference between resources.

The job of every farmer and management adviser is to place the available resources to best use. Most enterprises compete with each other for resources and the amount of any product that a farmer can produce is limited by a shortage of one or more resources. Farming resources are organised into productive units called farms or estates. The word 'farm' refers not to a single plot of land but to the whole business — all productive resources used in the business and under the farmer's control. It may include several plots of land. The concept of farm size, even in terms of land area, is difficult to determine in subsistence, shifting cultivation systems as in, say, Indonesia or much of Africa, because apparently unused land may have reverted to bush fallow and will be used again. Secondly, a household head's wife or wives may operate separate, fragmented plots independently. Thirdly, grazing land is often used communally. The size of the farm business therefore depends on the total quantity of all resources used in it.

NATURAL RESOURCES

Accurate definition of natural resources is difficult. They include all the 'free gifts of nature' not made by man, that existed before economic activity began. Natural resources include not only land but also the resources of the sea, minerals, soil fertility, water and climate. Land is defined traditionally as the 'original and indestructible properties of the soil'. It provides space, ability to be built upon or it may be used for other purposes, such as farming.

The difference between natural resources and capital is not always clear. Land is often regarded as a form of capital as it has usually been made more productive by man. Many of the services that land can provide to man need investment of capital and labour to obtain and keep. For example, land must be cleared, manured and protected for erosion. To this extent, it is a produced rather than a natural resource. Natural resources stay so only if left untouched. It seems that a wild tree stops being a natural resource when it is felled! Rainfall is a natural resource but what if it has been caused by shooting a rain-seeding rocket into a suitable cloud? Although there is no clear division between natural and other resources, the concept is kept for convenience.

Land productivity varies with environmental factors like terrain, soil fertility and length and quality of the growing season. Its full potential can be reduced by bad husbandry. The success of a farmer in developing a stable and profitable farming system depends largely on his concern for the fertility of his land. On most inefficient, low-income farms, land often fails to give its full potential because too few other resources are applied to it.

Natural resources, especially soils and climate, greatly affect the way in which land is used, the crops grown and the livestock kept. The enterprises chosen will be those for which the farm has the greatest comparative advantage in terms of soil, climate, markets and freedom from pests and diseases. They will give the greatest returns to the scarcest resource.

If land is short, the farming system must be intensified by using more labour, capital and skill on the limited area. Useful intensive enterprises include poultry and pigs fed on bought feeds — the product of someone else's land. The distinction between intensive and extensive farm businesses lies in the capital and labour inputs per

hectare. Area alone therefore is an inadequate measure with which to compare firms. The enterprise combination of the whole business should be intensive enough to earn an adequate profit and also be adapted to the ecological environment. Poor knowledge of the basic properties and needs of land and their relation to climate has caused rapid land damage in some areas, for example, over-grazing of natural grasslands in East Africa and salinisation of irrigated land in arid areas.

Large area farms tend to predominate in plantation agriculture and in countries with a high average income per head. They vary from intensive systems, such as sugar production, to extensive ranching. Large area farms are often associated with abundant and relatively cheap labour, and shortage of capital to fully develop them. They are often run by companies to try to obtain extra development capital, failing which the farming system may remain extensive. As farm size rises, output per hectare sometimes falls but costs per hectare may also fall owing to economies of scale in labour and machinery use.

Some farmers believe that vast areas, irrespective of quality, represent wealth in somewhat the same way as subsistence farmers regard large cattle or buffalo herds as a store of wealth. However, a small area of well-watered, fertile land, which a farmer can afford to develop fully, is likely to give a higher, more stable income than a large area of arid land.

Peasant farming is characterised by small-holdings where the land farmed tends to be limited to the area that can be cultivated by the family using a few simple hand tools. Advantages of small area farms include, first, the possibility that land will be used more intensively, resulting therefore in a rise in national production and, secondly, maintenance of a high rural population that can live off the land even in hard times; although we all know that famine occurs in peasant farming communities.

LABOUR

Labour is the group of productive services provided by human physical effort, skill and mental power. It is the work input of people — not the people themselves. Labour is the tool with which capital and managerial skill are used to extract profit from the land.

There are many different types of labour input, varying in the effort and skill needed. Labour is not homogeneous; a seasonal cotton reaper needs different skills from a man who can care for 50 dairy cows or maintain a combine harvester. It is often useful, however, to refer to the amount of labour as if it were homogeneous. Apart from some cash-cropping areas, little regular labour is hired in peasant farming.

Labour input is usually measured in man-days or, sometimes, man-hours. These represent the input of work of an average man in a working day or hour.

Although capital, to some extent, can be substituted for labour by mechanisation, personnel management will repay detailed study. Each unit of labour hired is a human-capital investment whether it is a farm manager or a manual labourer. While the quantity of available labour in the tropics can be large and its daily cost low, it may be of poor quality and unreliable. This can make the cost of each effective labour input high in relation to output. Labour thus is often the biggest single cost on commercial farms in the tropics and its cost is rising constantly. Well-trained, intelligent workers are needed to handle valuable livestock and equipment. In a few minutes an untrained poorly motivated worker can ruin equipment worth his wages for many years!

Although some enterprises, like certain forms of pig and poultry production, need little land, labour is needed for all enterprises. Labour differs from land and capital because, like time, it cannot be stored. If not used when available, it is lost forever. Wasted labour does not collect in heaps like wasted materials. It escapes, like invisible gas, and disappears. Excess labour costs can go unseen for years unless management knows how to detect its waste. Employees, therefore, have real worth to an employer as ultimately they affect profit. Consequently, it is important to maximise labour productivity by giving time and effort to training and motivation.

CAPITAL

Capital consists of assets, resulting from past human effort, that are available to earn future income. These assets have either been produced on the farm or bought and have not yet been consumed but are still being used in production. Two important features of capital are, first, that its creation means giving up current consumption possibilities. Saving is required therefore to amass capital for future use. Secondly, capital normally raises the productivity of both land and labour. This is the reward for the sacrifice needed to create it.

Many capital assets are used in farming. They include livestock and both durable and consumable goods. Durable assets consist of buildings, roads, fencing, soil and water conservation works and any other permanent improvements to land; also machinery and tools. Consumable assets include crops and livestock in hand, materials in store and cash.

Capital consists of so many different items that is can be misleading to treat it as a single resource. Money is normally needed to obtain or produce capital assets and their value is usually measured in money terms. However, in planning capital use, it often helps to separate the different types of assets and measure them in physical terms. Machinery resources can be measured in available machine hours just as labour resources are measured in man-days. Buildings may be measured in cubic capacity or floor space, livestock in terms of animal units, crops in hectares and stocks in terms of physical amounts.

In ordinary speech, capital is often confused with money. It is important, however, to realise that money is not itself productive until it is invested. However, commercial farmers will usually keep some capital in money form until it is needed for investment in the business.

Some capital is needed before any production can occur. Shortage of farm capital often limits production and profit. There is often too little available to develop all the land or employ sufficient labour for profit maximisation. If a farm is under-capitalised relative to its production potential, as is common, farmers must 'ration' their use of capital. The capital a farmer controls is a measure of his access to other resources, as it normally can be used to buy land or hire labour and thus influence his scale of production. Farm management problems increasingly have become those of capital use and therefore capital should be seen as a managerial tool.

Large commercial estates in some countries have access to much foreign-owned capital. Conversely, large estates in southern Africa are generally financed with local capital. In peasant farming, shortage of capital apart from hoarded cattle is one of the factors that often prevents farms being developed to their full economic potential.

Much capital used in businesses is borrowed and interest is paid for its use. Charging of interest serves two useful purposes. First, it is an incentive to individuals and organisations to lend funds. Secondly, provided interest rates are fixed only by supply and demand, they should help to encourage the most efficient allocation of available capital. Those investments with the highest potential return can afford to pay more interest and will thus attract capital first.

Long term	Medium term	Short term
Fixed	Partly-fixed	Working
Land and improvements to it, e.g., buildings, fencing, dams, roads, some perennial tree crops.	Plant and machinery, some building, tools, some perennial crops, breeding livestock.	Seed, fertiliser, fuel, chemicals, feed, repairs, labour, rent trading livestock, growing annual crops, cash.

Figure 3.1   Types of Capital

As shown in figure 3.1, the capital used in a business may be divided into long, medium and short term. Both long-term and medium-term capital provide a flow of services over time; most medium-term assets last for up to 10 years whereas many long-term assets last much longer.

Fixed capital consists of permanent, durable assets that are fixed to the land such as tubewells. Partly-fixed capital is usually called 'semi'-fixed capital but it is not exactly 'half' fixed. This consists mainly of capital invested in movable assets for several years and therefore unavailable for day-to-day farm finance. These income-earning assets are eventually consumed in the production process.

Working capital sometimes is called circulating or floating capital. It consists of assets that are used quickly to produce a regular flow of income. This capital is completely consumed in the production process — normally within a year. Working capital is available for day-to-day running of the firm and its amount tends to vary with seasonal trading.

TIME

Perhaps the single most important skill, according to many practising managers, is that of managing time. Time, like labour, is lost forever if not used when available. It is an inflexible resource as it cannot be stretched. It seems to exist in unlimited future amounts, but waiting until more is available means delay. It is important for managers to practise time management because the timing of

decisions is often as important as their accuracy. There is no virtue in making decisions too quickly or too far ahead but neither is there any value in delaying them too long.

Planning the use of time needs more awareness of time; how much is available and how is it being used? Some managers have found it useful to record their own use of time. They often have been surprised by the waste disclosed.

Effective managers learn to find both long and short periods of unbroken time. Long, free periods are especially useful for jobs needing sequential thought like reading or writing reports and planning. Short free periods are useful for keeping on top of routine, small jobs like signing vouchers, correspondence and keeping charts and records up to date.

An obvious way to find long spells of undisturbed time is to work outside normal working hours. This may be necessary at peak periods but, if it becomes a regular habit, it could lower long-term effectiveness and imply poor work organisation or lack of delegation. A possible solution is simply to be unavailable on certain days or between certain hours. It should be made clear when you can be reached by telephone and trade representatives should be seen only by appointment. The statement that 'My door is always open' is not only often untrue, but can also be inefficient. Perhaps it should be 'My door is always open between 10.00 and 11.30 each morning'? Most daily interruptions, for many managers, relate to small matters. As each one occurs, the manager should think how it could have been avoided and then modify procedures so that such disturbances, such as noise, do not recur or are minimised in future. It is important to improve the *quality* of working time.

BUSINESS SIZE

Ideally, any measure of business size should include all resource inputs but it is hard to combine different inputs into a single measure. In a specialised business with a single product, total quantity of output could be used. Sales figures alone might be inadequate as some farms may sell wholesale and others retail. Other measures that have been used include hectares, net assets and number of employees. In Britain, the size of the business is often assessed in terms of standard labour requirements for the farming system on a given farm. The labour input needed will vary, however, both with yields and the degree of mechanisation. A further measure used to assess business size in industry is the number of levels of management in the organisational structure. A 'small' business may be regarded as one needing only one top man who can know everything that is going on without consulting records.

Small, family-run, farm businesses have the advantages of easy supervision, especially with mixed farming, and less dependence on the money economy. They are protected from the full force of the market, for example, as labour costs are less important.

Large businesses also have several advantages. They have more bargaining power in dealing with large-scale, commercial firms and can be more creditworthy. They can, for example, make full use of quantity discounts by buying in bulk. Scope for economic mechanisation is raised because the fixed costs of machinery and other overheads can be spread over more output. Lastly, management should be better as the farmer can spend more time on managerial work by delegating other work to staff.

The disadvantages of large farm businesses include poor management which sometimes arises if the business becomes too complex for an average man. Also, capital may limit development especially on big area farms.

It is often wondered if there is an optimum farm size. There cannot be one for all farms, however, as they engage in different types of production, with different methods and managerial skill. There possibly is an optimum size for a particular business with respect to the *total* quantity of substitutable resources used. Farm boundaries are usually fixed and therefore if land is limiting, business size can be increased by using more labour, capital or skill. The farmer should expand the proportion of intensive enterprises, raise yields and produce high-quality products.

In general, the business should be big enough to fully use each full-time worker and the most costly, single piece of capital equipment, for example, a tractor or battery house. It should be small enough for the farmer to manage it efficiently.

SOME USEFUL ECONOMIC CONCEPTS

Whatever the business, knowledge and use of the following basic economic concepts may help to maintain success or to avoid failure.

COMPARATIVE ADVANTAGE

This concept explains why any economic unit, whether farm, district or nation, tends to specialise and trade in those products for which its relative advantage is greatest or its relative disadvantage is least. Relative advantage may refer to climate, closeness to a market for perishable products or low population density.

A farmer who is less efficient than another in all enterprises may still have a comparative advantage for any product in whose production he is relatively more efficient. This is best shown by an example

			Relative advantage of	
	S'd Nama (A)	S'd Fakhry (B)	A(A/B)	B(B/A)
Cassava (t/ha)	6.8	6.4	1.06	0.94
Sorghum (t/ha)	2.0	1.4	1.43	0.70

Although S'd Nama has an absolute advantage for both products, he has a relative advantage for sorghum and S'd Fakhry has a relative advantage for cassava, i.e. 1.43>1.06 and 0.94>0.70.

If we assume that both farmers have 10 ha of land and need both products, then S'd Nama could produce 68 tonnes of cassava or 20 tonnes of sorghum. S'd Fakhry could produce 64 tonnes of cassava or 14 tonnes of sorghum. Now, if each planted half his land to each crop, they would get:

	S'd Nama	S'd Fakhry	Total
Cassava (tonnes)	34	32	66
Sorghum (tonnes)	10	7	17

However, if S'd Nama produced only sorghum and S'd Fakhry only cassava, there would be a greater total production.

	S'd Nama	S'd Fakhry	Total
Cassava (tonnes)	—	64	64
Sorghum (tonnes)	20	—	20

Consequently, although S'd Nama has an absolute advantage in both products, it could make sense for him to concentrate on sorghum production and supply S'd Fakhry in return for his cassava requirements.

Comparative advantage explains why, when market availability improves, farmers will change from producing mainly for their own needs to producing deliberate surpluses of chosen products for sale. They gain by choosing those products that give them the highest returns and buying some of their own needs from outside.

OPPORTUNITY COST

It often is appropriate to define the cost of a course of action in terms of the value of the most profitable alternative course that must be forgone. This cost may be very different from the market cost. It is not the money paid but the loss of income that otherwise could be obtained.

If a farmer is considering an investment of $1000 in a machine, the cost of that $1000 is an interest rate of say 15 per eent if the alternative to the investment is to use the $1000 to reduce an overdraft. However, it is 25 per cent if the alternative is some other investment costing $1000 and expected to yield 25 per cent return. This is the opportunity cost of the capital.

Opportunity cost is important in 'make or buy' decisions such as whether to breed livestock or buy weaners and the opportunity cost of a farm investment may well be the most the funds could earn outside farming altogether. For example, when planning small tropical farms, it often is necessary to use an opportunity cost for family labour. If unemployment is widespread, this will be below any minimum legal wage rate but it may vary with the season of the year.

With the best resource allocation, the returns to all resources will be at least as high as their opportunity costs; otherwise profit would not be maximised.

COMPLEMENTARY ENTERPRISES

Enterprises are complementary when transfer of resources to one raises the production of both. They aid or contribute towards each other's success by providing materials for each other, e.g. maize stems providing support for relay-cropped climbing beans; temporary pastures and following grain crops; paddy rice and fish production.

SUPPLEMENTARY ENTERPRISES

An enterprise is supplementary when its production can be raised without changing the production of any other enterprise. It does not need any limiting resource but can make productive use of spare resources that would otherwise be wasted. For example, a farmer's main enterprises may be dairy farming, maize and air-cured (bur-ley) tobacco. These may have used all the land but have left some labour and capital spare. Introduction of a seasonal porker enterprise with bought-in weaners would need no extra land but could use labour in slack months, skim milk, and the tobacco-curing barns would suffice as temporary·housing.

Other supplementary enterprises often include batches of poultry, fish, fruit, beef cattle to exploit otherwise idle grazing land or ducks in Viet Nam that scavenge fields for fallen rice after harvest; and in peasant farming, non-farming enterprises such as beer-brewing, hunting and trading, to help to even cash flow over the year.

It is essential, however, to calculate the scale on which a supplementary enterprise can be introduced without uneconomically raising overhead costs. What may be profitable as a sideline may well be more trouble than it is worth if allowed to expand so as to compete with the main enterprise. Supplementary enterprises to boost farm profit should be considered only when the main enterprises are being efficiently and profitably run. Too often, when panic-stricken by scarcity of funds, there is a strong impulse to embark on something new in the hope that it is some kind of magic potion.

ECONOMIES OF SCALE AND SIZE

Economies of scale are obtained only if *all* resources are increased at the same rate, that is, when no resources are fixed. This is rare in farming except when small farmers combine together to form co-operative or collective farms, as has occurred in Iraq.

Economies of size are more common and occur when growth in the scale of production raises production costs less than proportionately to output. As a result, long-run average production cost per unit of output falls. These economies arise through spreading fixed, overhead costs over greater output. There are both technical economies that arise within the farming process, such as better use of indivisible items like combine harvesters and specialist staff, and marketing economies that arise through large-scale buying and selling.

Diseconomies of size eventually arise owing to problems of supervision and communication.

MARGINAL ANALYSIS

Knowledge of marginal analysis is far more than desirable for decision-making — it is essential. A marginal change is a very small rise or fall in the total amount of some variable, such as output or an input. Marginal analysis is analysis of the relations between such changes in related economic variables.

Marginal analysis and marginal costing, as used in management, arose from the basic ideas of production economics but these have been slightly changed. A basic, economic idea is that of marginal change in something or the ratio of a marginal change in one thing to a marginal change in another. The economic principle of marginal analysis states that it is profitable to continue to supply inputs to an enterprise, for example, fertiliser or seed, so long as the return from adding each unit, for example, each kilogram or bag of input, exceeds the cost of that unit of input. This basic approach is behind much of the thinking of production economics conventionally taught to agricultural students. It includes ideas like optimum level of production occurring when marginal cost equals marginal revenue, diminishing marginal returns, the 'law' of equimarginal returns and so on. Fascinating as these concepts may be, they are of limited use to managers.

The manager's and accountant's approach, though less pure than the economist's marginal cost and marginal revenue theory, is more practical. It recognises that it is impossible in the modern firm to balance marginal costs and marginal revenues to maximise profit over some theoretical time period. It also allows for the 'cost of costing', the time available for making calculations, the level of accuracy needed and the diverse aspects of the modern firm.

The marginal principle refers to the extra output or input produced or needed by an enterprise. This can be measured either in physical or financial terms. In describing the return obtained from one more unit of input or one more unit of output, we can refer either to marginal physical product or marginal revenue. This is the extra total physical product or total revenue made by one more, or the last, unit of input or output. As regards inputs, we can refer either to the extra physical quantity of resources needed to produce one more, or the last, unit of output or to the marginal cost of producing one more unit of output or of supplying one more unit of input.

MARGINAL COSTING

The common word 'costing' causes more muddled thinking and wrong decisions than any of the complex phrases like discounted cash flow, programme planning, inventory control, breakeven analysis or other techniques. There are many errors and wrong ideas connected with costing. Managers speak of costs and costing without making it clear whether they are referring to actual costs or estimates; variable costs or common costs, etc. The easiest way to avoid confusion about costing is to ignore all figures. Many small firms consider that, as their prices are fixed by the market, all they need do is prevent obvious waste of labour and materials and keep their overheads down. Their managers try to justify this by pointing out that they are making profits. Their error is that they do not realise how much more profit they might have made if only they had known their cost structure.

Despite the fact that marginal costing has been known for 50 years, few firms use the technique. Those that do, do not disclose its benefits to their competitors. Marginal costing is much misunderstood. Contrary to popular views, it is not a magical way of making a profit while selling at prices below cost. It is simply way of looking at costs by classifying them by their behaviour instead of by their functions. Marginal costing is based on the idea that, at any given level of output, the only meaningful costs are those that would change if the level of output changed. While many costs vary with the volume of output, others, such as many overhead costs, are constant over a range of output levels. Marginal costing therefore is basically a study of cost:activity level: profit relationships and depends on finding which costs would change if the level or type of activity changed. It mainly gives managers information about the effects on costs and revenues of short-term changes in the volume or type of output although it may include study of long-term problems.

Conventional, full cost accounting, known as absorption costing, charges *all* costs to operations, processes or products, whether they are variable or not. The aim is to find the average cost of production and profitability (or net margin) of each enterprise or product sold. It thus entails splitting and assigning to enterprises the common costs of the whole business. Cost accountants tend to make a once and for all classification of costs as variable or common and often fail to distinguish between the long and the short term. Cost accounting, with its allocation of all cost to products and estimation of a profit for each was tried for 50 years in British farming but there were great problems in allocating all costs to enter-prises. It had little value in farming and was dropped as being too costly and, more important, misleading sometimes because the enterprise profits so obtained depend on an arbitrary allo-cation of common costs. Enterprises may seem to be unprofitable when, in fact, they are contribut-ing to profit. Many firms have failed through faulty decisions based on absorption costing. The whole business should be considered and not only separate enterprises. Common costs should be regarded as an indivisible lump whilst variable costs may be allocated to separate enterprises.

Table 3.1   Example of Absorption Costing

	Milk	Enterprise Maize	Cotton	Total
	($'000)	($'000)	($'000)	($'000)
Gross output	30	30	17	77
Material costs	18	13	7	38
Labour costs	7	8	1	16
Common costs	7	7	4	18
Total costs	32	28	12	72
Profit (loss)	(2)	2	5	5

Table 3.1 summarises the absorption-costing approach to a business producing three products — milk, maize and cotton. The common costs of $18,000 have been allocated to enterprises in proportion to their gross outputs. The mistake of charging all costs to products or processes can be seen from this example. It seems that the whole farm profit comes from cotton because the profit from maize is exactly offset by an equal loss on milk production. This makes it seem that whole farm profit could be raised merely by dropping milk production. However this would reduce total profit because the common costs allocated to milk would remain. As shown in table 3.2, total profit would fall from $5000 to nil and the implication now is that maize production should also be dropped.

One of the main merits of marginal costing is its simplicity because common costs are not allo-cated to cost centres (products or enterprises). As the technique tries to find the extra or marginal costs of producing any one item, it is necessary to distinguish between those costs that change with a change in output and those that, within limits, stay almost constant. This distinction is the basis of marginal costing whose supporters think it

Table 3.2   Effect of Discontinuing Production

	Milk	Enterprise Maize	Cotton	Total
	($'000)	($'000)	($'000)	($'000)
Gross output	—	30	17	47
Material costs	—	13	7	20
Labour costs	—	8	1	9
Common costs	—	11	7	18
Total costs	—	32	15	47
Profit (loss)	—	(2)	2	0

illogical to include in product costs those com-mon costs that will be incurred whatever the level or type of production. They suggest that these items should be regarded as costs of being in business and what is important is to maximise income within a *given* common cost structure.

The main point about marginal costing is that it regards profit differently from ordinary financial accounting and absorption costing. In these sys-tems, profit is regarded as the difference between total income and total costs. In marginal costing, this concept is modified. Marginal cost is the change in total production cost that occurs if the volume of output is changed by one unit at a given level of output. It is the avoidable cost of producing an extra unit of output. A unit may be a single article, a product or an enterprise. It relates to the change in output being considered.

Table 3.3 and figure 3.2 show the marginal costing approach to the situation shown in table 3.1. This shows that milk is contributing towards covering the common costs. Marginal costing directs attention to the need to maximise total contribution. Maize has the same gross output as milk but provides almost twice the contribution. Cotton has less output than maize but provides the same contribution. This clearly shows that turnover alone does not determine profit. The aim should be to maximise contribution from the resources available.

Table 3.3   Example of Marginal Costing

	Milk	Enterprise Maize	Cotton	Total
	($'000)	($'000)	($'000)	($'000)
Gross output	30	30	17	77
Material costs	18	13	7	38
Labour costs	7	8	1	16
Total variable Costs	25	21	8	54
Contribution	5	9	9	23
Common costs	—	—	—	18
Profit	—	—	—	5

An easy way to understand marginal costing is to think of enterprises as pouring their contributions into a tank representing the common costs. Thus no profit, or strictly speaking, net farm income, is made until the tank is full. Any overflow is profit; any gap between total contribution and total common cost is a loss. By separating common and variable costs, marginal costing overcomes the problem of allocating the overheads. Instead, it concentrates on maximising contribution, that is, the difference between gross income, or value of production, and all variable costs. This provides a fund to cover the common costs of the business; the surplus (or deficit) is the profit (or loss).

Marginal costing is useful in practical management. It is a particularly powerful tool for raising short-term profits in a multi-product business. Information gained by marginal costing is especially useful for decision-making as it concentrates on those revenues and costs that are relevant to assessing possible alternative courses of action by measuring their marginal effect on total profit. Information is given on those costs and returns over which enterprise (section) managers have direct control so that their individual responsibility for success or failure can be assessed.

The likely effect on profit of changes in volume of output becomes clear; similarly, the effect of alternative production methods, such as hand or machine planting can be assessed. Marginal analysis and costing also helps in making decisions about enterprise combination (product mix); such as introducing a new enterprise or changing the proportion of existing ones. Assessment of sub-contracting processes, such as buying-in stockfeed concentrates instead of growing them on the farm, is easier. Another field where marginal costing can help is in fixing selling prices when the market for a particular product is uncontrolled. Provided the product will make *some* contribution towards covering common costs, it will pay to accept an order at a stated price. This technique can be very useful when seeking business during a trade slump. Profit maximisation is not always possible. In fact, during a slump, the business may have to fight hard simply to minimise losses. Therefore any sales that cover marginal costs and contribute towards common costs will help. In this way, losses are minimised.

Marginal analysis and costing was first introduced into farming from industry in western Europe in 1956 by advisers, such as Liversage, to the relatively small farm businesses there. Consequently, its main use is in the small, multi-product farm. It is also of most use in the planning and control of annual crops and of livestock enterprises where the production cycle is no longer than a year.

Like most techniques, marginal costing should be used carefully, otherwise it could lead to financial ruin. It needs to be used in the correct circumstances and with a knowledge of its limitations. One problem is that a particular item may contain both common and variable cost elements and, in theory, should be split so that marginal cost can be found. This is often difficult and time-consuming. Another snag concerns pricing. Short-term pricing, as already stated, can be

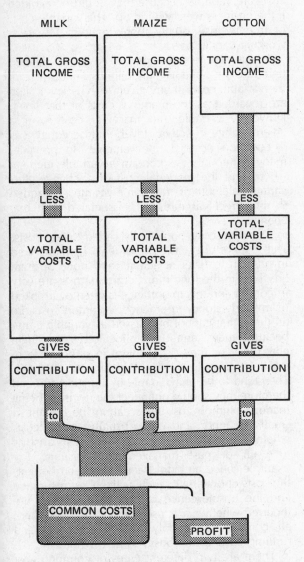

Figure 3.2 The Marginal Costing Concept of Profit

based on the product's contribution. However, in the long run, it is essential to have a reasonable margin from sales over and above *total* costs. Only by allocating common costs to products is it possible to see what price is needed to maximise long-term profit.

Not all accountants favour marginal costing. On one hand there are accountants who always avoid it. On the other hand, some accountants regularly use it. However, most accountants, while realising the value of marginal costing, also realise its limitations. Accountants responsible for financial control of large-scale company businesses do not usually use marginal costing to analyse company accounts. Reasons for this include the typical, necessary caution of accountants and the difficulty of combining the requirements of accounts for external publication with those for internal management use. Moreover, most company accountants have an industrial background where full or absorption cost accounting can be satisfactory.

The time when one set of costs was satisfactory for all uses is past. Costs must be adapted and arranged to meet the needs of a particular problem. This means that both orthodox and marginal costing have their uses. Marginal costing will not solve all problems but used properly it is a valuable aid to efficient management.

VARIABLE AND COMMON COSTS

One of the most important and most often misunderstood concepts in farm management, basic to many planning decisions and techniques, is the separation of variable and common costs. Provided they can be separated, there is no need to use the full costs of any product or enterprise as some of them will be constant whatever the farming system. Not only does this simplify planning, it avoids the almost impossible job of allocating common costs to enterprises.

Variable costs are often also known as operating, prime, direct or on-costs. They are all those costs that vary in roughly direct proportion to the level of activity, for example, to the area planted, the number of livestock or the volume of output. As shown in figure 3.3, they will rise steadily as the level of activity increases but remain constant for each hectare, head or unit of output. To a production economist, this is obviously oversimplified as it ignores, for example, both economies of scale and diminishing returns. Neverthe-

less, the concept is useful in practice. Besides varying in proportion to activity level, a variable cost must be specific to a certain enterprise. Most costs that vary with changes in the production pattern, scale of activity or yield will tend to come into this category. As shown before, an enterprise's variable costs are deducted from its gross income to find its contribution. Unfortunately, when marginal costing was introduced into farming, after it had been in use for about 30 years in industry, the descriptive term 'contribution' was changed to 'gross margin'. Features of variable costs are that they are usually avoidable if the enterprise is dropped and they can be controlled directly. In the long run, clearly, all costs are variable because they could be avoided if the business were wound up. The variable costs of an enterprise usually approximate closely to its working capital needs.

It may seem that variable costs, as defined here, are the same as marginal costs and, indeed, the variable costs at any given activity level often are regarded as the marginal costs at that level. However, variable and marginal cost, over a given change in activity level, will be equal only if common costs do not change. In practice, many common costs contain a variable element so, to find the marginal cost of a change, this variable element of common costs must be added to the direct variable costs associated with that change.

Common costs are usually called 'fixed' costs. This is unfortunate as only some of them are fixed in relation to level of activity and none of them are fixed in the long run. Common costs are very important and this importance is often overlooked or misunderstood. In practice, common costs are all those that cannot be classed as variable either because they cannot be allocated to a single enterprise, or they do not vary in direct ratio to small changes in the level of activity, or both. They tend to be unavoidable in the short run and therefore they should not affect the decision being made. Common cost items cannot be bought in small amounts and they usually incur costs whether they are used or not. These costs arise as a result of past management decisions. For example, once an extra tractor has been bought, the costs of *owning* it, such as licences, insurance and the obsolescence part of depreciation, are incurred whether it is used or not. Most, but not all, common costs apply to the maintenance and running costs of the business as a whole.

There are two different types of common cost — 'fixed' and integer costs. 'Fixed' costs, like

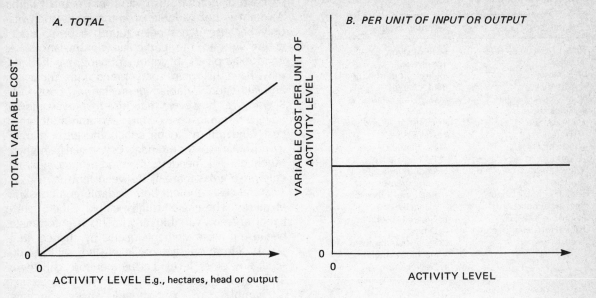

Figure 3.3   Relation between Activity Level and Variable Costs

rent, general overheads and interest on past borrowings, as shown in figure 3.4A, are constant, in
the short run, with any change in activity level.
They build up over time whatever the activity
level and are incurred even if nothing is produced. For this reason, they are sometimes called
period costs. In the long run, there are of course
no fixed costs. Integer costs are those that rise in
steps at different levels of activity but stay constant unless a fairly large change is made. An
example is the cost of owning, but not that of
using, a hand-operated cotton knapsack sprayer
and tailboom. One of these machines can cope
with about 4 ha of cotton. Consequently, in a
decision about expanding the cotton area from
5 ha to 7 ha, this integer cost will not change.
However, an increase from 7 ha to 9 ha would
need another machine and this extra integer cost
would be part of the marginal cost of such a
change. Other examples of integer costs include
the need to hire extra staff to cope with more
office work or supervision.

'Level of activity' can be regarded in different
ways. It may refer to the yield level per hectare,
per head or per enterprise or to the area cropped
or the number of livestock kept.

The higher the ratio between variable and
common costs, the greater control a manager has
over total costs.

It is important to specify the time period to
which marginal costing applies as the classification of a certain cost item will change with
time. 'Fixed' costs are fixed only at a given time
but not over a period of time. They vary with time
owing to changes in prices, market shortages, etc.
In the long run, no costs are fixed and in the short
run, they are nearly all fixed. After all, in the long
run, the farm could be given up and no costs
would arise. Again, once seed and fertiliser have
been applied to the soil, their costs become fixed
and will not vary with yield. The casual labour
cost for cotton reaping, however, will vary with
yield until the last stage of production. In practice, cost variability studies are based normally on
one year or one growing season.

In management decision-making, the costs that
are taken as common or variable in any particular
situation depend on both the time period
involved and on the sizes of the change being
considered. Integer costs will stay constant only
over a limited range of activity level. The size of
change being considered, therefore, will help
determine whether a particular cost should be
classed as common or not. There is no rigid division between variable and common cost items.
As shown in table 3.4, for example, tractor and
implement running costs and regular labour costs
may be classed depending on whether they are
avoidable or not and whether suitable physical
use records, by enterprise, are available.

Table 3.4   Typical Variable and Common Costs in Tropical Farming

Variable costs	Common costs
Raw material costs	Loan repayments, e.g.
– Seed	mortgages
– Fertiliser	Interest on existing loans
– Chemicals	Bank charges
– Livestock purchases	Managerial salaries
– Stock feeds	Rent
Contract hire	Administrative and office
– Transport	expenses
– Machinery	– Accountancy fees
– Labour	– Post and telephones
Veterinary expenses and	General overheads
dip fees	– Car expenses
Specific casual labour	– Road and other rates
Specific production	– Licences
insurance	– General insurances
Packing materials, selling	– Liming
costs, levies	Costs of *owning* fixed assets –
	machines, buildings
	– Depreciation[a]
	– Licences, insurance

[b] Costs of *using* fixed assets
 – Fuel
 – Repairs and maintenance
[b] Regular and non-specific, casual labour
[b] Electricity

[a] Depreciation contains both common and variable elements. To calculate the marginal cost of a change, these must be separated.
[b] These costs may be treated as either variable or common according to circumstances and availability of physical records.

Whether a particular cost item is regarded as variable or not will also vary from farm to farm. An item is not variable or common as such but only because it has been found to be so upon enquiring into the particular circumstances. A good example is in cotton spraying. Mr Kaunda may have his own sprayer and thus incur an annual, non-variable depreciation cost. Mr Sinyangwe, however, may hire a sprayer and so have a variable cost of hire. Assuming all other costs and returns to be equal, the gross margin (contribution) of Mr Kaunda's cotton will be higher. Much care is needed, therefore, in comparing enterprise gross margins between farms.

Some costs contain both variable and constant elements. These are usually called either 'semi-fixed' or 'semi-variable' costs. This is unfortunate because if these elements occur, by chance, in a 50:50 ratio at one activity level, they will not do so at any other level. Let us therefore call them 'part-variable' costs.

Examples of part-variable costs include depreciation, irrigation water, electricity and telephone charges. Depreciation due to age or obsolescence is a 'fixed' cost whereas that due to wear and tear varies with use. Water, electricity and telephone costs often include a fixed, minimum charge, or rent, and a charge that varies with use.

Sometimes, if a part-variable cost item is large, it *may* be worth dividing it into its common and

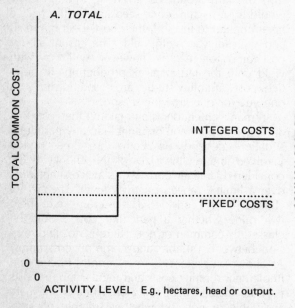

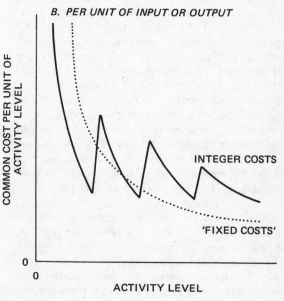

Figure 3.4   Relation between Activity Levels and Common Costs

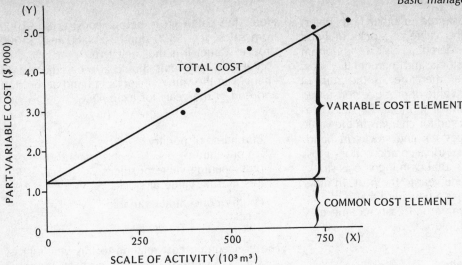

Figure 3.5   Separation of Common and Variable Elements in Part-variable Costs

variable elements so that the marginal cost of a proposed change can be estimated. There are several ways of splitting the two parts of part-variable costs. All give a rough division as precision is rarely possible. Graphical methods are often used. Statistical methods, like least squares regression, are also possible. Suppose, for example, the information in table 3.5 on irrigation water charges and water use is available over six quarters.

Table 3.5   Irrigation Water Charges and Water Used

	Irrigation water charges ($'000)	Water used ($10^3 m^3$)
Jan.–Mar. 1987	3.0	352
Apr.–June 1987	3.5	402
July–Sep. 1987	3.5	503
Oct.–Dec. 1987	4.5	554
Jan.–Mar. 1988	5.0	705
Apr.–June 1988	5.2	805

These values are plotted on graph paper, as shown in figure 3.5., and the line that fits the points most closely is drawn through them. Where the line cuts the vertical axis is the 'fixed' cost element. The variable cost element is related to the slope of the line; more precisely, it is the tangent that the line makes with the horizontal axis. Linear regression gives the equation $Y = 1.37 + 0.00496 X$. This means that the common cost element of the irrigation water charge is estimated at $1370 each quarter and that changes in level of activity cause a variable cost element of $4.96 for each thousand cubic meters of water used ($0.00496 \times \$1000$).

GROSS MARGINS

The concept of contribution from marginal costing has been used widely in farm management since about 1960. Within agriculture, it is usually called gross margin or, sometimes, gross profit.

The basis of gross margin analysis is that the farm is seen as a group of independent, productive enterprises, centred on the farm unit, which provides common services and the necessary co-ordination. Strictly speaking, gross margin is the difference between value of production and the marginal cost of that production. For practical purposes, however, it is taken as the surplus (or deficit) remaining after variable costs have been deducted from value of production or gross income. Table 3.6 gives some examples of enterprise gross margins derived from surveys of largescale, mechanised, commercial farms in Zimbabwe. Enterprise gross margins sort a shapeless mass of costs and returns into recognisable lumps. Provided common costs stay constant, then any increase in whole farm gross margin will raise profit by exactly the same amount. Consequently, provided an enterprise has a positive gross margin, it may be worth keeping even if its total costs, including overheads, exceed the value of production. However, it would be even better to replace it with an enterprise having a bigger gross margin.

Enterprise gross income is the total value of production and may differ from sales income. It also includes produce consumed on the farm, given as gifts or transferred to other farm enterprises. Changes in the value of stocks in hand must also be considered. This is important especially with livestock and perennial crops. Annual gross income, therefore, is the sum of produce sold and the value of produce consumed by the farm household or farm labour, less livestock purchases, adjusted for changes in the value of growing crops, livestock and stocks in hand. With annual crops, like tobacco and cotton, gross income usually equals total cash income as stocks are rarely kept from one year to the next. In other cases, however, such as a beef herd that is changing in size, gross income and cash income may vary widely. For example,

	$	$
Closing value of cattle	9000	
Cattle sales	1000	10000
Opening value of cattle	10000	
Cattle purchases		
Cattle gross income (deficit)	500	10500
		(500)

Here, the cattle gross output consists of $1000 from sales, less $500 purchases and less $1000 drop in value. It is thus negative.

Variable cost items should also be adjusted for changes in the value of stocks in hand at the start and end of the year, for example,

	$	$
Purchases of poultry concentrates	1200	
plus opening value of stocks	50	1250
less closing value of stocks		100
Poultry concentrate variable cost		1150

The concentrate cost relating to this year's gross output is $50 less than the cost of purchase because $50 worth of purchases remains in hand for next year's production.

It is important to account for transfers of products from one enterprise to another on the farm so that valid comparisons can be made on the relative profitability of enterprises. If a product is

Table 3.6   Examples of Gross Margins for Mechanised, Commercial Farming

	($ per hectare) Enterprise			
	Flue-cured tobacco	Maize	Rainfed cotton	Rainfed groundnuts
Yield	1580 kg	5820 kg	1894 kg	1180 kg
Gross income	1915	400	617	433
Fertiliser	141	100	57	27
Chemicals	110	11	107	12
Fuel	49	21	18	15
Tractors and implements	46	21	18	15
Machinery contract	1	2	22	—
Labour	301	49	128	93
Building repairs	11	—	—	—
Coal	73	—	—	—
Seed	—	16	8	40
Insurance	56	—	3	—
Packing	22	15	7	—
Transport	17	12	10	—
Levies/selling costs	78	2	9	—
Other	26	5	19	18
Total variable costs	930	254	406	220
Man-days	385	55	141	170
Gross margin/hectare	$985	$146	$211	$223
Gross margin/man-day	$2.75	$2.65	$2.88	$1.31
Gross margin/$ variable cost	$1.06	$0.57	$0.52	$1.01

(Whole farm common costs = $27000)

saleable, such as grain fed to stall-feeders, an in-calf heifer transferred from the rearing to the milk-production enterprise or a weaner pig transferred from a breeding to a fattening enterprise, it should be valued. It is valued at market price less what it would have cost to market it. This is its opportunity cost and should form part of the gross output of the giving enterprise and be one of the variable costs of the receiving enterprise. This is most important for making decisions such as whether to buy replacements or to rear them on the farm, whether to have a fattening enterprise or to sell weaners, etc.

Grazing and fodder crops, such as silage, are not usually saleable so they have no gross income or gross margin, only variable costs. These are allowed for by adding them to the variable costs of the consuming livestock enterprises. If they are consumed by more than one livestock enterprise, their variable costs should be split in proportion to the quantity used by each consuming enterprise. Satisfactory gross margin costings clearly need fairly complete and accurate physical records and financial accounts. This is particularly so because errors can multiply alarmingly with changes in activity level.

It follows from the nature of variable costs that there should be a constant relationship, within a given enterprise, between gross income, variable costs and gross margin, whatever the size of that enterprise, provided technical efficiency, costs and prices stay constant. This helps when gross margin analysis is used for planning. It forms the basis of most analysis and planning procedures and enables a practising farmer to understand his business better.

Gross margin planning helps to answer questions such as, 'What would be the likely financial effect of:

- changing the scale of production of an enterprise or dropping it?'
- buying in stockfeeds instead of growing them?'
- a change in prices or costs?'
- a change in a contract?'

It is often possible to plan with gross margins alone and ignore common costs. In most parts of the tropics peasant farmers have few common costs. They do not pay rent, or wages to family labour, they have hardly any buildings or equipment and borrow little capital. Most of their costs are variable so most of the total gross margin is family income. Increasing enterprises with high gross margins will usually raise profit at the expense of those with low ones.

Marginal analysis techniques are useful especially when the production cycle is a year or less. When it is less than a year, the actual period for, say, fattening a pen of cattle is used. This enables costs and returns to be directly related to a particular crop or batch of livestock.

If land is the scarcest resource, whole farm gross margin will be maximised by including enterprises with the highest gross margins per hectare. However, labour or capital is often the most limiting resource and the farmer should then try to maximise gross margin per man-day or per dollar spent. Table 3.6 shows that if land is most limiting, enterprises should be maximised in the order of tobacco, groundnuts, cotton and maize. However, if capital were more limiting than land the order should be tobacco, groundnuts, maize, cotton and, if labour is most limiting, cotton, tobacco, maize, groundnuts. Planning for the highest gross margin per unit of the scarcest resource is a useful short cut but any changes in common costs, capital or labour requirements must be considered.

The farm business will be unviable unless it earns a sufficient whole farm gross margin to cover its common costs and earn an acceptable profit. The target whole farm gross margin needed should be decided early in the planning process. This means listing all expenses to be met other than enterprise variable production costs. The whole farm gross margin target must cover:

- *General farm overheads*. These can be obtained from previous years financial statements. They include rent, rates, accounting fees, depreciation, office expenses, interest, bank charges and unallocated labour and machinery costs.
- *Tax liability* for either income or company tax. An estimate should be obtained from whoever prepares the accounts.
- *Repayments* on long- and medium-term loans. It is vital to allow for these annual cash expenses.
- *Capital costs* for buying new capital assets that will be depreciated slowly over the years.
- *Development costs*. The cost of present development plans must be met.
- *Basic living expenses*. This will vary from farmer to farmer as his personal situation, for example, insurance premiums, school fees and alimony, must be considered.

The sum of all these indicates the whole farm gross margin needed to meet all commitments. Any farming plan that fails to meet this target,

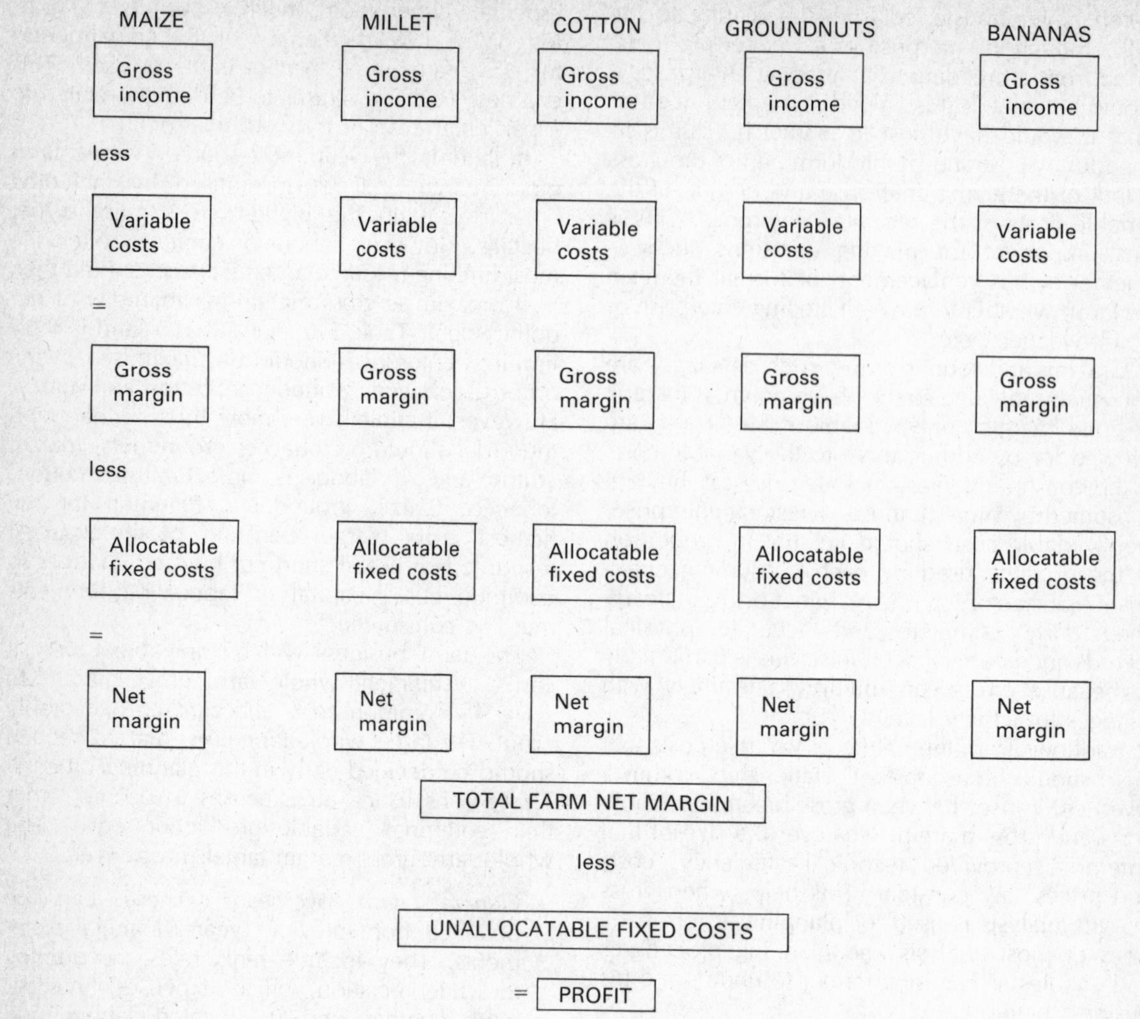

MAIZE	MILLET	COTTON	GROUNDNUTS	BANANAS
Gross income	Gross income	Gross income	Gross income	Gross income

less

Variable costs	Variable costs	Variable costs	Variable costs	Variable costs

=

Gross margin	Gross margin	Gross margin	Gross margin	Gross margin

less

Allocatable fixed costs	Allocatable fixed costs	Allocatable fixed costs	Allocatable fixed costs	Allocatable fixed costs

=

Net margin	Net margin	Net margin	Net margin	Net margin

TOTAL FARM NET MARGIN

less

UNALLOCATABLE FIXED COSTS

= PROFIT

Figure 3.6 Net Margin Analysis

with a bit over for the unavoidable bad season, will be unviable.

Gross margin analysis enables the detailed results of different enterprises to be compared on one farm. It also enables a particular enterprise result to be compared with that of other farmers to help find possible technical weaknesses. It cannot be used however to answer questions such as 'How profitable is my farming compared to Reuben's over the river?' Reuben may have a high total gross margin that is swallowed up by excessive common costs whereas your total gross margin may be lower but be offset by even lower common costs. Because of differences between farms in yields and inputs used, the safest use of gross margins is in comparing different enterprises on one farm rather than in comparing one enterprise on different farms with different types of management.

Gross margin planning can be dangerous unless used carefully. Common costs must also be considered. This becomes more important in commercial farming as the size and complexity of any proposed change rises.

NET MARGINS

An improvement to the gross margin concept has been proposed recently by Professor A.K. Giles of the University of Reading's (U.K.) Farm Manage-

For Each Enterprise

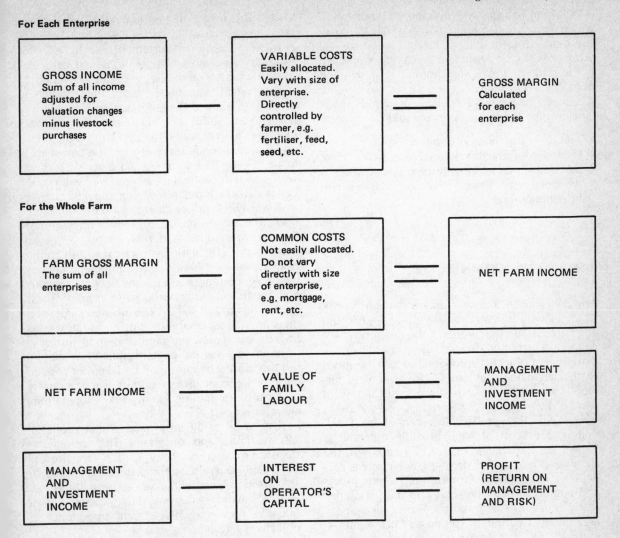

Figure 3.7   Gross Margin Analysis

ment Unit. This is to subtract from each enterprise's gross margin all common costs that are clearly allocatable (or assignable) to a particular enterprise or, in the case of a group of similar enterprises such as annual grain crops, to that 'sector' as a whole.

These common costs are those that are absorbed by that enterprise or sector — whether or not they would 'go' if the enterprise did. The result is known as a *net margin* and is shown diagrammatically in figure 3.6. This net margin concept relates specifically to the whole enterprise situation. Nothing would be gained — except misleading conclusions — by expressing it 'per unit of production'. While helping to increase understanding of how each enterprise has

contributed to the whole farm profit, net margin still does not measure the ultimate profit from each enterprise. To try to do so is elusive, pointless and dangerous.

NET FARM INCOME

Net farm income is sometimes called net income or net profit. It is the income from the business that pays for the farmer and his family's physical and managerial effort and interest on his own capital invested in the business. It includes the value of farm products consumed by him and his family so it is not necessarily cash income.

Net farm income will disappear completely if whole farm gross margin is too small to cover common costs. In such a case, a net loss will occur resulting in reduction of capital employed in the business unless the shortfall is made good from outside sources.

The alternative ways available of increasing net farm income, or reducing losses are:

- reduction of common costs
- reduction of enterprise variable costs
- increased level of activity provided this raises whole farm gross margin by more than any rise in common costs.

RISK AND UNCERTAINTY

Any decision or course of action is liable to risk or uncertainty if several possible outcomes could arise from it, in other words, if the outcome is not certain. As strictly defined in economics, risk exists when objective probabilities can be given to the alternative outcomes. However, few farmers can predict risk in any statistically valid way. Where there are alternative outcomes to which objective probabilities cannot be fixed, that is, the form of each possible outcome is known but the chance of it occurring is not, the situation is one of uncertainty. Uncertainty is particularly important in farming and affects inputs and outputs in both the short and the long term. Many situations that in practice are called 'risky' are, on strict definition, prone to uncertainty — not risk. Expectations are often based on subjective judgement which will vary between farmers in any given situation. In this way, a 'quasi-risk' situation is set up and planning data can then be quantified. Risk, but not uncertainty, can be insured against.

Having defined the theoretical difference between risk and uncertainty, to satisfy purists, we shall now call them both risk. All production involves taking risks as cost and effort are incurred before the final outcome is know. Decision-making in situations of risk is an important, managerial skill and basically is what makes planning so necessary. The further a manager plans into the future, the more risky is the outcome.

Many types of risk affect yields, inputs and prices. Variations in yield are caused by the natural hazards of farming such as crop failure, due to drought, pests and diseases, and storage losses. This means that an individual farmer does not know in advance what his output will be. However, good management such as, for example, timeliness of operations and early detection of pests and diseases, can do much to reduce yield variations. Input uncertainty is especially marked with labour and machinery inputs. Availability of casual labour or contract machine hire cannot be forecast far into the future and seasonal conditions will affect the time it takes to do a job. Bad weather can both slow down work and reduce the days when it is possible to do certain work. Input prices can quickly change, as when OPEC prices change for diesel and fertiliser. The same applies to product prices unless they are controlled and pre-planting prices are announced. Technological change also brings risk. As new methods and equipment become available, current practices may become outdated and inefficient. Other risks arise from sickness, injury, death and the actions of others. Few farmers provide for disability or death so unexpected tragedy can cause the family, despite insurance payouts, to suffer undue hardship and, maybe, to sell the farm. The farm may be taken for residential development or for building a road, railway or airstrip, or mining claims may stop farming operations.

The result of all these risks is that profits are also variable and uncertain. This complicates decision-making as planning data vary not only over time but also unpredictably. Farmers are risk-avoiders and will forgo some income each year to avoid occasional large losses. Consequently, they may place great importance on objectives like income stability.

Personal assessment and beliefs about the form and probabilities of outcomes, and the effects of these, vary. However, because risk is common to all businesses, and particularly farming, much attention is paid to reducing it. It has been described already how peasant farmers often reduce risk by producing less than the possible average output each year so as to maximise output in the worst seasons.

Risk-bearing ability is the ability to survive unexpected low income and unforeseen costs and still stay in business. This varies between farmers as they differ in resource availability, especially in capital reserves, or savings, and off-farm income. Ability to withstand risk falls as the risk increases and rises as assets increase. However, a wish for security is normal in all farmers — whether subsistence or commercial.

Farmers may practise some or all of the following methods to improve their ability to withstand the effect of risks:

- building up capital
- various methods of stabilising income
- getting a good credit rating
- lowering living standards
- less borrowing
- insurance against natural hazards

Various risks reduce income and hence the ability to repay debts. The more reserves can be built up, the easier it is to meet unexpected losses or costs. Every investment should be assessed on its ability to cover costs.

Several measures help to stabilise income in high risk areas. These include choice of reliable enterprises. Diversification of enterprises often helps as different enterprises react differently to seasonal conditions. This is shown in figure 3.8 for commercial farmers in Zimbabwe. It should be remembered, however, that over-diversification tends to stretch managerial skill too widely for efficient production.

Flexibility, that is, adjusting the business to fit new circumstances, also helps to stabilise

income. It may refer to cost, time or products. Cost flexibility is achieved by keeping overhead costs low in relation to total costs. This can be done, for example, by renting instead of buying or having low- rather than high-cost machines and buildings. Time flexibility refers to the length of the production cycle. For example, tree crops and breeding herds involve long time periods and inflexibility compared with annual crops or broiler production. Product flexibility means ability to adjust the product to suit changing conditions. For example, ranchers can sell either weaners or slaughter stock. Crops like millet can be stored until the price rises; tomatoes cannot.

Other ways of stabilising income include contracting prices in advance, to reduce risk of market and price changes, and insurance. Some risks can be assessed accurately when many farmers are considered, whereas the risk to a single farmer is hard to predict. For example, fire losses over many farm buildings are similar from year to year and, if a certain charge or premium is paid by sufficient farmers, it will be enough to cover fire losses and pay the costs of running the insurance scheme. By insurance, a private company or state

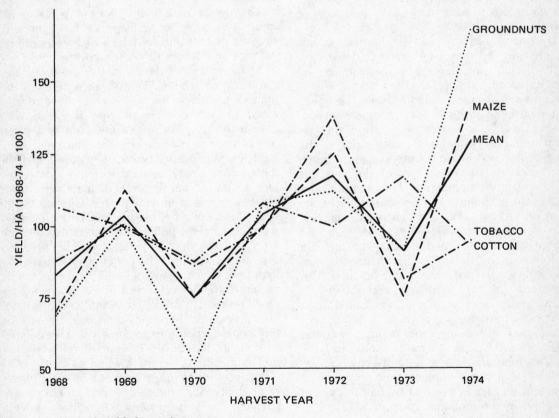

Figure 3.8 Annual Yield Fluctuations

organisation guarantees to pay a certain sum in the event of a disaster in return for a relatively small, annual premium. Farmers can thus exchange the risk of a large loss for a known cost. They can choose between profit maximisation in the short run and security in the long run. Some risk reduction, of course, can be undertaken by farmers themselves, for example, use of night watchmen and guard dogs, fodder banks, irrigation and other provisions against drought.

Whenever possible, events that could ruin the business should be insured against. The amount of insurance that a particular farmer should have depends on:

- existing risks and the effect these could have on the business
- his income and ability to buy insurance and, at the same time, pay other necessary farm and family expenses

These factors vary between farmers and each changes for the same farmer at different times. Every farmer should study these points for himself, getting advice from reputable insurance companies or brokers. Once taken out, insurance should be revised regularly as conditions change. It is dangerous to have inadequate or ineffective cover.

The main risks that can be insured against include fire, lightning, storm and hail damage, theft, liability for damage caused to or by vehicles, death, sickness and accidents. Fire is one of the main risks insured against. The policy should be for an adequate amount, otherwise payment is likely to be limited to a share of the loss equal to the proportion of the total value of the property that was insured. Farmers can do much themselves to reduce the risk of fire and this could reduce their premiums. For example, fire-breaks can be built, at an oblique angle to the usual dry season wind, through natural grazing areas. Ten-metre wide fireguards can be cleared around cribs and stacks, smoking can be forbidden in certain places, thatched roofs can be treated against fire and wire mesh can be put over the flues in tobacco curing barns.

Damage to crops from wind and hail is common. Hail insurance is usual in tobacco production and is becoming more popular in cotton production. A typical scheme is shown in table 3.7. Premiums are reduced for growers who have made no claims for compensation in previous years and damage is assessed by hail inspectors appointed by the insurance company.

Everyone is always liable to claims from the

Table 3.7 A Typical Scheme for Hail-damage Insurance in Tobacco

Max. gross compensation/ha	Annual premium/ha
$	$
3000	175
2750	161
2500	146
2250	131
2000	117
1500	88
1000	58

public or employees. Both should be fully covered with provision for all legal costs of defending or settling a claim. The sort of risks covered by such a policy would include damage done by fire spreading from the farm, poisoning by farm produce and any injuries to staff or the public. Personal liability, as an individual, is not covered by a farm policy. Cover for this should also include legal costs and have no indemnity limit. One policy will usually cover the whole family and this cover may be included in the household policy that insures the farm home and its contents.

The owner of a motor vehicle is liable for injury to people or damage to their property because of accidents for which he is responsible. All uses and drivers should be covered.

Many types of life assurance are available. Ordinary or whole life assurance requires an annual premium for a lump sum at death. However, the commonest type is endowment assurance that gives an assured sum at a given age or after so many years, the sum being fully paid if death occurs before it matures. Endowment policies may be used to repay mortgages. Life assurance satisfies one of a farmer's main needs which is to provide for his wife and family if he dies. Tax relief usually can be claimed on premiums and the policy is an acceptable security for a bank overdraft and often for a loan from the insurance company. The policy may guarantee a loan from the company of up to 95 per cent of its present surrender value and this is unaffected by credit restrictions on banks. When an endowment policy matures, the lump sum could be used to buy a paid-up policy or an annuity. If a life annuity is bought, the buyer is guaranteed a regular income for life but nothing when he dies.

It is possible to insure against temporary illness or disability. Such policies or medical aid society payments help the insured when ill or provide specified sums for permanent injury.

Table 3.8   Illustration of Rising Risk with Falling Per Cent Equity in Bad Years

| | Per cent equity | | | |
	100	75	50	25
Owned capital	12000	12000	12000	12000
Borrowed capital	0	4000	12000	36000
Total capital	12000	16000	24000	48000
GOOD YEARS (20 % return on capital)				
Total return	2400	3200	4800	9600
Interest on loans at 10 %	0	400	1200	3600
Profit left	2400	2800	3600	6000
Return on owned capital	20.0 %	23.3 %	30.0 %	50.0 %
BAD YEARS (5 % return on capital)				
Total return	600	800	1200	2400
Interest on loans at 10 %	0	400	1200	3600
Profit left	600	400	0	(1200)
Return on owned capital	5.0 %	3.3 %	0	(10 %)

A good credit rating, developed over the years, makes it easier to borrow in both good and bad times. This comes with honesty, responsibility, and a history of satisfactory loan repayment. It helps in a bad season. Lower living standards also increase risk-bearing ability because costs fall.

Raising the proportion of owned capital used increases risk-bearing ability partly because it lowers the risk of loss in bad years, as shown in table 3.8, and partly because borrowing ability depends so much on assets available as security. Table 3.8 shows the principle of increasing risk. In profitable years, loans raise the return on owner's equity but, in poor years, this drops as the proportion of loan funds rises. Hence, credit exposes farmers to possible losses by increasing risk.

GLOSSARY

Assurance cover against some event, such as death, that must occur sometime, but when is uncertain.

Broker a person joining buyers and sellers in a well-organised market.

Cover protection against risk of loss.

Indemnity compensation for loss.

Insurance cover against some event, such as fire, that may never occur.

Premium payment for insurance cover.

Slump a big or quick drop in price or demand, e.g., a depression — often causing unemployment.

RECOMMENDED READING
Books

Austin, B., Time, The Essence (British Institute of Management Foundation, 1979).

Bannock, G., Baxter, R.E. and Rees, R., Dictionary of Economics (Penguin, 1972).

Barnard, C.S. and Nix, J.S., Farm Planning and Control, 2nd ed. (Cambridge University Press, 1979).

Brech, E.F.L. (ed.), The Principles and Practice of Management, 3rd ed. (Longman, 1975).

Bridge, J. and Dodds, J.C., Managerial Decision-making (Croom Helm, 1975).

Capstick, M., The Economics of Agriculture (George Allen & Unwin, 1970).

Harrison, J.E., Profit from Farming (Rhodesian Farmer Publications, 1969).

Ministry of Agriculture, Fisheries and Food, The Farm as a Business, Book II (HMSO, 1963).

Ministry of Agriculture, Fisheries and Food, Terms and Definitions used in Farm and Horticultural Management (HMSO, 1983).

Redding, W.J., Managerial Effectivenes (McGraw-Hill, 1970).

Sizer, J., An Insight into Management Accounting (Penguin, 1969).

Stone, J.A., Accounting for Management in Agriculture (Angus & Robertson, 1968).

Upton, M., Farm Management in Africa (Oxford University Press, 1973).

Vause, R. and Woodward, N., Finance for Managers (Macmillan, 1981).

Periodicals

Giles, A.K., 'Net Margins' Farm Management (1987) 6.6.

Sizer, J., 'The Terminology of Marginal Costing', Management Accounting (August 1966). 'Marginal Cost: Economists v. Accountants', Management Accounting (April 1965).

Upton M., 'A Development of Gross Margin Analysis', J. Agric. econ. (1966), p. 111.

EXERCISE 3.1

As the new manager of an estate recently opened from virgin bush in a previously undeveloped area, for commercial maize production, you must decide how much fertiliser to apply to the first maize crop. The only information available on the likely local response of maize to fertiliser is the mean of three years' trials on a research station in a similar ecological area 80 km away. This is as follows:

Kg fertiliser/ha	Kg maize grain/ha
0	2714
200	3724
400	4556
600	5130

You expect fertiliser response in commercial farming to be half that got on the research station. According to your estimates, the variable cost of harvesting, shelling and transport will be about $9/tonne of grain.
(a) What is the best fertilisation rate if fertiliser costs $140/tonne and grain fetches $65/tonne?
(b) You have heard that fertiliser price may rise by 30 per cent and the marketing board may raise the price for grain to $75/tonne. If this occurs, what would the best application rate become?

EXERCISE 3.2

Mr Seushi has never fertilised his maize. However, he has just seen a field trial on his neighbour's plot and he wants to use fertiliser next year. The results on his neighbour's plot were as follows:

Sulphate of ammonia (kg/ha)	Yield of shelled maize (bags/ha)
0	15
150	24
300	30
450	35
600	39

A pre-planting price of $3.85/bag of maize has been set for next season but, if he stores the maize and sells to neighbours six months after harvest, he expects to get $6.00/bag. Storage would cost 10¢/bag. Sulphate of ammonia costs $7/50 kg bag but it may rise to $10 next year.

Mr Seushi wants to plant 1.5 ha maize next year and he could find up to $70 for fertilising it. What advice would you give him?

EXERCISE 3.3

You own a car that you bought on hire purchase and the following information is available.
(a) Purchase price — $4500
(b) Expected sale value after three years — $950
(c) Fuel consumption — 9.00 km/litre
(d) Fuel cost — 28¢/litre
(e) Tyres — a new set costs $190 and should last 20 000 km
(f) Licence and insurance — annual cost $180
(g) Regular servicing every 5000 km costs $30
(h) Extra repairs and maintenance – 0.95¢/km
(i) Hire purchase repayments over two years are: $187.50/month capital, plus $52.50/month interest
(j) Distance covered — existing travel is 1500 km/month

You are asked by the company you work for to do 300 km/month paid travelling on company business. They offer you 8¢/km. Should you accept or not?

EXERCISE 3.4

The company accountant has been analysing some figures on five 50 kW tractors, of the same make, sold this year. The following data were obtained.

50 kW tractors

Tractor no.	New purchase price	Price received on sale/trade-in	Age at sale (years)	Total hours worked
	$	$		
1	5600	5000	1	10
2	4000	850	4	4000
3	4475	1340	3	4500
4	4500	825	3	6000
5	5625	3500	1	2500

Find out:
(a) the obsolescence cost of depreciation each year for this type of tractor (using a graphical approach)
(b) the wear and tear cost of depreciation for *each* tractor per hour
(c) which of the two is 'fixed' and which is variable?

EXERCISE 3.5

On starting as a section manager for Kanyama Ranches Ltd, you extract the following information on the herd that you have just taken over:

No. on 1.10.86	Type	Value per head $	No. on 1.10.87
14	Bulls	300	13
350	Cows	225	345
125	Heifers 2–3 yrs	210	127
130	1–2 yrs	125	137
120	Steers 2–3 yrs	220	125
127	1–2 yrs	150	110
255	Calves	95	300

Purchases (1986–87)
25 Steers at $4625
20 Heifers at $3000

Sales
  70 Cull cows at $1700
  65 Cull heifers at $14300
  145 Steers at $34250

Births
310

Deaths
  1 Bull
  5 Cows
  10 Heifers, 2–3 yrs
  3 –   1–2 yrs
  10 Steers
  10 Calves

(a) Caluculate the gross income of the herd in 1986–7.
(b) Using the following conversion factors, work out the average herd size in Animal Units (AUs).

Type	AUs
Bull	1.27
Cow	0.91
Heifer 2–3 yrs	0.73
– 1–2 yrs	0.54
Steer 2–3 yrs	0.82
– 1–2 yrs	0.64
Calf	0.36

(c) For one month each year the herd is fed on the residues of maize produced for grain for both sale and feeding to it later. Without maize residues in this month, the herd would need to be grazed on a neighbouring estate at a cost of 35¢ a head. How much should be added to both the maize enterprise gross income and the beef enterprise variable costs for the value of these maize residues?

EXERCISE 3.6

Mr Ode planted 7.5 ha of crops in 1987–8. This consisted of 3.0 ha maize and 1.5 ha each of tobacco, groundnuts and cotton. During this season, he sold 64 bags of maize at $3.85 a bag and consumed nine bags. There was no change in stocks of maize in hand at the start and end of the year. During the year, Mr Ode sold 1100 kg of tobacco at 35¢/kg and 1450 kg groundnuts at 16.5¢/kg. He had 350 kg groundnuts in store on 1 September 1987, and 275 kg on 1 September 1988. He used 90 kg groundnuts on the farm during the year. Cotton sales amounted to $312.

Seed costs were as follows: maize — $20; groundnuts — $45; cotton and tobacco — free. Each enterprise received the following fertiliser: maize — 700 kg sulphate of ammonia and 700 kg 20:20:0 compound fertiliser; tobacco — 550 kg 20:20:0; cotton 350 kg 20:20:0. Sulphate of ammonia cost $7/50 kg bag and 20:20:0: cost $10/50 kg bag. Other costs were $3.50 for insecticides to control stemborer in maize and $40 for cotton-spraying chemicals. The cotton sprayer cost $50 and had an expected life of four seasons. The 16 tobacco-curing barns cost $20 each to build and are expected to last for eight years. Implements for ox-draught cost $75 and should last for six years.

Calculate enterprise gross margins, the common costs of the farm as a whole and Mr Ode's net farm income.

EXERCISE 3.7

The following information, on page 54, has been obtained for a typical peasant farm of 3 ha growing 2.0 ha maize, 0.4 ha groundnuts, 0.4 ha cotton and 0.2 ha vegetables for sale.

INPUTS/HA

Crop	Seed (kg)	Fertiliser (kg)	Chemicals (litres)	Casual labour (days)	Mech. unit hire (hours)	Transport (tonne km)
Maize	25 at 42¢	100	—	5	5	45
Groundnuts	90 at 63¢	—	—	80	7.5	—
Cotton	12 at 55¢	50	10	(3.5¢/kg)	5	75
Vegetables	2.5 at $4.20	150	30	—	10	—
Unit Price		$140/tonne	$1.40	$0.56	$3.50	5¢/tonne km

GROSS INCOMES

Crop	Yield/ha	Price/unit yield
		$
Maize	20 bags	6.30
Groundnuts	10 bags	21.00
Cotton	1500 kg	0.42
Vegetables	2000 kg	0.28

(a) Calculate the enterprise and whole farm gross margins.

(b) Given the following area average, suggest possible ways of improving both the technical efficiency of enterprises and the enterprise combination.

AREA AVERAGES/HA

	Maize	Groundnuts	Cotton	Vegetables
Yield	30 bags	5 bags	600 kg	2500 kg
Price	6.30	17.50	0.35	0.31
Gross income	$189.00	87.50	210.00	775.00
Seed	14.00	21.00	5.50	14.00
Fertiliser	21.00	—	3.50	70.00
Chemicals	0.70	—	22.50	42.00
Casual labour	5.60	28.00	35.00	7.00
Mech. unit	20.00	31.50	21.00	45.00
Transport	4.55	—	2.50	—
Total var. costs	65.85	80.50	90.00	178.00
Gross margin	$123.15	7.00	120.00	597.00

(c) Consider the possibility of bringing fire-cured tobacco, which has a gross margin of $500/ha, into the farm system. The total area of suitable soil for tobacco is 2 ha and tobacco can be grown on the same land only once every four years. One curing barn, costing $20, is needed for each 0.1 ha. The farmer cannot spend more than $70 on building curing barns or use more working capital than he is now. Tobacco has a working capital requirement (total variable cost) of $140/ha.

4   Financial planning

One of the most important responsibilities of a manager is that of financial planning and control which may be summarised as:

- Setting financial objectives.
- Planning to achieve those objectives.
- Managing the business or enterprise to ensure that the plans are achieved.

Most of the activities of managers have an effect on his business finances.

Financial management means obtaining funds for the business and making the best use of them. Funds are needed to buy the business, obtain fixed and part-fixed assets, like buildings and machinery, and to provide working capital, especially to pay wages and other expenses, until cash starts flowing into the business. The growing capital intensity in farming poses financial problems. One of the most serious problems is keeping enough funds flowing through the business. It is essential when starting to farm to decide how much finance is needed. Few farmers can start with sufficient finance of their own and inadequate finance can prevent adoption of the potentially most profitable farming system.

Before changes in farming system are made, farmers should estimate the capital needed. Sufficient finance to cover costs must not only be available, it should be planned and a sound plan needs accounting information for realistic forecasts. Estimating funds needed takes time and patience but it is not difficult. It should assure a farmer that his plans will yield a reasonable profit without danger of running out of funds. Not only the growing amount of capital needed in farming makes availability of credit so important; a business must get finance when it is needed. If there is a regular income from sales and a regular outflow of payments for production costs, no serious problem arises. However, most farm products take months or years to produce and the farmer must live and meet production costs until they are sold and bring in income.

Shortage of cash helps new farmers develop economical methods but, still, more farmers start with too little capital than with too much. Inadequate capital usually results either in failure or in delayed payments and excessive interest charges. Although it is clearly wise to keep borrowing down, it is better to incur extra interest charges than to work with insufficient capital. A business is under-capitalised if it has too little funds to finance sufficient turnover for maximum profitability. Using more capital, or reducing turnover, would raise the return on capital.

The main signs of an under-capitalised business are:

- Small-scale, uneconomic production units.
- Extensive land use owing to shortage of working capital.
- A continuously large bank overdraft.
- Purchase of small, uneconomic amounts of inputs and loss of quantity and cash discounts due to shortage of funds. For example, buying fuel at a local garage instead of bulk storage on the farm.

The following may help to remedy under-capitalisation:

- Better management of working capital.
- Leasing of capital assets, like land or machinery, to free tied-up funds for working capital.
- Reduced scale of activity while either raising funds from outside the business or distributing profit.

SOURCES OF FINANCE

Farming is so important to the economy of most tropical countries that a supply of adequate farm finance is essential for national well-being. The circumstances likely to need further extra finance can be divided into two main groups arising from:

- Previous trading, e.g. unprofitable trading, overtrading or inflation.
- Expansion.

When the capital needed has been estimated, it must be found. Each farmer should arrange finance from the best available sources. Many possible sources of finance, both internal and external, are available besides borrowing. The commonest of these are shown in table 4.1. Capital may be obtained as a gift or inheritance or it can be saved by keeping profits within the business. These are the traditional methods of patrimony, matrimony or parsimony. Also, capital assets, like dams, often can be made by direct use of labour.

With a new business, a farmer can work out his financial needs and build a capital structure to suit them. It is unusual, however, for a farmer to start afresh. Even if he is starting his first farm, the choice of holdings available is probably small and his personal funds are likely to be limited. Financing will usually be a joint venture between himself and financial institutions like the capital market and commercial banks. There are two basic sources from which extra finance can be obtained. These are:

- Increases in liabilities to owners or outsiders.
- Decrease in assets by, for example, sale of assets, leasing instead of buying, sale and leaseback arrangements and control of stocks and debtors.

Borrowed funds have become more important in farming. Luckily, many credit sources are available to most commercial farmers, to satisfy the rising demand both for investment in fixed assets and for more working capital. Need for more funds became more urgent with the growing need to modernise and develop farms. Farmers were often wise to get credit, that is, use of someone else's savings. As shown in table 4.1, sources of finance are classified as internal and external and by the time for which funds are used. Farmers normally need short-, medium- and long-term capital in their businesses at the same time and no one type can be used to best effect without the others. However, varying the use of one type may change needs for the others. If, as often happens, each type of capital comes from a dif-

Table 4.1   Summary of Sources of New Finance

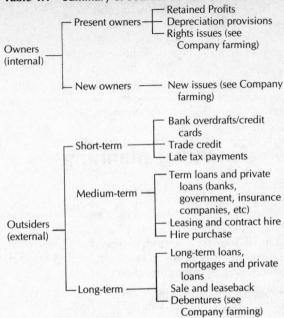

ferent source or sources, it is hard to get the best amount of each. Nevertheless, it is usually wise to use several sources rather than to rely entirely on one. In most tropical countries, arrangements for business finance in commercial agriculture are confusing and need reorganising.

Credit means having use of or possession of goods or services without immediate payment and a loan is money borrowed, at an agreed interest rate, for an agreed time. Because of the fear and distrust of borrowing long felt by many people, unnecessary mystery surrounds the whole subject. In fact, there are few tricks in finance. A dollar borrowed is a dollar that must eventually be repaid with interest. Any extra return, after interest, is profit for the borrower. Credit, as such, is neither good nor bad. It is the use made of it that determines its value. Credit is often unfairly blamed for all sorts of social ills when those ills are really caused by its misuse.

Farmers have some problems in borrowing. They usually want small sums for long periods but lenders prefer to lend large sums for short periods. Also farming tends to have more risks than most other industries and farmers often have little suitable security.

SHORT-TERM FINANCE

Shortage of short-term finance at fair interest rates is one of the main farm problems in less developed countries. Short-term capital consists of the current liabilities of the business and it is far more under the farmer's control than long-term or personal capital. Short-term funds are needed to maintain liquidity through a normal year's trading to bridge the gap between, for example, sowing a crop and selling the harvest, buying store stock and selling them fat and to provide something to live on in the meantime and to pay the tax man.

This short-term or seasonal credit is required to cover seasonal peaks in working capital needed for growing crops, wages, purchase of feeders and other short-term assets that are completely used up during production. If it pays to borrow the money for, or invest one's own capital in, these assets, the principal plus interest can be repaid out of income from seasonal sales. Hence, loans associated with short-term assets are called self-liquidating because their original cost is repaid out of income. Many, but not all, short-term funds are renewable. Similar, fresh advances may be got almost automatically, at least if results were as planned.

The main sources of short-term finance are:

- Overdrafts or short-term bank credit.
- Trade credit.

Much company financing consists of self-generated internal finance in the form of retained profits, depreciation and delayed tax payments. External sources include temporary loans and trade creditors. Most short-term, external finance comes, either directly or indirectly, to farmers from commercial banks.

Retained Profits and Depreciation Provisions

Farmers find much new investment money out of income, as many of their wives, lacking a deep freeze but seeing a new tractor on the farm, will agree. Increasing capital investment this way depends on the profits earned, the farmer's standard of living and the taxation rate.

Most capital in farming belongs to farmers themselves, especially if they inherit or have saved for many years. The most important source of finance is income from within the farm business; if past income has been enough, investment funds may come from savings. Farmer's own capital is supplemented by loans, trade credit, etc. Self-financing is safe because a business financed this way has no external, financial obligations. Whatever its origin, personal capital is increased only slowly by retaining annual profits.

In general, financing current operations and capital investment out of savings is wasteful as idle funds are carried for much of the year. It is also too slow to take advantage of chances for expansion as they occur. Retained profits are rarely enough to satisfy the capital needs of a growing firm. Shareholders expect good dividends and, if a large sum is needed quickly for development, a farmer can hardly delay the project until sufficient savings accrue from profits. Careful farmers can provide extra business funds by spending less on personal and family living. This is commendable but, with a low income, even a careful farmer cannot save much. Some new techniques need so much capital that, even with high saving rates, small farmers generally need more capital than they can save. With inflation, great skill is needed even to maintain personal capital in real terms. Total capital needed is seldom obtainable during a lifetime and inheritance tax may prevent passing of wealth from one generation to the next. Furthermore, taxation may make it hard for a business to amass sufficient reserves for working capital if these are taxable.

Profits kept in the business are rarely in liquid or realisable form. They are usually tied up in machinery, fixed equipment and livestock. The farmer's net worth and, likewise, his credit-worthiness grows as his land and fixed assets appreciate but he may still be short of ready cash unless he sells some of his land or uses it to secure a loan.

The cost of using retained profits for finance is too often ignored, even though it is hard to measure. It has an opportunity cost which is what it could earn if invested in its best alternative use either on or off the farm. The cost of using his own capital varies between farmers and depends on the investment opportunities open to them.

Reinvestment of funds arising from depreciation charges is also a major source of short-term finance provided profits are earned and then reduced by an internal depreciation charge.

Overdrafts

Commercial banks exist to make profits for their shareholders by borrowing money at lower rates of interest than they charge to borrowers. They undertake all kinds of ordinary banking business through a highly developed financial network which provides a unique way of collecting short-term funds. The extent of the branch and agency network of the big clearing banks allows them to collect surplus funds from people and institutions.

Long banking experience shows that few depositors demand repayment of their balances at any one time. This means that banks need only a small fraction, usually about 12.5 per cent, of their assets in liquid form at any time. Thus, for every customer depositing $100 in a bank, the bank can invest or lend about $87.50 to borrowing customers as there is only about a 12.5 per cent chance of a depositor demanding repayment. This ability to multiply the amount of credit available is one of the most important parts of the supply of finance, known as the 'credit mechanism'.

Banks fulfil the task of putting potential lenders in touch with potential borrowers. They obtain very profitable funds from current account balances, on which normally they pay no interest, and deposit accounts on which interest is paid and that, in theory, need up to a week's notice of withdrawal.

Because special Land Banks exist in many countries, commercial banks have played a smaller part in agricultural finance than they would otherwise have done. In developed countries and for industrialists in less developed ones, commercial banks, sometimes wholly or partly nationalised, are the main source of credit. They also usually provide most of the production credit in commercial farming but play little part in the smallholder sector. This is because of the management problems of lending small amounts to many farmers and the general lack of collateral security.

In deciding lending policy, bankers must consider the general credit situation and compromise between their wish both for liquidity of assets and for maximum profitability. These objectives conflict; the shorter the loan period, the greater the banker's liquidity but the less he will earn in interest. This necessary compromise is called the 'banker's dilemma'.

Most bank lending is by agreed overdraft on current account. A current account, always used for ordinary business purposes, is a running account used as needed for payments in and out — withdrawals being made by cheque. Cash and cheques received from debtors are paid in regularly whilst payment of creditors' accounts and various other expenses is made by cheques which are eventually received by the bank and debited to the customer's account.

Many banks charge their customers a ledger fee or commission for managing their accounts, sometimes even if they are in credit. Such charges are sometimes called account charges. Service charges normally relate to the manager's management time with a customer who makes an appointment with him.

Bank lending takes two forms — term loans, which are fixed amounts repayable in equal instalments over a specified period, and overdrafts. The latter are most common and a popular, cheap means of finance. An overdraft is merely a loan facility on a customer's current account allowing him to draw cheques up to an agreed overdraft limit, greater than the amount deposited at the bank, for a limited time.

Because banks lend funds deposited by their customers and withdrawable at short notice and because they may be told by government to make less advances in a credit squeeze, credit given by them is traditionally short-term and on demand — except by default! However, farm overdrafts are in practice often treated as long-term borrowing. Borrowing short and lending long causes most bankers' problems. Banks do not like and may not allow overdrafts to be used permanently. In a credit squeeze, a bank may refuse to renew overdraft facilities or may lower the limit when this is least convenient to a farmer. An overdraft is not a contractual loan.

Bankers like fluctuating balances, accounts that swing regularly from credit to debit and back again rather than 'solid' or permanent overdrafts. They expect farmers' overdrafts to be cleared yearly at marketing time. However, it is often possible to renew seasonal credit and sometimes raise the limit when the next production cycle starts; depending on last year's results and next year's plans. Although an overdraft is theoretically repayable on demand, customers are usually, in practice, given at least some months' notice if repayment is demanded.

A bank overdraft provides seasonal capital to replenish stocks, pay current expenses, take advantage of good financial opportunities and provides a cash reserve for emergencies. The seasonal needs for finance of annual crops and other types of batch production, like certain pig and poultry enterprises, are well suited to borrowing on current account. Also, in hard times when yields are low or disease strikes, farmers with a past record of good profits and in whom the bank has confidence can often rearrange their overdraft limits to cover the interruption in cash flow.

An overdraft is nothing to be ashamed of but rather a sign of the bank manager's confidence in a farmer. Its main advantages are cheapness and flexibility. A farmer simply agrees overdraft limits with the bank and can vary his use of the facility according to trading needs. Another advantage of overdrafts, to many farmers, is that they can use the credit however they wish despite the fact that they are arranged in terms of cash flows and supposed to be used as agreed. If a business is financially sound, or even if it is short of liquid assets but can offer reasonable security and good management, banks are one of the easiest and quickest sources of short-term finance.

Cheapness of bank overdrafts is because interest is charged daily on the amount drawn — not on the agreed limit. The interest rate charged varies with the security offered and the borrower's credit-worthiness but it is normally around 1½– 3½ per cent above the base rate. Life cover may be required and this should be added to the cost of borrowing and banks occasionally attract adverse comment by not detailing their charges or by charging for interviews between managers and clients. While an overdraft may be agreed at, say, 2½–3 per cent above the bank's base rate, changes in this base rate add some uncertainty and increase costs if it rises. Interest is a tax-deductible business expense, usually paid quarterly.

Despite the many advantages of bank borrowing facilities, it is unwise continually to use these resources to the full as it removes the relief otherwise available when temporarily embarrassed financially by factors outside the farmer's control.

Commercial bank managers usually know less about farming than land bank inspectors. This may not be a disadvantage. A veterinary surgeon need not be a technically sound agriculturalist to cure a sick animal; a bank manager need not be one either in giving financial advice or assessing a farmer's credit-worthiness. There is now, however, a growing tendency to employ expert agricultural advisers at head offices.

As negotiating bank credit is usually a personal matter between a farmer and his local bank manager, the farmer's reputation and honesty is important. The wide network of small branches in rural areas enables close personal connections to grow between the bank manager and his farmer clients. The manager's permission and approval should always be got before overdrawing a bank account, however short the period or great the security. It is also most important that any agreement — often verbal — should be honoured and overdrafts be repaid as agreed.

Credit Cards

A credit card lets the holder obtain goods and services on credit from certain suppliers up to an agreed credit limit; a certain minimum repayment being made monthly to the issuer of the card. Cards are issued after the applicant's credit-worthiness has been assessed. Most credit cards are issued by companies controlled by commercial banks. There is generally no joining fee or annual charge and the holder pays no interest if he clears his credit within 25 days of receiving the monthly statement. If this is not done, however, interest can be quite high — about 25 per cent a year is now common. Also, use of credit cards often stops the customer getting a cash discount – the supplier pays this to the credit card company! The main advantage of credit cards is that they save using cash but they are costly for extended credit. You can, however, get up to eight weeks or more interest-free credit.

Trade credit

This is credit allowed by one business to another that has bought goods from it. Trade credit has its disadvantages. First, the farmer's bargaining power is weakened if he has to trade with a particular merchant even though it is not to his advantage to do so. Secondly, extended trade credit can be costly. The merchant himself often uses borrowed funds, on which he is paying interest, so he must recover this interest from late payers. He may do this by adding a surcharge to the sale price, this is by raising the price of all goods enough to cover the cost of the credit given. He may have two prices — one for cash and another for purchases on account — or he may include the cost of credit in his sale prices but offer a discount for prompt payment.

If the cost of open-account credit were better known, it might be used less. The real cost of trade credit is shown by the cash discount offered for prompt payment, the difference between cash and credit prices.

Buyers on account generally do not pay for goods supplied on credit until they get a monthly statement from the supplier. This is sent as soon as possible after the end of each month and gives details of transactions during that month. Figure 4.1 shows a typical statement on which the settlement terms are stated. It starts with the closing balance from last month's statement and shows all transactions during the month. Finally, it shows what is due to the supplier at the end of the month. Discount is an allowance or deduction from the quoted balance offered to induce the debtor to pay promptly. In this example, 6 per cent can be deducted from the balance outstanding if payment is made within 14 days and 4 per cent if the account is paid by the end of April. Without a discount for early payment or surcharge (interest) on late payment, prompt settlement is unlikely. When part-paying a supplier for several purchases, pay an exact number of invoices and not a round sum or account so that you know what has been paid for. By supplying goods or services without immediate payment, the receiver of the credit gets capital to help run his business from another firm. The trader or merchant who allows credit has to go without the capital that he allows customers to use. Immediate cash payment for goods is rare because traders run open accounts for their customers' convenience. Customers cannot get cash on an open account — only goods or services.

Delayed payment for credit transactions may seem out of place as a way of getting short-term finance but it is nevertheless common. Many small farmers would have problems without trade credit which is even more useful to them when bank credit is limited. Trade credit is often informal and not backed by a stop order. It is convenient and, because of their wide local knowledge, traders are often useful links between lenders and borrowers. Lending, of course, is not the main business of traders and they do not lend formally to farmers. They simply await payment of bills because extended credit attracts business.

Credit is often given by general dealers, auctioneers and merchants such as stockfeed manufacturers. The latter, for example, may supply feed on credit, interest free, so a farmer they know well can carry through a two-to-three-month cattle-fattening project and pay for it on sale of the fat cattle. Items for which trade credit is usual are seasonal inputs like fertiliser, seed, feed and labour rations, payment being delayed until sale time. Farmers receive trade credit for often only four to six months. Even so, it can be useful and is often renewable when the next production cycle starts. For seed and fertiliser in rainfed farming this is usually a year ahead so there is a gap of several months between paying for one lot of goods and buying the next. During inflationary times, it pays to be in debt to your suppliers. Extract the longest credit period that you can.

It is surprising how much trade credit sometimes costs. Clearly, if a trader offers a discount of 4 per cent for payment by the end of the month, there is no incentive to pay before then. However, if payment is delayed until a month after the discount period has ended, which often happens, then the 4 per cent discount lost means that

STATEMENT

MWANZA AUTO-ENGINEERS LTD
PO BOX 247, MWANZA.   'Phone: MWANZA 0-2596

Messrs Karoyi Tobacco Enterprises,
PO Box 196, Kitale.   31 MARCH 1986

			Debits	Credits	Balance
February	28	Account rendered $			146.47
March	4	Cheque		135.48	
		Discount		10.99	0
	5	Goods	30.00		30.00
	8	Goods	42.53		72.53
	12	Returns		8.67	63.86
	29	Goods	74.21		138.07

The last amount in the balance column is the amount owing.
Terms: 6% − 14 days; 4% − 30 days.

Figure 4.1   Example of a Monthly Statement from a Supplier

for the extra month of credit the effective cost is about 48 per cent a year. If payment could be delayed for another two months without further penalty, the cost would fall to about 6 per cent a year. However, this is still likely to cost more than an overdraft. The penalty for further delay, in any case, will probably rise. Similarly, a farmer may find he could buy a machine for $1000 and get three months' credit on an open account. He could also get that machine for $950 in cash. If he buys on credit, he is paying $50 for the use of $950 for three months. This is equivalent to an interest rate of about 21 per cent a year ($\frac{50 \times 4 \times 100}{950}$). The cost of trade credit should be compared carefully with that of other credit. It often pays to borrow from a bank and take advantage of cash discounts rather than take trade credit. However, cheaper credit can often be agreed between farmer and trader if a properly organised plan of buying and selling is arranged in advance and such an agreement is strictly honoured by both parties.

Trade credit has advantages as a source of short-term finance as it is free unless discounts are lost or surcharges imposed. It helps every well-managed firm to take as much free trade credit as possible while keeping the trader's goodwill and leaving a small safety margin should extra finance be needed at a difficult time. Returning to the example of a 4 per cent discount for payment within 30 days, it would clearly pay the farmer to buy at the beginning of each month and pay within 30 days of receiving his statement at the end of the month. In effect, he thus gets almost two months' free credit. However, where extended trade credit involves a surcharge or loss of discount, it is usually better for the farmer to avoid it.

Stop orders and co-operative credit

A stop order is a cheap, formal type of trade credit. A farmer gets goods on credit and gives the trader a lien over income from his produce as security. As a lien, a stop order is a right or claim that the trader has to certain of the farmer's property. A stop order over, say, the proceeds of a crop is a legal document telling the organisation through or to whom the produce is sold to pay the proceeds, up to a certain amount, direct to the lender. This possibility of deducting debts from the price of the indebted farmer's produce is often adequate security for repayment of a short-term loan.

The cost of registering stop orders is low and the system avoids high interest charges. These are often no more than bank lending rates. However, there are disadvantages in both the ease of registration and because it can be unfair to other creditors who will not take stop orders. Firms with stop orders are often paid for much later purchases than those from other firms who remain unpaid at the end of the season.

Various commercial organisations, such as marketing organisations, co-operatives, fertiliser, pesticide and machinery suppliers, provide goods either wholly or partly without payment, the outstanding balance being registered against crop proceeds. Stop orders are registered in sequence and when tobacco, for example, is sold, the marketing body pays them. Payments to the grower start only when all stop orders registered against tobacco are paid.

Co-operative credit is of growing importance. Farmers who belong to a co-operative society can often get credit for inputs against liens over produce for sale through the society. This credit may be unsecured apart from a stop order. Farmers can usually get their inputs cheaper through a co-operative that buys in bulk. Credit is normally limited to goods and services but cash loans are sometimes made. This is most common in countries where governments like to direct available credit for small farmers through co-operatives. The co-operatives have the advantage of close personal knowledge of members and may take collective responsibility for repayment of government credit directed through them to members.

Late tax payment

Although the amount of tax payable and when it is due is not easily changed, the fact that it is payable many months after the profits being taxed were earned provides a useful source of free, short-term

depends on the firm's financial year and this is within its control.

It is possible to delay paying tax until a final demand comes, often several months later. Authorities usually charge no interest until the final demand expires so this is a free loan. (It is not unknown for farmers to wait even longer, paying interest and using this as a form of credit!) Clearly the tax bill must eventually be paid so delayed payment is not a secure loan, merely a reduction in alternative short-term credit.

MEDIUM-TERM FINANCE

Medium-term finance consists of funds tied up from one to about seven years in medium-term assets that are eventually used up in the production process.

These assets are partly fixed capital items like vehicles, tractors, implements, temporary buildings, medium-term improvements and breeding livestock. Finance may also be needed for opening new arable land, conserving it or for range improvement. Loans for medium-term assets are partly self-liquidating as they are repaid out of net income over several years.

The main sources of medium-term finance are:

- Term loans – 2–7 years.
- Leasing of equipment – 3–5 years.
- Hire purchase – 3–5 years.

The main features of term loans are that:

- They are made for a fixed period.
- Normally they are secured.
- The interest rate is either fixed or linked to a base rate.
- Such loans may be offered by banks to replace part of a continuing overdraft.

Private loans

Families still supply much farming finance, relatives and friends providing most of the credit used in tropical, peasant farming. Gifts and loans are important in extended families, providing a way of transferring income from person to person in changing times of want and abundance. Many commercial farms are also family businesses, passed from father to son, so new entrants to the industry are often finaced by relatives or through executors, trustees or friends. For example, a farmer may leave his estate to be split equally between his widow and children. Only one of the therefore might arrange a mortgage with which he buys out the other heirs over a period during which he pays them interest on their share of capital in the business.

Private sources form a simple and very common way of raising capital. They are more common than often thought because they are private. A relative or friend will often sell a farm to a young man and allow the debt to be repaid over several years. Prompt repayment or repayment at commercial rates may not be enforced. Private investors may be persuaded to finance sole traders or partnerships by supplying capital on loan. Such business is mostly done on a personal basis but it is still best to draw up proper written agreements about any investments made or cover the terms of repayment and recall.

The desirability of private sources of farming finance varies greatly. Private loans from relatives and friends have the advantage of secrecy and, sometimes, little or no interest. Parents, brothers or sisters simply leave funds invested in the farm. However, there are several disadvantages. One problem with this informal financing is that it may be withdrawn at short notice, especially if a lender or borrowing partner dies, although it is normally of a medium- or long-term nature. Family sources are often inadequate and farmers usually need extra finance after a crop failure, a fall in prices or other misfortune. Most private lenders are less experienced in lending than the staff of lending institutions. They may depend on interest and principal repayments for their living and need quick repayment. Also, they may be unable to renew when the need arises — even if they wish to do so. Some individual lenders may fully understand farming and extend credit according to farming needs; others may have less understanding and insist on impracticable terms. These points are not meant to scorn this source of funds but to indicate some important limitations of private financing. Institutional sources are more likely to have sufficient funds to allow renewal if they think it is justified.

Bank loans

Many banks now provide medium- and long-term credit. Besides overdrafts and personal loans, bank advances to customers also include bank loans where a fixed sum is placed to the credit of a customer's account and interest is charged on this fixed amount, even though cash repayment reduces the outstanding amount. During times of inflation and credit restraint, farmers have become more aware of the need to plan replacement of wasting assets like machinery and specialist buildings. If credit is needed, a loan account with an agreed reduction plan may be the best solution. This will often exist together with the varying short-term needs financed by overdraft and reductions should be planned to occur when current account surpluses are expected.

A term loan, arranged for two to five years, or more for a specific purpose, cannot be recalled before it matures. Banks may make loans for working capital up to, say, five years but it is usually unwise for them to agree to such long credit unless the minimum debt during a year's trading (hardcore element) is removed in five years. Fixed-term loans help if the business is starting a major new enterprise or expansion and needs capital for buildings, machinery or livestock. Though a bank may refuse to grant a loan to one person, it may do so if someone else stands surety as guarantor, agreeing to repay the loan should the borrower fail to do so.

The size of a bank loan and its interest rate depends on the bank's estimate of the customer's character, security and the purpose of the loan. The advantage of a formal, medium-term bank loan is that it guarantees the farmer use of funds for an agreed time, despite any credit squeeze, provided an approved repayment plan is followed and satisfactory audited accounts are produced each year.

Leasing and contract hire

Leasing and contract hire have become important because it is the use of equipment, not its ownership, that is vital for profitability. They enable enterprising farmers to expand by getting equipment to increase output, cut labour costs and raise profit. This way of getting use of capital by leasing equipment is similar to renting land. Plant and machinery, belonging to a leasing company, is rented to the user. A lease hire agreement is a contract whereby the owner of an asset (the lessor) allows someone else (the lessee) to use it for a stated time while he keeps ownership. A decision to lease depends upon comparing the costs of direct purchase (interest, depreciation and maintenance) with the rental charge.

Leasing has several attractions, the main one being that funds that would have been tied up in outright purchase of equipment can be used more profitably elsewhere, as in day-to-day running of the business. Costly capital equipment can be used without capital outlay so there is no need to tie-up existing capital, wait until enough is available or borrow extra funds. Overdraft facilities are left free for the normal job of financing short-term needs and bridging finance. If the equipment is profitable, after paying leasing charges, it is better to lease than to wait until sufficient funds are available for outright purchase. Another big attraction of leasing concerns tax liability. If equipment is bought, only depreciation and interest are tax-deductible business expenses. However, all rental and leasing charges are tax-deductible but any capital allowances or investment grants are lost. As depreciation can be charged only according to a fixed scale, it is generally a less valuable tax allowance than leasing costs. Other advantages of leasing include unreduced borrowing powers, use of the latest equipment, less repair and maintenance costs and ability to plan the payments in advance. A lease agreement is not shown as a liability on the balance sheet and can thus improve balance sheet ratios and borrowing powers. On the other hand, the equipment is not a business asset. No security is taken over already owned assets and leasing facilities cannot be withdrawn like an overdraft. When leased equipment wears out or becomes obsolete, a new set can be leased so it will be modern with low repair and maintenance costs. Leasing helps reduce the effect of inflation by ensuring fixed costs for a known period. Reliance upon external capital and the danger of change in the balance of control of the business is reduced. The popularity of leasing contracts for farm machinery, buildings and livestock is increasing.

Much confusion exists because leasing covers both operating leases (usually a finance lease) and contract hire.

Plant is normally hired under a finance lease for a primary period of 2–7 years during which the lessor (finance house) recovers all principal and interest payments. Interest is often less than the minimum lending rate or than that on hire purchase agreements or even bank loans. It may be fixed or vary with the finance house's base rate. Payments are usually monthly but sometimes may be negotiated to match expected cash flows. Insurance, spares, repairs and maintenance are the responsibility of the lessee and replacements for broken-down plant are not available.

The lessor retains ownership of the plant throughout the term of the agreement and is solely responsible for its purchase and disposal. The lessee has no *right* to buy at the end of the primary period. In practice, however, he may often

be allowed to arrange for the disposal of the asset and be credited with 90–100 per cent of the sale proceeds. He then has the choice of buying the asset for its residual value through a third party, or to re-rent for a secondary period of at least 15 years at a nominal cost. Clearly, buying at the end of the primary period makes a finance lease somewhat like hire purchase.

The difference with contract hire is that at the end of the primary period the equipment remains the property of the lessor, is returned and is not sold for the benefit of the hirer. However, the lessor is responsible for repairs and maintenance and will usually replace broken plant if necessary.

Besides leasing new plant and equipment, it is possible for farmers to sell existing assets to raise finance and lease them back from the new owner. As this usually applies to land, it will be dealt with under long-term finance.

Hire purchase

Instalment credit, as hire purchase is often called, is usually considered a medium-term financing method. Like a stop order, hire purchase is basically formal trade credit in which a deposit is paid and the rest of the purchase price, plus fixed interest, is paid in regular instalments over time. With hire purchase, vehicles, plant or machinery and sometimes livestock are hired for a period and then usually become the property of the user.

A hire purchase agreement is any contract whereby goods are sold on condition that, despite delivery of the goods, ownership of them passes only in terms of the contract and the purchase price must be paid in instalments.

The main features of hire purchase are that:

- A finance house buys the asset and hires it to the user with an agreement for the user to obtain eventual owership.
- The user agrees payment terms — usually an initial deposit and at least two instalments.
- Ownership remains with the finance house until final payment is made; so the asset cannot be sold or used to secure a loan.

A hire purchase agreement can apply only to identifiable, movable goods. These two qualities are necessary because the seller under certain conditions may recover his asset if the agreement is broken by the buyer. The buyer may also end the agreement, under certain conditions, and return the asset to the seller. In either case the seller must be sure that he is getting back his own goods and not similar ones.

Used wisely, within the buyer's means, hire purchase finance lets farmers use modern equipment that will often raise profits. It helps small or new businesses with insufficient funds to buy outright and those who cannot get a bank loan. Liquidity can be retained during an expansion project when assets are in unrealisable forms so hire purchase may be justified for buying assets with high earning power like machinery for contracting.

Minimum deposits and maximum repayment periods are usually set by governments for various goods. The deposit is the amount due as the first instalment, often expressed as a percentage of the cash price. This deposit must be paid on or before delivery of the goods to the buyer. The maximum repayment period is usually two or three years— less than that of banks and other sources. A buyer may decide to pay a bigger deposit than the legal minimum and may pay the balance before the maximum time allowed.

The seller does not have to accept only the minimum deposit nor to extend payment for the maximum time allowed. He may set terms according to current trading practice, competition and his estimate of the buyer's credit-worthiness and other relevant matters. The seller, however, must not accept less deposit than the legal minimum nor extend payment over longer than the legal maximum or the agreement will be invalid.

With hire purchase, unlike credit sale, the goods remain the seller's property until the last payment is made when ownership automatically passes to the buyer. A seller rightly regards the goods as security during the life of the agreement. A buyer cannot claim partial or conditional ownership of them during this time. Any financial interest he may have in the goods can be realised only when the agreement ends and he gets ownership. Only with the consent of the seller may the buyer sell or dispose of the goods during the life of the agreement and, in this event, the seller must receive the balance outstanding immediately. If the agreed instalments go unpaid, the hire purchase company may be able to repossess the goods although they may not be worth the balance of the debt. Normally, however, there are certain rules protecting borrowers against seizure of goods. Most finance houses prefer voluntary surrender of the goods and formal ending of the agreement as this ensures that the goods are sold to best advantage, thus minimising any loss for the buyer. Nevertheless, fail-

ure to pay instalments can still be costly. This occurs in poor seasons because hire purchase payments seldom rank high in the list of farmers' stop orders.

When buying on hire purchase, farmers usually like to pay seasonally after harvest rather than by monthly instalments. Payments should be arranged to coincide with expected cash inflows and a basic principle of hire purchase financing is that the value of the goods should always exceed the balance owed. With seasonal payments, the finance house may get nothing for up to 12 months while the goods are in use and falling in value.

If, owing to crop failure or delayed marketing, a farmer cannot pay an instalment when due, he should tell the finance house at once; otherwise they will activate their 'collection procedure'. If advised of payment problems in advance, the finance house usually will consider sympathetically a request for time to pay. In doing so they will compare the current market value of the asset to the balance outstanding under the agreement. Should the market value be the lesser amount, they will probably ask for sufficient payment to create some equity in the goods. However, no agreement can be extended beyond the maximum legal repayment period.

Buying capital items on hire purchase has advantages and disadvantages similar to those of leasing with the extra advantage that it may be one of the few ways for a firm without a proven record to raise finance. Funds are readily available — perhaps too much so! Hire purchase is a way of preserving liquid capital by financing purchase of capital equipment out of current earnings. Goods are available for immediate use without heavy, initial capital outlay. Other advantages are that the instalments cannot be called up before they are due, charges cannot be raised during a current agreement and the asset being bought appears on the balance sheet.

Borrowers on hire purchase eventually get ownership of the asset. They therefore can claim any initial allowance and depreciation as tax-deductible business expenses. Interest charges are also allowable but not repayment of principal as this is really capital growth.

Need for hire purchase facilities in business often arises but this is unfortunate. Many farmers have used them unwisely — especially for car purchase. A sound business that can gain the respect and confidence of bankers and get all its finance from them will pay less interest than one buying assets on hire purchase. Lending institutions often err by pushing farmers into hire purchase. No programme backed by a bank should make a farmer use hire purchase to fully execute it. This simply raises his costs and reduces his chance of success. A disadvantage of hire purchase compared to outright purchase, besides its cost, is that the liability outstanding shows on the balance sheet. With leasing, of course, the asset does not show on the balance sheet at all.

Let us look at an example. Suppose a tractor is bought on hire purchase over three years in equal monthly instalments at a nominal flat interest rate of 9 per cent each year. It costs $6000 and a 20 per cent deposit is needed so the sum borrowed is really $4800. Details would be as follows:

Principal	$4800
Interest at 9% per year for 3 years	$1296
Total owed	$6096($4800 + $1296)
Monthly repayment	$169.33 $\left(\$\dfrac{6096}{36}\right)$

Monthly repayment of principal

$133.33 $\left(\$\dfrac{4800}{36}\right)$

The average principal owed during the whole agreement period is the amount paid after 18.5 instalments, or $2466.67, so the real interest rate is 17.5 per cent, as shown in table 4.2.

$$\text{True Ann. Percentage Rate (APR)} = \frac{1296 \times 100}{2466.67 \times 3}\% = 17.5\%$$

The general formula for this calculation and from which table 4.2 was derived, is:

$$\text{Real APR} = \frac{\text{Flat rate (\%/yr)} \times 2 \times \text{no. of instalments}}{(\text{no. of instalments} + 1)}\%$$

Table 4.2 Effective Annual Percentage Rate Paid on Advances Repayable in Equal Monthly Instalments

Annual flat rate	Real annual interest rate (%) on advances repayable over:		
(%)	1 year	2 years	3 years
8	14.8	15.4	15.6
9	16.6	17.3	17.5
10	18.5	19.2	19.5
11	20.3	21.1	21.4
12	22.2	23.0	23.4
13	24.0	25.0	25.3
14	25.8	26.9	27.2
15	27.7	28.8	29.2
20	36.9	38.4	38.9

Had the farmer negotiated this same agreement with repayment in three annual, post-harvest instalments, the interest rate would reduce to 13.5 per cent:

$$\text{Real interest rate} = \frac{9 \times 2 \times 3}{4}\% = 13.5\%$$

His annual repayment would have been $2032 ($6096/3). He would owe $4800 in the first year, $3200 in the second and $1600 in the third year. The average credit taken would therefore have been $3200.

$$\text{Real interest rate} = \frac{1296 \times 100}{3200 \times 3}\% \; 13.5\%$$

Though the nominal interest rate may seem attractive, the real one may make farmers look for cheaper sources of finance because hire purchase finance is costly.

Too much equipment, and thus finance, is sold by salesmen who care only about selling, and getting commission on each sale, and do not bother if the purchase is wise or not. Admittedly, this is the buyer's and not the salesman's fault but the attitude of hire-purchase salesmen is naturally different from that of say a trader who must estimate his client's credit-worthiness.

The main disadvantage of hire purchase as a source of finance is the cost. It is almost double that charged on a bank loan or overdraft both because suppliers of hire purchase finance are usually financed by banks themselves and because of the poor security and high risk involved. An unfortunate aspect of using hire purchase for development is that sound buyers have to pay for the losses made on unsound ones.

Hire purchase and 'personal loans' from banks are credit sources where interest charges are calculated as a 'flat rate' on the initial amount borrowed (the principal). Before signing any agreement, the true interest rate should be assessed.

It is easy to get hire purchase finance. Although a farmer could go straight to a finance house and tell it what he wants, it is more usual for him to select what he needs from a dealer who will arrange repayment terms, complete the hire purchase agreement between himself, as seller, and the farmer, as buyer, and then sell the agreement to a finance house which thus gains total possession of the goods and the contractual rights in the agreement. The finance house will then pay the rest of the purchase price, selling price less deposit already paid, to the dealer. The farmer then owes the finance house the instalments due under the agreement.

The larger finance houses are mainly controlled by commercial banks who provide the necessary funds. Industrial bankers are those small finance houses that attract savings from the public by offering higher interest rates than they could get saving with commercial banks.

Insurance companies

These are important, organised sources of both medium- and long-term funds which they will invest in agriculture. They grant loans on their own life or endowment policies if these have a surrender (cash-in) value and are not already charged to support other debts.

Life offices allow policy holders to borrow up to about 90 per cent of the cash-in value of their policy at a low rate. Only interest need be paid as the principal is deducted from the proceeds of the policy when it matures.

Some insurance companies give mortgages for up to 60 per cent of *their* valuation of the farm. They normally insist on the borrower covering his loan with a non-profit endowment policy. This tends to be costly credit but it does ensure that if the borrower dies his dependants are safe.

Government and quasi-government organisations

In some countries, governments have set up special organisations to supply credit to farmers. Some of these arose during emergencies such as drought, depression or both. Land banks often lend money cheaply for buying land, improvements, equipment or livestock, holding the title deeds, often, as security. There are often other state credit sources run by departments of agriculture, natural resources, land, water development, settlement, etc.

The government is often the only non-private credit source available to small, peasant farmers. Interest rates are often subsidised and full security rarely is insisted on. On irrigation and other agricultural settlement schemes capital may be provided as improvements to the land — irrigation works, roads, buildings and so on. Interest and depreciation charges may be recovered through the rent charged. In big agricultural development projects, short- and medium-term credit is usually given in kind.

LONG-TERM FINANCE

There is no generally accepted definition of how long 'long-term' is. Normally, however, it means 10 to 30 years and is used to buy long-term assets, like land, that are not used up during production. Properly managed land is at least as productive after a production cycle as it was at the start. This is reflected on balance sheets by not depreciating land value each year. Therefore investment in land is sound if net returns exceed the interest, or opportunity cost, of capital invested in it. Thus, loans for land purchase are called non self-liquidating. Besides buying farms, long-term finance is often used for initially equipping them, developing and improving land, buildings, fencing and water supplies.

Long-term finance sources include:

- Long-term loans granted by finance houses or institutes, e.g. mortgages.
- Sale and leaseback.
- Long-term loans in the form of loan stock, e.g. debentures.

The advantages of long-term loans include certainty of interest charges and use of funds for a known period. A great benefit during inflation is that the real value of principal repaid is far less than the real value originally obtained from the lender. The problem connected with long-term loans is that interest must be paid on a fixed sum irrespective of whether it is all needed. Should interest rates fall, the borrower keeps paying the higher rate until the loan expires. Again, most lenders need security in the form of a right over some or all business assets. Long-term finance is rarely flexible. Only when major new purchases are in hand or the loan has existed for many years is it practical to increase a long-term loan.

Mortgages

These were discussed in chapter 2. They are a form of long-term credit, available to landowners, using a mortgage on land or buildings as security. The loan is used either to buy the farm, erect buildings or to carry out other improvements or repairs. Payment on a mortgage may be only interest at a low rate but it usually includes both interest and gradual repayment of principal in the form of an annuity.

From Appendix A.4, $15,000 borrowed on mortgage for 20 years at 9 per cent, needs annual repayment of $1643.21 ($15,000 × 0.109547) for 20 years. The early repayments consist largely of interest and the later ones largely of principal repayment as shown in table 4.3.

Annual interest payment is tax-deductible but not principal repayments, which have to be found from taxed income.

Table 4.3 Mortgage Repayment Data. $15,000 borrowed at 9% for 20 years

Year	Annual payment	Interest	Principal repayment	Principal owed at year end
	$	$	$.	$
1	1643.21	1350.00	293.21	14706.79
2	1643.21	1323.61	319.60	14387.19
3	1643.21	1294.85	348.36	14038.83
4	1643.21	1263.49	379.72	13659.11
5	1643.21	1229.32	413.89	13245.22
6	1643.21	1192.07	451.14	12794.08
7	1643.21	1151.47	491.74	12302.34
8	1643.21	1107.21	536.00	11766.34
9	1643.21	1058.97	584.24	11182.10
10	1643.21	1006.39	636.82	10545.28
11	1643.21	949.08	694.13	9851.15
12	1643.21	886.60	756.61	9094.54
13	1643.21	818.51	824.70	8269.84
14	1643.21	744.29	898.92	7370.84
15	1643.21	633.38	1009.83	6361.09
16	1643.21	572.50	1070.71	5290.38
17	1643.21	476.13	1167.08	4123.30
18	1643.21	371.10	1272.11	2851.19
19	1643.21	256.61	1386.60	1464.59
20	1596.40	131.81	1464.59	0
	32817	17817	15000	

Leaseback

In a sale and leaseback or 'renting back' agreement, the former owner of a capital asset, like a farm or estate, sells it outright, often at tax-depreciated value, to an investor such as a financial institution, pension fund, insurance company or sometimes to a wealthy private person. The original owner then leases the asset back from its new owner. If it is a farm, he then becomes a rent-paying tenant.

Large investors like to hold some of their assets in land because they believe it to be a safe investment and a good hedge against inflation. They get a guaranteed rent income and the chance of much capital growth owing to a rising land value. The price paid for the farm seldom exceeds 85– 90 per cent of its current, freehold market value, as vacant possession is not gained, and investors usually expect a return of about 5 per cent a year on their investment.

The main advantage to farmers of arranging sale and leaseback is that capital tied up in owning or buying land is freed for more effective and profitable use in the business; for example, to increase working capital and aid expansion thus raising profits. They remain farming as before. Leaseback is also a way of getting the tenancy of a farm being offered for sale with vacant possession.

If mortgage payments are being made, this so drains available funds that excessive borrowing may be needed to finance day-to-day farming. If a 150 ha farm were bought on a $30,000 mortgage in 1985 over 30 years at 8 per cent, the buyer would be paying about $2664 ($30,000 × 0.0888, see Appendix A.4) each year and would expect to do so for another 11 years or so. However, the market value of the farm may have risen to, say, $66,000 and expansion and investment on the farm may need a further $25,000 in loans on the security of the greater land value. Thus, besides the forced 'saving' of $2664 a year in mortgage payments, interest charges of say $3000 may be payable. Even if the business is prosperous, such costs may leave little disposable income. Selling and leasing back the farm could improve matters. The $57,750 ($66,000 × 87.5%) raised would almost repay both the mortgage and the other loans. However, the advisability of a sale and lease agreement for land, if a buyer can be found, depends mainly on personal circumstances. The advantage of freeing tied-up capital must be set against certain disadvantages.

If land is leased-back, the former owner loses the benefit of any rise in land value and his borrowing power drops because of loss of a balance sheet asset. Rent becomes a new, fixed cost. The benefits of land ownership or purchase should not be given up too easily for hard cash and each owner-occupier should assess his position carefully before taking any final step. Many farmers whose position apparently justifies sale and leaseback would do better to sell and leave farming entirely. Farmers, however, can sometimes keep some equity in land they sell and lease-back.

Company farming

A company's source of long-term capital is called its capital structure. Its choice between different sources depends upon their cost, the type of business it is, its past and expected earnings, taxation and other factors. Most company finance comes from issue of ordinary (equity) shares. These earn no fixed dividend, unless they are 'deferred' ordinary shares, because this depends on the net profit earned by the company. A company's capital structure depends on the number and type of shares it issues and its reliance on fixed interest funds (gearing). Each type of share seeks to attract a particular type of investor. Limited liability helps company promoters collect capital from several sources but the capital of a private company must come from sources known to its promoters.

Assuming that future capital needs are known, policy questions arise on how much equity capital should be sought from sources besides the main owners. This raises the danger, particularly for small businesses, of loss of control and the need to share profits. There is no one best way of getting long-term company finance. Certain general rules exist but basically the best method depends on the business involved, its size, growth prospects and the wishes of its owners or shareholders.

The original capital invested by shareholders is used to run the business and a dividend is paid to shareholders according to profits made. Shares carry part-ownership of the business so an investor, if he has many shares, can influence company policy.

The main types of company finance are debentures, preference shares and ordinary shares.

DEBENTURES

Debenture stock is issued by companies to get loan capital. They recognise the debt, guarantee to repay it 10 to 40 years ahead and meanwhile to pay interest, usually twice a year, at a fixed rate. Debentures are a form of bond giving evidence of a loan raised on the security of company assets. They are not part of the share captial so debenture holders are not members of the company and they can influence company affairs only if their interest is not paid. Debenture-stock holders are simply creditors.

The fixed interest on these securities, issued by limited companies in return for long-term loans, is usually a little lower than that on preference shares. However, there is little risk for investors as it must be paid before any share dividend, whether the company makes a profit or loss, if it is to stay in business. Interest paid on debenture stock is tax-deductible.

PREFERENCE SHARES

These were once a common way of raising capital; the name arises from the fact that any funds available for dividends go first to preference shareholders, at a fixed interest rate, with only the rest going to ordinary shareholders. Preference shares are now seldom issued because the dividends are not tax-deductible but these shares are simliar to a fixed-interest loan, interest on which is deductible. Preference shares, therefore

have been replaced generally by hard interest loans. Like debenture capital, however, they can help family farming companies get extra capital without danger of loss of control.

Preference shares rank for payment straight after debentures (if issued), and before ordinary shares, both when profits are distributed and if the company is liquidated. However, a preference shareholder, unlike a debenture holder, cannot claim any payment if no profit is made or distributed. If profits are made and distributed, he gets his fixed rate of dividend before ordinary shareholders take the rest. Preference shareholders usually have no voting rights unless their dividends are in arrears. Cumulative preference shares carry the right that unpaid dividends are carried forward and that these arrears also must be paid before on ordinary dividend can be paid.

ORDINARY SHARES

Also called equities or risk capital, ordinary shares are shares in the equity capital of the business entitling holders to the residual earnings after prior claims are met. If the company expects good, regular profits, it can issue debentures or preference shares, knowing that profits will cover the fixed interest charges. If, however, the business is risky with fluctuating profits, the company will not wish to commit itself to large interest charges in low-profit years and preference shareholders will not accept the risk of years without a dividend. In this situation, new risk-takers looking for large possible profits are desirable. In any case, there must be some ordinary share capital when a company is formed.

Ordinary shareholders own the company and usually have voting rights and therefore control proportional to their holdings. It is right that voting control lies mainly with ordinary shareholders because they bear most risk. Because of voting control, raising extra ordinary share capital from outsiders leaves the original owners less control of the company. Existing shareholders may not accept this. However, ordinary shares are sometimes issued without voting rights.

Financing by ordinary share capital has a special advantage to the business in that dividends are paid only when adequate profit has been earned; there is no fixed charge. Ordinary shares attract investors willing to take a risk in return for possible high dividends in good years.

The main disadvantage to companies of over-reliance on ordinary share capital is that exisiting shareholders cannot always find extra investment funds when needed and bringing in new ones affects control of the business. The only other disadvantage is that dividends are not tax-deductible. Investors must accept the risk that, if profits fall, they suffer first. They have last claim on both profits and on a share of the value of assets sold after liquidation. In bad years, they may get no return on their investment.

Gearing

The ratio of owned to borrowed finance strongly affects investment decisions. A yield of 3–4 per cent on owned investment of say $200,000 might be poor, in accounting terms, if there were not other benefits. In personal terms, however, it would provide quite a good income and a satis-factory living. Had the same investment been made with funds borrowed at 10 per cent, it would have to yield this before producing any personal income. The ratio between owned and borrowed funds needs thought whenever invest-ment decisions are made.

The ratio of long-term fixed interest loans (and maybe, current liabilities in farming with its long trading cycle) to owned capital shows how much the firm relies upon borrowed capital. This is called the firm's 'gearing'. A limited company's capital gearing is shown by the proportions of different shares and debentures it has issued. The ratio of the annual payment due on preference shares and debentures to profit available for dis-tribution is the capital gearing of the company. A delicate balance is needed between ordinary shares and fixed interest stocks, according to the company's circumstances, for a healthy capital structure. Gearing is high if a high proportion of assets are financed by long-term loans and the prior charges on issued debentures and prefer-ence shares absorb much of company earnings and may leave little for ordinary shareholders. It is low if these prior charges are low.

Table 4.4 shows that in the high-geared company a 75 per cent drop in profit, from $24,000 to $6000, would wholly remove the yield on ordinary shares and even preference dividend falls. In the low-geared company, however, preference shareholders are still well covered and even ordinary shareholders still get a dividend until profits drop by 96.25 per cent to $900. Thus high gearing can raise the yield per share on ordinary shares but it also increases risk, showing a common dilemma of business finance; the delicate balancing of risk and return both for

the borrower and the lender. A highly geared farm gains much from inflation. It borrowed maxi-dollars and repays them with mini-dollars — funds worth less in real terms.

If the return from the whole business is well above interest rates, borrowing greatly raises the return on owned capital. Indeed, the higher the ratio of borrowed capital, the higher the gearing and the greater the return on capital owned by the business. Used wisely, therefore, by a farming company confident that it can earn good profits, borrowing raises its growth rate. Using borrowed capital is risky of course and the higher the gear-ing, the greater the risk. If the proportion of fixed dividend capital gets too big, it may be hard to stay solvent during poor seasons because, if return on capital drops below interest rates, losses are also geared-up or increased.

LENDING POLICY

Having described the main sources of business finance, it is worth looking at the policy of

Table 4.4 Illustration of High-geared and Low-geared Companies

Company 'A' High geared. Dividends at Different Profit Levels

Profit	$6000		$12000		$24000	
Capital	Total $	Per share $	Total $	Per share $	Total $	Per share $
150,000 6% preference shares at $1 each	6000	0.04	9000	0.06	9000	0.06
15,000 ordinary shares at $1 each	NIL	—	3000	0.20	15000	1.00
	6000		12000		24000	

Company 'B' Low geared. Dividends at Different Profit Levels

Profit	$6000		$12000		$24000	
Capital	Total $	Per share $	Total $	Per share $	Total $	Per share $
15,000 6% preference shares at $1 each	900	0.06	900	0.06	900	0.6
150,000 ordinary shares at $1 each	5100	0.03	11100	0.07	23100	0.15
	6000		12000		24000	

lenders and the factors potential borrowers should consider. Lending and borrowing are two sides of the same coin — bad lending is bad for both lender and borrower and a sound advance is good for both.

Lenders are businessmen selling the use of funds. Potential borrowers should not be apologetic or shy in seeking credit. It is, after all, a lender's job to find borrowers for his, or his investors', funds. If a farmer has a sound use for credit, likely to enable him to repay it with interest, then an agreement is possible. However, financing farming is largely a problem of financing small farms with special risks and offers only a limited return to lenders. In estimating a potential borrower's credit-worthiness, lenders usually assess the risk involved on the three Cs of credit — competence, collateral and character.

COMPETENCE

Competence means repayment capacity which depends largely on the profitability of the use to which borrowed funds are put. Is the loan for something that will depreciate or is it for land or livestock likely to rise in value? Can the applicant repay out of ordinary income while keep liquidity?

A banker is more likely to lend if he knows how and when he will be repaid. He knows that temporary overdraft reductions will follow sale of products and that, to keep the farm running, inputs must be bought at the right time of the year. He therefore regards reductions caused by sale of products as only temporary. Permanent reductions come from after-tax profits, from after-tax proceeds of sale of assets — though such sales usually affect later profits — and, lastly, from capital brought in from outside the business.

It is not hard to assess an applicant's managerial competence to meet his financial obligations from information got from a well-designed questionnaire and from his remarks during discussion. Applicants of good character usually can assess for themselves how much income can safely be committed for repayments, avoiding over-optimism and choosing the best repayment period from those possible. The main consideration is that external finance should help and not be too great a load. Bankers usually consider it unwise to lend in hard times when there may be repayment problems; hence the 'umbrella bankers' concept — he lends you an umbrella when the sun shines and asks for it back when it rains!

A lender will assess all the local knowledge he can get and all the financial information obtained. Many factors are considered, such as the applicant's personal and profit record, his proven reliability, past prosperity and other commitments. Banks and commerce may give information on his credit standing. The main source of financial information, however, is the profit and loss account and balance sheet which are both needed for studying the financial standing and soundness of a business. A small-scale, 50-year-old farmer in difficulty is hardly likely to change, with credit, into a dynamic, 55-year-old entrepreneur!

Audited, financial statments are the most important tools used to vet loan applications and analyse the business for stability, flexibility, liquidity, profitability and growth. Balance sheets showing steadily rising net worth (owner's equity) impress lenders more than plans. A list of annual profits may not be accepted as evidence of success as net worth may have fallen through drawings taken or dividends distributed.

Banks seldom lend more than one-third of a borrower's net worth and may allow even less credit. This is not only due to risk of loss but also because of the high finance charges more credit would load on the business. As a rough guide, total finance charge should not exceed 15 per cent of gross income.

If past performance and technical ability as shown on the financial statement is only fair, little notice will be taken of future plans. If funds are scarce, a banker's decision to grant a loan or not will rely about 80 per cent on past performance and credit-worthiness and only about 20 per cent on future hopes and plans. This does not, however, mean that realistic, detailed budgets, setting out costs and returns, are unnecessary. Any competent lender wants to know where his money is going and the borrower should assure himself with a sense of realism, of the potential profitability of the credit and avoid taking on commitments that may be hard to meet. A clear proposal showing how the funds would be used and how both interest and principal would be repaid on a monthly or quarterly basis, is helpful therefore.

COLLATERAL

Sometimes called security, collateral is anything offered as promise of loan repayment, to be given to a lender in case of non-payment. There is always a risk that a loan will not be repaid.

Indeed, if a project fails or the loan is used unproductively, the borrower may be unable to repay. Hence a lender generally wants some security against total loss of his funds. Such assurance may be simply personal knowledge of, and trust in, the borrower. This is often enough within a family or village community and even much bank lending to farmers lacks formal security, relying on the farmer's credit-worthiness.

Whatever the security offered, lenders regard it only as something to cover unforeseen risks. The loan itself should be justified on its merits if both lender and borrower are to gain. Apart from pawnbrokers, lenders never lend hoping to obtain the security. Collateral cannot be ignored but normally it is considered less important than character or competence.

The security of a loan, or lack of risk, greatly affects the cost of credit. If land is bought on long-term loan and the deeds are given as security, interest will be low. However, a high-risk, short-term loan will be costly. The less the security and the lower the credit rating the more credit will cost.

Available security is limited if a farm is rented or under communal tenure and cannot be mortgaged. As most peasant farmers own few movable assets and do not legally own their land, they can offer little security for long-term loans. A prospective borrower, with insufficient security of his own, may be able to find a guarantor who agrees to bear the risk of the loan and repay it should the borrower not do so himself. A lender will often accept this if the guarantor is well established and has ample, liquid funds to meet the liability if necessary. Individuals as guarantors are best avoided as default of the borrower can cause them hardship and result in great bitterness.

Collateral security can take many forms. A mortgage may be provided over fixed property like land, buildings or perennial crops such as tea or cocoa. Notarial bonds over loose assets are not good security unless the lender can claim them *before* insolvency. Marketable stocks and shares and assurance policies are often pledged and stop orders on crops and produce also give security. Retained ownership of livestock and machinery bought on hire purchase is itself security for the credit.

CHARACTER

Prospective lenders need to know the farmer's technical training, past business record and reputed character. Personal reputation is important in farming as businesses are mainly small family units. Even though this may be hard for newcomers to the industry, it is reasonable. Before granting credit to a new borrower, lenders may ask for trade or bank references; that is the names of firms and banks with which the borrower has dealt before.

The applicant's sense of responsibility to the financial obligations of a contract can be assessed during negotiation. Good appearance and frank replies to questions may help gain a lender's confidence.

In practice, character or integrity is the most important consideration of lenders. Nobody of good character will over-commit himself or knowingly accept duties that he probably cannot fulfil. He is likely to tell his creditors if anything affects his ability to repay so as not to harm his credit standing.

In general, the importance of the three Cs can be summed up as follows:

- If character is good, risk is low.
- If character is only 'fair' or is doubtful, risk is high even if competence is good.
- If both character and competence are good, risk is slight.
- Competence and collateral, whether considered together or apart are not sufficient evidence on which to estimate risk.

BORROWING POLICY

Where capital comes from depends much on business size and willingness of the owners to use outside capital.

One of a farmer's main jobs is to allocate capital to enterprises so it will give the best return and maximise profit. This really means that his task is finding the most profitable farming system with the limited capital available. It does not really matter whether farmers own or borrow the capital used provided any borrowed funds return more than enough to repay the loan with interest.

Productive use of credit is normal in efficient farming. Most good farmers now run seasonal overdrafts and farmers' attitudes towards borrowing have changed. The old idea that borrowing is morally wrong, and that admitting to a bank overdraft is almost a confession of sin, has gone. Borrowing does not mean running into debt or that the business is making a loss.

There are, however, dangers in over-reliance on borrowed funds. A farmer too dependent on credit may be unable to free himself from debt when return on capital is falling and, even with skilful farming, may be below the interest rate.

There are certain basic principles that should be clearly understood when considering seeking a loan. Plans and budgets should be both conservative and as accurate as possible, all credit sources should be investigated and any arrangement should be in writing. In particular, the written agreement should include sound arrangements for repayment. If possible, these should include a lower charge if the borrower repays faster than first planned. It also helps to allow for reduced repayments if in temporary difficulty. Whenever possible, an automatic repayment method, like a banker's order, should be arranged.

The main factors a person or business seeking new funds should consider are:

- Cost.
- Period of the loan.
- Convenience.
- Flexibility.
- Degree of control required by the lender.

The annual charge for finance obtained from the owners of the business is different in nature from that obtained from outsiders. Loans from outsiders carry an explicit cost, usually stated as a percentage rate, that the borrower is contractually obliged to pay. With owner-supplied finance, whether in the form of retentions, depreciation provision or ordinary share capital, there is no explicit cost as there is no obligation to repay. There is however, an implicit 'opportunity' cost which is the income surrendered by the owners as a result of their financing decision.

A managerial task is to seek that mixture of sources of funds that will be cheapest. In general terms, the greater part of the finance needed to run a business at its normal level should be obtained from long- or medium-term sources. This ensures the continued availability of finance; usually at known, fixed rates.

A small part of normal requirements and amounts to meet temporary peaks, may be financed from short-term sources. Profitable, expanding businesses often face temporary shortages of working capital. The increasing need for short-term funds, to finance growing stocks, livestock, debtors and machinery, will be hindered if they have been wrongly used to buy assets that should have been bought with long-term funds. It is better and safer to have some spare short-term

funds, including borrowing power, than to fully use all possible sources. Short-term funds normally should be used only to buy liquid, current assets so that they can be repaid when necessary. Long-term finance should be used for long-term investment in fixed and part-fixed assets such as land, buildings, plant and machinery.

Whether borrowed funds are invested in fixed assets or used to raise working capital, worthwhileness and profitability are the chief tests for finding the best use for them. Factors borrowers should consider can be classified simply into the three Rs — returns, repayment capacity and risk-bearing ability.

RETURNS

Borrowing is justified whenever the use to which the loan is put produces more than sufficient income to repay it with interest. This simply means ensuring that returns are likely to exceed the cost of the investment, so the loan shows a profit. The main justification for borrowing is that it enables a farmer to buy things that will provide greater future returns than their cost. So a farmer should decide, as accurately as possible, what future returns he expects from borrowed funds and what future costs are likely to be. For each form of asset — long, medium and short term — the income expected during the asset's life is compared with its cost.

The greatest quick return is usually gained from short-term, production credit for seasonal inputs like seed, fertiliser, feeds and sprays. Peasant farmers often use loans for consumption or pleasure, rather than productive investment, so they get no income from which to repay them. Because of such unproductive borrowing, some farmers become indebted and dependent on moneylenders. Small farmers should not borrow for investment like buying a tractor or erecting boundary fencing if it would cost less to cultivate by hand, or contractor, or to hire a herdsman.

REPAYMENT CAPACITY

This means whether the cash flow from the investment will be such that the agreed loan repayments can be met. An investment may yield a profitable, long-term return but the borrower may be unable to repay when due if he has insufficient cash at the time. When assessed only on their returns, most farm investments are profitable.

Most farmers can easily pay the interest on borrowed capital, but they often find it hard to repay the principal.

Good financial management improves repayment capacity and uses as much credit as is profitable, thus raising net income. The following all help improve repayment capacity:

- Extending repayment time. The longer the time, the lower each annual payment and the greater the chance of finding it.
- Planning repayments to coincide with income. This is one of the main reasons why cash-flow planning aids efficient business management.
- Planning and running the business to minimise overhead costs and thus raise net income.
- Stressing enterprises with quick turnovers. Related to this is maximum use of self-liquidating loans. For example, renting rather than buying land, contract hire of equipment and cattle fattening rather than breeding.

The cost of finance affects both return and repayment capacity. As credit terms vary a lot, all possible sources should be compared critically. Shopping for capital is the same as any other buying; both good and bad bargains exist.

The interest charged for using someone else's money is highly important to borrowers. They want funds as cheaply as possible. Lenders, of course, want the best possible return for investing funds in another business. Market and economic forces interact and fairly stable interest rates result. A base rate of interest is set by major banks from which all other rates are fixed. The usual bank rate is 1.5–2 per cent above this base rate. In practice, loan costs vary from nothing, from a friend or relative, to 50 per cent on more, according to the loan period and the lender's risk.

Depending on the amount borrowed, interest can be more important than many borrowers realise. This is shown in table 4.5.

Table 4.5   The Effect of Interest Rate on a $30,000 Loan amortised over 20 years

Interest rate/year	Total interest paid
%	$
8	31140[a]
9	35700
10	40500
14	60600

[a] $30,000 × 0.1019 (see Appendix A.4) = $3507 each year.
$3507 × 20 = $61,140 total repayments over 20 years.
$61,140 − $30,000 = $31,140 total interest paid.

Farmers are sometimes unaware that stated interest rates often differ from the real cost of borrowing. They should check the true cost carefully before deciding anything. Differences between apparent and true finance charges are due to the different ways of calculating interest. The three commonest ways are the flat interest rate, interest on the unpaid balance and the discount method.

As shown in table 4.2, if interest is charged on the principal which is being repaid in regular instalments during the loan period, the true interest rate is nearly twice the flat rate.

If interest is paid on the unpaid balance, it is paid only on the amount still owed. Assume $1000 is borrowed for a year at 14 per cent interest on the unpaid balance and $500 is repaid after six months, the rest being repaid at the year end. Interest would be:

$1000 for 6 months at 14% — $70
$ 500 for 6 months at 14% — $\underline{\quad 35}$

$105

Table 4.6   Calculation of Interest on Borrowed, Short-term Funds on a Per Annum Basis

	Amount borrowed each month	Interest charges at 10%/year
	$	$
Aug.	7600	760
Sep.	2300	211
Oct.	3100	258 [a]
Nov.	650	49
Dec.	800	53
Jan.	550	32
Feb.	740	37
Mar.	700	29
Apr.	2200	73
May	850	21
June	1200	20
July	900	8
Total	$21590	$1551

[a] $\dfrac{\$ 3100 \times 10 \times 10}{100 \times 12}$

This is equivalent to 10.5 per cent on the original loan.

If funds are borrowed for a short time and interest charges are quoted as a per annum rate on actual drawings, the real interest rate may seem less than the nominal one quoted. This is shown in table 4.6. The first column shows the actual amounts drawn in particular months for example, by a bank overdraft at a quoted rate of 10 per cent a year. $7600 was borrowed in August, $740 in February, etc. The second column shows the interest the farmer pays at 10 per cent a year. Total interest for the year comes to $1551. If this is compared with the total amount borrowed of $21,590 until the end of July, the true interest rate may seem to be about 7.2 per cent a year. The reason for this *apparent* difference lies in the words 'per annum'. When calculated on outstanding balances, the total amount borrowed during the year, $21,590, is the cumulative total after 12 months. Hence the administrative rate of 10 per cent is charged only on the actual amount borrowed at any time. It *is*, however, the real, effective rate.

In the discount method, interest is deducted in advance. Assume $1000 is borrowed for a year. If the lender discounts the loan in advance at 12 per cent, he will make the loan for $1000 but provide only $880 so $120 interest is charged. Twelve per cent interest a year on $880 is only $105.60 and the true interest rate is 13.6 per cent a year.

Taxation also changes the real cost of interest to a farmer from its apparent rate. Interest is tax-deductible so if the farmer is lucky enough to make a taxable profit the real interest cost is reduced. Consider the example given in table 4.7. Mr Harvey finances all his working costs himself. Mr Belafnte, however, borrows most of his. The profit of both farmers is $10,000 before tax in Mr Harvey's case and before both tax and interest in Mr Belafonte's. Assuming their after-tax profits are $8152 and $5132.50 respectively, the difference of $2992.50 is the true cost of the $45,000 borrowed — not the apparent cost of $4275. Therefore the real interest rate is:

$$\frac{2992.50 \times 100}{45,000} \% = 6.65 \% - \text{not } 9.5 \%$$

Clearly, if a farmer pays no tax, this reduction does not apply. Conversely, the higher the profit and tax bracket he is in, the lower is his real interest rate.

Besides interest, the period of the loan is also important. It greatly affects the wisdom of borrowing owing to its effect on the annual repayment and therefore the return needed to meet both interest and redemption. This is shown in table 4.8. Time, therefore, is important to a borrower. He may need the flexibility that lower annual payments provide and be prepared to pay more in total for the loan. This is especially true for farmers with limited capital who could face

Table 4.7   Effect of Taxation on the Real Rate of Interest

	Mr Harvey	Mr Belafonte
	$	$
Own finance	60000	15000
Borrowed finance		45000
Profit before tax and interest	10000	10000
Interest at 9.5%	—	4275
Profit before tax	10000	5725
Tax: $2500 allowances; balance as in table 2.4		
Disposable income after tax	1875	592.50
	$8125	$5132.50

Table 4.8   The Effect of Repayment Period on Total Interest and Annual Cost of $30,000 borrowed at 10% Annual Interest

Years of loan	Total interest	Annual repayment
	$	$
10	18810	4881
15	29175	3945
20	40500	3525
25	52650	3306
30	65490	3183

trouble in poor years if the annual repayment commitment is too high.

If a depreciating asset, for instance, a tractor, is bought on credit, the loan should be repaid well within its expected useful life, that is, before it needs replacing. Non-depreciating assets, like land, should be financed over a period that will not retard full development of the business itself. The effect of taxation is often underestimated and too high a repayment rate is agreed.

As shown in table 4.2 one of the commonest ways of repaying a loan is by amortisation. This means that the debt is slowly repaid in equal instalments consisting partly of interest and partly of capital. When this method is used, the principal repayment rises over the loan period as less interest is paid on the lower outstanding balance. This method, however, seldom is likely to be the cheapest and, if possible, use should be made of a life assurance policy as collateral. Besides repayment of the loan on death, this has other advantages.

The idea is that the loan stays outstanding for the full term and interest, at the agreed rate, is paid on it. This interest is tax-deductible. A life policy is then taken out to mature when the loan must be repaid and the proceeds are used for this. Tax relief can also be claimed on the policy premiums. If the policy is with profits, there may be a tax-free surplus when it matures. Such a policy may, however, cost too much although some lenders are satisfied with a with-profits policy for say 70 per cent of the sum lent. In this case, a with-profits policy, linked to the units of a well-managed unit trust, may well be better than amortisation if the lender will accept it as collateral.

RISK-BEARING ABILITY

This is concerned with the reserves that the business has available to offset the risk involved in using credit. As shown in tables 3.8 and 4.3, the risks involved in borrowing rise as equity ratio falls. This may discourage farmers from making large changes in their farming systems if they would need heavy borrowing.

SMALLHOLDER CREDIT

Owing to the special problems of extending credit to small farmers, a case is often made for less developed countries to set up separate credit schemes for them. Credit was identified as a pressing need for Indian small-scale farmers and from 1951 to 1971 rural credit co-operatives grew from supplying 3 to 23 per cent of India's agricultural credit. Rural moneylenders' share of the credit market fell from 75 to 50 per cent. Most third world countries do have agricultural banking and credit institutions and programmes to ensure that external funds are used to best advantage and increasingly to bring local capital into circulation, but these funds are not always available to smallholders.

Small farmers need more service and closer supervision than more commercialised, larger farmers and less rigid policies on collateral, deposits and repayment timing. However, institutional and manpower constraints enable few countries to set up separate schemes. A practical solution is to keep programmes and accounts for small farmers separate from the others so that different procedures can develop if necessary. For the same reasons, multipurpose co-operatives

should hold their credit accounts separate from other activities.

Many governments and more and more academics blame shortage of agricultural credit as a main block to achieving adequate rises in production and rural living standards. Skilfully used credit is often the key to modernising farming. Need for more institutional credit in less developed countries is now realised and reflected in current plans for agricultural financing by the World Bank and other international agencies. It forms a component of World Bank agricultural lending but many small farmers are still held back by being unable to get credit easily. This is not only a financial constraint; it retards adoption of innovations.

The main reason for providing institutional credit to small farmers is usually said to be the exorbitant interest charged by private moneylenders and traders. It is also said that lack of capital both causes poverty and slows national growth. The amount of private credit available is said to be inadequate and borrowers may have to sell their produce to lenders on unfavourable terms that discourage productive effort.

Most credit granted to small farmers in less developed countries is short-term, seasonal credit for buying current inputs like seed, fertiliser and pesticides. This stress on seasonal, production credit may cause neglect of other development needs like equipment, livestock and perennial crop establishment. In Malawi, however, the Government Loans Board concentrates on oxen and ox-drawn equipment.

There are many problems in granting credit to smallholders. It is much more complex than lending to large farmers; the many small loans and the spread of borrowers over vast areas makes smallholder credit more costly. Each credit assistant can handle few accounts. The high cost of extending smallholder credit to individuals also arises because administrative costs — investigation, office needs, servicing and collection — are similar whether the loan is large or small. Total administrative costs rise as average loan size falls, its time shortens and as clerical work grows to cope with many small borrowers. Peasant farming contains traditional features, mainly social and institutional, that make large-scale extension of credit difficult. A farmer's capacity to use credit effectively is often rated too highly by both new potential borrowers and unwary lenders. Difficult communications make contact with borrowers time-consuming and costly and also raise the risk of irregular payment of interest and principal. The duty to repay loans to large organisations is often taken lightly as experience has often shown that there are few penalties for defaulters.

The main problems in running any smallholder credit scheme are the administrative ones met in granting many small loans and the even greater ones of getting repayment later. Security is limited if land is owned communally. However, the critical problem is not so much security for loans to smallholders (as often stated by bankers) but the profitability of managing many small advances (as usually stressed by economists).

Some people think it is cheaper, nationally, for governments to subsidise input prices rather than to incur the cost of setting up a credit system. However, peasant farmers may have no spare cash to buy inputs at any price and subsidies distort resource allocation and are politically hard to withdraw.

Although credit provision can do much to help develop peasant farming, there is a danger that too much may be expected of it and it becomes regarded as *the* tool to quickly raise output. Credit is not essential for agricultural development; merely an accelerator. Like a car's accelerator, it can cause disaster if misused. Ability to use credit productively is low; it can be a liability — not an asset! Persuading simple peasant farmers to accept credit, unless they know how to use it skilfully, would do them a disservice. Credit alone cannot change a poor farmer into a good one! Even at low interest rates, credit will not automatically raise output or the incomes of the rural poor. Only in the modernising sector of peasant agriculture will more capital yield a high return. In the traditional sector, development will come only after new information and inputs have made capital more productive. Credit should be a crutch to help farmers raise their productivity and living standards, not a financial cross to which they will eventually be nailed!

Where relevant inputs are physically available and market prices are profitable, many of the problems that providing institutional credit is designed to solve soon disappear. Various studies — many done in Nigeria — show, for example, that peasant farmers have often successfully incentives well through on-farm savings and informal money-lenders, especially for investments with no large, lumpy expenses.
well through on-farm savings and informal money-lenders, especially for investments with no large, lumpy expenses.

Credit never stimulates innovation and greater output if farmers are unwilling to adopt better methods or have poor incentives to do so. If the necessary incentives are there, credit is seldom essential for adoption. Productive credit, in other words, has little importance in the early stages of small farm development and improvement. If national development policy places most importance on the smallholder farm sector, there is little evidence that absence of credit is a limiting factor, at least in the crucial take-off stage, and it is doubtful if more credit would help much without other support services.

Smallholder borrowers should be chosen on their credit-worthiness, like any other borrowers, but the assessment of credit-worthiness may differ. Few of those urgently needing loans are credit-worthy in the usual sense. The criteria need not be as strict as for larger borrowers. The main ones for small-farm credit should be management ability and repayment capacity, not the amount of collateral offered. The farmer's managerial competence should not be overrated nor should it be assumed that provision of credit will automatically improve it or that lack of credit is necessarily limiting output. Important factors are the status of the applicant in the local society, technical feasibility of the proposed enterprise on his own farm and the cash flow expected from the investment.

In Central Africa it was proved, without doubt, by years of field trials that right planting time, plant population and weed control are essential for high maize yield and, without these, fertiliser is a waste of money. However, correctly used, fertiliser could raise yield from 1800 kg/ha to 9000 kg/ha — and this was economically profitable. However, some farmers could not repay seasonal fertiliser loans as maize yields were too low to cover fertiliser cost, because the basic husbandry practices mentioned above were not adopted.

Care is also needed with credit for livestock production that, unlike annual crops, is usually based on longer time periods and normally with more conservative farmers. Credit is often given now for stock purchase, fencing, pasture improvement and disease control. Pasture and genetic improvement is a long-term project that should be based on proven data. Fencing alone will not necessarily improve grazing — especially boundary fencing — so stress should be on fencing to assist rotational grazing and separation of different types of stock. The fastest and cheapest improvements come from adopting better husbandry practices like feeding, bull selection, early castration of bull calves not needed for breeding, control of breeding and weaning times, provision of water and pest and disease control. Apart from water provision, these are cheap and do not load borrowers with repayments beyond their capacity to earn.

Sources of credit to smallholders in less developed countries vary with economic and political conditions. The traditional credit channels are a mixture of non-institutional sources such as relatives, friends, moneylenders, input suppliers and marketing and processing enterprises. As shown in table 4.9, non-institutional sources may lend more than institutional ones.

Many distribution systems have been tried such as specialised agricultural banks, credit co-operatives, commercial lenders, government direct-lending programmes and specialised government agencies such as land reform and settlement authorities. Government-sponsored credit is often the only type available to small-holders in less developed countries. Credit co-operatives have often failed because of the high costs and risks of granting many small loans. No single approach is better than any other and the best way to get credit to small farmers depends mainly on local conditions in each country or even in regions within a country. Although different programmes have been used in various countries, few governments have tried alternative systems within their country.

Table 4.9   Sources of Agricultural Lending in Selected Countries

	A	B	C	D	E
India	8000	26	4	51	19
Philippines	1245	16	26	51	7
Thailand	912	7	1	36	56
Colombia	433	27	69	3	1
Costa Rica	157	70	0	20	10
Sri Lanka	150	20	0	45	35

A   Total loans to farmers (US $million)
B   Loans from public institutions (%)
C   Loans from private institutions such as commercial banks (%)
D   Loans from commercial lenders like landlords, storekeepers, middlemen and moneylenders (%)
E   Loans from non-commercial lenders like friends and relatives (%)

SOURCE: Agricultural Credit, World Bank (1974).

For credit institutions to maintain and expand their services, there must be a continuing inflow of capital both from repayments and by adopting policies and developing facilities to attract savings. Farmers traditionally have made savings in kind (stocks of produce, livestock, equipment), in jewellery and in cash.

However, these types of savings, particularly hoarded jewellery and cattle, are available as funds only if the owner sells them. To mobilise cash savings, at least, they should be drawn into a lending organisation. Although usually moderate in amount, rural savings can often help greatly in expanding credit availability for rural development.

Credit is unlikely to be used productivity unless it is combined with other services and pre-requisites that work efficiently at the small-farmer level. No agricultural credit programme can be effective unless it is combined with effective extension services and other forms of support. Credit and extension staff must work closely together. Extension workers know the background and farming activities of the applicants and can help with loan appraisals. They should have no part, of course, in regulatory matters like fore-closure or collecting repayments or they will become ineffective as extension workers. Production credit given together with technical advice and help is called supervised credit. Credit and techical help are complementary. The credit ensures that the farmer can finance new tech-niques and these, in turn, provide a sufficient rise in income to repay the loans with interest. Close supervision helps to ensure that credit is used productively.

Other prior needs for successful use of small-holder credit are availability of profitable inno-vations and markets, timely delivery of inputs in a complete package, a land tenure system enabling farmers to benefit from improvements they make to the land and recognition by farmers of the importance of business character as a basic for credit. That is, prompt payment of interest and instalments, with notice of any delays, and the development of skill to handle borrowed funds wisely.

Credit in kind is often more suitable than giving it in cash. This relieves farmers of further, possibly strange, transactions and provides the institution with some assurance that the credit is used as intended. Credit in kind should be in the form of a composite package of inputs suited to the area to be planted. There is little point, for example, in giving hybrid maize seed on credit to a farmer who will not use fertiliser or fertiliser on credit to a cotton grower who will not spray his crop against pests. Composite packages should be issued at a set price that should include internal costs and credit charges — without quoting interest rates. For example, a composite package for a hectare of tobacco could be 125 kg compound fertiliser, 125 kg sulphate of ammonia for top dressing and provision of centrally grown seedlings. For this the farmer might be charged a fixed rate of, say, $60 which includes a credit charge of $6.65.

Credit, especially credit in kind, must be timely; inputs must be available when and where they are needed. As smallholders seldom have adequate storage, too early delivery often results in losses and wastage of inputs like seed and fertiliser. Delivery after the best time for use is, of course, pointless. Many smallholder development projects still fail because farmers receive inputs too late.

Farmers usually should have to pay cash for some of their inputs, thus contributing towards the cost of the investment for which they are borrowing. This increases their incentive to make it successful, increases the lender's security and spreads credit funds wider. All procedures should be as simple as possible. There must be a legal basis for using credit, fair to both borrower and lender; but it should be possible to borrow quickly with few formalities. Of more concern to the farmer than the interest charged is ability to get credit promptly, without complex procedures and with convenient repayment conditions.

The phrase 'the availability of credit' often really means 'the availability of cheap, subsidised credit'. The low interest rates for smallholder credit in many tropical countries are well below the rates of return readily available on most low-risk investments. It is widely believed that interest rates charged by local moneylenders are excessive and they are often over 100 per cent a year. Concern over these rates has led to unrealistically low rates for institutional credit. Smallholders do not need credit subsidised at 6 per cent a year for projects that can yield 20 per cent or more. They would be better off paying a realistic interest rate on easily obtained credit.

If there is only one wealthy trader in a district who is the only credit source, he can extort excessively high interest rates. Thus there may be a need for government to provide rural credit, on grounds of social justice, as an alternative source. Even then it is wise not to remove the private credit source. With two or more sources of credit, each one bids for the farmers' custom by making its interest rates and way of extending credit as

reasonable as good financial practice allows. Consequently, farmers are less likely to think they are being exploited by the lender with whom they deal, whether he is a private person or an official.

For short-term loans, local moneylenders may attract farmers despite high interest rates; the main attractions being the lack of need for collateral security and the ease of getting such loans which may be made on the spot without any formal contract. Local moneylenders, many of whom are traders, often make small cash advances on the understanding that the borrower will sell his crops to them. Besides giving the lender a chance to recover his loan after harvest, this also assures him in advance of a supply of produce.

Charging an interest rate below that needed to cover the cost of obtaining funds and administering loans means subsidisation which few less developed countries can afford for long. However, governments may need to subsidise farm-credit schemes at first and can raise interest rates once farmers realise that use of credit can increase their profits. The interest rate should then cover the cost of capital plus administration. New technology that could triple a farmer's production and, with favourable prices, multiply his net income severalfold, would enable him to pay 25 per cent on his loans if necessary. Commercial lenders now seem to charge much more than institutional ones. Weighted mean annual interest rates from table 4.8 are:

Public institutions	10%
Private institutions	13
Commercial lenders	26

The real interest rate for public institutions may be negative — less than the rate of inflation. This usually low interest rate for institutional credit helps larger farmers most as they get larger loans.

The average repayment rate for institutional lending in the tropics varies widely — from 18 per cent in Jordan to 98 per cent on the Lilongwe Land Development Programme seasonal credit scheme in Malawi — but is about 70 per cent. Ninety per cent recovery is considered high but no scheme is likely to be viable unless it exceeds this. High recovery rates are particularly important with revolving funds. This is aided if community pressure shames defaulters and if there is a fixed, automatic penalty for failure to repay fully on time — not interest on overdue balances.

Although loan repayment should be each borrower's responsibility, and should be assessed according to his expected repayment ability, there is strong community spirit in most rural areas and this can be used effectively to raise the recovery rate from borrowers. If transactions are made in public for all to see — quite opposite from conventional financial practice — and the whole village or co-operative society realises the possible results of a single default, either in damage to their reputation or some form of penalty, group responsibility can help to ensure repayment. It becomes hard for a borrower to withstand the pressure of his friends and avoid his obligations.

The surest way of getting repayment is for the credit institution to market the farm produce and deduct the repayment from proceeds. A stop order can be obtained on a particular product. This is easiest to administer if there is only one wholesale buyer, like a marketing board or state corporation, and for products like tea, cotton or sugarcane that must pass through one of a few processing plants. The chances of a borrower by-passing the maketing body with which he has a stop order arrangement are then much less.

Annual crop loans are sometimes granted to farmers through recognised, responsible farmers' organisations such as farmers' clubs, that accept group liability. Reduced interest rates may be offered to those clubs that are self-accounting. Village credit associations in South America have mobilised unexpected amounts of local savings and granted loans on a self-help basis.

Providing an efficient credit service to smallholders needs many trained, well-motivated staff and incentives for performance. Credit agents should understand the technical and management problems of farming. The number of accounts a credit agent can handle depends on dispersion of borrowers, transport, support from extension services and commercial firms, local participation and back-up accounting services.

USE AND MISUSE OF CAPITAL INVESTMENT

Financial management means assessing the best use for funds besides obtaining them. Farmers have become more aware of capital problems — of how to obtain and use capital — and investment appraisal has recently tended to dominate farm management literature, just as comparative analysis, budgeting and finally gross margins did during the fifties and early sixties.

To the extent, however, that existing investment is past investment and is often fixed, the element of farm capital that is studied most closely is probably that small share of it that is newly invested. In theory even past investment is constantly being freshly invested because, if it is decided to leave it alone, it is in effect being reinvested in its present use. Some farmers do, of course, move out of or retire from the industry and others often make large changes to their farming system and, therefore, to their pattern of investments. Most farm capital, however, tends to stay where it is.

There are two main questions about capital — how much to use and where to use it. In practice, these are closely related. The best amount of capital to use is usually determined by expected return, its real cost and the amount of risk at a given level of management skill. In comparing the return on possible new capital investments with interest rate, both farm and non-farm opportunities should be considered.

Farmers with unlimited capital will increase investment in their businesses until the returns from further additions fall below outside opportunities; they will then invest extra funds off the farm. However, decisions are usually dominated by limited capital owing to shortage of credit and to competing uses for capital within the business.

Basically, investment is a matter of choice. The choice of investment opportunities on many farms is endless — as are the returns that these investments offer. The range of investment between different farms and farming systems varies widely and investment policy varies not only between farms but often even for different items on the same farm. The best policy to follow depends on the importance of the asset for production, the need for it and the capital available for investment. Common capital investments are buildings, machines and irrigation schemes.

Capital investment decisions are probably the most important ones in shaping the future of the business. Not only do they lock up much money, they also largely determine the farm's future profitability and, probably, production methods for many years ahead. Good profits usually arise from correct investment decisions made in the past. Well-planned marginal investments on existing, well-managed farms can still provide very good returns.

Capital investment in farming has two general restrictions: the slow turnover rate of most farm businesses and the high proportion of overhead costs. Both these factors limit the possible return on capital. Slow turnover rates need high margins from each production cycle for a satisfactory overall rate of return. Most business-conscious farmers agree that farming rarely yields as high a return as many other business investments. Some argue that rising land values more than offset the relatively low annual returns on capital. While it is reassuring that farm value is rising, it should be remembered that this rise in value can be realised only when it is sold. To sell a farm, built up with much time, care and effort, is hardly the average farmer's main ambition!

Probably the most important effect of moderate or poor investment performance is failure of the farm to realise fully its profit and growth potentials. Even farmers who do not borrow heavily, by misusing capital resources, may get less profit than they could. Despite the importance of sound investment policies, the methods many, or even most, farmers use to evaluate projects are crude and often misleading. Two examples of such methods are 'average return on capital' and 'pay-back'.

In peasant farming areas, scarce capital is often misused owing to financially sub-optimal choice of investments. Investment in low yielding assets, such as an expensive house, a flashy car or a boundary fence, is understandable from a social or pride of ownership standpoint but adds little to output.

Investment appraisal is a branch of business economics that sets out rules, methods and criteria to help practical decision-makers to choose between alternative investments. All investment appraisal methods initially need at least a partial budget to estimate the extra profit likely to arise from the investment. Comparison of competing investments is often best handled by partial budgeting. This will show whether or not the investment is likely to significantly raise profit and what it could earn if invested elsewhere — on or off the farm. The steps needed to prepare a partial budget include establishing the input: output situation, assessing total capital needs and calculating the return on capital. If an existing enterprise is being expanded, assessing the physical co-efficients is easier if farm records are available but the effect of the increased scale of activity must be considered.

Financial assessment, therefore, should involve not just setting down the capital needed but also a partial budget to show the extra profit expected. Determining the extra capital needed may require detailed plans, such as an irrigation layout. The real cost of capital, and interest rate to be used in profit calculations, is its opportunity cost. This is normally higher than bank borrowing rates and it

will vary between farms. This opportunity cost of capital is what it could earn in its best alternative use at the same level of risk. Possible outside investments include building societies, first mortgage debenture stocks, shares and unit trusts. If 20 per cent return could be obtained, any farming investment should yield more than this.

Another element influencing investment decisions is the effect of taxation which tends to make capital investments more attractive. This is designed by governments to encourage investment by usually reducing the capital needed.

Farmers need detailed calculations to estimate how much capital they need to follow their proposed plants. They should ensure there is sufficient capital; if not, they must amend their plans, adjust the proposed pattern of payment and sales or try to obtain more capital.

Just as important as the rate of return, or profitability on capital investment is its timing and estimating the amount of capital needed at different periods. Consequently, besides a partial budget to estimate profit, a cash-flow budget is needed to assess the ebb and flow of funds and estimate any need for credit. This helps greatly when credit agencies are approached. The effect of the investment on the balance sheet should also be estimated by considering depreciation and the reduction of liabilities.

To sum up, there are three basic considerations in capital investment:

- What will be the capital cost — interest or opportunity costs.
- What will it earn — usually assessed by partial budgeting.
- Will liquidity and solvency be kept — normally assessed by cash-flow budgeting.

There is no completely satisfactory, comprehensive way of planning capital use in farming as farmers' needs and objectives vary. Only occasionally are formal investment appraisal methods applied. Despite the recent interest in these techniques, the more complex ones, like internal rate of return and net present value, are rarely used and even the simpler 'pay-back period' and 'rate of return' methods are used in a rough and ready way. Investors still usually follow a particular course because, instinctively, they *think* it right or just *want* to do it.

These mainly instinctive reasons for investing can be sound — especially if they are influenced by past investment that (with due regard for future cost and price trends) has proved sound. By themselves, however, such methods lack the objectivity of formal investment appraisal and must always be suspect for this reason. While this situation continues, some more systematic but agreeable kind of appraisal could help to prevent unintended misuse of capital. The following guidelines should help to improve the likelihood that a potential investment is worthwhile.

See what you can do without

Can you raise income without extra capital, for example, by improving crop yields or margin over concentrate feed costs? There may be many ways of raising profit without using more capital; some may even release capital. These should be fully studied before making any new investment.

Concentrate on short-term, directly productive investments

Although commercial farming, unlike many other industries, is characterised by a high ratio of fixed to working capital, it is important to keep these two items in balance. Over-investment in fixed assets may leave insufficient funds to carry through current production plans. A basic aim should be always to keep capital as available as possible by keeping as much as possible as productive or working capital that can earn a return and remain realisable. Sufficient liquidity must be kept to pay interest, loan repayments, living costs and tax.

Long-term assets, like buildings, roads and fixed equipment commit funds for a long time. They help and maintain production but show no direct return as they generate costs rather than income. Perennial crops and breeding cattle usually have a slow turnover and thus need much capital. Priority is best given to assets that offer short-term prospects of generating income that can later be reinvested to buy longer-term assets. Examples include water supplies, annual crop inputs, pullets, work animals and implements. These all give prospects of quickly generating cash surpluses. This is especially important if borrowed funds are hard to get which tends to be so for young farmers who need them most.

It is sometimes hard to classify whether an asset is productive or consumer capital in peasant farming where there are close links between farm and household.

How long must you wait for a return?

Investment always means outlay now for a return in the future. The further ahead that future, the more uncertain any forecast of the return will be. Other things being equal an early return is better than a later one.

Could the investment be cheaper?

There are two sides to cheapness: the interest paid on borrowed capital and the amount of capital used. It is easy to pay too much attention to the former and not enough to the latter. The question to ask is, 'How little capital do I need?' not 'How much?'

Concentrate on existing enterprises

Investment within an existing farming system, by specialisation or by using up spare resources, usually offers more return than introducing a new enterprise. With an existing enterprise you are fairly sure about the return and your managerial skill; so rather expand or intensify existing enterprises than start new ones. Too often, when suddenly alarmed by shortage of funds, farmers tend to start something new, expecting it to be some sort of magic potion. It is, however, usually not worth trying to revive any enterprise giving a poor return now.

If a new venture, are all the facts available?

This is important both technically and financially. Do not rely on hearsay or vague reports; rather let others pioneer innovations first! Start on a small scale and cost it carefully before making any large investment.

Keep the investment as flexible as possible

This makes adapting to changing times easier. One advantage of the capital resource is its fluidity and adaptability for different uses. Once it is invested in a fixed asset, ability to change course in future is limited — thus hindering adaptation. Give investment priority to appreciating assets like breeding livestock. These can be sold whenever funds are needed for reinvestment. Permanent improvements can be cashed only by selling the farm. Multi-purpose and cheap, rather than specialised, buildings and equipment make the investment flexible enough to allow for alternative uses if necessary.

Plan repayments as well as interest

Unlike many farmers if borrowing to invest in a

depreciating asset, plan full repayment as well as interest payments within its expected life.

Require higher returns from risky investments

Farmers must decide how much capital they will use before knowing crop yields and product prices. They may, therefore, quite rationally decide to use less capital than what would probably be most profitable. They may be cautious and forgo some possible profits if things turn out well rather than risk a loss if they turn out badly. The higher the risk, the higher should be the expected return.

Will labour really be saved?

If you are investing to save labour, where is the saving? Will the wages bill really fall or will extra output arise from productive use of freed labour elsewhere?

Initially ignore tax concessions

Do not be unduly tempted by tax allowances when making initial decisions. Work out tax adjustments, which are liable to change, later. It is best to ensure a profit before considering tax savings.

Avoid personal bias in calculations

Do not expect technical perfection or be over-optimistic. Too often there is a tendency to rig the figures to get the answer wanted.

RETURN ON CAPITAL

The rate of return on capital employed is used either to indicate overall business efficiency or to assess the likely profitability of an individual investment. This crude calculation is one of the most common measures of the worth of particular investments. Managers are expected to make the best use of the resources they control so the profit generated by business assets is clearly an important indicator of business efficiency. Also known as the book or accountant's rate of return, simple rate of return on capital employed recognises that it is not only profit that is important in business but also the amount of capital used to produce it. Profit is regarded not by itself but as a return on capital used because one of the main objectives of most business firms is to earn a satisfactory return on capital used.

The ratio most commonly used is profit/capital employed. This is the percentage of average profit, over the expected useful life of an asset, to the initial investment (or, sometimes, to half the initial investment). The annual return on capital is the total annual gain or benefit from using it less all extra costs incurred including depreciation and repairs and maintenance. Rate of return draws management attention to the main objective of most businesses — making the most profit possible with the capital available.

The basic data needed for calculating or estimating return on capital are capital employed and profit for the period for which the ratio is calculated. This is not as simple as it seems as there are many ways of calculating both profit and capital employed. Great care is needed in interpreting figures of rate of return as they can mislead unless both profit and capital employed are clearly defined.

The profit or return figure varies greatly with treatment of interest on borrowed capital, tax and depreciation. It may refer to expected return in the first year of an investment, average annual earnings over its expected life or those of a peak year. The profit to be related to capital employed may be profit before tax, profit after tax, profit before or after deducting interest on long-term loans, profit available for distribution to the owners, management and investment income, net cash flow or marginal return. This gives at least eight alternatives.

Return on capital is strongly affected by the amount of borrowed capital used and the interest paid for it. A farmer working almost entirely on borrowed capital with say $4000 of his own invested in the business and who makes $1000 profit has a return on his capital of 25 per cent. This does not mean that he is more successful than his neighbour who has earned $40,000, using only his own capital, showing a return of 20 per cent on $200,000. In calculating rate of return, interest charges are best excluded from costs.

It is hard to allow for tax payments in the calculation. If the business is run by a sole trader or partners, profits should be calculated before tax as income tax varies so much with the personal situation of the owner(s). However, if the business is a limited company, tax is easy to work out and profit may be expressed after tax.

Return is always expressed after allowing for depreciation. However, alternative depreciation methods can give different results. Depreciation may reflect actual wear and tear or it may be the notional depreciation used for tax purposes. Land value appreciation is usually disregarded; machinery may be undervalued and stocks written down. All these features tend to raise the apparent return on capital. Return on capital should be calculated on the current replacement value of fixed assets less adjusted depreciation. To measure the rate of return on assets employed without considering current prices is likely to mislead and could spur unjustified expansion.

The method of calculation is as follows:

	$
Machinery – original cost in 1983	20000
Accumulated depreciation – 4 years at 10 %	8000
Book value in 1987	12000
Current replacement cost in 1987	30000
Depreciation for 4 years at 10%	12000
Depreciated current replacement value	18000

Capital employed is sometimes taken as either the initial sum invested or the average over the period of the investment, that is (initial amount + residual value)/2. Again, the capital may be total business assets, total assets less current liabilities, owned capital, fixed capital, working capital or marginal capital. Usually it is net assets, that is, total assets less current liabilities, such as an overdraft or other creditors. If capital is defined to exclude loan capital, then the return measured is that on equity capital.

If there is little difference between the amounts of capital employed at the start and end of the accounting period, it is usual to calculate it from the latest balance sheet. If there are large differences, the average amount of capital employed is generally better.

The only calculations of capital return relevant to decision-making in farming are those involving the total capital in the business and returns on marginal capital expenditure. An example of the former is given in table 4.10. The profit/equity ratio measures business profitability solely from the standpoint of the company's members or equity shareholders.

Table 4.10   Example of Calculation of Rate of Return from Annual Financial Statements

Balance sheet

	$		$
Owner's capital	90000	Fixed assets	110000
7.5 % loan	40000	Current assets	40000
Current liabilities	20000		
	150000		150000

Profit and loss account

	$
Profit from trading	23000
Less: loan interest	3000
Net profit	20000

Rate of return on capital employed = (23,000 × 100)/(150,000 − 20,000) = 17.7% per annum.
(Rate of return on owner's equity = (20,000 × 100)/(90,000 = 22.2%)

Table 4.11 gives a simple example of the return on marginal capital expenditure on an investment project within the business.

Table 4.11   A Simple Investment Project

	Year 0	Year 1	Year 2
Net cash flow	(200)	110	121

The direct rate of return is calculated by dividing total profit over the economic life by the amount of outlay – (231 − 200)/200 = 15.5%. The more usual simple, or average rate of return results from dividing average annual profit by the initial outlay – (231 − 200)(2 × 200) = 7.75%.

Rate of return on capital is simple to calculate and easy to understand. It considers the profit of a project over its working life and is useful for comparing the profitability of alternative investments or different enterprises. If the project is fairly short-term and simple, with the whole investment made initially (in year 0) and producing an equal, annual income over its life, an accurate answer will be obtained. Should return vary much in different areas of the business, it would clearly pay to invest in and expand those enterprises giving the best returns, provided the risks involved are considered. If the

Table 4.12 Illustration of Failure of Simple Rate Return to Distinguish between Constant, Decreasing and Increasing Cash Flows

Project	Year 0	Year 1	Year 2	Year 3	Year 4
A	(200)	70	70	70	70
B	(200)	100	80	60	40
C	(200)	40	60	80	100

relative return *between* enterprises is fairly accurately assessed, lack of precision *within* individual enterprises may not matter much. Precise absolute rates are often unnecessary.

Profit/assets is probably the most useful single ratio for showing long-term trends in business efficiency over the years. Average, annual return can be compared with what could be earned by investing off the farm.

Use of simple rates of return to compare alternative investment projects is open to many serious criticisms. First, it does not consider the timing of the capital outlays and earnings and so it ignores the time value of money. As shown in table 4.12, it also ignores the fact that earnings from alternative projects can occur at different rates. Options with the same *total* returns, initial outlays and useful life are identically ranked even if the returns of one of them accrue earlier and are worth more because of the greater reinvestment possibility. Each of these projects has an average annual rate of return of (280 − 200)(200 × 4) or 10 per cent although the high early surplus produced by project B could be reinvested to earn interest. While it seems sensible to choose between competing projects by comparing the average, annual return on capital invested, it is hard to express, as a single figure, the varying returns expected over the projects' lives.

Rate of return is a good indicator and a crude screening device but a poor management tool as it seldom explains *why* it is high or low. Because of all these problems in estimating rate of return, time-discounting methods are increasingly used for appraising capital investments.

If a whole business has a falling annual rate of return, this may result from falling profit margins or overcapitalisation − too much investment for the profit generated. A low return on total capital can happen merely because farm value exceeds its agricultural worth owing to its amenity or potential development opportunities. A business earning a long-term rate of return lower than its cost of capital is usually using resources inefficiently. Although this seems obvious, many businesses regularly produce a return on capital below the bank rate and seem content to do so. One reason for this is the assumption, widely held in small businesses, that constant hard work and an assured market will automatically ensure more profit. There is too little regard for the need to ensure that capital is used efficiently in producing extra income. On starting a farm with limited capital, it is important to invest in those enterprises likely to give the greatest rate of return on the investment.

Some farmers get at least as much return on their capital as they could get in any alternative investment; others can barely service their borrowings and support a low standard of living. It is the latter who 'live poor to die rich'. They hope that when they cease farming, the greater sale price of the farm will enable them to meet their commitments and have a surplus on which to retire.

It is hard to define a satisfactory return on capital employed. Perhaps the lowest rate should be the cost of capital raised externally to finance the project, say, 15−17 per cent before tax. Obviously, an owner of a business does not always seek an immediate adequate return on capital employed since, in their early years especially, businesses are run with a view to future growth. In the long term, however, a return on capital greater than the going rate on investment markets must be achieved if the owner is to get a fair reward for his effort and enterprise.

The return on total capital used on the farm depends upon the return on working capital and the proportion of total capital used for running the farm instead of being tied up in land. The higher the proportion of capital that is working, the better the rate of return is likely to be, provided the business is not overcapitalised.

PAY-BACK

This criterion is still often used by managers as a crude decision tool for comparing investment projects. It is simply the time taken for expected profits to fully recover the initial outlay or investment; that is, the time taken for cumulative net cash flow to become zero. Depreciation, interest and tax charges are usually ignored.

If an extra $700 income each year can be earned by spending $4900 on a new machine, the pay-back period is seven years. Table 4.13

Table 4.13 Illustration of Pay-back Periods for Alternative Investment

	Project D	Project E	Project F
	$	$	$
Year 0	(8500)	(8500)	(8500)
Year 1	2000	4000	1000
Year 2	2000	3000	1000
Year 3	2000	1000	1000
Year 4	4000	1000	3000
Year 5	1000	1000	4000
Year 6	0	1000	1000
Year 7	0	1000	1000
Year 8	0	1000	1000
Cumulative net cash flow	2500	4500	4500

shows three alternative investment projects each needing an initial capital outlay of $8500 but giving rise to different patterns of net cash flows.

In comparing alternative projects, the quicker the pay-back, the better the project and it is usually accepted if the pay-back period is less than some maximum decided by management. In the example above projects D and E are considered equally desirable as they both pay back in year 4 and they are both better than project F which does not pay back until year 5.

Pay-back has its advantages as an investment criterion. It is a simple cash flow easily understood by farmers. If a business is short of cash, it is essential that investments are changed into cash as soon as possible and this method acknowledges the value of early returns and keeping liquidity. By selecting investment projects with the quickest pay-backs, the time value of money is allowed for to some extent. Because the method considers only the years in which cost is recovered, estimates are not based over long time periods and so tend to be more accurate than other methods in which the whole life of the asset is considered.

Unfortunately, the pay-back criterion has several disadvantages that limit its use for appraising long-term capital investments. Full regard is not taken of the time value of money, that is, that $1 tomorrow is worth less than $1 today. Study of projects D and E in table 4.13, shows that options with the same pay-back period are ranked the same even if one of them yields extra earnings after the period, or more in earlier years during it, and thus has more reinvestment possibilities. The whole useful life of the project is not considered. Thus project D is favoured over project F although the cumulative net cash flows are $2500 and $4500 respectively. Profits earned during the working life of the asset after recovery of its cost are ignored. Thus, this test can rate a project giving a quick return above another that takes longer to get established but gives higher returns in the long run. Pay-back also ignores the timing of cash flows within the pay-back period. No distinction is made between projects D and E although project E is clearly more desirable as it recovers much more of the initial capital outlay in the first year or two. If a project has high early net cash flows, these can be invested to earn interest.

Despite all these disadvantages, pay-back is useful as a crude screening tool. Projects needing small capital outlays that pay back early in their expected life are probably worth while and can usually be accepted without spending time on more complex study. The technique also helps to isolate especially poor projects that have a long pay-back time, put funds at risk for a long while and thus do not warrant further study. It therefore helps in the initial appraisal of large risky capital projects. Pay-back is an adequate criterion when quick recovery of funds is important in a risky project; it saves time by automatically accepting or rejecting many possible investments.

Simple rate of return and pay-back are common, crude methods of investment appraisal that ignore the time value of money. Modern techniques of investment appraisal allow for the time value of money. They are called discounted cash-flow techniques and will be described in chapter 11.

CAPITAL BUDGETING

Budgeting is needed for all progressive, successful concerns — whether a farm, store, hotel or even for personal affairs — to estimate whether a proposed change is justified by the chance of raising profit. All budgeting is forecasting; it means trying to state now what will occur in future. Inevitably, this is uncertain but it is better to budget intelligently than not to budget at all but simply hope for the best. One of the main purposes of budgeting is to create, in the whole management team, the mental attitude to seek constant improvement in operating performance by relating management decisions and actions to their effects on performance. There are usually many alternative ways of spending scarce funds so a choice is necessary.

In its widest sense, a budget is a forecast of income and expenditure in the future in contrast to an account that records financial transactions as they occur. It is a plan for the future in terms of money, quantity or both. Budgeting is an aid to planning rather than a planning method itself as it is used to estimate in financial terms the likely results of an already made plan. Complete budgets, partial budgets and cash-flow budgets will all be described.

Many business run into serious financial trouble by taking on commitments beyond the financial resources available to them. As capital resources are usually limited and as it takes a long while to recover the cost of investment in plant and machinery from operations, thereby leading to inflexibility, capital budgets should be carefully combined with long-term planning. No farming system should be changed without estimating its likely financial results by detailed budgeting — not on the back of a cigarette box! Although much planning is still done informally, budgeting reduces the chance of any important effects being overlooked. In large organisations, business planning and control has become highly developed and smaller firms should realise that the same techniques are available and just as necessary. In small firms, the issues are usually clearer.

Budgeting is not a trick for confusing traditionalists more interested in technical efficiency than business efficiency! It should help both and is essential for financial planning and control.

Budgeting has many advantages. A capital budget may reveal the need to adjust a farm plan or, at least, to retime it. If the maximum finance for a proposed system, designed to maximise profits, cannot be obtained, then the budget should suggest where adjustment is needed. This may require, for example, delayed buying of new equipment or, if allowed, payment of tax. It can help to postpone the introduction of subsidiary enterprises into the system. Whatever changes need making they are certainly best made before the plan is executed. A capital expenditure budget enables a manager to decide his expenditure and to control capital expenditure.

The commonest use of budgets is to help decide between one farming system and another or between changes in an existing system. Budgeting helps to check proposed plans by translating technical data into financial terms. Many farmers have regretted making changes without budgeting; they have adopted a change and found that, instead of rising, profits have dropped. Guesswork is rarely adequate for profit planning. In any case, one change can seldom be considered alone. If alternatives are available for consideration then a separate budget for each is a suitable way to decide which one should be adopted.

Budgets have the following main advantages:

- They force management to produce clear plans for the future. This should mean that objectives are examined and guidelines produced for operating managers. This prevents waste as it directs spending towards a definite purpose. Managers must study markets, products and methods, thus finding ways to strengthen and expand the business.
- They help to control management activity since performance can be compared with the budget. This comparison gives managers a guide to their performance, showing where the system needs adjustment. They set a target for each business activity and aid control-by-exception by highlighting the variances between forecast and actual results.

A budget may be for a short time ahead or for several years. Capital expenditure is often planned five to ten years in advance so that, for control purposes, it should be broken down into convenient periods like years or months. Detailed budgets are usually made for a year ahead but outline plans may be produced for several years. The period for which it is reasonable to budget depends on the type of business, the length of the production cycle and the ease or difficulty of forecasting future market conditions. This time period need not be the same for each section of the business. A long-term budget is often prepared as a general objective and this is added to more detailed, short-term budgets. The short-term budgets are compared period-by-period with the long-term one so that the manager can see if the business is progressing as planned. When necessary, the long-term budget should be adjusted in view of experience.

The starting point for any budget is the plan of action determined by top management. These proposals are then translated into quantitative and monetary terms to show how the desired ends can be reached within the limits of available resources. Budgeting may need little more than the often crude mental calculations that most farmers already make but it usually needs more formal methods and many farmers seem to lack confidence in preparing formal budgets. These will therefore be explained later in some detail.

It is essential to build into a budget performance and price data relevant to the particular business. Many farmers, but too few, can provide this information from their records. Data specific to the farm concerned are important because of the difference between farms (and farmers). It is often tempting to use data obtained from government costings or from commercial firms in budgets simply because they are often available in an immediately usable form. This temptation should be resisted as no standard data, especially of labour and capital needs, can replace a farmer's records of his own recent physical performance, combined with his best estimates of future prices and costs. Use of average data is likely to result in wrong decisions. It is reasonable to assume that 500 kg/ha of a standard type of fertiliser will give the same response on coffee under identical conditions. However, similar climate, soil type and cultivar alone do not constitute identical conditions. Owing to varying management skill affecting, for example, time of pruning and fertiliser application, plant population, weed control and pest and disease control, fertiliser response varies widely with the same cultivar, climate and soil type.

If new enterprises are being planned for which your own data are unavailable, then standard data must be used. These, however, should be adjusted to conform to the standard of management shown by existing enterprises. If, for example, burley tobacco yields are only 60 per cent of standard and it is proposed to substitute cotton for burley, it would probably be unwise to expect standard cotton yields. Care is also necessary in using input: ouptut data based on research results as most of this work is conducted under ideal management conditions that few commercial farmers could, or should, provide. These responses should be reduced.

Budgets and plans are pointers to — but not guarantees of — success! They are seldom entirely accurate because so many factors, like rainfall, are beyond a farmer's control. There is also uncertainty about future costs and prices so reasonable 'guesstimates' of future trends are needed. It is possible to forecast the future, but not to determine it!

Before detailing specific types of budget, such as complete, partial and cash-flow budgets, it is worth describing depreciation in some detail.

DEPRECIATION

Depreciation is provision for writing off the cost of fixed and part-fixed 'wasting' capital assets,

expected to last over a year, over their expected working life. It is a financial estimate of the annual loss of value of these assets and an amortisation of past capital costs allowed as an annual tax-deductible expense.

Capital expenditure represents a deferred cost. It is the cost of obtaining assets for use beyond the current accounting year and that must be written off, by means of annual depreciation charges, over the accounting periods that benefit from their use. Only part of their cost is charged to the year of purchase. In this way, these items are brought into the balance sheet each year at a realistic book value. The depreciation charge in financial accounts is therefore merely a book entry representing the spreading of the cost of a fixed or part-fixed asset over its working life. Capital assets are eventually used up or destroyed and they must usually be replaced. In other words, a current replacement cost is associated with any new capital asset brought into the business. This replacement cost of medium- and long-term capital does not arise each year. Thus if the productive life of irrigation risers is estimated as eight years, then a replacement cost will recur every eight years but not in the intermediate years. It is often more convenient to think of this replacement cost as a series of annual costs equal to the total eight-yearly expense. This imaginary annual replacement cost is called depreciation.

Not all fixed assets depreciate in value. Land usually appreciates and buildings, livestock, seasoning timber and perennial crops usually gain value over at least part of their life.

Depreciation provision is important because the reported annual profit depends, among other things, on the depreciation method used and the assumptions made about the life and residual value of the asset. Allowance for asset depreciation is always made before calculating profit because consumption of capital assets is one of the costs of earning the revenues of the business and is allowed as such, according to special rules, by the tax authorities. Depreciation provision is an imprecise measure of the loss of value of an asset and there are many problems in estimating and interpreting it. However, to ignore it would suggest that no economic change had occurred with respect to the capital asset. A fair estimate, based on known assumptions, is usually better than no information at all. Depreciation is a way of keeping funds within the business to help to meet the cost of renewal; otherwise those funds might have been distributed as dividend or drawings. It is somewhat flexible and can be adjusted, within limits, in the best interests of the business.

The word 'depreciation' covers many concepts and has several meanings. To some, it means physical deterioration; to others a fall in value; and to others, an allocation of costs. Some accountants argue that depreciation is simply a way of sharing the cost of a capital asset farily over the time that it is likely to last. It is an annual charge made to recover a prepaid cost over the asset's economic life. Other accountants argue that the asset was bought to provide services so, to maintain earning capacity, sufficient funds should be kept in the business to finance replacement of the asset at the end of its economic life. This immediately raises the question of whether depreciation should be based on original or replacement cost.

CAUSES OF DEPRECIATION

There are three main reasons for depreciation: wear and tear with use, obsolescence and gradual deterioration with age. Physical deterioration is hastened by use and running maintenance will not keep assets in their new condition. This is what is meant by wear and tear. It is loss of value caused by ordinary use as distinct from damage resulting from carelessness or accidents.

In accounting, obsolescence is the ending of an asset's useful life for reasons other than deterioration. Functional obsolescence can cut short useful economic life long before physical life of an asset is over. An asset is obsolescent when it is going out of use but is not yet completely unusable. It is scrapped because it has become out-dated and needs replacing with a more efficient, newer version. This often necessitates scrapping machinery that is still in good condition. Items held in stock may become out of date and of no further use, such as spare parts for a scrapped machine.

Depreciation due to natural deterioration over time, which can rarely be wholly stopped by preventive maintenance, is often caused by corrosion or perishing of parts. Chemicals can lose their effectiveness if stored too long. Poor storage conditions increase the rate of deterioration over time. For example, letting cement get damp or storing bags of grain on a concrete floor without dunnage. Accelerated deterioration is also caused by accidents, bad handling and rats.

Age and obsolescence are most important with buildings and they tend to limit the life of machinery in years. Wear and tear with use is often the most important aspect of machinery depreciation on large farms where machinery is used intensively. This tends to limit its life in hours. Machinery depreciation is then no longer an annual fixed cost, regardless of use, but can be taken as a constant variable, running cost per hectare or tonne.

To calculate depreciation it is necessary to know or be able to estimate the cost of the asset, how its value is likely to fall, it useful life and its residual value, if any, at the end of that life. It is hard to know how long an asset will last before it is scrapped but an attempt should be made to estimate its life because it is logical to write it off over this time. The productive life of a capital asset can be prolonged by regular, careful maintenance so that annual depreciation falls. However, repairs and maintenance also involve cost so that a balance should be found between these and depreciation costs. Economic life varies widely with specific type of asset, maintenance and use. A very rough guide is: building — 30 years; plant and machinery — 10 years, with a residual value of 10 per cent; lorries — 200,000 km with 10 per cent residual value; cars and Land Rovers — 100,000 km, with a residual value of 10 per cent.

METHODS OF DEPRECIATION

Several methods are used in practice for calculating depreciation. The only really accurate one is to find the loss in market value. By definition, depreciation is loss of market value over a period of time — usually a year. It is thus reasonable for depreciation to be estimated by finding out the market value at the end of the period and deducting this from the original price or last estimate of its value. Unfortunately, this is seldom practicable. It depends on the secondhand market which varies both with place and time. The main methods used in practice are:

- the straight-line or fixed instalment method
- the reducing balance method
- the sum-of-digits method
- the annuity method
- use-adjusted methods
- revaluation

There are other ways of providing for the cost of replacement by investing in equities, securities or an insurance policy but I shall not describe these.

Straight-line depreciation is simply the estimated total loss of value of the asset divided by its estimated useful life on the farm. Depreciation is written off in equal annual instalments that are included as an expense in the profit and loss account. The estimated residual (salvage or scrap) value of the asset is deducted from original cost and the balance divided by its expected life. This method is simple, easy to work out and favoured by most accountants. Furthermore, the time when the asset will be fully written off in the books is known in advance. Straight-line depreciation is normally used in budgets where concern is mainly with average annual cost. The name comes from the line obtained when depreciation is plotted graphically as in figure 4.2.

A disadvantage is that, even though annual depreciation is constant, the repairs and maintenance costs of many fixed assets increase as they get older. The combined effect of these costs and straight-line depreciation is therefore to increase the total charge against profits each year. This is shown in figure 4.3 for an asset bought for $10,000 with an estimated life of 10 years and a scrap value of $1000. Repairs and maintenance are assumed to cost $400 in the first year and then to rise by 20 per cent each year. Straight-line depreciation is really suitable only for very durable assets, like fences and buildings, that need the same maintenance throughout their lives and are not liable to large changes or improvements in design.

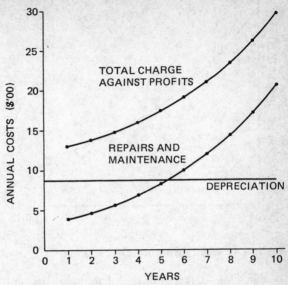

Annual depreciation = $ (10,000 − 1,000)/10

Figure 4.3   Combination of Straight-line Depreciation with Repair and Maintenance Costs

With the reducing or diminishing balance method of depreciation a fixed percentage is written off the falling, written down or book value each year; not off the original cost as with straight-line depreciation. Annual depreciation thus falls as the asset ages. Using the previous example of an asset costing $10,000 and taking an annual depreciation rate of 20 per cent, the effect of the reducing balance method on its annual written down value (book value) is shown in table 4.14 and figure 4.4.

From an accounting standpoint, it is desirable to spread the costs of fixed assets fairly evenly each year to avoid undue variations in profit. Some accountants argue that the straight-line method does not allow for the higher repair costs

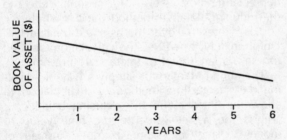

Figure 4.2   Straight-line Depreciation

Table 4.14   Calculation of Depreciation at 20% on the Reducing Balance

Year	Value at start of year	Depreciation	Book value at year end	Year	Value at start of year	Depreciation	Book value at year end
	$	$	$		$	$	$
1	10000 (cost)	2000	8000	6	3277	655	2621
2	8000	1600	6400	7	2621	524	2097
3	6400	1280	5120	8	2097	419	1678
4	5120	1024	4096	9	1678	336	1342
5	4096	819	3277	10	1342	268	1074

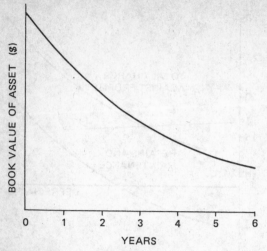

Figure 4.4   Reducing Balance Depreciation

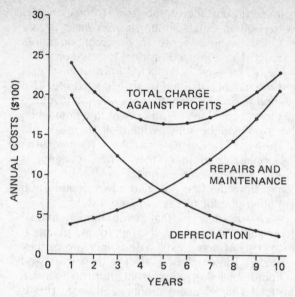

Figure 4.5   Combination of Reducing Balance Depreciation with Repair and Maintenance Costs

in the later years of the life of many assets and that it is therefore better to write off larger sums as depreciation in earlier years. Reducing balance does this.

Note that with this method the value of an asset, in theory, is never completely witten off. There is always some small amount remaining so the asset stays on the books until it is scrapped or sold. In practice, the rate is applied until the salvage value is reached and then no further depreciation is taken. It will also be seen that almost twice the percentage must be used with this method (20 per cent against 10 per cent) to write down an asset to a required figure over a given period.

Figure 4.5 shows that the reducing balance method gives a more even combined annual charge against profit by making smaller depreciation charges in later years when repair and maintenance costs, as well as the risk of obsolescence, are likely to rise.

Depreciation by reducing balance is allowed by the tax authorities in most countries for certain capital items, for instance, vehicles and machinery, that lose value fastest in the early years of their life. It reflects the way in which the market value of a machine drops after purchase more accurately than straight-line depreciation.

The sum-of-digits depreciation method is popular in North America. It is a compromise between the straight-line method and the rapidly diminishing curve of the reducing balance method. Annual depreciation is calculated as a fraction that drops each year so the total depreciation is entirely covered during the asset's useful life. The denominator is the sum of the indi-

vidual years of expected useful life and the numerator is the years of life still left at the start of the year.

Table 4.15 again uses the example of an asset bought for $10,000 with an estimated life of 10 years and scrap value of $1000.

This is probably the most realistic way of depreciating most assets. It has a similar effect to the reducing balance method but the whole net cost is written off over the asset's estimated life. Unfortunately, however, the calculations needed for, say, a large fleet of tractors are tiresome.

The annuity method is simply a way of adjusting the straight-line method of depreciation to allow for 'interest' on the reducing balance. It adjusts the assset value to the time value of money. Using the same example, instead of charging $900 a year, depreciation would be $1464.30 each year ($9000 × 0.1627 — see Appendix A4).

Use-adjusted methods consider the use of an asset rather than passage of time. The expected loss of value of the asset over its useful life is divided by the output that it is likely to process or produce. Alternatively, the amount to be written off may be divided by the hours that the machine is likely to work or, if it is a vehicle, by its estimated life in kilometres. The depreciation charge will therefore vary each year according to use. This method considers only wear and tear; both obsolescence and gradual deterioration with age are ignored.

Table 4.15 Calculation of Depreciation by Sum of Digits

Year	Fraction	Annual depreciation
		$
1	10/55	1636
2	9/55	1473
3	8/55	1309
4	7/55	1145
5	6/55	983
6	5/55	818
7	4/55	655
8	3/55	491
9	2/55	327
10	1/55	164
55		9000

Table 4.16 Effect of Inflation on Conventional Depreciation

Year	Depreciation provision	Price index	Purchasing power of depreciation provision
	$		$
1980	1250	100	1250
1981	1250	110	1136
1982	1250	122	1025
1983	1250	135	926
1984	1250	149	834
1985	1250	165	758
1986	1250	182	687
1987	1250	200	625
	10000		7241

Some accountants and many economists like to revalue assets at replacement cost each year and calculate depreciation from the new cost of an equivalent asset. This method requires annual revaluation of all the firm's assets in the light of current conditions. It seems logical to revalue them each year and write off the loss of market value. However, this method leads to the sort of error that we wish to avoid. If assets are revalued each year, trading profits may not truly reflect operating profits as they will vary simply because of sudden changes in the market value of assets. A solution to this problem would be to revalue each asset yearly and enter a value as correct as possible on the annual balance sheet but to charge the profit and loss account with a 'rent' representing what the firm would have had to pay for use of the asset if it had been hired. For example, rent could be charged according to what it would have cost to lease. It is as if the business has set up a hiring department to hire assets to an operating department at an economic rent and take credit for any gains or losses made in the capital value of the assets themselves.

Depreciation provisions are seldom sufficient to pay for eventual replacement because prices rise so quickly. This is especially so in the present inflationary times and replacement can cause a serious cash shortage. The steady fall in the real value of money, which lasted for many years and will doubtless recur, means that depreciation based on an asset's original historic cost will not retain enough funds in the business to replace it. Consequently, 'inflation' or 'current-cost account-ing' has become popular in management accounts to avoid overstating and distributing fictitious profits, understating costs and thus the tendency to increase personal drawings and living standards by consuming capital.

Inflation can be allowed for merely by raising the depreciation rate to amass funds equal to expected replacement cost. However, firms are increasingly employing for internal use, formal methods of inflation accounting based on replacement cost. In practice, of course, the cost of replacing plant and equipment rises not only because of inflation but also because, with tech-nical progress, plant and equipment must often be replaced by more advanced models.

A machine bought for $10,000 in 1980 with a working life of eight years may cost $20,000 to replace in 1987. A true provision for depreciation needs two separate entries in the accounts. The first is the normal method of depreciation designed to spread the cost of $10,000 evenly over the eight-year period. The second is to spread the extra $10,000 over the same period and this is shown as a separate item in the accounts with its own title such as 'Provision for increased replacement cost of plant and machin-ery'.

Table 4.16 shows that although the farmer may feel he has recouped the $10,000 originally invested, he has recovered only $7241 in pur-chasing power and has overstated his profits by 2759 current dollars.

Table 4.17 shows how this problem could be treated in inflation accounting to ensure that total depreciation provision is sufficient to finance replacement.

To decide which depreciation method is best for management accounts, two things should be considered; which method best approximates the loss of value of a particular asset and ease of

Table 4.17   Current Cost Accounting. Asset bought for $10,000 with an Estimated Trade-in Value of $2,000 after 8 Years. Sum of Digits

Year	Fraction	Annual depreciation	Book value at year end	Replacement cost at year end (A)	8 Year trade-in price (B)	Total depreciation (A−B)	Annual depreciation	Book Value at year end
		$	$	$	$	$	$	$
1980	8/36	1778	8222	11000	2200	8800	1956	9044
1981	7/36	1556	6666	12200	2400	9800	2127[a]	8117[b]
1982	6/36	1333	5333	13500	2700	10800	2217	7200
1983	5/36	1111	4222	14900	3000	11900	2294	6306
1984	4/36	889	3333	16500	3600	13200	2406	5500
1985	3/36	667	2666	18200	4000	14600	2383	4817
1986	2/36	444	2222	20200	4000	16000	2173	4444
1987	1/36	222	2000	22200	4400	17800	2244	4400
		8000					17800	

[a] $(9800 × 15/36) − 1956.   [b] $12,200 − (1956 + 2127)

Table 4.18   Comparison of Conventional Depreciation Methods. Machine Costing $9,000. Estimated Scrap Value after 8 Years — $1,000

Year end	Annual depreciation			Book value		
	Straight line	Reducing balance (25 %)	Sum of digits	Straight line	Reducing balance	Sum of digits
	$	$	$	$	$	$
1	1000	2250	1778	8000	6750	7222
2	1000	1688	1556	7000	5062	5666
3	1000	1266	1333	6000	3796	4333
4	1000	949	1111	5000	2847	3222
5	1000	712	889	4000	2135	3222
6	1000	534	667	3000	1601	2333
7	1000	400	444	2000	1201	1666
8	1000	201	222	1000	1000	1222
	8000	8000	8000			1000

calculation. Straight-line depreciation is easiest to calculate but most farm assets depreciate faster in their early years than this shows. Reducing balance and sum of digits give similar results as shown in table 4.18.

Depreciation is not a current cash expense. The only time that a flow of funds occurs is when the asset is bought or sold. Annual depreciation is a charge against the year's profits to amass funds so the asset can be replaced when necessary. However, it is unusual to simply put depreciation funds aside each year. They may be used for other purposes because it is usually a waste of resources to keep idle cash reserves for replacement. Depreciation provisions are a source of funds for reinvesting in the business in any suitable way to raise its profit. When the time comes to replace a capital asset, the fact that the business has grown, and presumably has become more profitable, should make it easier to raise extra finance.

The depreciation rates used to produce management accounts for internal business planning and control may vary widely from the standard rates used in financial statements prepared for the tax authorities. Depreciation is accepted for tax purposes as a business expense but it must be calculated according to certain rules that may not reflect the real loss of value charged in management accounts. For example, revaluation is not usually acceptable in conventional accounting and if it is done, to give a more realistic profit, it may be deliberately changed by the tax authorities. Tax calculations depreciate machinery faster than usually recommended for management purposes. Because depreciation affects taxable income, the profit and loss account and the balance sheet, various objectives may conflict when it is calculated. However, a much undervalued inventory can seriously weaken accounts for management use.

Accelerated depreciation is a practice by which, for tax purposes, a firm may write-off an asset before its expected useful life is over, with possibly a much larger allowance in the first year than in later years. Depreciation rates are adjusted by governments to encourage certain types of investment. For example, the Rhodesian authorities tried to encourage farmers to build better labour housing and more tobacco-curing barns by allowing them to be written off at 20 per cent and 33 per cent straight-line respectively. These assets will obviously last more than five and three years.

COMPLETE BUDGETING

Complete budgets are concerned with the whole farm system so all expenses and receipts likely to be incurred are included. In planning a farm system from the start, or in comparing the potential of existing and alternative farm systems, the initial selection and choice of scale of enterprises could be done in several ways. These vary from simple intuition to strict programme planning, depending usually on the complexity of the situation and the detail needed in the plan. Programmed solutions usually need professional advice but, if a farmer prepares his own budgets, less formal, though still fairly precise, methods are available to plan the farming system.

Complete budgeting is used most for the following situations:

- When a plan for a 'new farm' or 'new farmer' is needed — whether the farm is virgin land or already partly developed.
- When a large, basic change is being considered that would affect most, perhaps all, the farm costs and receipts. Examples include conversion of a dairy farm into a pig unit or a change from tobacco to beef and maize. Labour and capital needs are sure to change with different farming systems and these cannot be dealt with adequately by partial budgeting. Because of the many possible interactions of a large change in plan, it is often better to

prepare a complete rather than a partial budget.
- When the profit potential of an existing farm needs to be assessed either when tendering for a farm tenancy or for later use as a check on actual performance.

Thus, complete budgeting helps farmers in the following ways:

- To forecast future profits as precisely as possible. This means budgeting a profit and loss account for some future period.
- To estimate the future capital needs of the business when seeking credit.
- To assess future tax commitments.
- To set up and work a system of budgetary control.
- To compare the likely financial effects of a proposed large change with the present system. The latter is compared with alternatives and these with each other. The present system provides the base for comparison.

Farmers taking over a new farm often have an unrealistic idea of which farming system would give *them* the best profits. Often the only plan seems to be to follow the system of the previous occupant, forgetting that he might have bought or rented the land cheaper than the current cost. High interest and repayment charges may make a more intensive system essential now.

It should be remembered when preparing a complete budget that there is a limit to every farmer's managerial ability. If the plan requires expansion of a current enterprise on an existing farm and present management achieves better results that those expected from 'standards', then the farmer's own results should be used. If current performance is below the normal standard, but it is expected that future results will improve, then expected output should be set as near the standard as seems realistic. The following aspects should all be considered: expected output, bearing in mind rotational constraints; enterprise size; inputs needed and input and output prices.

I do not suggest that all farmers should adopt the theoretically best farming system found by complete budgeting. A farmer's preferences for different enterprises and his skill in managing them will influence the system. Nevertheless, if this theoretical best plan is not adopted, it still provides a standard with which to assess the profit likely to be lost by adopting another one.

Complete budgets include the entire farm system so all physical data, costs and returns (including those in kind) must be estimated. If adequate farm records and accounts are available, it may be unnecessary to prepare a new budget for the existing system. However, when a new system is started, there may be no data to use. In this case it is necessary to compare budgets for alternative systems to find the one that best suits existing resources and objectives. Once a master budget is prepared, preparation of alternative ones usually need far less work as much of the same data can be used. These succeeding budgets arre often simple adaptations of the main one and partial budgeting, as described later, may suffice.

Table 4.19   A Typical Complete Budget

		$	$
Output:	Milk	37400	
	Beef	5100	
	Goats	200	
	Maize	36400	
	Cotton	10600	89700
		$	
Costs:	Bought feed	16800	
	Bought seed	4300	
	Fertiliser	6300	
	Wages	14400	
	Fuel	6800	
	Repairs and Maintenance	13600	
	Miscellaneous	6800	69000
Net farm income:			20700

Table 4.19 shows a typical summarised complete budget. Assumptions, both physical and financial, about potential performance are built into each figure but the detailed items needed to construct the budget are not shown.

PARTIAL BUDGETING

Managers and consultants often need a quick way to assess the financial effect of a proposed change in policy or prices in a basically satisfactory farm business where the overall farm organisation is unchanged. Partial budgeting not only enables them to assess the effect of small changes, such as buying a sprayer instead of hiring one or adding a few more sows to the herd. It can be used also to assess the likely financial effect of fairly large changes such as disposal of a dairy herd and substituting beef and cash crops on the freed land. Of all planning techniques, only simple par-

tial budgeting receives much attention from mos farmers.

Partial budgeting is a marginal analysis technique as it looks at the *changes* in costs and receipts, and thus net farm income, likely to resul from a marginal change in farming system. There are, of course, many possible changes in a farm system that need only partial reorganising of the business.

Whenever a change is considered or suggested, the first question is usually 'Would it pay?' The situation is often complicated because there are usually several ways of changing the farm business and the question becomes, 'Which would pay best?' Partial budgeting helps to answer such questions by applying logic to the situation so that the likely effects of proposed policy changes are examined objectively in terms of expected effect on net farm income. It is a simple technique equally useful in both private and business affairs, and should be used more often.

There are two main situations in which partial budgets help.

CHANGE IN THE COMBINATION OF ENTERPRISES — (PRODUCT SUBSTITUTION)

This could mean complete substitution of a new enterprise for an existing one, introduction of a supplementary enterprise, without displacing any existing one, or changing the scale of an enterprise. Farmers often face problems such as whether to keep pigs or poultry in existing buildings, to intensify a certain part of the business and raise profit by increasing turnover or whether to dispose of cattle and grow more crops. These are some of the commonest questions facing any farmer when he plans for the next year or two.

Partial Budget to Estimate the Effect of . . .

Losses	Gains
Income lost:	New income:
New costs:	Costs saved:
Net gain	or	Net loss

Figure 4.6   Standard Layout for Partial Budgets

CHANGE IN PRODUCTION METHOD

For example, buying a new machine or building either to raise output or to reduce costs. This, in effect, means finding the most profitable of two or more alternative policies such as whether to invest capital in a machine and thereby reduce labour costs, to invest in a grain drier or larger tractor or to feed cattle in pens instead of grazing them.

Partial budgets thus assess the likely effect of future policies or changes in part of the farm system on whole farm results. They are quicker and easier to make than complete budgets which are not needed when only fairly small changes are considered. However, a partial budget is sometimes too weak a tool and complete budgeting is necessary. For example, introducing a first tractor to replace most hand labour, will affect most existing inputs and outputs and modify the whole farming system.

As much detail and effort is needed to calculate each item for partial as for complete budgets if the same precision in the result is needed.

Partial budgeting simplifies decision-making for many problems by giving the most precise possible forecast of the financial effect of a proposed change. This should prevent unprofitable changes being made and the budget also serves as a target against which to compare later performance.

It is important in partial budgeting to be systematic, to head the budget clearly and to state clearly any assumptions made. Four basic questions must then be answered:

- what new costs would arise?
- what former costs would be saved?
- what former income would be lost?
- what new income would arise?

Provided each question is answered, there should be few errors in the budget. In its simplest form, when a new enterprise or method is introduced without substitution, only new costs and income arise. However, if there is substitution the former costs and income lost must also be considered. It helps to use the standard layout in figure 4.6 and headings even if some headings are not needed for a given budget.

The two sides of the budget must balance. If total gains on the right exceeds total losses on the left, the balancing figure is an expected rise in net farm income (net gain) and is placed on the left. If, on the other hand, total losses exceed total gains the balancing amount is a net loss on the right.

The importance of being systematic means more than merely having a standard layout for partial budgets. As with complete budgeting it is essential to clearly specify the proposed change stating what is involved and when it occurs. It then helps to go through the following three stages:

- find out the present situation
- calculate the situation after the proposed change
- complete the partial budget

I find it helpful to insert a description of all cost and income items likely to change *before* fixing values to them.

To clarify the use of partial budgets, let us look at a few examples.

Example 1

Headman Kaunda is thinking of cutting his maize area by one hectare and substituting a hectare of tobacco. He would have to hire more casual labour to cope with the tobacco in the peak reaping and curing period and to build curing barns. The likely financial effect of this proposed substitution is shown in table 4.20.

Example 2

Mr Chok is considering changing from hand-milking his dairy herd and produces the partial budget shown in table 4.21.

Table 4.20   Partial Budget to Estimate the Effect of Substituting 1 ha Tobacco for 1 ha Maize

Losses	$	Gains	$
Income lost		*New income*	
36 bags maize at $5.20/bag	187.20	900 kg tobacco at 50¢	450.00
New costs		*Costs saved*	
Fertiliser — 7 bags at $10.50	73.50	22.5 kg maize seed at 20¢	4.50
Specific casual labour	54.00	Fertiliser — 5 bags at $7.50	37.50
Depreciation of 7 curing barns at $3.50 each	24.50	Fumigant for storage	3.30
Net gain	156.10		
	$495.30		$495.30

Table 4.21   Partial Budget to Estimate the Effect of Introducing Machine Milking

Losses		Gains	
Income lost		New income	
New Costs	$	Costs saved	$
Electricity, $10 per week	520	Four less dairy labourers	1900
Replacement of liners, etc. $7 per week	364		
More cleaning materials	180		
Annual machine depreciation	270		
Annual interest on capital	190		
Net gain	376		
	$1900		$1900

Whether this change is adopted or not may depend on the farmer's estimate of any likely change in milk yield or quality.

Example 3

Mr Perreira is thinking of installing a grain drier for his 170 ha maize crop. He normally produces 62 bags (91 kg)/ha. With the drier he would harvest the whole crop earlier than usual. The lowest quotation for a suitable plant was given by Agro-Industries (Pvt.) Ltd for $2904 installed. It would also cost Mr Perreira $1000 to erect a suitable building. The expected life of the drier is five years and the estimated variable cost, including labour, of reducing grain moisture content by 6 per cent is 54 cents/tonne.

Agro-Industries claim that earlier harvesting and drying has proved to raise yield by 7 per cent or more by reducing losses due to lodging, termites and vermin — both two- and four-legged! They also say that farmers should raise yield by 5 bags/ha and reduce land preparation costs by ploughing two months earlier when the soil is still moist. Being shrewd, Mr Perreira halves these claims though he agrees that they *could* be true.

Using this information, he wants to assess the worth of the proposal by partial budgeting. Naturally, the capital costs must be changed to annual costs. Assuming straight-line, that is, average, annual, depreciation for five years, his annual depreciation will be $3904/5 − $781. However, he must also consider the other costs of ownership such as interest on capital invested, repair

Table 4.22   Partial Budget to Estimate the Effect of Drying Maize from 170 ha

Losses	$	Gains	$
Income lost		New income	
New Costs		3.5% increase in crop at $65/tonne[a] (due to earlier harvesting)	2270
Annual depreciation	781		
Annual interest	195		
Annual operating costs[d]	558	Interest at 10% on payment received 2 months earlier[b]	1119
Shelling and transport on extra yield at $9/tonne[e]	662	Increased yield from earlier ploughing estimated at 2.5 bags/ha[c]	2514
		Costs saved	
Net gain	3707		
	$5903		$5903

[a] $(170 \times 64.5 \times 91 \times 3.5 \times 65)/(1000 \times 100)$

[b] $(1700 \times 64.5 \times 103.5 \times 91 \times 65 \times 10 \times 2)/(100 \times 1000 \times 100 \times 12)$

[c] $(170 \times 2.5 \times 91 \times 65)/1000$

[d] $(170 \times 64.5 \times 103.5 \times 91 \times 0.54)/(1000 \times 100)$

[e] $[(170 \times 2.5 \times 91)/1000 + (170 \times 64.5 \times 91 \times 3.5)/(1000 \times 100)] \times 9$

and insurance. Repair and maintenance costs are included in the variable cost per tonne and he does not intend to insure the drier. With regard to interest on his investment, he will charge 10 per cent a year on the average investment of $1952 ($3904/2) so average annual interest is $195. Therefore, the total annual ownership cost is $976 and to this Mr Perreira must add variable operating costs to find his total annual cost.

He does not expect to lose any income or save any costs by installing the drier. On the other hand, he does expect extra income for, besides higher yields, he expects to receive payment two months earlier and thus reduce the interest paid on his seasonal overdraft for working capital.

Table 4.22 shows that, according to Mr Perreira's calculations, it would pay him handsomely to install the grain drier if he can obtain finance for it.

As the farm is not being completely reorganised, some or many costs and returns will not change; these are ignored. Only those costs, whether variable, 'fixed' or integer that are likely to change are considered. In other words, we consider only marginal costs and returns. To make an informed decision about transferring, for example, 10 ha from sorghum to cassava, financial information about the cowpea crop or the telephone account is irrelevant. Again, expanding the maize enterprise from 95 ha to 105 ha would involve extra costs only for seed, fertiliser, pesticides, fuel and probably some types of labour. However, doubling the maize area to 190 ha would probably also raise many common costs like tractor and implement depreciation, interest, storage space and skilled labour, such as drivers. Therefore, it is essential to know for any given proposal which costs and returns will change and to estimate the amounts of those changes.

Special care is needed, in dealing with items such as buildings, machinery and labour costs in partial budgets, to include only those items that will change in the annual profit and loss account. The capital cost or purchase price of buildings and part-fixed capital assets like machinery must not be included in partial budgets but only the annual charges resulting from such investments. These are mainly average depreciation, interest and repairs and maintenance on any *extra* capital assets bought. If a proposed change does not need investment in extra buildings or machinery, the annual running costs of existing buildings and machinery should be excluded from the budget.

Labour costs are easy to deal with in partial budgets. On small peasant farms most labour comes from the family with perhaps a few regular labourers. Regular labour cost is therefore unlikely to change with minor reorganisation although casual labour cost may change. However, on large commercial estates in the tropics, the regular labour gang tends to be large and to change rapidly. Consequently, even small changes can alter regular labour costs and this should be forecast in partial budgets.

In costing capital in a partial budget, interest is often charged at the current bank overdraft rate. With high inflation the consequence is to give results that are heavily biased against capital investment. The simplest solution to this problem is to charge a 'real' interest rate instead of a 'nominal' one. This is calculated as follows:

$$r = (n-i)(1+i)$$

where r = real interest rate
       n = nominal interest rate
       i = inflation rate

Thus a nominal rate of 20 per cent interest associated with an inflation rate of 15 per cent implies a real interest of

Table 4.23 Partial Budget to Estimate the Effect of Substituting 1 ha Tobacco for 1 ha Maize

Losses		Gains	
Income lost	$	*New income*	$
Gross margin of 1 ha maize	141.90	Gross margin of 1 ha tobacco	322.50
New costs		*Costs saved*	
Extra fixed costs — depreciation of 7 curing barns	24.50	Fixed costs saved	—
Net gain	156.10		
	322.50		322.50

$$(0.2-0.15)/(1+0.15) = 0.0435 \text{ or } 4.35\%$$

Another important principle in partial budgeting is to avoid personal bias because if some change is wanted strongly, there is often a tendency to load the figures to get the required answer. Use only hard facts in the budget itself. Personal preferences may affect the final decision *after* you have estimated what they are likely to cost in terms of lost profit. Output and prices should be estimated realistically as over-optimism often causes trouble when the plan is executed.

Costs, yields and prices used in partial budgets, whenever possible, should be based on a manager's own experience. However, if he expects next season's costs or prices to change, this should be allowed for.

Partial budgets may be set out in full or in gross margin form. Use of gross margins simplifies them for enterprise substitution as it deals in convenient packages and it is unnecessary to use the longer four-question approach because variable costs have already been deducted. Use of gross margins, however, is only a crude form of partial budget and it can give misleading results. This is because gross margins themselves may change at different scales of activity and they ignore common costs that do change with different policies. This is shown in table 4.23 — cf. table 4.20.

CASH-FLOW BUDGETING

Money has been likened aptly to a blood supply, circulating to all parts of the organisation. Too little can prove fatal, so every wise manager controls his cash flow. The health of any enterprise depends upon sufficient money flowing freely. Clotting, i.e., failure to flow freely, occurs for many reasons. A common one, however, is excessive leakage into stocks — of raw materials, work in progress and finished goods — and into debtors who are slow to pay.

Managers are more effective if they understand clearly the way in which money circulates within their own business. They should determine which activities need large money flows or soak up money for a long time. They should then examine the effects of their own operations on these critical activities and consider how to improve the flows.

Unfortunately, cash-flow budgets have many different names such as cash budgets, cash statements, cash-flow plans, cash-flow profiles and capital diaries. In chart form, they are often called cash-flow charts or capital profiles.

There are four main elements in successfully managing working capital: namely stock control; work-in-progress control, for example, capital tied up in annual crops; debtor control and cash budgeting. The importance of this area of financial management cannot be overstressed. Management of working capital is vital in all firms and no small firm is likely to last long if it neglects it. Balancing the supply of and demand for any resource is an essential part of management control. This is especially true of balancing the supply of and the demand for cash. Cash planning is especially important in financial planning because a manager needs cash to execute his plans and policies. Two essential criteria for running a business successfully are:

- Profitability, which greatly aids long-term cash flow
- Liquidity, or short-term cash flow, which helps the business keep running and to achieve its long-term cash flow.

A third criteria often is growth.

However excellent business potential is, it cannot be exploited if there are insufficient funds to finance it. Cash budgeting is therefore most important as use of money involves the cost of interest or dividends so it is essential to use it as effectively as possible. This is usually realised in hard times when credit is tight. However, in the past, generally buoyant trading conditions allowed firms to grow and make profits without paying much attention to using their funds to best advantage. That attitude has now changed. Managers have noticed the apparent paradox of being short of cash while making good profits. Cash flows, therefore, are now usually planned carefully in contrast to the old idea that if a firm is making profits the cash position will look after

itself. Liquidity can be reduced by either higher output or by management decisions like buying new plant or high stock levels in relation to sales; or simply by giving too much credit or investments not giving the expected profits. Maintaining liquidity requires management of cash.

A cash-flow budget is a forecast of the case position of the business for the immediate period ahead. It is similar to a bank statement in that it sets out the cumulative cash position as transactions occur. It is designed to show the future ebb and flow of cash and to show when funds should be borrowed and also when there is likely to be surplus cash. It is merely a forecast of the expected flow of money forecasted profit and loss account as it gives no estimate of future profit because it excludes depreciation and valuation changes. You make a cash budget by adding to your expected cash balance at the start of each period all the cash receipts expected during that period and from this total subtracting all the cash payments you expect to make.

Cash-flow planning is one of the most important planning techniques available to farmers. There are great benefits as the budget clearly shows the pattern of income, expenditure and repayment of creditors. Cash budgets predict expected inflows and outflows of cash and show whether cash resources will meet commitments. Cash shortage causes critical situations to develop, such as failure to meet current liabilities.

Cash budgeting is used not only for the whole business but also for individual investments within it. Indeed, it is one of the main tools used in investment appraisal. If money must be borrowed to finance a capital investment, it is important to know exactly what the cash commitment is likely to be. A cash-flow budget shows the peak cash need of a project and therefore any borrowing needed to finance it. The demand that would be made upon cash resources by buying a new machine or hiring another assistant is fully shown if a special cash-flow budget is prepared related only to those extra flows that will arise from the decision. This indicates the best type of financing to obtain the amount needed and when.

Cash-flow budgets have the following uses:

- They ensure that managers properly plan their credit needs and repayment programmes so that sufficient working capital is available for effective trading and the capital expenditure programme.
- They raise the probability of bank managers agreeing to loan or overdraft facilities.

● They aid financial control. The budget shows what things should look like at any given time. By comparing the actual flow, against the budgeted ones, the health of the business is constantly reviewed and differences can be investigated immediately.

To forecast short-term cash flows start with the bank balance at the end of the last period, estimate the cash likely to flow in and out during the current period and so forecast the balance at the end of the current period. This is continued for several periods ahead. Having planned the farm programme, expected physical inputs and outputs are valued and allocated to the expected time periods. Receipts and expenses are usually split into operating items and capital ones. The latter are those with a life of over one year. Expense items, sometimes called negative cash-flows, usually exceed receipt items in number. Operating receipts are usually broken down by enterprise. The budget normally has two basic parts — the budget or plan and the actual performance. Both parts are needed to gain full value from the exercise.

During each time period, the net cash flow is total cash inflows less total cash outflows. It shows the likely effect of trading during that period on total cash availability. The cumulative net cash flow is the forecast cash availability at the end of each period.

The time period used in cash-flow budgeting varies from a week to a year or more. Large, long-term investment made by firms with ample finance are usually appraised by yearly periods. However, this time period is too long for small businesses or for the daily running of any business. Yearly profit figures, balance sheets and cash flows are of little use as management control tools because they appear too seldom and are usually out-dated when available.

Cash-flow budgets are best made for a year or, better, eighteen months, broken into monthly periods. They should be revised at least twice, and preferably four times, a year so that adjustments are made for price, cost or policy changes. Preparing a cash-flow budget is a continuing job. Budgets may extend for several years, to find the peak cash needs, when large changes are planned, for example, from annual crops to cocoa. The overall budget is often split into quarters but months are recommended as quarterly periods can hide particular peak cash needs; much money may have to be spent at the start of a quarter and most receipts may come near the end. A rolling budget for 12 months ahead by

month, and quarterly for the next 2–3 years, is recommended. If it is considered too onerous to make monthly forecasts, two-monthly ones are an acceptable compromise. The business must keep solvent throughout the year, not only at the end of each budget period. If funds are really tight, it may even be necessary to forecast weekly to see if you can avoid the danger point.

Each expected movement of cash in or out of the business is allocated to the month when payment is expected to be received or made — not the month when the transaction took place. This means that credit terms and the time lag between receipt of goods and payment for them must be considered.

All *cash* payments and receipts expected during the budget period must be included — whether current, operating, or capital transactions. Captial expenditure, loan repayments, interest and tax payments should all be included; also receipts of loans and gifts or from sale of capital assets. It is also important to include payments made by banker's order for which no cheques are drawn. The cash-flow budget usually contains more than farm items alone. Personal drawings or dividends usually come from the same source and should therefore be included as they reduce the cash left for running the business.

Income and expense items that do not involve movement of cash, like depreciation, valuation changes and other internal transfers, are excluded. Although these are important, from an accounting standpoint, or preparing the balance sheet and calculating profit, they produce no cash flows.

Having described cash-flow budgeting, let us look at some typical examples. Figure 4.7 shows the annual, forecast cash flow for a small, peasant farm. Figure 4.8 gives a cash-flow budget for the six months ending 31 December 1986, for a company starting with a balance in hand of $3990. The receipts and payments arising from credit transactions have been phased to account for one month's credit allowed to customers and granted by suppliers. Let us assume that:

(i) Credit transactions are as follows:

	Purchases	Sales
	$	$
June	9800	18600
July	9300	20000
August	9000	20700
September	9300	21300
October	12800	26600
November	14600	27900
December	13800	29300

    (ii) Monthly salaries and wages to $8000

   (iii) Other monthly cash expenses are estimated as $2100 to the end of September and $2700 afterwards

   (iv) A dividend of $5300 is due in August

   (v) Capital investment in a new machine of $5600 is planned in December

The dividend payment in August means that an overdraft will probably be needed from its payment until the end of September. After this the cash balance should grow enough to meet the capital expenditure in December.

The information needed for cash-flow budgeting comes from operating budgets, budgeted balance sheets and discussions with staff. The prime need is a record of past costs and income, giving dates. This can come from previous cash flow plans. The next need is a budget for each enterprise preferably also based on past records. Using the planned areas of crops and scale of livestock interprises, it is then possible to forecast likely expenses and income in each budget period. If no records are available, the data must come from estimates, budgets for overhead costs, quotations for buying capital assets, etc. A cast budget unites the cash elements of all other budgets so it is prepared only after the other budgets are available.

		Oct.	Nov.	Dec.	Jan.	Feb.	Mar.	Apr.	May	June	July	Aug.	Sept.
Cash inflows (receipts) (+)													
Opening cash balance $696													
Maize			261		480								
Tobacco									174	383	17		
Groundnuts												63	90
Orchard									52	52	52		
Livestock		44							14	17	17	17	
TOTAL INFLOWS		44	261	0	480	0	0	0	240	452	86	80	90
Cash outflows (expenses) (−)													
Seed		26											
Fertiliser		188											
Biocides								7					
Labour		24	24	24	24	24	24	24	24	24			
Capital costs									1832				
Personal drawings		94	21	21	21	21	21	21	35	21	21	21	21
Other		7								70	35	38	
TOTAL OUTFLOWS		339	45	45	45	45	45	52	1891	115	56	59	21
Net cash flow	$	−295	+216	−45	+435	−45	−45	−52	−1651	+337	+30	+21	+69
Cumulative net cash flow	$+696	+401	+617	+572	+1007	+962	+917	+865	−786	−449	−419	−398	−329

Figure 4.7   Example of a Cash-flow Budget for a Small Arable Farm

		July	Aug.	Sept.	Oct.	Nov.	Dec.
Cash inflows: ($)							
Opening balance	$3990						
Income		18600	20000	20700	21300	26600	27900
Cash outflows: ($)							
Purchases		9800	9300	9000	9300	12800	14600
Salaries & Wages		8000	8000	8000	8000	8000	8000
Other Cash Expenses		2100	2100	2100	2700	2700	2700
Dividend		—	5300	—	—	—	—
Capital Costs		—	—	—	—	—	5600
		19900	24700	19100	20000	23500	30900
Net cash flow ($)		(1300)	(4700)	1600	1300	3100	(3000)
Cumulative NCF ($)	3990	2690	(2010)	(410)	890	3990	990

Figure 4.8   Cash Budget for the Six Months ending 31 December 1986

Forecasting future receipts and expenses is the key to cash control. Cash-flow budgeting helps ensure that there is sufficient cash to achieve and maintain planned profits and that there is always sufficient liquidity to meet planned commitments. By showing seasonal cash needs, it enables credit to be arranged well ahead. Unless a contingency allowance has been made, always ask for a facility at least 10 per cent above the peak borrowing need. This also indicates the amount of bank interest payable on a fluctuating overdraft. Anticipation and suitable planning of borrowing and loan repayments help reduce the cost of credit while maintaining profit.

Among the many advantages of cash-flow planning are the following:

- As with labour and tractor use planning, the manager must think ahead and co-ordinate his activities; policies must be defined.
- It shows the likely timing of peak cash needs.
- It shows whether capital expenditures can be financed internally or not.
- It reveals opportunities for adjusting purchases and sales to reduce peak cash and credit needs and to minimise tax liability.
- It reveals the availability of cash so that advantage can be taken of cash discounts or surplus cash can be invested.
- Interest charges can be forecast accurately.
- A cash-flow plan can bring peace of mind to those who have borrowed heavily.

Perhaps the most important benefit of cash-flow planning is budgetary control. By comparing actual costs and income each month with the forecast ones, variations from plan are quickly seen. Unforeseen changes, such as inflation, may have upset the original calculations and these should be studied for future use. All other significant variances need examination and remedial action if possible.

Recording actual figures beside the forecasts is a check on slow-playing debtors, outstanding bills and on rising costs. As there are normally more cost than receipt items, this tends to cause over-attention to the costs of the business at the expense of the output. For example, a cow's yield together with milk prices are far more important components of profit than most single cost items but, with feed, fertiliser, dairy stores, medicines, etc., there are far more costs than returns.

Cash-flow budgets are only estimates and are therefore open to error. Also, a favourable cash balance at the start and end of a month is no guarantee that an overdrawn situation will not arise during the month. It is, however, likely that this could be avoided by withholding certain payments until the month end.

Cash-flow budgets reveal any possible ways of adjusting the timing of purchases and sales to reduce peak cash need and bring it within any overdraft necessary. One must be prepared to modify the plan if necessary. On most farms there are times of the year when expenses exceed income though purchases can often be delayed or sales advanced to reduce the deficit and, in turn, loan requirements. It is useless making a forecast just to know in advance that disaster is coming; the whole aim is to be able to do something about it.

The timing of receipts and expenses can greatly reduce loan requirements and therefore annual interest charges. For example, if a farmer expects a $3000 profit from $15,000 gross income, his costs of $12,000 will need varying amounts of initial capital, depending on the flow of receipts. For an enterprise with equal monthly receipts and equal monthly expenses, he would need only $1000 of initial capital because monthly receipts would be enough to pay the following month's cash expenses. However, if the receipts from this enterprise accrued only at the end of the year, he would need $12,000 of initial capital.

If a cash-flow budget forecasts availability of excess cash, this would be an underuse of resources so any sum exceeding normal needs should be invested to earn return and not be left idle. A temporary surplus could be invested in a bank deposit account or in the money market at seven days' notice. It is important that short-term investments should be readily realisable when needed. If a permanent surplus is forecast, it should be decided if more internal investment is desirable to expand or re-equip the business or if external investment would be more profitable. Too much cash in hand will lower the overall return on capital and, if it is borrowed, raise interest charges.

If a farmer thinks a new tractor will be needed during the year, that land improvements, like bunding, will be undertaken and that some more cows must be bought, construction of a cash-flow budget enables him to consider the relationship of these decisions. He can judge whether there will be sufficient cash to do all these things at once and, if not, which to do first, do partially or postpone. In other words, budgeting is a discipline that forces decision-makers to consider the effects of alternative policies.

OVERTRADING

Overtrading is much more serious than under-capitalisation. Though occurring most often in small fast-growing businesses, it may occur for short periods in other businesses because in farming there are usually seasons when expenses exceed receipts.

Briefly described, overtrading means trying to maintain or increase the scale of operations with inadequate cash to finance current operations. It is a situation in which the business, though perhaps trading profitably, has expanded beyond the level that can be supported by its finance structure. This is likely to be due, at least in part, to inadequate financial control. Basically, it is a situation of capital being invested in medium- and long-term assets and leaving the business short of cash for its day-to-day running. It means shortage of cash rather than of assets. Cash means funds available for immediate, unrestricted use. It consists mainly of money in current bank accounts. This differs from profit, much of which may consist of a rise in the value of assets.

The main signs of overtrading are one or more of the symptoms of under-capitalisation plus inability to pay bills on time. Overtrading may occur in quite promising firms; it does not imply inefficiency but rather misguided financial management. It shows in balance sheets and profit and loss accounts as a rise in bank borrowing, an increase in trade creditors and an increase in both current and fixed assets without a corresponding rise in sales. In some businesses, growth can produce more sales and profits together with a fall in net current assets. This is also a sign of overtrading and should be regarded as a danger signal. Liquidity is dangerously reduced. A firm is liquid if a high proportion of its assets are held as cash or other liquid assets. The amount of a firm's liquidity, in this sense, indicates its ability to pay bills quickly enough to satisfy creditors and avoid bankruptcy.

The main cause of overtrading is rapid growth as when a new enterprise is introduced or an existing one expanded to leave too little working capital. Management may wish to raise output to take advantage of good trading conditions. More inputs may be bought (raising the figure for creditors), more people employed, more production costs are incurred and more fixed assets may be required for expansion. Overtrading can also be caused by holding products on the farm in the hope of getting a better price later or by taking earlier and larger deliveries of inputs in order to gain rebates or discounts. While a business is expanding, cash inflows always lag behind outflows and cash resources fall because so much working capital is tied up in work in progress. The position will worsen if returns are less than expected or if the time taken to build the enterprise up to full capacity is underestimated — a common fault in farming. There are cases of buildings being erected to expand a livestock enterprise but leaving too little cash behind to buy the extra animals to produce the planned returns when expected.

Shortage of cash is both a cause and a sign of overtrading. Too much business can be as fatal as too little unless the growth is covered by more cash or by short-term loans.

A similar problem is that of the farm owner who has taken a large mortgage to buy the farm. If the mortgage forms a high proportion of business liabilities, he may find it hard to borrow enough to provide sufficient working capital to run the business for maximum potential profit. The mortgage repayments which, in effect, are forced savings transferring cash into the long-term asset of land, may be too high to leave sufficient cash for other essential purposes. This situation is improved, however, by any fall in the value of money. The borrower repays in depreciated money at fixed, historic interest rates. It is also improved when the early payments are treated as mainly interest rather than capital repayment and are therefore tax-deductible.

Overtrading may be caused, and is certainly stimulated, by inflation; especially in farming where there is a long time lag between investment and the resulting rise in output. It can be brought about by higher tax demands on inflated profits since taxes must be paid in cash although they are levied on profits that may consist mainly of an increase in physical assets. Higher personal drawings caused by the higher cost of living also make direct demands on available cash. Also, more working capital is needed to finance higher input costs and it costs more to replace fixed and part-fixed assets at their new prices.

The effects of overtrading all arise, directly or indirectly, from shortage of cash, not shortage of capital assets. When a business has too little working capital to finance its operations safely, it may be unable to pay its creditors and is in danger of a sudden cash crisis and bankruptcy. Firms with sound trading profits may still be forced out of business if cash is unavailable to pay debts on time.

Farm businesses usually experience exceptional

trading for short periods during the year and overtrading during these periods is no problem if prior steps are taken to obtain more short-term funds. Efforts to correct overtrading should be like those described for undercapitalised businesses. It can generally be avoided by giving up potential profit for liquidity. To cope with it, it must be recognised as it develops and funds must then be tightly controlled.

Tight control on the cash flow is especially important with seasonal production systems. Well-managed firms can anticipate overtrading arising by making cash-flow budgets for six to twelve months ahead, breaking the budget down into monthly periods. This sort of budget not only enables management to see when and how much extra credit is likely to be needed, it also helps in deciding how to time capital expenditure.

All businesses need cash to pay their creditors and it must be seen to be available even if bills are not paid immediately upon receipt. Although much money in the bank suggests poor management because greater use might be made of it, nevertheless liquid assets ensure smoother relations with creditors and also make management ready if a take-over opportunity arises. Ready cash is very useful when firms are growing.

INVESTMENT IN BUILDINGS

From an economic standpoint, investment in farm buildings is worthwhile only if it will raise profit, by raising income, saving costs or both, enough to cover the extra annual building costs. Like land, buildings should be financed out of capital or from medium- or long-term loans.

The main building costs are depreciation and interest. Buildings are usually depreciated, on a straight-line basis, over their expected economic life. The economic life of specialised buildings can be less then their physical life and they are best written off within 10 years. Durable, general-purpose buildings can have an economic life of up to 20 years.

Other building costs are repairs, maintenance and insurance. Repair and maintenance vary with the quality and expected life of the building and the use to which it is put. The annual cost is generally from 1 per cent to 3 per cent of initial cost, say 1.25 per cent, after the fourth year of operation. Annual insurance premiums are usually about 0.2 per cent of value but, if wooden or filled with combustible material, this

could well rise to 0.4 per cent. Buildings, however, should be insured for replacement cost — the inflated cost of replacing them should total loss occur. Unless this is done, insurers seldom pay the full cost of claims for partial damage but reduce their payment in proportion to the under-insurance. Replacement cost should include any cost for removing debris and architect's and surveyor's fees needed for rebuilding.

Appraisal of a building investment is normally done by partial budgeting with the 'real' interest on capital included. Tax relief on the investment and extra tax on any extra profits should also be considered. It is hard sometimes, however, to quantify the benefits of certain buildings, for instance, workshops, which should reduce machinery repair and maintenance costs and lengthen their working lives. An alternative to partial budgeting is to omit interest on capital, estimate the return on it and compare this with the return on alternative investments.

If a farmer decides to invest in a building, capital availability will influence his decision on whether to erect a cheap structure, likely to last no more than 10 years, or a more costly and durable one. Farmers often rightly choose general purpose instead of specialised buildings to reduce initial costs and provide more flexibility. If possible, buildings should be designed to save labour by suitable siting of doors, water taps and stocks of feed and other inputs.

Suppose, for a given purpose, a farmer may either invest $9000 in a building likely to last eight years or $13,000 in one likely to last 16 years. His first response might be to choose the more durable building as two cheap ones would cost over $18,000 over the 16 years. However, the decision is more complex than it seems. He could invest funds now to yield the replacement cost of the cheaper building in eight years. $3635 invested now in fixed securities at 12 per cent interest would yield $9000 in eight years — see Appendix A.2 — so total investment in the two, cheaper buildings would be $12,635. Moreover, if capital invested in the business is earning 18 per cent, the farmer needs to invest only $2394 now to produce the required $9000 in eight years. This is a total investment in the cheaper buildings of $11,394. These days, however, the problem is complicated further by inflation. Assuming an annual inflation rate of 9 per cent, the cheaper building will cost $19,292 (see Appendix A.1) to replace in eight years' time. At 12 per cent interest, this would require investing $7792 now and, at 18 per cent, it would need

$5132. In the latter case, total building cost over 16 years would be $14,132 — more than that of the more costly building now. However, the less permanent building may still be chosen as it leaves the farmer more flexibility if he wants to change his farming system or benefit if he wants to change his farming system or benefit from improved building design after eight years.

INVESTMENT IN MACHINERY

WHY MECHANISE?

Analysis of yields in various countries indicates that a power level approaching 0.4 kW/ha is needed for efficient farming. However, machinery and labour are closely related, alternative power sources and should be considered together when decisions are made on using machines to replace hand labour. Mechanisation is so country and product specific that general policy recommendations are impossible.

A defect of technology — or rather of technologists — is that bigger is thought better. The right question for a manager is not, 'Isn't there a bigger tool for the job'?; it is, 'What is the simplest, smallest, cheapest and easiest tool that can do the job?' Another simple rule is that the machine must serve the work. Work does not exist for the sake of the machine but for the sake of production. This rule is constantly broken by today's computer users. They become fascinated by the capacity, speed and memory of the latest type of computer. Consequently, when the new computer arrives, a frantic search begins to find work for it to do. In the end, it is used to produce masses of information that nobody wants, nobody needs and nobody can use.

The basic justifications for mechanising a job are that it should raise productivity and thus profit beyond that previously achieved, or because insufficient labour is available. Correct selection and use of machinery can raise profit in three ways — by lowering cost per unit of output, by reducing risks or both. In practice, profiability is usually influenced by many factors that should all be considered. Not all of them can be measured in cash terms, for example better working conditions and less drudgery or worry during busy times. Thus farmers can sometimes justify investing in machinery that will reduce overall profit.

The most valid reasons for mechanising are then as follows:

Greater income	— increased output
	— increased yield/hectar because of more efficient cultivations
	— greater timeliness of work
	— higher product prices
Reduced costs	— lower labour costs
Reduced drudgery	
Labour shortage	— particularly at peak periods
Reduced risks	

Mechanisation can raise the productivity of land, labour or both. For example, irrigation will probably increase the return to both. However, mechanisation does not always raise yield per hectare. It is usually thought that higher yields occur with tractor use but the evidence is not convincing. This is most likely to occur if the greater power produces better soil conditions, increases rooting depth and allows more timely cultivations. A ripper, for example, can break a 'hoe pan' and a wide range of implements enables varying conditions to be handled. If timeliness is important for certain jobs, such as planting and harvesting, and labour cannot cope in time, mechanisation probably does raise yield per hectare. On large irrigable farms, tractor use has enabled double cropping in some areas where it would have been impossible without mechanical power. The important effect of timeliness on crop yields is particularly marked in the sub-arid tropics with their short growing season(s). In some countries physical soil conditions allow only a few days between ploughing (after the rains begin) and planting. As present animal power systems can plough only about 0.25 ha/day, cultivated area is often limited by lack of power. If mechanisation makes possible a greater crop area, the fixed costs of the machine will be spread over a larger area.

Some machinery can raise product prices through improved quality; for example, thermostatically controlled, automatic stokers for flue-cured tobacco-curing barns. However, mechanisation of jobs like groundnut shelling and cotton reaping can reduce both quality and price.

Reduced costs can also raise profit by, say, making labour more productive; with increased yields, by improving the conditions for higher output; with reduced yields, by cutting costs by more than the value of the yield lost.

Replacement of labour by costly machinery needs careful thought and calculation. It is, of course, unavoidable if adequate labour is unobtainable, as in Iraq. Farm work is highly seasonal and with growing urbanisation some labour shortage is experienced at peak periods on large farms. In industrially advanced countries mechanisation was a response to the rising cost of labour which was itself caused by industrialisation. In most tropical less developed countries, the rural population is likely to increase greatly in the next decades owing both to high birth rates and low rates of industrial growth. The problem in these countries, therefore, is not how to replace labour but how mechanisation can increase labour use and rural employment. The aims of mechanisation in crowded, rural areas are not to increase cropped area or save labour but to raise yields per hectare, improve timeliness and lessen drudgery. However, when mechanisation and large-scale commercial farming are brought to relatively uncrowded areas, they can raise both output and employment. New land can be opened for cultivation and new jobs created. In countries with a land surplus, mechanisation may allow its exploitation either by raising average farm size, by developing new settlements, with tractor-hire provided, or by creating large mechanised holdings.

Saving of labour does not always save cost if farms are small and have only a small regular work force. Saving part of a worker's time may not cut the minimum number of men needed to run the farm. Nevertheless, with good planning, saved labour can often be used productively; and though it is only when a worker's labour can be saved throughout the year that big cost savings result, smaller labour-saving benefits are worthwhile if the saved labour can be used profitably elsewhere.

High labour and machinery costs are often found together on the same farm, although mechanisation was justified by an expected drop in labour cost. However, even if mechanisation will not raise profit, it may be introduced to reduce drudgery by shortening the working day, leaving more time for leisure and thus helping to attract and keep staff by raising their morale, if this is possible at an acceptable cost. Many jobs can be mechanised to lessen drudgery — like handling grain and fertiliser in bulk rather than in bags and machine-milking straight into a bulk tank.

Reducing risks pays off only when a known hazard can be overcome. For example, there must be a likelihood of drought before considering installing supplementary irrigation and a drought must occur for the equipment to pay off. The sum worth spending on supplementary irrigation can be fixed only when the frequency of droughts is known. The same considerations should influence decisions to use chemical-application machinery like sprayers.

A less obvious gain in changing from animal-power to tractors comes from the extra output of land formerly used for grazing draught animals. In intensive farming, as in Pakistan, this gain can be large.

There are many dangers in unselective mechanisation. Many farmers tend to invest heavily in modern farm machinery without being able to raise output enough to justify the investment. Desire for speed for its own sake or to be among the first to have a new model and the search for technical perfection can easily lead to over-capitalisation.

In less developed countries, capital-intensive mechanisation often reduces employment or productive self-employment and speeds the drift to the towns. Introducing labour-saving machinery into such areas is often unwise and sows seeds of social discontent. Farmers may be too ready to rest from drudgery at the cost of the greater productivity intended. Mechanisation also tends to favour larger farmers, reduce the permanent labour force and increase the use of casual labour. Moreover, much arable land is too steep or has too many surface hindrances for animal, or mechanical draught, but this land can be productive with hand labour.

Tractor mechanisation in tropical, smallholder farming has usually failed for several reasons. These include the short life of tractors and equipment owing to corrosion and severe lack of skilled labour to maintain and run them efficiently; poor management; cost; low use; equipment unsuited to tropical soils and varied ecological conditions and problems connected with poor land clearance and size and siting of fields. Tractor introduction can also cause loss of soil structure and fertility by soil exposure and erosion.

When considering mechanising tropical small-holder farming, it is not enough for agricultural engineers to know how to keep the wheels turning. They should also understand the restrictions to adoption/adaptation of mechanised equipment by small farmers. In no other aspect of tropical agricultural development — with the possible exception of large irrigation schemes — has so much money been wasted as in premature and unplanned farm mechanisation. Engineers and economists differ greatly in their advice on mechanisation policy for less developed countries. For example, economists often use catchy phrases like 'premature mechanisation' and the 'paradox of mechanisation in labour-abundant economies' while agricultural engineers often push the idea of a 'mechanisation ladder'.

In summary, the case for tractors in small-holder, tropical farming should be critically reconsidered country by country because of the general failure of tractor schemes, widespread underemployment and unemployment and the potentially bad effect of mechanisation on employment and foreign exchange. There is a limited place for unsubsidised tractor mechanisation in some areas. However, there is also more scope for animal-powered mechanisation than is commonly thought, especially with rising oil prices.

What seems necessary in smallholder farming is *selective* mechanisation seeking new methods to suit small farms. Most motor-powered farm machinery was designed mainly for the labour-scarce, captial-abundant developed countries. Selective mechanisation should aim at raising output per hectare not at reducing labour per hectare other than during the busiest times of the year. This need not affect employment adversely if limited to work that cannot be done quickly or efficiently by hand or to relieve labour peaks. Farming is characterised by a few tasks that need much labour for quite short periods and long periods of low labour demand. A farm's cropping pattern has to be adjusted to suit the labour supply in the few peak periods so mechanisation of these tasks may allow more intensive cropping over the whole farm. When seasonal labour peaks are identified, engineers and agronomists should direct their research to breaking them and economists should devise policies to encourage selective mechanisation of these peaks, without mechanising the other farm jobs.

Selective mechanisation should not consider only field, or even farming, jobs but all the power needs of farm families. Lorries, wheelbarrows, irrigation pumps, water carts, tree-stump pullers, building-block making machines, tube-well borers and simple processing machinery, like threshers and mills; and well-designed hand tools, like swan-neck hoes, can release much labour for direct production.

Appropriate technology means any manufacturing technology, based on modern science, that is adjusted to its environment. Less developed countries should make the most use of available resources which usually means developing some technology between the labour saving of developed countries and the capital saving of hand methods.

The need is urgent to develop research ability *within* the tropics to design mechanical techniques suited to the resource endowments of each country and of varying ecological zones within countries.

In some ways, Japan is a model for use of appropriate technology. In farming, for example, the first technical innovations involved use of manure, improved plant breeding, weeding and building of drainage and irrigation works — all capital-saving methods. Use of cultivators and other labour-saving machinery began only when industry's need for labour made it necessary.

More attention is needed to improve the design of hand tools, containers and hand-powered machines to raise labour productivity. Keeping hand tools in good condition can raise labour efficiency. Hoes, axes and knives should be kept sharp and have tight-fitting handles.

Tropical governments have tried many tractor mechanisation schemes and hire services. Tractor mechanisation has been linked usually to large-scale farming schemes like state farms and land settlements. Many of these large mechanised farming systems soon failed or had so many problems that they were run down as much as politically possible. Some of the better-known failures include government plantations in Sierra Leone, state farms in Ghana, farm settlements in Southern Nigeria, Nigeria's Mokwa scheme and settlements in Tanzania during its First Plan. Tractor hire schemes were set up by governments in Tanzania, Ghana, Morocco, Zambia, Nigeria, Malawi and other countries. These were usually inefficient as their financial profitability depended on subsidies. However, these subsidies are often unnoticed by technical agriculturalists, politicians and donor countries. Also, tractor schemes need much foreign exchange for new equipment, spares and fuel and loss of jobs due to tractor schemes is often ignored.

For these reasons, it seems strange that tractor schemes are still promoted and subsidised by so many tropical governments and aid agencies. The reasons for this continuing appeal include:

- Prestige. Tractors and mechanised farming are equated with modern farming in developed countries.
- Inadequate appraisal of mechanisation projects from a national viewpoint.
- Tied aid. Support for mechanisation often

results from the 'tied-aid policies' of donor states.

- An alternative to 'unresponsive' tropical peasants. Despite much research showing that peasants are 'economic men', many civil servants, especially technical agriculturists, still unfortunately see mechanisation as a shortcut to the tedious job of helping peasants improve their farming systems.
- Timeliness. Rainfall patterns and hard soils that allow only a few days between ploughing and planting in some sub-arid, tropical countries *might* justify mechanisation despite widespread rural unemployment and underemployment, for example, the dry zone of Sri Lanka.
- Opening new land.

The main reasons for government tractor hire services making a loss are low demand for the service, low charges and high repair costs owing to working in poorly stumped fields. Data from Western Nigeria in the early 1970s showed that the hirer's extra costs were not met by his reduced costs and extra income. Tractor hire is a costly addition to the package of farm practices. For economic use of a tractor hire service, improvements are needed in the size and shape of fields, road access, marketing of produce, timeliness in tractor service, availability of spare parts and use of yield-increasing inputs like fertiliser, better seeds and biocides. Governments should usually help by clearing land, training drivers and setting up workshops.

Obviously many things need considering before investing in machinery. In assessing these, the financial effect of the machine upon the farm business should come first in a farmer's mind — What will be the increase in return?; What will be the extra annual cost besides the initial capital outlay? Both flexibility and the chance of more profitable alternative uses for scarce capital should be considered. Work quality is also important.

In a partial budget to decide whether it is worth replacing hand labour with a bought machine, the following items should be considered:

Losses	*Gains*
Income lost	New income
Value of any fall in yield or quality	Value of any rise in yield or quality
Gross margin of enterprises dropped or reduced in size	Gross margin of enterprises introduced or increased in size

New costs	*Costs saved*
Fixed costs of new machine	Fixed costs of any replaced machinery
Net rise in fuel and repair costs	Net fall in fuel and repair costs
	Regular labour costs saved
	Casual labour costs saved

Replacing labour by machinery may cause many changes in farming systems through the resulting effect on seasonal labour needs and hence cropping possibilities. These changes may be so great that they can be fully forecast only by complete budgeting.

ANIMAL DRAUGHT

There is a sound case for farmers who have moved partly towards commercial farming to use draught animals for power. The step from hand-hoes to mechanical power is great, so the ideal power supplement for farmers changing from subsistence cultivation to commercial farming is often draught animals. In general, animal power seems to be a more practical way of helping smallholders than tractor mechanisation in much of the tropics at their present stage of development and with great underemployment problems.

Some people regard animal draught as the only way to solve the problem of intensifying tropical, smallholder farming. Others consider it an almost essential half-way stage leading eventually to the final goal of some form of engine-power. Others, whose number seems to be rising, think that animal power is now largely out-dated. There are many variations between these different, conflicting opinions. In fact, there is no universally most suitable equipment for tropical small farmers. It is worth, however, encouraging and developing animal draught cultivation where it exists already. Small, family farmers in the tropics can often afford no other sort of mechanisation.

Draught animals used in the tropics include oxen, water buffalo, camels, horses and donkeys. Work in Zimbabwe and Pakistan has shown that, for maximum effectiveness, a pair of oxen is needed for about every 4.5 ha of tillage area. But this varies with soil conditions and whether crops are planted on ridges or on the flat.

Use of animal power has many advantages over hand labour and its introduction is progressing well in sub-arid areas as it is within the technical ability and pocket of many peasants. Old draught animals can be fattened to supply either extra income or a better diet. Where population density is low, bringing in animal draught can reduce underemployment by increasing the cultivated area. This has been shown in the Gambia and parts of Tanzania when both the animals and the farmers are well trained. In West Africa, in particular, there is much land available. Thus, in the non-irrigated cotton area of Mali, a farm family cannot cultivate by hand more than 3 ha, of which 0.5 ha is cotton. With draught animals, this can rise to 5–10 ha of which 1–2 ha is cotton.

Other advantages include higher labour productivity while simultaneously reducing the drudgery of work and gaining prestige, at low cost. Better ploughing and weeding can raise yields and carts make transport for short distances much easier. In Malawi, carting has been shown to be much more important than any other task both as a user of ox time and as a source of income.

Use of animal draught enables cropping to be combined with keeping of livestock. This enables manure to be used both to raise crop yields and to improve soil structure and, if land is available, it may enable a stable, ley-farming system to be introduced. Also, in countries or areas where animal power is justified, local manufacture of implements — carts, ploughs, ridgers and cultivators — can provide jobs and save foreign exchange.

One of the main advantages of animal draught is said to be a saving of hand labour. For example, hand labour takes about 54 man-hours to prepare a hectare of land in Malawi whereas a pair of oxen takes 32 hours. However, the oxen need both a driver and a ploughman, using 64 man-hours, so there is no labour saving in land preparation. The oxen simply ease the task. In this study, possession of oxen saved no labour except in carting produce and manure. Nevertheless, there are potential labour savings on many tasks, such as land preparation, weeding and groundnut lifting, if farmers and animals are better trained and minimum tillage techniques are used to replace conventional ploughing.

Animal draught as a source of power also has disadvantages. Its critics often state that animal-powered mechanisation simply increases the area cultivated and produces a weeding problem. This criticism is usually true in areas with low population density. Many studies show that farmers who adopt animal-powered land preparation face labour shortages at weeding time. Opening more land than can be managed properly later raises the risk of erosion. Animal disease is not serious in sub-arid areas, which are those best suited for animal draught, but it can become so in more humid areas. Where land is scarce, animal draught may require some scarce arable land to be used for fodder growing. This could add to the poverty of already poor, small farmers by reducing their staple food production.

Draught animals are usually in poor condition at the start of the rains in the sub-arid tropics and soils are always hard to work. At the end of the rains, grazing quality falls so cattle are kept through the dry season on crop residues left in the arable land. Land is therefore unavailable for ploughing until it is baked hard in the dry season when oxen are too weak to work it. This leads to poor, shallow ploughing at the wrong time of the year, late planting and loss of potential yield. Sufficient supplies of animal feed are essential during ploughing time which is usually soon after the rains break. This is when feed supply is always lowest and it is also a hard time for people. Serious shortages of animal feed occur mainly in areas with an average rainfall of below 800 mm and a long and severe dry season.

Use of draught animals is often very low. This results in low output by the poorly trained animals when they do work. The working season is short, often only three or four months of the rainy season; for eight or nine months the animals are not used, only fed, and this puts a strain on the farmer. What is needed is a whole-farm, integrated system to use draught power throughout the year — ploughing, carting, pumping, etc.

Use of draught animals is often prevented by land shortage, livestock diseases, excessively heavy soils or unstumped land.

Training draught animals is easily done by farmers themselves if they have a stock-keeping background like the Nguni tribes of Southern Africa. However, where livestock introduction is new, ox-training units are sometimes set up to train both oxen and farmers. This has been done in Ghana, Mali and Malawi. The difficulty of using a standard cultivator with poorly trained oxen often prevents use of this implement to ease the weeding bottleneck in Malawi where smallholder's crops are planted on ridges.

Insufficient attention is given to improving equipment for animal draught. Past government efforts to encourage animal draught usually included only token spending on research and development of animal-powered equipment compared with encouraging use of animals simply for ploughing. There is much need to expand research on weeding devices for attaching to animal-powered toolbars, and on developing a cheap groundnut lifter. Much existing animal-drawn equipment is inadequate and unsuitable for the conditions in which it must be used. Implements are poorly designed, cover the ground slowly and cause too much loss of moisture. Field trials in Botswana, an arid country, show that cultivation implements like chisels, sweeps, planters with press wheels and flat-bladed rather than tined inter-row hoes are far more efficient than the equipment available now. These are essential for successful introduction of a better cropping system there.

Introduction of animal power incurs high initial and opportunity costs but repair and maintenance costs for equipment are low. Spares are usually available at fair prices and local blacksmiths can do any necessary repairs. However, the cost of the oxen themselves is tending to rise as such animals can be profitably fattened.

Work in Malawi shows that field length is important for efficient use of oxen for cultivation. On average, an increase of 30 m in field length gave a 55 per cent increase in rate of ploughing and a 24 per cent increase in rate of ridging. A square field of 0.8 ha was ploughed 40 per cent faster than a square field of 0.4 ha. The most important aspect of work output in carting was the speed at which the oxen pulled the loaded cart. A slightly higher speed, obtained with a good team of oxen, gave much higher work rates. It is essential that the oxen are well fed and strong enough to move fast.

This Malawi study found that the average working day of an ox team was only five hours. Mean speed was 1.48 km/h when ridging and 3.13 km/h when ploughing. This resulted from the greater width and working depth of the ridger. These work rates enabled 0.44 ha to be ploughed or 0.85 ha to be ridged each day. Speed of carting was 3.18 km/h when carting maize, 2.74 km/h carting manure and 5.01 km/h when empty. These speeds reflect the different loads being pulled. Assuming maize is 240 kg/m^3 unshelled and manure is 930 kg/m^3, the average mass of a cart of maize is 385 kg and of manure is 1290 kg. Using these assumed densities, maize and manure were carted at rates of 541 kg/h and 3890 kg/h respectively over a distance of about 400 m.

At the average maize yield, manure application rate and distance from field to store found in that study, 1.76 ha maize could be cleared in a working day and 0.47 ha could have manure spread on it in a working day with one cart and a two-ox team.

One should not conclude that animal draught cultivation is the sole, ideal and final answer to increasing the power available in tropical, smallholder farming. It is a useful way of helping this sector, especially where population density is low, but other methods are still needed. Research on ways of providing power is needed just as much as that being conducted into solely agronomic problems. This could result in a special form of power mechanisation — possibly somewhere between animal-draught cultivation and a 50 kW tractor! However, as long as we still seek an answer (and this type of research has only recently started), it seems that one of the most effective ways to aid agricultural development in sub-arid areas is through greater use of draught animals.

MACHINE PURCHASE

The need to understand the economics of using farm machinery is growing. Labour and tractor work together make up 30–60 per cent of the production costs of most crops, so efficient use of these resources is important for financial success. The problems farmers have in machinery selection and management are becoming more than technical. Managers must make many decisions about the ownership and use of machinery like how many tractors, lorries, combines and so on; the size of equipment; whether to own, rent or use contractors and whether to own new or used machines. Many of these decisions depend on the balance between machinery, land and labour combined in the best proportions. This is a special problem with machinery, which often comes in large single units.

I know few farmers who buy equipment, including tractors, in a planned way. This is not surprising as most farmers buy farms just the same way, without preparing proper budgets showing their planned farm programme will meet the combined needs of costs, capital repayment and living expenses. All new investments should be assessed as carefully as possible as any large investment will affect the business for a long while. A bought machine is likely to be on the farm for several years. The farm must live with the results of its investment decision and, if the machine is inadequate and is scrapped and sold, cash will have been wasted. Decisions on choice of machine cannot be separated from those on whole-farm planning. In particular, seasonal labour supply, timeliness and capital availability affect the choice.

Selection and use of farm machinery usually depends more on partially understood considerations than on the merits of individual machines. There is no single best way to farm or to mechanise; there are always several equally right ways. Selecting machinery from that already used is likely to be successful only if all known factors are considered. The greater the knowledge and depth of planning, the more accurate the budgeting and the greater the confidence in the decision.

While it is easy to condemn excess machine capacity and over-mechanisation, optimum choice of machinery is highly complex and would still be so even if full and certain information were available about the way in which yields, costs and prices change over time.

One of the most important considerations affecting machine costs for each unit of output is the amount of use made of the machine. Using a machine near the limit of its technical capacity helps to spread the common costs of depreciation, interest, taxes and insurance. As the variable cost of each unit of output stays fairly static, however many units are produced, average cost per unit falls as use rises.

Size of enterprise affects all aspects of mechanisation and it is easier to mechanise economically on large farms than on small ones. Most farm implements and machines have only a short working season and cannot be fully used on small farms where the area to be covered or the size of the job is small. Few machines can be reduced in size and price to suit small farms so, even if the smallest available type is chosen, investment in machinery per hectare on small, commercial farms tends to exceed that on larger farms. It can be excessive unless some of the work is done with equipment not owned by the farmer. As mechanisation progresses, advanced types of field equipment tend to become too costly for economic use on small farms, unless contractors are used.

Investment in machinery can have widespread effects throughout the business so these should be assessed and evaluated systematically. Some of these effects will be obvious — for example, coping with a faster output and the effect on other jobs on the farm — but others can be less obvious. If the machine only partly mechanises a job, other machines may be needed. Building layout may need changing or cultivars, and therefore yields, may change. The basic method used to help decide whether further spending on machinery is worthwhile is partial budgeting, bearing in mind capital availability and the return from alternative investments when capital is scarce. All the likely results of an investment should be considered in terms of effects upon costs and income and these should be compared.

The margin of return over costs, related to the capital needed (that is, purchase price plus working capital) gives a measure of the rate of return. This rate should be compared with the interest the farmer would pay if he borrowed the capital or, if he would use his own capital, with its opportunity cost. The whole task is concerned with expected future costs, prices and yields and therefore involves risk and uncertainty. Farmers should try to estimate how these uncertainties are likely to modify the return calculated and this modified value should determine the final decision.

It should now be clear that careful evaluation of all the costs and returns connected with machinery investment is needed. Let us first look at the costs. These must be estimated as a basis for every analysis of any decision and, clearly, ownership costs should not exceed the cost of contract hire. It is important to consider what production methods other farmers use as they might have lower production costs through using more efficient plant and machinery. This could enable them to reduce selling prices and so capture a larger share of the market; or increase profits by keeping their present share of the market with higher margins achieved through lower costs.

One of the main problems of machinery ownership is keeping down total machinery cost per unit of output. Farmers often want to know what it would cost to use a particular machine; sometimes they may be thinking of changing from one model to a better one; they may wish to compare the cost of using their own machinery with that of contract work. For whatever reason costs are needed, the principles are the same and, if these are understood and certain basic facts and assumptions are agreed, it is simple to adapt the costing system to suit any special need.

For convenience, the costs of machine ownership are normally considered to consist of two different groups — one common and one variable. The common group consists of overhead costs which do not change however much a machine is used. They are fixed in total but fall per hectare and per unit of output as use rises. Overhead costs include depreciation (to a large extent), interest, housing, insurance, tax and licences. These are best considered on a yearly basis. They are those costs incurred by owning a machine *whether it is used or not*. The variable group consists of those operating, or running, costs that vary with the use made of the machine. They vary in total according to annual use but are almost constant per unit area or output. Running costs include labour, spares, repairs, maintenance, fuel and lubricants. When added together, overhead and operating costs make up the total costs of machinery ownership.

Depreciation is usually the largest single, annual cost of machine ownership. With most costly farm machines, depreciation occurs regardless of use and is therefore considered to be a fixed, overhead cost. Many farm machines are so little used each year that depreciation is little affected by annual use. In practice, however, there is much in favour of considering depreciation charges to consist of two parts, one of which (obsolescence) is independent of annual use while the other, caused mainly by wear and tear, is influenced greatly by the amount of use made of the machine. Depreciation that occurs over time, regardless of use, varies much with the machine's design and the care taken in housing and maintaining it. For example, unprotected sheet metal will corrode quickly if not kept dry and costly rubber tyres will be spoiled if left too flat while the weight of the machine is on them.

Besides depreciation with time, obsolescence can be important with little-used equipment. Some machines are so little used that they become obsolete long before wearing out. Obsolescence is especially important with machines that are being rapidly improved. Such machines may be worth little or nothing when a better model is available. It is always wise, therefore, to estimate the useful life of equipment with caution. Machine obsolescence can be caused by factors beyond the farmer's control. Any specialised, little-used machine should not be given an estimated life longer than, say, 12 years unless the farmer is fairly sure that he will still need and use it in the distant future.

Machinery loses value as soon as it leaves the showroom. The depreciation rate varies with the kind of equipment, the amount of work it does and the care taken in servicing and storing it. It can be estimated in different ways. For some purposes, for instance, calculation of expenses to set against income for tax purposes, it is necessary to use certain fixed, depreciation rates that ignore the annual use of equipment and anything else that might affect its useful life. Depreciation, as described here, is the difference between original cost and trade-in value. Real depreciation rates therefore depend largely on the state of the second-hand market which varies between times and places. Second-hand prices may, therefore, be a guide to the size of fall to be expected but farmers must make their own estimate in the case of new types of machine.

Depreciation of new machines should be estimated as realistically as possible. The real annual fall in value is the sum for which a machine could be sold at the start of the year less what it would fetch at the year end. Loss in market value is the best estimate of depreciation.

Several factors should be considered when estimating the annual depreciation of a particular machine. Good management generally requires that the rate of writing-off capital invested should not depend only on the working life of the machine. The assumption that 'depreciation life' is a fixed small number of years, with no allowance for annual use, is clearly so false that it should be made only in exceptional circumstances. There is therefore much to be said for a compromise, with depreciation life related to annual use so long as no machine is depreciated over more than 12 years, even if annual use is very low and physical life is likely to be much longer.

For comparing total annual machine costs the simple 'straight-line' method is usually most suitable, except for particular types of highly developed machines. For these there are valid arguments in favour of using less simple ways of assessing depreciation. Wtih machines like tractors and combine harvesters there is a large fall in resale value during their early life, regardless of use. If depreciation is assessed at a constant high percentage of the falling value, the result better matches resale values and is nearer the truth than figures obtained by the straight-line method. It is considered therefore that, while the straight-line method is adequate for use where the long-term effects of using machines are concerned, the reducing balance method is better when it is important to obtain a realistic assessment of the value of a few important items of equipment at various ages, for example, in considering when replacement is most economic on a given farm.

To estimate the annual cost of depreciation by

Table 4.24   Estimated Useful Life in Years of Power-operated Machinery in the Tropics Related to Annual Use in Hours

Equipment	50 or less	100	150	200	250	300
			Annual use (hours)			
Ploughs, ridgers, cultivators, rolls	12+	12+	12	11	10	8
Harrows, grain drills — without fertiliser boxes, grinders, rotary hoes and slashers, combine harvesters, pick-up balers	12+	12+	11	9	8	7
Planters	12+	12+	11	9	7	6
Mowers, forage harvesters	12	10	8	7	6	6
Hay rakes	12+	12	10	7	6	5
Sprayers, fertiliser distributors, grain drills — with fertiliser boxes	8	7	7	6	6	5

Tractors	500	750	1000	1500	2000	2500
Small — 30 kW	12+	10	8	5	4	3
Medium — 45 kW	12+	11	8	6	4	3
Large — 60 kW	12+	12+	10	7	5	4
Electrical motors	12+	12+	12+	12+	12	12

the straight-line method, a farmer must decide the useful, economic life of the machine, that is the time during which it seems economically wise to keep it, considering the chance of obsolescence and the question of mechanical reliability. Ideally, farmers should consider both likely depreciation and repair costs and decide the useful life of the machine as being when the total is lowest per unit of use. The two items should be considered together. For example, it is often assumed that the average working life of a medium-powered, wheeled tractor is about eight years. In fact, however, many tractors over twice that age are still working, merely because they are well cared for and little used.

If the average life of a tractor must be estimated, 10,000 working hours is a better guess than eight years. This may represent a working life of eight years at 1250 hours a year or one of 12 years at about 800 hours a year. With care in repairs and maintenance, the tractor may keep giving good service well after 10,000 hours. This applies especially to tractors with high annual use. For example, the life of a tractor that does 2500 hours a year and is well cared for is likely to be at least five years. Little-used tractors, conversely, can become obsolete before they have done 10,000 hours.

The annual or hourly depreciation cost is obtained merely by dividing capital cost of the equipment, less trade-in value, by its estimated useful life. The hourly depreciation cost stays constant throughout the machine's life and, although it is not exactly true, there is little error if

repair costs are also averaged. For example, using table 4.24, the annual depreciation charge for a medium tractor expected to work about 1200 hours a year (life about seven years) and costing $4200 will be $600 or 50¢ an hour. Similarly, the life of a forage harvester that does 150 hours a year is set at eight years so, if it costs $5000, annual depreciation is $625 and the hourly charge is $4.17. Simple figures like these help to stress the use of costings. If the machine could be productively used for 250 hours a year by working for a neighbour, the annual depreciation over its lower estimated life of six years rises to only $833.33 and the hourly charge falls to $3.33. If its working rate is 0.4 ha/h, contracting would thus reduce depreciation per hectare from $10.42 to $8.33 ($3.33/0.4).

Although farmers can easily recognise certain machinery costs — they can hardly fail to notice the cheques written for fuel, oil and repairs — they should remember that the main costs of owning machinery are depreciation and interest.

When the cost of doing farm work in different ways is compared, interest on the funds invested in machinery should be charged. Clearly, farmers who invest their own funds could earn a return on them by investing outside farming. They could also possibly use these funds to produce a higher return from farming by investing them in, say, more fertiliser. Because the capital cost equipment is written off regularly by depreciation provisions, the capital invested falls from the purchase price in the first year to nothing when the machine is fully written off. Using straight-line depreciation, the average investment is clearly half the capital cost and it is therefore usual to charge interest on half the capital cost over the whole life of the machine. The interest rate should vary according to whether owned or borrowed funds are to be used. In the former case, it should be the opportunity cost, in the latter, the 'real' interest rate. Consequently the interest rate correct for a given farmer depends on the specific choices he has for capital investment.

Other overhead costs include insurance, housing, taxes and licences. Insurance charges are easy to obtain by getting a quote from a broker. In general, equipment should be insured for its replacement cost less accumulated depreciation. This is because any claim is likely to be settled in this way. Annual premiums on machinery are usually about 0.35 per cent of the insured value; 1.25 per cent for tractors.

Housing costs can be estimated by costing the building used and then allowing for the space occupied by the machine. If no shelter is provided higher depreciation may be warranted. Licence fees and road taxes are easily costed by checking with the local authorities.

Labour costs are sometimes forgotten when making investment decisions but the gains from mechanising often result from saving scarce labour. Similarly, most machines need operators so the normal wage paid to these staff should be added to direct machine costs. The number of hours charged should normally include wasted time and time used travelling to and from the job.

Annual machine running costs include repairs and maintenance, spares and servicing. These costs tend to rise as a machine gets older if it does the same work as before and is not allowed to become a spare bit of equipment kept only to raise capacity in hard seasons. Repair costs rise irregularly over time and are closely related to the amount of use. They vary between models and with the care with which the machine is used. Only experience can indicate likely repair and maintenance costs. A farmer's own records of repair costs should indicate the probability of costs arising at particular times (especially for things like tyres). Without past records, the best guide is probably advice from machinery dealers.

Maintenance costs tend to be similar for large and small machines of the same types. Repairs tend to be similar for, say, a large and a small tractor — as a mechanic's labour is often the main part of the cost and varies little with machine size.

As with depreciation and interest, it is best in making investment decisions to average repair and maintenance costs over the whole life of machines if this is feasible. Average, annual repair costs are often quoted as a fraction of the original cost of machinery. There are two main snags in using such figures. One is that repair costs clearly vary with use. The other is that when two machines or implements for doing the same job are compared, there may be a choice between a cheap, poor machine and a costly, reliable one. In such a case, repair costs tend to be inversely rather than directly proportional to capital cost. This problem is hard to avoid but the relationship between annual use and cost can be allowed for and table 4.25 gives estimated costs for different machines and amounts of use in average tropical conditions.

The way to use the estimates in table 4.25 can best be shown by an example. Suppose a pick-up baler costs $1500 and it is expected to be used for 150 hours a year. Average annual repair and maintenance cost over its estimated useful life of 11 years (see table 4.24) is $(5.5 \times 1500)/100 = \$82.50$ a year.

It is well known that wear on cultivation implements depends largely on soil conditions and some adjustment is necessary for severe conditions. The figures given in tables 4.24 and 4.25 are for use only when no better ones are available. They are based on data published in Britain, the United States, South Africa and Zimbabwe. Where records, kept over several years, are available on your own farm, these would be better. If the tables given are used, adjustments are needed if a high purchase price reflects high quality and durability, for example, by use of special bearings or specially hardened wearing parts. Adjustment is also needed if low price reflects a rapid likely rate of wear and tear.

Fuel and lubricant costs can be estimated from past records or from machinery specifications coupled with the on-farm cost of fuel and lubricants. Fuel use is usually quoted 'per hour' so for a 'per hectare' figure it is necessary to estimate the rate of work and adjust the hourly figure.

The sources of the return to offset the cost of investing in machinery may be easy to see; for example, extra output, higher quality output or less labour. If labour is saved without reducing the labour force, then that saved labour must be used productively. However, a particular machine may save labour only in the sense of saving time, leaving the labour force unchanged. A larger combine could shorten the harvest by a few days. What are the benefits in such cases? If a farmer reckons them to be 'peace of mind', he

Table 4.25   Estimated Average Yearly Cost of Repairs and Maintenance, as a Percentage of Purchase Price, Related to Annual Use in the Tropics

	Annual use (hours)				For every extra 100 hours, ADD
	50	100	150	200	
Equipment	%	%	%	%	%
Ploughs, cultivators, toothed harrows, rotary cultivators, mowers	4.0	7.0	9.5	12.0	5.0
Combine harvesters — p.t.o. driven, pick-up balers	3.0	5.0	5.5	6.5	1.5
Disc harrows, fertiliser distributors, grain drills — with fertiliser boxes, sprayers	2.5	5.0	7.0	8.5	3.5
Hay rakes, planters, forage harvesters	2.5	4.0	6.0	8.0	4.0
Grain drills — without fertiliser boxes, milking machines	2.0	3.5	5.0	6.5	3.0
Combine harvesters — self-propelled and engine driven	1.5	2.5	3.0	4.0	1.5
Grain driers & cleaners, rolls, hammer mills, feed mixers, threshers	1.5	2.0	2.5	3.0	1.0
	Annual use (hours)				For every extra 100 hours, ADD
	500 %	750 %	1000 %	1500 %	%
Tractors	6.0	7.7	9.3	12.6	0.7

should try to estimate the value of this by comparing the new costs and returns with those of machines that would 'just do'.

Machinery investment is rarely discussed for long without considering income tax allowances which many see as a strong incentive to replace machinery more often. This view arises mainly from wrong understanding of the way the allowances are usually given. In practice, taxation allowances have little bearing on buying decisions. Investing in new machinery to reduce tax without being sure of greater profit could ultimately leave the business weaker. Farmers are too often attracted by the large capital allowances often available early in a machine's life.

Taxation may, however, affect the best time to buy new equipment, justified by its likelihood of raising overall profit, in businesses where profit, and therefore tax payable, varies between years. If a large pretax profit is likely, this should be clear by the tenth month of the financial year and it is then possible to invest quickly in plant or machinery such as better office equipment, to reduce that year's taxable profit. Re-equipping in good trading years often makes it possible to reduce average, annual tax payment. Two rules should guide decisions on this subject. The first is to invest in machinery when there is a sound

commercial need to do so and to regard any tax allowances as less important than the main aim of farming efficiently. The second is to re-equip at the end of a tax year, rather than wait until the start of the next, so that tax relief is obtained as soon as possible.

Taxation rules vary between countries and also over time. Investment decisions involving tax matters are therefore best taken after consulting an accountant.

MACHINE SELECTION

Choosing a major piece of plant — a machine, vehicle or new building — is an important management decision; creating a stream of effects that will continue for several, perhaps many, years. Selection should be based upon the widest possible choice of designs.

There is constant argument about the relative merits of complex, specialised machinery and simpler equipment. The former are naturally inflexible, prone to obsolescence and costly but they are supposed to do better work. The simpler equipment is cheaper and is often very flexible. The need to reduce tractor size to make it more suitable for small farmers who cannot depend on large-scale production to keep costs down, prompted development of the single axle or two-wheeled walking-type tractor. This design omits certain features of the normal tractor but offers farmers something at reasonable cost even if they have to forgo some of the advantages of a standard machine. This approach has succeeded on small farms, especially in rice-growing areas. Experience in the sub-arid tropics shows, however, that the human effort needed to work these machines is high and, in hard conditions, this limits daily work time. These machines are still often too complex for small farmers in the tropics.

Expecting machine choice always to be ideal is over-optimism. Perfectionists in anything are normally frustrated. Nevertheless, it is possible to follow simple, logical procedures for selecting machinery to save costs, reduce risks or both. Most information needed for decision-making should be obtained by farmers themselves because the final responsibility for decisions is entirely theirs. It is possible, however, to guide them with a few general principles for selecting and using machines.

A large factor in machinery selection is the maximum time available for a certain job; for example, planting 40 ha of maize in 10 days or 120 ha of cotton in 7 days or perhaps ploughing 40 ha in a fortnight on an irrigation project. Weather records should show how many days are usually suitable for mechanical work on these jobs. One can then judge the adequacy of different machine sizes. Thought should also be given to the worst possible conditions that could limit available time more than usual. The urgency of following operations should also be considered. A big problem is that available hours cannot be known exactly nor can the rate of work; both vary with seasonal weather and allied factors.

When a farmer has decided to mechanise a job, he must decide how many machines and what size to buy. He may have a choice between several different models and many considerations enter into the decision. First, he must calculate the work rate needed by dividing the area or output to be handled by the hours or days available for work. Next, he must select the number and size of machines capable of this.

Although we think of capacity as the size of a machine, it really depends on two things — the work rate achieved and the time available to complete the job. Both vary with site and season. Machine capacity is thus highly variable and, consequently, deciding the best capacity for a job is hard. Another problem is that the capacity of

successive new machines keeps rising. This gain comes from larger machines with higher work rates and from better design which allows work in worse conditions, thus extending the time available for work. However this greater capacity arises, it raises capital cost.

Another problem adds to the complexity of the matter. The capacity of any machine depends on the capacity of all other machines used for the job. Thus a whole range of machinery has to be assessed.

Many farmers do not know how much land is covered in an hour by tractor-drawn implements. Rate of work depends on the width of the machine, its speed and how efficiently it is used. This formula helps:

$$\text{Effective work rate (ha/h)} = \frac{\text{Implement width (metres)} \times \text{working speed (km/h)} \times \text{field efficiency (\%)}}{1000}$$

For example, if a tractor and implement(s), with a 2-metre working width, travel at 6 km/h at a field efficiency of 70 per cent, the effective work rate is 0.84 ha/h — $(2 \times 6 \times 70)/1000$.

In assessing the work rate of a farm implement, its width, as quoted in its specifications, should be modified to the average width used when working. There is always some overlap in tillage and cultivation work; with implements like mowers and harrows it is almost impossible to use the full width without occasional misses. The overlap needed to cover a field thoroughly depends on speed, ground conditions and operator skill. In combine harvesting, it is not always possible to use the full width of the header because of crop density — even at the slowest possible speed of the machine. Row-crop planters and similar implements use their full-rated width — unlike most other field implements. The rated width of a row-crop machine is found by multiplying the number of planting or seeding units by the spacing of these units. Working width is assumed to include a half space beyond each outside unit.

Working speed cannot stay constant because all the time the machine is working changes in draught are caused by changing circumstances, for example the power available from the tractor, variations in land slope, the rolling resistance, which can vary from point to point in the field, and the traction of the tractor, which can vary according to the soil or vegetation the tractor is working on. Table 4.26 shows how average working speed can be obtained.

Table 4.26   For Obtaining Speed of Tractor

Pace off 30 metres. Time the tractor over this distance a few times and average the time. Then read the speed in km/h from the table below.

Seconds to travel 30 m	Km/h	Seconds to travel 30 m	Km/h
9	12.1	19	5.75
10	10.9	20	5.45
11	9.90	22	4.95
12	9.08	24	4.54
13	8.38	26	4.19
14	7.78	28	3.89
15	7.26	30	3.60
16	6.81	35	3.16
17	6.43	40	2.72
18	6.05	45	2.42

Whereas theoretical spot work rate is calculated simply by multiplying effective width by working speed, this rate is rarely if ever attained owing to various losses.

Spot work rate (ha/h) = machine width (m) × working speed (m/h) ÷ 10 000

The percentage of time lost varies with the time spent on lubricating and fuelling the machines; regular checks and adjustments; breakdowns; turning at headlands; removing blockages; adding seed and fertiliser and breaks taken by operators. Time lost in turning, idle time and machine adjustment is normally proportional to the operating time of the machine. Other time losses, such as clearing blockages of filling seed and fertiliser boxes or spray tanks tend to vary with area rather than operating time. Time lost in harvesting crops tends to vary with yield, area and length of haul of the harvested crops. Time losses of this type gain in importance as the width or speed of the machine rises. This is because they then become a bigger proportion of the total time needed for each hectare of work.

Turns at ends or corners of a field take up much time, especially in short fields, irrespective of whether the machine works back and forth or around the area. Time studies show that rectangular rather than square or irregularly shaped fields raise the effective work rate.

Field efficiency when working various machines is roughly shown in table 4.27.

Many factors affect the working rate of field operations and each one should be studied carefully to minimise lost time and so raise the effective rate. Routine maintenance, timely repairs and following the instructions in the operator's manual are essential for economic and efficient machine use.

Table 4.27   Average Field Efficiency of Machinery in the Tropics

	%
Subsoilers, rippers, disc ploughs, harrows, cultivators, rolls, rotary slashes and mowers, tractor hoes	7
Mouldboard ploughs, finger-bar mowers	6
Grain drills and combine harvesters	6
Combine drills and forage harvesters	5
Fertiliser distributors, pick-up balers	5
Planters	4
Crop sprayers	4

Table 4.28   Approximate Draught or Power Needs of Various Field Implements in Sub-arid Tropical Conditions

Subsoilers, rippers	20 kg/cm depth up to 30 cm depth; 43 kg/cm from 30–4 cm depth; 65 kg/cm over 45 cm depth.
Ploughs, disc or mouldboard	0.42–0.70 kg/cm² furrow section (increase with working depth)
Heavy disc harrows	370 kg/m width
Light disc harrows	220 kg/m width
Tined cultivator 25 cm deep in loose soil	70 kg/tine
Cambridge roller	75 kg/m width
Heavy 'notched' or spiked roll	120 kg/m width
Planter	70 kg/row
Fertiliser distributor	30 kg/m width
Rotary slasher	5 kW (at drawbar)/m width

Field efficiency does not allow for mealtime or stops other than those expected with serviceable equipment and an able driver. One-and-a half hours off work in every eight may be expected from a tractor and implement unit. This may often be too little if, through poor organisation, the planned stop becomes just a rest break while the driver finds and refills an empty grease gun or spends the time refueling without tending the machinery. Failure to realistically estimate the maximum capacity of machines, obvious as it may seem, is the commonest fault made in planning machine use.

To estimate implement needs it is necessary to find out not only their field efficiencies but also their power demands. Table 4.28 gives a rough idea of the power needed by various implements in the sub-arid tropics.

As the demand for tractors and implements depends mainly on the size of a job and the time available to do it, it is first necessary to select a suitable tractor/implement combination to do a articular critical job and then to use this equipment as efficiently as possible for other work.

For example, suppose the main enterprise is 25 ha maize and it is decided that:

ploughing must be 25–30 cm deep and must be finished in eight weeks
• planting must be done within ten days

We estimate that the tractor and implements will be available for 50 working hours each week; this provides a safety margin allowing for repairs, bad weather, illness and so on.

Ploughing 125 ha in eight weeks (400 hours) needs an effective work rate of 0.31 ha/h. From table 4.27, we see that the field efficiency of a disc plough is about 70 per cent and by calculation we find that for a net ploughing rate of 0.31 ha/h with a disc plough at an effective field capacity of 70 per cent we need a plough width of 89 cm at 5 km/h or 74 cm at 6 km/h. The width of disc ploughs is normally 25 cm per disc, therefore three discs would have a working width of 75 cm and, at this width, we must work at 5.9 km/h.

Now we must decide what tractor would be needed to pull this plough. We have already decided that ploughing depth must be between 25 cm and 30 cm and from table 4.28 we find that soil resistance will be 0.42–0.70 kg/cm^2 furrow section — say 0.60 kg/cm^2.

Total area of furrow section
= width of cut × average depth
= 75 cm × 27.5 cm = 2062.5 cm^2
. Draught of plough
= total furrow section × soil resistance =
2062.5 = 0.60 = 1237.5 kg

Enquiries at your local farm machinery dealers show that at your altitude above sea level, a drawbar pull of 1237. kg at 5.9 km/h for ploughing under firm field conditions needs a tractor with a maximum belt power of about 45 kW at sea level.

A similar exercise can be done for the planting — assuming 90 cm rows and speeds not more than 6.4 km/h. From table 4.27 we find that the expected field efficiency of a planter is only about 45 per cent. To finish this work in the required 10 days we must have an effective work rate of 12.5 ha/day or, in a 10-hour effective day, 1.25 ha/h. By calculation we find that, at the maximum speed we want to use, we need a working width of 4.34 m. As the limiting factor is likely to be the lifting capacity of the tractor hydraulics or practical considerations of size, the largest unit is likely to be a four-row model with an effective working width of 3.6 m. It would be wise therefore to have an extra minimum size, two-row planter for a second tractor.

From table 4.28 and discussion with the dealer, it is found that at your altitude a 15 kW tractor would be sufficient for a four-row planter but since it would have to carry considerable weight with a mounted planter and, since you already have a bigger tractor for ploughing, you should use it. If the extra tractor for the two-row planter is available only for part of each day, because of other commitments, it may be best to have a three- or four-row planter. Alternatively, some of the maize could be hand-planted.

Surveys of machinery use on farms in many tropical countries have shown how much tractor power is used on average on various types of farms for growing the main crops. Standards used vary slightly with the area and the many types of farming carried out in it; so if specific rather than national standards are available and suitable, they should be used. Standards are useful for checking mechanical efficiency. In the absence of personal experience and in areas where standards drawing on farmers' collective experience are available, they may be used to suggest how a particular farm should be equipped to effect management plans. However, it seems best to use practical experience for choosing equipment and to use any efficiency standards for checking results. This is because those standards available have a restricted range and are soon outdated through improvements in both power units and equipment. It is undeniable, however, that farmers should know roughly the work capacities of the main power units and machines under their conditions and it helps if these are known both in terms of area each hour or day and of likely output each season. The main snag is that output varies with climatic and soil conditions, field size and topography and size and exact type of both the equipment and the power unit.

Farm machinery comes in several sizes so farmers must choose a machine from a range of similar but different sized models. Larger machines may give flexibility, greater timeliness and thus higher yields through, for example, less grain shedding and maybe lower costs for grain drying or having to hire a contractor to complete the harvest in bad seasons.

Choosing which size of machine to buy is especially important with tractors whose size usually fixes the size of implements. This decision

Table 4.29   Typical Tractor Sizes and Hypothetical Costs

Main type	Power group	Approx. power range (kW)	Capital cost	Capital cost/kW	Mean rate of work (Ha/h)	Annual overhead cost	Operating costs per	
							hour	ha
			$	$		$	$	$
Market garden	Motor hoes	0.75–2.2	250	170	—	40	0.5	—
	2-wheel GP	2.2–7.5	750	155	—	110	0.15	—
Wheeled, rear-wheel drive	Small	11–22	3000	182	0.6	450	0.55	0.9
	Small—medium	23–24	3800	133	0.8	540	0.72	0.9
	Medium	35–45	4300	108	1.0	650	0.79	0.7
	Medium—large	46–60	5500	105	1.4	900	1.00	0.7
	Large	61–75	8000	118	1.8	1160	1.47	0.8
	Huge	Over 75	12600	168+	2.2	1900	2.32	1.0
Wheeled, four-wheel drive	Medium—large	46–60	8000	152	1.6	1200	1.47	0.9
	Large	61–75	10800	160	2.1	1620	1.99	0.9
	Huge	Over 75	12600	168+	2.5	1900	2.32	0.9
Track-layers	Medium agricultural with linkage	52–67	13800	232	2.5	2100	2.54	1.0

may be between one large tractor or two smaller ones. Choosing the tractor size and the range of implements must obviously be related to farm size, the crops to be grown, capital availability and the cost of labour. This is difficult because tractors are used for many jobs and have a different work rate and capacity for each one. Huge tractors (over 75 kW) are costly per unit of power and also in matched equipment. They are unlikely to be worth while except on large areas run with an unusually small work force or where they can help solve problems caused by periods of peak labour demand. Even then, alternatives like contracting should be considered.

One of the problems in studying tractor economics is that tractors vary greatly. Engine power, which was mostly 19–22 kW in the early 1950s, doubled to 38–45 kW in the next 20 years and is still rising. Typical tractor sizes and hypothetical but relative costs are given in table 4.29.

In estimating the capacity of a tractor for heavy work, a useful guide is how many plough furrows of moderate depth can be pulled in medium-textured land at about 4 km/h. Typical figures for wheeled tractors are:

Tractors	Furrows	kW
Small	1–2	11–22
Small–medium	2–3	23–24
Medium	3–4	35–45
Medium–large	4–5	46–60
Large	5–6	61–75
Huge	6–8	Over 75

Table 4.30   Relationship of Tractor Size and Area Worked to Total Annual Tractor Cost

		Hectares work/year		
		1000	2250	3500
Tractors		$	$	$
Small	Overhead cost	450	450	450
	Operating cost	920	2070	3220
		(1370)	2520	3670
Medium	Overhead cost	650	650	650
	Operating cost	790	1778	2765
	Total cost	1440	(2428)	3415
Medium—large	Overhead cost	900	900	900
	Operating cost	710	1598	2485
	Total cost	1610	2498	(3385

(   ) indicates cheapest tractor size at each level of use

Table 4.30 shows how costs such as those in table 4.29 can be used to estimate the total annual cost of different sized rear-wheel drive tractors over different areas of work. In this example small, medium and medium–large tractors have the lowest total annual cost at 1000, 2250 and 3500 hectares of total work a year respectively.

On large farms it is often worth having more than one size of tractor to enable the tractor to be suited to the job in hand. However, this is often a case of choosing suitable implements and gears rather than of tractor size.

The normal, medium, rear-wheel drive, four-wheeled tractor, most common in world farming and thus mass produced, has a relatively low cost per kW — see table 4.29 — besides many other advantages like versatility and low fuel use. Its main disadvantage is that annual capital cost per hectare is high. This is also true of small conventionally designed models of about 14 kW.

Small farmers would probably benefit from specially designed small four-wheelers of about 10 kW. The essential difference between these and small conventional types is that they are specially designed to eliminate some of the sophisticated elements of the smaller conventional models and reduce cost/kW. Such specially designed versions have been made in Uganda and Swaziland. The basic aim is to produce a simple tractor, assembled locally, in quantities to meet local demand, using mainly mass-produced parts and for sale at a price suited to six- to eight-hectare farms.

It is well known that small tractors have two common inescapable failings. First, traction is limited because of wheel slip and their light weight and, secondly, small engines are notorious for their relatively high fuel use. Fortunately, both problems could be minimised. For instance, traction problems could be partly overcome by adopting tined implements, with lower draught requirements, instead of using ploughs for basic tillage — especially in extreme conditions. Fuel consumption could be lessened by substituting diesel engines for petrol ones when sufficiently cheap engines of this size are available.

The purpose of specially designed small tractors is not to compete with draught animals or with conventional wheeled tractors but merely to fill a special, maybe temporary, gap for farmers with no access to draught animals and holdings too small to justify buying a standard wheeled tractor.

Deciding how many tractors are needed on a farm involves factors such as labour supply, capital availability and timeliness — all related of course, to the cropping and stocking of the farm. At its simplest, however, the number of small–medium size tractors can be estimated as roughly one for the first 800 hours of annual work and another for each further 1200 hours.

Farmers can choose between two or more smaller machines or one large one. As the time available for various jobs limits the capacity of any one tractor, it may be more efficient to have two smaller tractors instead of one large one. However, it may be financially sounder to work overtime or use contractors than to buy another tractor that would be idle for much of the year. Nevertheless, adequate machine capacity reduces risk. People talk of a reserve of power, a stand-by tractor, giving economists nightmares about over-capitalisation, when they really mean that the machinery they have enables the necessary work to be done at peak periods. If the machinery selected has raised profits, then the selection must have been about right. Whether it was as right as it might have been is another matter.

The theoretical approach to the number of tractors needed is to estimate the total tractor hours needed each year. This is imprecise for two reasons — the time needed for a job varies between years and the hours a tractor works depends mainly on the demand. If it is necessary to work a tractor day and night, this can be done.

One of the problems in estimating tractor needs is that peak demand tends to be forgotten; also tractors can be joined to attachments. Sugarcane production is a good example of the latter as mechanical loaders are joined to tractors. These may work only four hours a day but are unavailable for other work.

Most farmers realise the difficulty of selecting suitably matched tractors and implements. This means balancing mechanical, agronomic and economic matters and can be done only by considering all of these together. It is important to examine what already exists on the farm and whether it can still be used, then consider the suitability of new equipment, its simplicity and reliability. Generally machines should not be considered separately but as part of an entire system. As fieldwork becomes more mechanised, the time required for each job needs closer study and more accurate job co-ordination or there will be much waiting time that raises cost. For example, in combined operations like harvesting, bulk handling and silage making, it is important to match the capacity of various machines to avoid bottlenecks and standing time.

The working environment includes climate, soils and their likely reaction to rain or a long drought, the machine operators, their attitudes and mechanical skill — or lack of it. Even if it helps simply to decide the tyre size needed to stop a fertiliser distributor sinking into sandy soil, a thorough knowledge of the environment helps in planning to mechanise even the simplest job.

One way to improve the working environment is to improve your knowledge of machines and train your operators. Lack of this knowledge is one reason why many farmers ask for a demonstration on their own farm before buying a new machine. It is not necessarily the machine they are unsure about but the effect local conditions could have on it.

With the working environment, the end product, the costs and finance clearly known, selection of individual machines simply means finding the most suitable one to fit into this framework. For example, work conditions always determine the needed machine strength and the capital available determines the initial price that can be paid.

Most farmers practise what might be called 'informal' planning of tractor (and labour) use and I suspect many successful managers are adept at this. However, if a farmer is changing his farming system or wishes to review it, it is necessary to make a 'formal' plan. Tractor (and labour) plans have been used successfully on many tropical farms.

The procedure is simple. For each enterprise in turn the operations are listed, like land preparation, planting, cultivating and so on. The size of tractor needed, the number of tractor hours (and labour days) are also recorded. The timing of each operation is listed as it is often critical.

Annual work plans for each tractor are plotted on graph paper. In practice, it is wise to start with planting and then fit in the jobs before and after planting in sequence. This is because the planting date is so often critical. Because of the seasonal needs of most crops there is a tendency for scarcity of machinery at peak times of the year and underuse of it the rest of the time. As it is usually impracticable to plant crops regularly each month, the alternative is to delay or advance the timing of less critical jobs so that tractors are used as evenly as possible. The timing of each job is usually recorded relative to other jobs but sometimes by calendar date. Sometimes the relationship is critical. For example, tobacco should not be planted within two weeks of fumigation and this should be recorded.

Table 4.31 and figure 4.9 are good examples of a calendar of work and tractor use plan from a Zimbabwe estate growing maize, tobacco, groundnuts and soya beans.

For each crop in turn the jobs are listed. For every job the method is recorded together with the type of labour to be used (this list is also used for labour planning) and the tractor to be used. The expected output for the operation is recorded using the farmer's own estimate of what he will achieve. With a crop that the farmer has not grown before, work study data or survey results may be used.

Separate plans are made for each tractor as in figure 4.9. Provision is made for six working days a week (48 hours tractor week) and for four weeks in each month.

In practice, there may be more work to do than there is time available and it may be necessary to work seven days a week instead of the normal six. A farmer may need to work tractors on double shifts to complete the work. These periods will be shown on the plan with the shift in hours.

Tractor plans often reveal, weeks or even months ahead, when problem times are likely to occur. This may necessitate buying or hiring another implement if it is clear from the plan that the farmer cannot cope with the one he has. The plan may even indicate a need for a new tractor or reveal problems in the planned combination of enterprises.

It is important to allow for tractors to haul trailers at harvest or to fetch fuel for curing tobacco; or tractors needed to power a maize or groundnut sheller.

A well-drawn tractor plan can enable a farmer to arrange his work so that he relies less on his older tractor whose breakdowns had reduced working rate before.

The modern diesel engine's characteristic of relating its fuel consumption to the output of power seems to mislead some farmers into thinking that a 55 kW tractor taking rolls of wire to a fencing gang is working economically. It is *not*, nor is a 75 kW electric motor driving only a feed mixer. Nevertheless, surplus power is rightly thought an advantage if the surplus is limited to 20–30 per cent of maximum working capacity but there is no sense or profit in planning any 45 kW tractor operation in which something like rolling or drag-harrowing is done separately.

Tractor efficiency can be raised and cost much reduced by doing two or more jobs together whenever it is practicable. More thought could profitably be given to multiple operations, like milling and mixing, planting, pre-emergence band spraying and harrowing or ploughing and rolling, because each job combined with another should lower costs through saving time, handling and fuel. However, care is needed to avoid reducing the efficiency of an operation by combining an unreliable machine with a critical task like planting. For example, if a planter by itself has a field efficiency of 50 per cent and it is combined with a harrow, the field efficiency of which is only 60 per cent because of sticky soil conditions, the combined efficiency will be only 30 per cent which could be unacceptable for planting. However, on a less critical job, like

Table 4.31 Calendar of Work

Operation	Method	Labour	Average output	Hectares	Tractor days	Tractor	Gen. lab. days	Comments and timing
Land preparation (1986)								
Ploughing	3 furrow, disc	D (driver)	1.2 ha/day	25	22	A or B	—	
Discing	3 m tandem	D	6.1 ha/day	65	11	A or B	—	
Ridging (tobacco)	Single row	D + 2	2.4 ha/day	20	8	A	17	At least 2 weeks before planting
Furrowing (maize)	2 row — with fertiliser	D + 2	4.0 ha/day	20	5	A or B	10	25 ha end Nov.
Herbicide	Mounted boom and disc	D + 1	6.1 ha/day	50	8	A	8	25 ha end Oct.
Cross disc	3 m light	D	6.1 ha/day	50	8	B or C	—	25 ha end Oct.
Land preparation (1987) (Aim to complete by end of September 1987)								
Ploughing (early)	3 furrow, disc	D	1.2 ha/day	70	57	A or B	—	After tobacco reaping
Discing	3 m light	D	6.1 ha/day	70	11	A, B or C	—	After ploughing
Ploughing (August)	3 furrow disc	D	1.2 ha/day	20	17	A or B	—	August
Discing	3 m tandem	D	6.1 ha/day	20	3	A or B	—	August
Tobacco								
Seedbeds	Prepare	4					100	Sow 10/8
	Watering	3	Full time					10/8 to end Nov.
Dig holes	Hand	17	0.2 ha/lab. day	20			100	Before 1/11
Water planting	Boom system	24 reg., 6 cas.	3.2 ha/day	20	18	A, B or C	144	6 days. 1/11 to 15/11
Cutworm treatment	One tractor	24	0.4 ha/lab day	20	2	A, B or C	50	Straight after planting
Refilling	One tractor (Cups)	24	0.4 ha/lab. day	20	2	A, B or C	50	One week after planting
Top dressing	Cups (×2)	17		40			100	End Nov. and mid Dec.
Cultivation	Mechanical (×2)	D	6.1 ha/day	40	7	A or B	—	Mid Nov. to end Nov.
Cultivation	Hand (×2)	24 reg., 6 cas.	0.1 ha/lab. day	40	—		400	Straight after mech. cultivation
Topping	Hand	6		20			60	10 days end Dec.
Suckering	Hand	12 women	Full time	20			60	Last week Dec. to end Dec.
Reaping	Max. 2 barns/day Often 1 barn/day (Kurt machines, horizontal string hanging)	8 reapers 4 waiters 1 foreman 6 at barns 4 curing 1 boilerman 4 women untying		20		A, B, C (mornings only for A and B)		Last week Dec. to mid March
Disease control	Power duster (white mould)	2	4.0 ha/day	20				Full time, mid Dec. to end Jan.

Operation	Method	Labour	Rate					Timing
Grading	Single stage	12 graders, 4 check graders, 1 final check, 6 women tying hands, 3 baling, 1 boilerman	550 kg/day	34000 kg				After reaping for 63 days
Maize								
Planting	Hand into furrows	24	0.4 ha/lab. day	20	—	—	50	Early Nov.
Covering	Spike harrow	D	10 ha/day	20	2	A, B or C	—	After planting
Cultivation	Mechanical (×2)	D	4.0 ha/day	40	18	A or B	—	Mid Nov. to mid Dec.
Cultivation	Hand (×2)	Variable	0.3 ha/lab. day	40	—		200	Mid Nov. to mid Dec.
Top dressing	Tractor	D + 2	6.1 ha/day	20	4	A or B	8	Mid Dec.
Stemborer treatment	Hand	22	0.3 ha/lab. day	20			67	Mid Dec.
Guarding (inc. other crops)		2	Full time					End Jan. to end of harvest
Harvesting	'Bang board', stationary sheller	12 reapers, 1 gleaner, 8 shelling, 12 W. gleaning	Approx. 1.2 ha/day	20	24	A, B or C		Mid July to earily Aug.
Groundnuts								
Planting	2 row machine	D + 2	2.6 ha/day	25	10	A or B	21	Early Nov.
Gypsum	Hand	18	0.4 ha/lab. day	25			62	Late Dec.
Loosening	Tractor, 2 rows	D + 2	4.0 ha/day	25	6	A or B	12	Early Apr.
Lifting and cocking	Hand	15	7.1 lab. days/ha				180	Early Apr.
Stripping	Contract	5						2000 bags of pods
Shelling	Contract	5	50 bags/day		10	C	50	After stripping
Soya beans								
Planting	Machine	D + 2	3.2 ha/day	25	8	A or B	16	Early Dec.
Cultivation	Mechanical (×1)	D	4.0 ha/day	25	6	A or B		Late Dec.[a]
Cultivation	Hand (×1)		0.1 ha/lab. day	25	—	—	250	Early Dec.[a]
Spraying	Mounted (×1)	D + 2	8.1 ha/day	25	3	A or B	6	Early Jan.
Harvest	Contract combine	5	12 ha/day	25			10	2 days Apr.
Harvest	Transport bags	5			2	C	10	2 days Apr.

[a] Not necessary if herbicide effective

dry-season ploughing, it may be acceptable to trade lower overall field efficiency for a drop in overall cost.

Supplying new machines is not the only job of a dealer. He should know his new machinery well enough to be able to advise on its adjustment in use and he should be able to back his sales with adequate spare parts and service. In assessing the reliability of a supplier, ensure his business is sound and profitable because, as many farmers have learnt to their cost, without a fair profit in any business there can be no continuity.

Farmers usually get the quality of service they pay for. Service is either adequate or dangerous; there is no half-way state. Unfortunately, one still suspects that a farmer who insists on making a busy serviceman travel 200 km from a depot and back again on an emergency call to repair his tractor's power take-off still has not read the instruction book. He probably thinks it is absurd that he is charged, say, $270 when it took the field serviceman less than a minute to put the p.t.o. drive into gear. Dealers also think it is absurd!

Economic use of machinery depends mainly

on using it to as near maximum capacity as possible. However, maximum capacity is often misunderstood. Farmers know that a tractor cannot work for 24 hours a day. Everyone knows there must be refuelling stops but in rush periods the equally needed checks and preventive maintenance routines, so carefully set out by manufacturers, are often overlooked. This is deplorable as the resulting damage and costly repairs often constitute up to a quarter of a tractor's running costs. Engines often wear out rapidly through inattention to air cleaners. Batteries filled too seldom and often from taps, lubrication oil levels too low and use of the wrong grades of lubricant often raise running costs unnecessarily. It is cheaper in the end if the maximum capacity of each machine is calculated on its possible work output after allowing for all service checks.

Machine replacement

One of the hardest policy decisions facing a manager is when to replace machinery. It is eventually necessary to replace worn-out plant and machinery to maintain production efficiency. Should it be replaced after a certain time, when it is fully written off in the books, should he adopt a fixed replacement policy every few years or wait until the machine no longer works? Frequent breakdown, due to worn-out plant and machinery, causes production delays, incurs the related costs of having assets idle and causes loss of valuable machine- and man-hours. There comes a time when the rise in repair costs is so great that it would pay to get a new machine but one never knows when a large breakdown will occur.

The plant selection problem becomes more complex when an existing piece of plant is being considered for replacement; on the basis that the task it does is likely to continue indefinitely. In such a case, the problem involves *two* choices:

- *Which item* in the market is best as a replacement.
- *When* should replacement be made.

The choice of the best time for replacement can make a significant difference to overall costs. As time passes, there are two factors that must be considered:

- The existing plant is becoming less effective, more prone to breakdown and more costly to maintain.
- Improvements in technology are resulting in more effective and efficient alternatives coming onto the market.

Practical selection policies need to consider both factors.

There seems to have been little investigation or formal research into farm machinery replacement policy and there is certainly a shortage of practical advice. The formal theory is fairly clear, though not necessarily valid, but using it in practice is difficult.

No single replacement policy can suit all farmers so the best policy will vary between farms. If

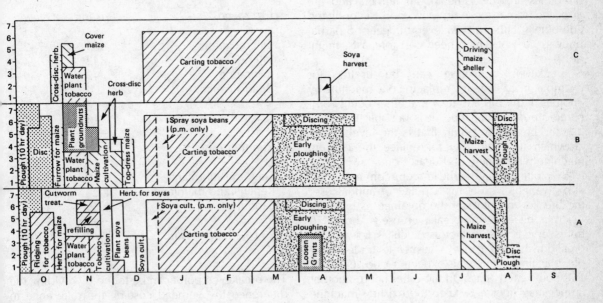

Figure 4.9 Tractor Plan

this were not so, there would be no second-hand or trade-in values. Each farmer should consider his own position and decide his own policy. On many farms there seems to be no replacement policy. Plant is kept as long as possible or until repair costs become obviously uneconomic. When replacement is no longer avoidable the only decision needed is what type of replacement to obtain. However, to ensure reasonable, long-term efficiency in mechanisation management, it is important to have a long-term plan for replacing the main items of equipment. Different investment policies may be followed by different farmers. The best policy depends on the importance of the machine for production, how much it is used and the funds available for investment.

Delaying tractor replacement postpones cash expenditure and this might be considered when capital is limited and there are alternative more profitable investment possibilities. Alternatively, the risk of breakdown during a critical period may encourage a policy of early replacement, especially as there is little difference in cost between early and late replacement.

How should a farmer or his advisors, amongst whom I include the machinery trade, decide whether or not to buy a new piece of equipment? There is, of course, no standard time in a machine's life when it should be changed. Repair costs vary between farms according to the farmer's ability to do maintenance work himself economically or to bargain for a low price with a repair firm and also upon soil type. However, the principles are the same for all farmers and are also the same for both new and secondhand equipment. Either a breakeven budget or a partial budget, both of which need cost forecasts, might be used.

Breakeven budgeting can be useful, for example, in considering replacing one machine by another. If the old and new machines being compared are of similar type, it is possible to calculate the breakeven point, that is the level of use at which it would pay to replace the existing machine with a new and dearer one.

By plotting total overhead (common) and operating (variable) costs for a range of annual use, the breakeven point can be obtained as in figure 4.10. In this case, the breakeven use is about 750 hours a year. Other factors, such as less risk of breakdown, may cause a farmer to trade-in the old machine even if he expects to do less than 750 hours each year. On the other hand, another farmer may decide to keep the existing machine although he expects to exceed 750 hours. If there

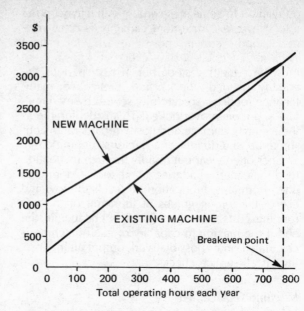

Figure 4.10   Estimated Breakeven Point for Existing and New Machines

is little apparent financial gain from the change, he may quite correctly decide to use the capital for something else.

Sometimes there is a clear choice between different machines, one having higher common costs but lower running costs than the other. Let us assume that cost estimates, averaged over a planning period of three years, are as follows:

Overheads	Existing	New
	$	$
Depreciation	485	742
Interest	342	644
Insurance	45	59
Road licence	74	74
Total each year	946	1519

Running costs		
	$	$
Fuel and lubricants	0.69	0.62
Repairs and maintenance	1.24	0.56
Labour	1.55	1.55
Total each hour	3.48	2.73

The breakeven point can be found by dividing the difference in overhead cost by the difference in running costs, for example,

Breakeven use (hours/year) = (1519−946)/(3.48 − 2.73) = 764 hours.

At any use over 764 hours a year it would be cheaper to get the new machine.

Breakeven analysis may also be used to compare the relative costs of owning or hiring a machine at different scales of activity, for example,

Machine annual overhead costs	= $1500
Operating costs/tonne	= $4.50
Contract charge/tonne	= $11.00
Breakeven tonnes	= 1500/(11.00−4.50)
	= 231 tonnes

All these examples demonstrate ways of making decisions about whether to replace machinery *now*. There remains the question of *when* should it be replaced.

The best way to assess what changes in profits will occur by replacing a machine is to rebudget the profit and loss account of the whole business. Partial budgeting, however, is a simple short cut.

Replacement analysis is a branch of operations research that seeks to analyse the problem of *when* to replace a firm's machinery and other equipment, given that the machinery gradually wears out or that new, more efficient models become available.

In planning the replacement time, it is worth trying to spread replacements fairly evenly. In an emergency, due to breakdown or unexpected improvement in a new model, replacement of a particular item may be advanced. Planning can help in making the right decisions on such matters as sharing the use of a machine.

Towards the end of any accounting year, if a farmer expects his profit to be unusually high, is a good time to buy replacement machinery. A machine costing $4000 may cut $4000 off the taxable profit and will reduce tax liability in that year. Table 4.32 shows a direct cash saving of $526 owing to careful planning and prompt action, as compared with waiting and buying replacements in a year of lower profit.

Several reasons might, either together or apart, necessitate changing replacement policy. These include the following:

- Change in the farm programme. If different enterprises are included, the importance of some operations may rise relative to others.
- Change in husbandry practices; for example, a change to minimum tillage might increase the importance of a sprayer and reduce that of a plough.
- Changes in the labour force. The need to mechanise when labour is hard to find.
- Need to replace equipment to gain advantage of improvements in machine design and manufacture. Such improvements may offer a direct operating cost advantage or they might give an indirect advantage by completing a job faster or better.

The main factors affecting the best replacement time for a machine are the following:

- Depreciation rate. (Expected future loss of market value.)
- Rate of increase in future repair and maintenance costs.
- Interest rates.
- Availability of capital.
- Annual use.
- Reliability.
- Improvements in the machine.
- Morale and psychological aspects.

The first five points are most important; the others being harder, if not impossible, to measure.

The lower the actual yearly loss of value, the more likely is early replacement to pay since it will involve less drop in value. The rate and way in which trade-in values fall as machines age determine annual depreciation. The faster trade-in prices drop in the early years, the less likely is early replacement to pay. A change will often make equipment obsolete. For example, shortage of labour on a large maize farm means that a sheller needing much labour to work it will become obsolete and purchase of a combine to reduce labour needs may be necessary.

The faster repair costs rise, the more advantageous early replacement is because the saving in repair costs is higher. Budgeting the repair and maintenance cost of existing equipment is perhaps the most difficult item to quantify. If a large repair bill can be avoided by trading-in without much effect on the price received, then obviously replacement is further favoured. Annual replacement of a machine may eliminate repair and maintenance costs entirely if these are covered by guarantee.

The lower the interest rate, the more early replacement is encouraged as the higher interest charges involved in this policy is then less weighty.

The more capital is available, the stronger the case for early replacement because there is less

Table 4.32   The Advantage of Capital Investment in High Profit Years

	1984–85	1985–86	
	$	$	
Profit Year to 30 Sep. 1985	12000		
Year to 30 Sep. 1986		4000	
Tax allowance on replacement machinery costing $4000 in Sep. 1985 at 100%	4000	—	
Taxable profit before personal allowances	8000	4000	
Approx. income tax payable (see table 2.4)	1275	150	Two-year mean = $712
Alternatively			
	$	$	
Profit Year to 30 Sep. 1985	12000		
Year to 30 Sep. 1986		4000	
Tax allowance on replacement machinery costing $4000 in Oct. 1985 at 100%	—	4000	
Taxable profit before personal allowances	12000	Nil	
Approx. income tax payable	2475	Nil	Two-year mean = $1238
Saving by replacement in high-profit year		$526	

chance of more profitable ways of using the extra capital tied up.

The more a machine is used, the more is early replacement favoured — because hourly depreciation is lower; trade-in value depending more on age than condition and more on condition than use. The more intensive the use, the lower will be depreciation cost per hectare, whereas repair costs per hectare will be fairly constant, in other words, the higher the repair bills that will be saved by early replacement. Also, the more the annual use the higher the possible losses that might occur through breakdowns and thus the greater the benefit from early replacement.

The less reliable a machine becomes with age, the more early replacement pays; the more intensive the use, the more important is reliability. It is often valuable to consider how important the reliability of a particular machine is to the prof-itability of the whole business. For example, many farmers keep one or two old tractors simply as motive power for transport. It is not essential to keep these in top condition. However, most farmers accept the need for fairly new 'ploughing' tractors as these are vital for operational efficiency. The greater the improvements in a machine the more early replacement is encouraged. They must be real, not imaginary or superficial, improvements, however, and result either in higher yields, better quality of work or lower running costs. Naturally, better versions will lower the trade-in value of older ones. Improvements in tractor design are frequent. One should try to distinguish between those that improve performance by, for example, a higher working rate or more drawbar pull, and superficial changes that do not raise the usefulness of the tractor. Useful improvements are not always reflected in higher prices but they may have a real advantage for individual farmers and may well prompt a decision to buy a new tractor.

Besides the factors mentioned there are others that are even harder to quantify. However, this does not reduce their importance. For example, there is the effect on the mental attitude of operators. New tractors, like new cars, get more attention and care in use. If having new machinery helps to keep good staff or the farmer's self-esteem depends on it, he may need to replace equipment sooner. Peace of mind, through freedom from breakdowns, is also included here, although too often this is not given by a new machine, especially in its first season.

Farmers vary in the way they balance all relevant factors to decide their machinery replacement policy.

A reasonable working life for a tractor is 10,000 hours and you should get 5000 hours without much trouble or costly repair bills. If tractors work really hard, say 1500 hours a year, there is often a strong case for trading-in every two years — before any large overhaul. Below 1000 hours a year, a four-year replacement rate is often suitable.

The aim is to decide when to replace equipment or machinery. If the criterion is that total cost averaged over the life of the equipment should be minimised, we could proceed as shown in the following worked example.

Worked example

A farmer wants to decide when to replace a machine that costs $8500 to buy. He estimates repair, maintenance and depreciation costs as follows:

Year	Yearly repairs and maintenance	Yearly depreciation[a]
	$	$
1	100	1409
2	200	1268
3	350	1127
4	600	986
5	800	845
6	1150	705
7	1550	564
8	1950	423
9	2450	283
10	2750	140

[a] Sum of digits. Scrap value of $750 after 10 years.

The average cost after one year, therefore, is repairs and maintenance + depreciation = $100 + 1409 = $1509 and, after two years is $(300 + 2677)/2 = $1488 each year.

We can conveniently tabulate the costs as follows:

Year (n)	Cumulative repair cost	Cumulative depreciation	Cumulative total cost (C)	Average annual cost (C/n)
	$	$	$	$
1	100	1409	1509	1509
2	300	2677	2977	1488
3	650	3804	4454	1485
4	1250	4790	6040	1510
5	2050	5635	7685	1537
6	3200	6340	9540	1590
7	4750	6904	11654	1665
8	6700	7327	14027	1753
9	9150	7610	16760	1862
10	11900	7750	19650	1965

Hence total cost averaged over the machine's life (C/n) is least when n = three years and the farmer should replace the machine soon after it is three years old.

By delaying the replacement of any machine, its capital cost is spread over more years but annual repair cost rises. The total of these two costs — capital cost and cumulative repair cost — is known as the machine's holding cost. This concept is that machinery costs can be separated into recurrent running costs, like fuel and oil, and cumulative capital, repair and maintenance costs. This latter is the holding cost.

Let us look at the previous example from the standpoint of holding cost.

Now, from the previous example, we know by subtracting cumulative depreciation from the initial cost, what trade-in price is likely each year end:

Year	Cumulative depreciation	Likely trade-in price
	$	$
1	1409	7091
2	2677	5823
3	3804	4696
4	4790	3710
5	5635	2865

Year	Cumulative depreciation	Likely trade-in price
	$	$
6	6340	2160
7	6904	1596
8	7327	1173
9	7610	890
10	7750	750

From these figures it seems that minimum annual holding cost is reached after seven years but that this could be reduced further by trading in any time before this. However, the maximum excess of likely trade-in price over the price needed to justify replacement also occurs after seven years ($1596 − 0 = $1596).

Instead of calculating holding cost per year, it could be calculated per hour of use to estimate optimum replacement time in terms of hours worked.

The concept of holding cost has the advantage of being fairly simple and easy to understand. However, it has two major drawbacks. First, it is highly theoretical and unless a farmer keeps detailed records of individual machines, he is unlikely to know exactly when the economic life has been reached. The holding cost concept is hard to use in practice. The practical or empirical approach involves observing different types of farmers and their replacement policies and seeing how these affect their machinery costs. Study of some case studies shows how different farmers cope with machinery replacement and how this affects their costs.

Table 4.34 shows relative machinery costs and tractor complements for four different types of farmers.

The unrestrained mechanics averaged a total cultivated area of 172 ha. For this they had four

Table 4.33   Total Holding Cost and Breakeven Exchange Prices for a Machine Costing $8500 New

Year	Initial cost	Yearly repairs and maintenance	Holding cost	Holding cost/year	Trade-in price needed to justify replacement
	$	$	$	$	$
1	8500	100	8600	8600	6707[a]
2	8500	200	8800	4400	5014
3	8500	350	9150	3050	3471
4	8500	600	9750	2438	2179
5	8500	800	10550	2110	1086
6	8500	1150	11700	1950	343
7	8500	1550	13250	1893	—
8	8500	1950	15200	1900	—
9	8500	2450	17650	1961	—
10	8500	2750	20400	2040	—

[a] Holding cost less 1893 n where n is the number of years.

Table 4.34   Relative Machinery Costs and Tractor Complements for Different Types of Farmers (Hypothetical)

	Unrestrained mechanics	Muddlers	Misers	Ideal
Cultivated hectares	172	170	83	232
No. of tractors	4	6	5	4
Hectares/tractor	43	28	17	58
Fuel cost/ha $	26	41	34	15
Tractor cost/ha $	46	29	98	4
Plant and implement cost/ha $	26	22	21	6

tractors in excellent condition. The area covered by each tractor averaged 43 ha and annual fuel cost per hectare was reasonable. However, tractor, plant and implement costs were excessive. Unrestrained mechanics are often so busy repairing machinery that the rest of the business suffers and profits therefore fall.

The muddlers seldom sat in their offices to plan operations. This resulted in too many tractors for a fairly simple farming programme. Their argument was that they needed extra tractors to cope with peak work loads. However, by spending more time planning operations, their machinery costs and tractor fleet could be much reduced.

The misers were so afraid of spending on new equipment that they kept tractors on the farm long after their economic life had passed. Tractor repair cost per hectare was far higher than for the other three groups in this comparison.

The fourth group approached the ideal. They had a realistic attitude to farming and spent much time and effort planning. They thus coped with the largest cultivated area with each tractor. Fuel, tractor repair and plant and implement costs were the lowest. A typical policy was to replace trac-

tors at such intervals that, of three of them, only one was older than three years. The fourth tractor, used for towing, was much older.

In choosing between various courses of action, one important principle applies. Costs that were incurred in the past, such as the purchase price of existing machines, are irrelevant. Accordingly, it is unimportant whether an item of plant being considered for replacement has been fully depreciated or not. All that should be considered are future costs that will be incurred with one alternative, compared with future costs if another is chosen. Housing and insurance, for most equipment, are fairly small costs and differ little with the age of the machine. Apart from interest on capital, the main costs are future depreciation, fuel, repairs and maintenance. Thus one economic objective might logically be to minimise the *future* total annual cost of these items, particularly if output differences are either negligible or small.

Leaving aside taxation, what should concern us is the current state of a machine, what we need it for and what it would cost to replace it. Consequently, the historic purchase price of the current machine already written off is irrelevant to the decision whether to replace it or not. Similarly, in our private lives when we consider whether or not to sell our car, what should concern us is the future performance we expect from the present car, how much we could sell it for, the performance we expect from a possible replacement and what it would cost. How long we have had the present car is irrelevant to the decision whether to keep or replace it!

All this is not to say that past history is necessarily unimportant, simply that it is relevant only in so far as it affects the state of affairs we are in when making a decision. An adequate description of the state we are in must include all the information relevant to possible future performance. It is only in this sense that the past history should affect present decisions.

In the earlier example, minimising average annual cumulative depreciation and repair cost indicated replacement after three years; minimising average annual holding cost indicated replacement after seven years. The first method is valid if we consider only *future* likely repair cost and loss of market value. The holding cost concept, with its consideration of initial cost seems to me to be invalid!

In this section I have tried to show some of the theories preached for deciding a machinery replacement policy. These theories are shown to have many practical drawbacks that make it hard for individual farmers to apply them. A general conclusion is that major repair of fairly old machinery and plant is rarely economic. Such studies as have been done show that there is little to choose, however, between fairly early and later replacement, provided machinery is used close to its capacity. In such a case, an impending large overhaul or repair bill, a good trade-in price offer or a good harvest, that is, high profits and therefore a high tax year, is enough to trigger replacement and often the balance is so fine that these are totally valid reasons.

A practical approach is to examine the replacement policies of other farmers and see how these affect their machinery costs. It seems from this approach that the best policy is to keep most equipment in new or nearly new condition. Keeping old machines and/or spending too much time and money repairing them causes excessively high machinery costs and lower farm profits.

Alternatives to buying new machinery

The growing demand for more capital for mechanisation is often thwarted by credit curbs and need to expand the productive enterprises on the farm. This has caused a search for other ways of gaining use of more machinery. Possibilities include buying used equipment, leasing it, use of contractors and sharing equipment with others. As machinery tends to be costly compared with other farm inputs, more thought is needed on lowering costs by using contract services or by joint ownership of machinery. These are both preferable to hire purchase and often offer a wide range of machinery and much lower costs.

Total machinery cost per hectare drops with increasing area owing to the spreading of common costs such as tax, insurance, housing interest on capital and depreciation. It is therefore often wise for managers with too little work to justify a given machine to either use a contractor or to co-operate with other farmers in joint ownership. These are ways of gaining the advantages of mechanisation without the problem of much greater capital investment. Important factors are:

- the importance of timeliness for a particular job
- availability of dependable service under acceptable conditions
- little or no loss of income compared with outright ownership
- large cost savings compared with ownership
- limited availability of capital and more profitable alternative investment opportunities

Generally, the more critical the timing of a certain job, in terms of either yield or quality, the less farmers are likely to substitute an alternative arrangement for ownership.

Buying used equipment is one way of reducing machinery investment. In general, buying used rather than new equipment results in higher variable costs per unit of output but lower annual common costs. Depreciation, interest and insurance are higher with new equipment; repairs, fuel and lubrication costs higher with used equipment. Used equipment is generally less reliable than new and breakdowns may reduce the quality or quantity of yields.

Because breakdowns are more likely with used equipment, the farmer's mechanical skill is important. If he is a skilled mechanic, he may be able to run used equipment without many delays. Repairing and servicing machines in the slack season is more important with used machines.

Breakeven analysis helps in choosing between new and second-hand equipment. Assume, for example, that a manager is thinking of replacing an existing old cotton picker with either a new one or a good used one. His cost estimates are:

	New $	Existing $
Present cost/value	10000	5700
Expected remaining life (years)	5	3
Common costs		
Annual depreciation	2000	1900
Insurance and tax	320	200
'Real' interest at 5.2%[a]	260	148
Total/year	2580	2248
Variable costs/ha	15	25

[a] adjusted for inflation

At 30 ha total annual cost would be $3030 for the new machine and $2998 for the old one. However, at 40 ha the respective total annual costs become $3180 and $3248 so the new machine becomes cheaper. The actual breakeven usage is readily found by dividing the increased common cost of the new machine by the saving in variable cost per hectare:

$$\text{Breakeven hectares} = (2580 - 2248)/(25 - 15)$$
$$= 33.2 \text{ ha}$$

Now, let us say this farmer's cotton crop is 50 ha. At first sight, it seems that the new machine is justified. However, for three years an extra $4300 is invested in the new machine. Total annual cost, *less interest*, is $3070 and $3350 for the new and used machines respectively. If this $4300 would return more than 6.5 per cent (3350 − 3070/4300) in real terms invested elsewhere, then the used machine is a better buy.

With growing mechanisation costs and more sophisticated machines, too large for one farmer, a machinery syndicate is an obviously attractive answer to a costly problem. By forming syndicates, farmers cannot entirely avoid investment in machinery as an initial capital outlay is needed to start the syndicate but, because the equipment is shared by several farmers, the total investment is also shared.

Plant hire was formerly associated mainly with the building and allied industries but it is now increasing in farming. There are two main forms of hire — contract hire and leasing. With contract hire, unlike leasing, the hire company is often responsible for all repairs and usually supplies a replacement if there is a serious breakdown. At the end of the hire period, the machine is returned to the hire firm. This scheme has the advantage that all payments are tax-deductible and that the farmer gains the benefit of reliable maintenance. It is especially worthwhile if a high-powered tractor can be fully used for cultivations and also for work like forage harvesting.

The main advantages of leasing lie in the choices available at the end of the leasing period. These may include continuing to lease the machine at a nominal charge or leasing a new one with the credit of a cash sum closely related to the value of the old one. After a three-year lease the remaining value may be entirely the hirer's. Such an option usually represents a substantial sum and appeals to farmers who care for equipment well.

Dealers who run leasing schemes tend to favour sale of the machine at the end of the lease and often offer an attractive bargain to obtain new business. Having recovered both capital and interest charges, the dealer is pleased to offer a farmer a good trade-in price if this will persuade him to lease a new machine. The farmer may then be happy to reduce the new leasing charge by deducting the trade-in offer from the cost of the new machine. All leasing charges are normally tax-deductible business expenses. Leasing is therefore an attractive way to obtain use of new equipment and it has quickly become popular with many farmers.

If you choose to lease a machine, you are usually free to negotiate the best terms with your local dealer; a leasing company usually then buys the machine and leases it to you.

Whether you choose contract hire or leasing, the company will probably want a banker's reference and, although it has the equity of the equipment to fall back on, it will usually inquire closely into your farm business if you want to lease fixed equipment like a grain drier. Leasing companies are careful to whom they lease and in the same way you should be careful who you lease from.

Machinery leasing has several advantages and is well worth considering as a way of providing necessary equipment on farms. First, and perhaps most important, is the fact that the needed assets are usually obtained fairly quickly and easily. Secondly, existing farm enterprises suffer less from capital shortage. This might heavily favour leasing on farms where growth of enterprises is already under way or planned. In fact, if released finance can be used, even indirectly, to expand productive enterprises, the return obtained is usually worthwhile.

A further point which may influence some is that a contract is entered into and this provides security because it specifies the amount and timing of the repayments. There is thus no danger that sudden recall of the finance could occur. In times of inflation, there is much in favour of the benefits of liquidity it can provide.

In deciding whether to lease plant or buy it, managers will consider both financial and non-financial factors.

From a financial viewpoint the manager wishes to enjoy the use of the plant or equipment as cheaply as possible, subject to availability of the necessary cash. He requires, therefore, a method that will enable him to compare directly the cost — and cash requirements — of the alternatives.

If we lease the equipment, we shall pay a stream of periodical payments for as long as we need the use of it. If we buy the equipment we shall pay a substantial initial price and then little or nothing else until we dispose of the equipment.

A stream of payments that are spread out in time cannot simply be added up but must first be discounted back to their value at the present time (see Chapter 11).

The decision to buy rather than to lease is taken if the payments 'saved' by buying represent a sufficiently large rate of return on the sum 'invested' in buying the equipment (when all payments have been suitably discounted).

Besides the factors of relative cost and the availability of cash, other considerations may sway the manager's mind on whether to buy or lease equipment, e.g.:

- Suitable equipment may not be available to lease.
- Suppliers of leased equipment can often provide replacement units for those suffering breakdown, accidental damage, etc.
- What is important is to secure the use of the equipment; its ownership is unimportant. However, it cannot be denied that some managers have a strong emotional preference for *owned* assets;

If the proportion of leased equipment is allowed to rise *too* high, it can affect the balance sheet unfavourably.

Farm contractors have come to stay. Each year new contractors open businesses in new areas. Use of contract services has the advantage of not tying up capital. Heavy peak demands on labour can be avoided and the contractor's specialised labour often does a better job. Specialist work is carried out by trained staff using complex equipment that most farmers could not afford to own as they have insufficient work for it to be economic.

Another advantage of using contractors is that costs can be forecast fairly accurately, because the contractor's charges are known. Regular contracting not only enables farmers to use their labour more effectively but, if planned ahead, can save machinery costs. Also, capital that would otherwise be tied up in physical assets is freed to finance the day-to-day farm expenses that offer higher returns. Some farming systems have marked seasonal changes in activity and it may pay better to hire a contractor during busy periods than to keep a large labour and machinery force all the year. By hiring a contractor during emergencies, like adverse weather delaying work, farmers avoid the need to buy extra machines or hire more men to insure against uncertain weather.

A contractor doing general, unspecialised farm work to help farmers cope with peaks and emergencies must be over-mechanised and over-staffed himself to cope with such peaks. The farm problem is transferred to the contractor.

Few commercial farmers today do not use contractors for one job or another. For some jobs, aerial spraying, ground application of toxic sprays, lime spreading, land clearing and drainage, this is often the most satisfactory and cheapest way of getting the work done. Combine harvesting and pick-up baling is also often contracted out.

New settler farmers find that by hiring a contractor they save the initial large investment in equipment or hire purchase; this enables them to get established before buying their own equipment. Contractors are often used to clear and rip virgin land and remove roots before it is ploughed, to remove termite mounds, build roads through wet areas, clean out and build dams, dig silage pits and many other jobs. A reason often given for using contractors for ploughing is, 'Why buy two big 45 kW tractors with costly tyres to be used for only two to three months' ploughing each year when two or even one 30 kW tractor can handle all the other work?' Many farmers use small wheeled-tractors and give their heavy work to a contractor. For instance, many cattle and maize farmers in the sub-arid tropics feed their maize residues to cattle in the land. They thus have to plough late when the ground is too hard for their own wheeled tractors to do a good job. A contractor may plough, to an average depth of 35 cm, and harrow 240 ha with two crawler tractors in 20 days. The farmer would have been running back and forth looking after four tractors and ploughs for two-and-a-half months and could not have averaged 35 cm depth.

Any farmer knows the importance of being able to rely on getting work done at the right time. Attainment of timeliness depends on close and continuing co-operation between farmer and contractor. Calling in a contractor as a last resort in emergencies is rarely satisfactory. Contractors can give good, economic service only if work is planned well ahead so that their specialised equipment is both reasonably fully used and ready on time for all customers. This requires advance booking of the service within specified periods followed by close co-operation between farmer and contractor during the working season.

Contract work is often unsatisfactory for jobs

that are critical from a time standpoint — except to supplement a farmer's own equipment. Furthermore, many farmers often want the contractor at the same time in emergencies.

If a group of farmers do not already have the skill to manage mechanisation, it can often be introduced to the group by hiring a contractor or by finding another member who is willing to work for the group on a contract basis. There are therefore situations where a contractor can provide better service than that obtained by sharing.

The difference between the cost of leasing and buying machinery on bank overdraft is often small but borrowing from the bank is usually cheapest. Leasing usually carries an interest rate about two-and-a-half percentage points higher than that on a bank overdraft. For small sums, this difference may be negligible, especially when taxation allowances are considered. If large sums are needed, it can be vital, however.

The more you consider the choice the more you are likely to decide that there is no single, universal answer as the decision usually depends on personal circumstances and taxation. However, to help you gain *some* idea of the differences between the possible ways of obtaining equipment, I will give five examples of how an $8000 machine could be obtained. The table refers to an average farmer — who does not really exist! He is assumed to pay tax at 30 per cent and to be able to benefit fully from a 100 per cent first-year allowance. In each case the machine is kept for four years and sold for $2000.

At first glance, financing the new machine by personal capital (table 4.35A) seems to be the cheapest way. However, is it really? Should there

Table 4.35   Alternative Methods of Buying a Machine

A. *Personal capital*

Year	Payments Capital cost	Cash savings Tax relief 100% first-year allowance	Net cash outlay
	$	$	$
1	8000	—	8000
2	—	2667	(2667)
3	—	—	—
4	—	—	—
5	—	—	—
Totals	8000	2667	5333
TOTAL, less trade-in value: $2000			$3333

Assumption: Trade-in value net of tax.

B. *Personal capital*

Year	Payments Capital cost	Opportunity cost	Total cost	Cash savings Tax relief	Net cash outlay
	$	$	$	$	$
1	8000	831	8831	—	8831
2	—	917	817	2667	(1750)
3	—	1012	1012	—	1012
4	—	1117	1117	—	1117
5	—	—	—	2667	—
Totals	8000	3877	11877	2667	9210
TOTAL, less trade-in value: $2000					$7210

C. *Leasing. Quarterly payments*

Year	Payments Leasing charge	Cash savings. Tax relief on leasing charge	Net cash outlay
	$	$	$
1	3200	—	3200
2	2560	960	1600
3	2560	768	1762
4	2560	768	1762
5	600[a]	768	(168)
Totals			$8156
TOTAL, less rental rebate on terminal value: $2000			$6156

Assumptions: Leasing £80/quarter/$1000. Three months rent in advance in first year.

[a] Tax on rental rebate.

D. *Bank loan account. Quarterly payments*

Year	Payments. Principal	Interest	Total	Cash savings. Tax relief on interest and depreciation[a]	Net cash outlay
	$	$	$	$	$
1	2000	1160	3160	—	3160
2	2000	840	2840	828	2012
3	2000	520	2520	636	1884
4	2000	200	2200	463	1737
5	—	—	—	306	(306)
Totals	8000	2720	10720	3483	8478
TOTAL, less trade-in value: $2000					$6487

Assumptions: Interest paid on the reducing balance at 16%. Trade-in value net of tax.

[a] Depreciation calculated by reducing balance at 20%/year.

not be a charge for the opportunity cost of using this cash if it could earn, say, 10 per cent net when invested at compound interest on fixed deposit off the farm? As soon as an opportunity cost for using owned capital is added in, even at only

E. *Hire purchase. Quarterly payments*

Year	Payments Principal	Interest	Total	Cash savings Tax relief	Net cash outlay[a]
	$	$	$	$	$
1	2000	1200	3200	—	3200
2	2000	1200	3200	3027	173
3	2000	1200	3200	360	2840
4	2000	1200	3200	360	2840
5	—	—	—	360	(360)
Totals	8000	4800	12800	4107	8693

TOTAL, less trade-in value: $2000 $6693

Assumptions: Interest charged at a 'flat' rate of 15% on total capital. Trade-in value net of tax.

[a] *Includes 100% first year allowance.*

10 per cent, it becomes the dearest way to buy the new machine (see table 4.35B) and leasing comes out on top followed by the bank loan and hire purchase. However, this is often untrue so you should consider your own situation before committing yourself to any particular method of finance. This applies especially to tax consider-ations. Apart from contract hire, ownership of the goods finally passes to the buyer and he can obtain any taxation allowances available just as if the goods were bought for cash. With a leasing contract, however, the finance house buys the goods and keeps ownership and, therefore, obtains any taxation allowances payable and these are reflected in the leasing rentals payable by the lessee (customer).

Breakeven analysis can help in deciding whether to buy a machine for a certain job or to use a contractor. Consider a groundnut stripper costing $1600, expected to last six years and to have a terminal value of $160. Interest and other common costs are estimated at $100 a year and running costs at $3 a tonne of pods. A contractor is now hired to do this job at $7.50/tonne. The breakeven throughput needed to justify buying the machine is easily calculated by dividing the common cost incurred each year by the saving on running cost per tonne.

	$
Capital cost	1600
Terminal value	160
Balance to depreciate	1440
Annual depreciation	240
Other annual common costs	100
Total annual common costs	340

Breakeven throughput = $340/(7.50 - 3)$ tonnes = 75.6 tonnes/year. This shows that at any annual throughput over about 76 tonnes, the total annual cost of owning the machine would be less than the contract charge.

In the mid-1950s a British farmer, aware of the increasing burden placed on farmers by the capital needed to mechanise small and medium-sized farms faced with a shortage of labour, combined with some friends and a commercial bank to form the first machinery syndicate. The basic aim of a machinery syndicate is to improve the economic efficiency of farm mechanisation by enabling the joint ownership and use of machines and by providing credit to finance their purchase.

Two or more farmers may jointly buy a costly machine if the volume of work on any one farm is too low to justify sole ownership. Essentially, such a syndicate is a partnership based on clearly stated rules.

The advantages in joint ownership of shared equipment include:

- less capital tied up in machinery on each farm
- use of the latest equipment
- efficient maintenance is easily arranged
- providing sufficient work to use large machines close to their capacity
- enabling a labour gang to be formed for tasks where one is needed
- reduced running costs

Joint ownership is a way of reducing capital investment and can improve the efficiency of capital use. Less capital is needed by each user as common costs per hectare or unit of output are reduced by spreading depreciation and interest over a larger throughput. Machinery syndicates offer a source of credit, not open to non-members, on terms that often compare well with those available to lone borrowers. Approved syndicates can often borrow money on favourable terms partly because the syndicate members agree to 'joint and several' legal liability and because of the precise arrangements made for repayment within a short time — such as four years.

At the end of, say, four years' service, the equipment which may have been the best avail-able when bought will probably have become outdated by a better version. The market value of the old machine owned by the syndicate is likely to be as much as it would have been had it done only about a fifth as much work for a single farmer. The latter could hardly justify replacing a machine that had done so little work; but the syndicate-owned machine would now be paid for and there will be no reluctance to trading it in and buying a new model that is not merely a replacement but is probably better. By using newer, better-developed machines, group members benefit from higher output owing to better performances and timeliness.

Experience has shown that maintenance of equipment in a well-run syndicate is usually better than under single ownership. Members usually ensure that the machine is in good working order when it leaves them. Moreover, the arrangements usually made with a dealer for regular inspection and an annual overhaul help to ensure optimum servicing.

Machines in groups are used nearer to their full capacity than most of those owned individually. However, one of the most important rules for successful running of a syndicate is not to overestimate machine capacity. Whereas it is clearly desirable to obtain reasonably full and economic use, nothing is worse than planning an over-ambitious programme that is not finished in good time. For example, in planning syndicate use of a cotton picker, it is best to plan for somewhat less than full use to ensure that all members' known needs can be met without high field losses during adverse conditions. The machine may, of course, be capable of more work in good conditions if extra contract work is found for it.

Machine sharing encourages labour sharing on gang work and other forms of mutual help in, for example, silage making. It is likely to lead to shared ownership of a range of complementary equipment such as, for example, labour-saving drying and storage facilities with a combine harvester.

Besides saving on capital investment, machine sharing also tends to reduce running costs by reducing the unit cost of work. Given sufficient goodwill and compromise this can halve the cost of some tasks. There are few other chances in farming today for cost reductions approaching this size.

Little systematic information is available on the importance of informal machinery sharing outside the syndicate movement. Sometimes, this is a simple sharing of cost and use; sometimes co-operation between relatives in buying machines for previously agreed periods of use on different holdings run by members of the same family; sometimes 'cross purchase' in which two farmers buy a different machine each and exchange them when needed; sometimes purchase of a machine by a farmer who has arranged to hire it to others for a set price.

Contract work done by a farmer might help to justify machinery purchase, especially if the extra work spreads the overhead costs to a worthwhile extent.

The reasons why many farmers prefer such informal arrangements to formal syndicates vary greatly. So do the reasons why others make no use of machinery sharing. However, surveys have found certain reasons why many doubt that the advantages of syndicates exceed the disadvantages.

Despite much publicity, machinery sharing or, more formally, syndicate ownership is less popular than might be expected. Small farmers can often obtain the same advantages by buying either smaller machines, if available, or second-hand ones; the machine is then available when needed and the full value of timeliness is gained.

Reasons farmers give for opposing sharing include:

- poor maintenance compared with privately-owned machines
- lack of timeliness of work if all members want a machine at the same time
- risk of disagreement
- loss of freedom
- problems in agreeing how to share costs
- finding willing and acceptable farmers with whom a good personal relationship can develop
- the distance between farms in many tropical countries

In practice, few group members have trouble with machine maintenance. Machines used by a group may seem less well maintained but this is usually wholly owing to their greater use not to anything else.

One of the commonest criticisms of equipment sharing is that all farmers often need the equipment at the same time and that, therefore, it is impossible for all to be satisfied. The problem with small farm groups, however, is similar to using the same machine on a large farm. Machinery used seasonally does, at first, seem unsuited for sharing but hay-making equipment, nevertheless, is often shared. Problems of organisation and time of harvest exist but these can be settled by group discussion. Provided the total work needed from a machine is within its capacity, there should be no trouble. It is important to agree a rota for seasonal machines and to change the order each year so that each member gets a chance of first use. Naturally there are a few times when machinery is not free when wanted and output suffers. Such problems also arise with farmers using their own equipment.

Some farmers tend to avoid sharing because they foresee agruments with unco-operative neighbours.

Sharing naturally causes some loss of freedom. Some farmers, who have hired or borrowed, say that by joining a syndicate they have gained more control over their operations and are more likely to meet their targets. Larger farmers seem to worry more about loss of freedom than smaller ones because they prefer to keep more work under their sole control. A striking feature of group operation of machinery is, in fact, the absence of trouble associated with sharing seasonal equipment, like combine harvesters, grain driers and coffee pulperies.

Experience has shown that the apparent disadvantages of the system are less than they seem at first and that they can be minimised by care and commonsense.

For successful machinery sharing, certain rather obvious conditions must exist. For example, it is usually essential for sharers to farm close to each other. Also, to avoid trouble, total planned use should be well within the capacity of the machine in a hard season and the method of allocation and responsibility for costs, together with the order of use, should be carefully detailed and understood.

The following points should be considered by all group members before agreeing to any machinery sharing arrangement:

The share of the initial cost that each member shall contribute

This can be assessed by area, volume of work or hourly use. A rule that each member of the group shall be liable for group debts is needed to obtain a low interest rate on borrowed funds.

The way in which members shall contribute to running costs and other expenses

It is desirable for the group to appoint one member or an accountant as its secretary; he should keep the account books, machine maintenance records and, if necessary, a meetings minute book. The group members should ensure, if they are borrowing money, that a bank account is opened in the group's name.

The right of members to use the machine

If practicable, this should include a rota stating the order in which members shall use the machine and the time for which each may keep it. Agreement should be made in advance on priority if two or more members need a machine at the same time. This can be decided only after discussion between members considering each's volume of work, soil type, crop cultivars and all the other factors that affect group operation. An annual work programme should be set up and the start of operation rotated amongst members so that each gets a fair share. The time that each member may keep the machine on any occasion should be fixed with regard to factors such as weather conditions.

Who shall operate the machine

With many machines performance depends greatly on the operator as it is only by good operation that full use can be made of them. The best performance is usually obtained where one man operates the machine on all farms but, in any case, a decision in principle on who operates the machine is needed at the start.

Who provides fuel and other operating inputs

This must be decided from the start. One simple solution for supply of fuel is that the machine is filled when it leaves any member's farm. Alternatively, a single source of fuel can be used, records kept on fuel consumption and charges made at the end of the season.

Who provides any extras

These can be use of a tractor or trailer or other equipment or labour used together with the main machine that is shared. It is important that adjustments are made for use of extras and these should be decided upon when group decisions are first made.

Who shall be mainly responsible for maintaining the machines

One member of the group should be responsible for machine maintenance. He may not do the work himself but it should be his responsibility to ensure it is done regularly and to report to the group regularly. It is also important to decide who stores the machine when not in use and on what terms.

Who inspects the machines

It is important for group operation to keep machines in good order and it therefore helps the group to have an outside opinion on the state of the machine. This could be given by a dealer. The group may arrange for the machine to be inspected once or twice a year.

Any maintenance bonuses to be paid to machine operators

Any maintenance bonus should be fixed at the start and it should depend on a favourable inspection report. This ensures that all group members and operators know the position.

Machine insurance

All machines should be adequately insured for working on any group member's farm.

Personal insurance

Each member should insure his workers against liability at common law while working away from his own farm. This is necessary so that all workers are fully insured when using machines on any farm within the group.

Formal machinery syndicates usually commit all participants financially. This can lead to problems. A formal agreement ensures financial and legal stability and gives protection on retirement, death, disputes or voluntary departures of members. There should be a prearranged method for orderly winding-up of the syndicate if necessary.

Most machinery syndicates have between two and four members. A balance should be found between cost and service for each member, which both tend to fall as numbers rise. The larger groups are mainly concerned with fixed plant, such as storage and processing equipment.

Few farmers working in groups have had trouble finding partners but members should be chosen carefully to ensure mutual trust between them. In such a close business relationship it is essential that members can work well together. For a successful group at least one member should have specialist knowledge of machinery.

Banks will often give syndicates four- to seven-year loans for purchase if members can together raise 20 per cent of the cost. Repayment is usually made twice a year and interest is low because banks usually demand a full guarantee from each syndicate member.

Sharing of both initial and running costs can be done in several ways. Each member's contribution to costs should preferably vary with his use of the machine. The trend is to divide equally the costs of long-life, low maintenance machinery, like monkey-winches and post-hole diggers, but to split cost according to use for more complex, high maintenance machinery like combine harvesters.

Machine sharing is most useful for equipment not needed at the same time by all members or that is used in jobs where timeliness is unimportant. Because sharing can cause farmers to lose yield or incur extra expenses and many problems in a bad season, there is a better case for sharing the use of fixed or portable barn equipment or machines used for jobs where timeliness is less important. Typical equipment of this type includes storage plants, milling and mixing equipment, concrete mixers, power saws, hydraulic scoops and rippers. Nevertheless, several British machinery syndicates have successfully worked for years with such highly seasonal machinery as balers, combines and forage harvesters.

GLOSSARY

Asset　anything the business owns that has a money value.

Base rate　the rate on which other interest rates, such as overdrafts, bank loans and bank deposits, are based.

Book value　the value of assets on a balance sheet. This is often purchase price and may be less than market value.

Bridging loan　borrowing to buy before receiving payment for a sale.

Budget　an estimate of future income and expenditure as opposed to an account of past financial transactions.

Creditor　one to whom a debt is owed.

Deposit account　a bank account in which deposits earn interest and withdrawals from which need notice.

Discount　normally a deduction from face value. A discount for cash is a percentage deductible from an invoice to encourage a debtor to pay quickly.

Discounted cash-flow　a way of comparing alternative investments. Comparison is made of the present values of the flows of cash expected from each investment during its life. (See Chapter 11)

Gilt-edged   stock exchange securities or investments that carry little risk about payment of fixed interest or repayment for the stock (unless undated) when due. Usually applied to bonds guaranteed by government.

Inflation   a process of steadily rising prices causing lower buying power of a given nominal sum of money.

Lease   an agreement between the owner of property (lessor), to grant use of it, and another party (lessee) for a stated time for a single premium or a stated rent. A recent development has been rapid growth of leasing arrangements in which title of the property passes to the lessee at the end of the lease for a nominal charge. In effect this is a form of hire purchase without a deposit and is subject to differences in tax treatment that may be worthwhile.

Liabilities   debts of all kinds.

Lien   The legal right to possess the property of another until certain lawful demands, such as payment of debt, are met. Thus, a banker has a lien on collateral security until a loan is repaid.

Liquidity   a business is liquid if a high proportion of its assets are held in the form of money or other liquid assets. The extent of the firm's liquidity shows its ability to pay its expenses soon enough to satisfy creditors and avoid bankruptcy.

Mortgage   a legal document transferring conditional ownership of assets as security for a loan until the loan is repaid.

Operational research   an inter-disciplinary field of activity that tries to find optimum mathematical solutions to management problems.

Principal   money that has been invested, excluding any interest it has earned.

Quotation   a statement of a price. A buyer may ask his supplier for a quotation for a certain order. When such a quotation is given it applies only to that particular transaction.

Reserves   amounts set aside from profits to meet contingencies (unexpected future expenditure) or for future investment.

Solvency   having enough money to meet all financial obligations.

Surcharge   an extra charge.

Surrender value   the case an assurance company will pay the holder of an endowment life assurance policy if he cashes it in before it matures.

Trade-in   giving a dealer an article in part-payment for a newer, more costly, similar article.

RECOMMENDED READING
Books

Anderson, R.G., *Corporate Planning and Control* (Macdonald & Evans, 1975).

Barnard, C.S. and Nix, J.S., *Farm Planning and Control*, 2nd ed. (Cambridge University Press, 1979).

Brown, J.L. and Howard, L.R., *Principles and Practice of Management Accounting*, 3rd ed. (Macdonald & Evans, 1975).

Butterworth, B. and Nix, J.S., *Farm Mechanisation for Profit* (Grandada, 1982).

Castle, E.N., Becker, M.H. and Smith, F.J., *Farm Business Management*, 2nd ed. (Macmillan, 1972).

Clery, P., *Farming Finance* (Farming Press, 1975).

Culpin, C., *Profitable Farm Mechanization*, 3rd ed. (Crosby Lockwood Staples, 1975).

Dyer, L.S., *A Practical Approach to Bank Lending* (Institute of Bankers, 1977).

Giles, A.K. and Stansfield, J.M., *The Farmer as Manager* (George Allen & Unwin, 1980).

Leeds, C.A. and Stainton, R.S., *Management and Business Studies* (Macdonald & Evans, 1974).

Mallinson, D., *Understanding Current Cost Accounting* (Butterworths, 1980).

Norman, L. and Coote, R.B., *The Farm Business* (Longman, 1971).

Oxford, R., *Be Your Own Accountant* (Nelson, 1975).

Sturrock, F., *Farm Accounting and Management*, 7th ed. (Pitman, 1982).

Taylor, A.H. and Shearing, H., *Financial Cost Accounting for Management*, 6th ed. (Macdonald & Evans, 1974).

Upton, M., *Farm Management in Africa* (Oxford University Press, 1973).

Vause, R. and Woodward, N., *Finance for Managers* (Macmillan, 1981).

Warren, M.F., *Financial Management for Farmers* (Hutchinson, 1982).

Watts, B.K.R., *Business and Financial Management* (Macdonald and Evans, 1981).

Wood, E.G., *Bigger Profits for the Smaller Firm* (Business Books, 1987).

Periodicals

Anon, 'Agriculture and its Finances', *Midland Bank Review* (Autumn/Winter, 1982).

Bright, G., 'The Current Economics of Machinery Leasing', *Fm Management* (1986) 6, 3.

Brooke, M.D., 'Machinery replacement policy', *Fm Management* (1980), 4, 3.

Camm, B.M., 'Choice of Borrowing', *Fm Management* (1985) 5, 10.

Crabtree, J.R., 'Machinery leasing', *Fm Management* (1980), 4, 3.

——, 'Machinery Finance Options', *Fm Management* (1984) 5, 7.

Craig, M., 'Why Leasing Boomed', *Management Today* (January 1979).

Evans, M., 'Methods of Machine Sharing, *Fm Management* (1986) 6, 4.

—— 'The Benefits and Problems of Machinery Sharing', *Fm Management* (1987) 6, 5.

Giles, A.K., 'The Use and Misuse of Capital: A checklist', *Fm Management* (1973) 2, 5.

Gladwell, G., 'Labour and Machinery Analysis and Planning'. *Fm Management* (1985) 5, 9.

Hedley Lewis, V.R., 'Leasing: To be or not to be', *National Westminster Bank Agricultural Digest* (April, 1984).

Jones, J.V.H., 'Equity Share Shortgages for Farmers — Do they have a future', *Fm Management* (1983) 5, 3.

—— 'Essential Information for Bank Borrowing', Fm Management (1985) 5, 12.

Hill, G.P., 'Current Cost Accounting', *Fm Management* (1978), 3, 7.

Howard, C.R., 'The Draft Ox. Management and Uses', *Zimb. Rhod. Agric. J.* (1980), 77, 1.

Lewis, R.W. and Jones, W.D., 'Current Cost Accounting and Businesses', *Farming J. Agric. Econ.* (1980), 31, 1.

Mitchell, W., 'Choosing between Alternative Fund Commitments', *Fm Management* (1979), 2, 11.

Preston, E., 'Under-Insurance: Caution is the Best Policy', *Rhod Fmr* (3 January, 1975).

Schlie, T.W., 'Appropriate Technology: Some Concepts, Some Ideas and Some Recent Experiences in Africa', *E. Afr. J. Rur. Dev.* (1974) 7, 142.

Sutherland, R.M., 'Costing of Capital in Partial Budgeting — Allowing for inflation and Taxation Effects', *Fm Management* (1980), 4, 2.

Uzureau, C., 'Animal Draught in West Africa', *World Crops* (May–June 1974).

Warren, M.F., 'Use of the Partial Cash Flow Budget', *Fm Management* (1982) 4, 11.

World Bank, 'Agricultural Credit', Rural, Development Series (August, 1974).

World Bank, 'Agricultural Credit', Rural Development Series (August 1974).

EXERCISE 4.1

Calculate the present value of the following annuities.

	Interest rate (%)	Annual flow starting after one year $	Number of years
(a)	9	2000	15
(b)	7	1600	10
(c)	12	3000	5

EXERCISE 4.2

Replacement of a machine is expected to cost $5500 in six years' time. What equal sum should be set aside at the end of each year and be invested at 9 per cent annual interest to provide the funds needed for replacement?

EXERCISE 4.3

Penaing Oil Producer's Co-operative Ltd is thinking of buying a new processing unit that would both reduce labour costs and raise the quality and price of the product. A & B are alternative plants and the following information is available from which the chairman asks you to decide which plant should be bought on the basis of payback.

	Plant A	Plant B
Estimated life (years)	6	8
Installed purchase price	$50000	$80000
Annual running costs	3000	4000
Extra annual cost of supervision	4000	6000
Estimated annual labour savings	25000	40000
Estimated extra annual sales (Taxation is 30% of profit)	10000	6000

EXERCISE 4.4

Mr Ndhlovu sprays his cotton with pesticides but does not use fertiliser.

He discovers that fertiliser use would probably increase his yield by 305 kg/ha at 18¢/kg.

However, the fertiliser would cost $30/ha and to get the required response he would have to add soluble boron to his pesticide spray mixture at a cost of $4.50/ha. Also, the higher yield would raise the cost of casual reaping labour by $6/ha.

Should Mr Ndhlovu fertilise his cotton?

EXERCISE 4.5

Mr Kanyemba grows 4 ha of groundnuts with an average yield of 900 kg/ha. Price for groundnuts is 20¢/kg for entire nuts but only 8.5¢/kg for damaged ones.

He usually spends $75 on casual labour to shell his nuts by hand and is considering buying for $90 a mechanical sheller that should last for five years. Mr Kanyemba could get a loan for this purchase over three years at 5 per cent interest a year.

The sheller would produce 35 per cent damaged nuts instead of none that he now gets with hand labour.

Should he buy the sheller?

EXERCISE 4.6

Chief Ode has 2.5 ha maize and 0.8 ha ground-nuts.

He is considering applying for a loan of $70 to obtain a plough ($25) and a ridger ($45). He already possesses two trained oxen and a cart.

ASSUME:

- Both the plough and the ridger will last for five years.
- They will both incur annual repair costs of $1.50.
- There will be no saving in labour costs.
- The $70 loan is for three years and is to be repaid in annual instalments at 8 per cent interest on the reducing balance.
- The ridger will save one hand-weeding of maize that is now done by casual labour at a cost of $4/ha.
- Both maize and groundnuts will be planted two weeks earlier than usual. This should raise yields by 450 kg/ha and 125 kg/ha respectively.
- Maize is worth $3.80 for a 90 kg bag and groundnuts 18¢/kg.
- Extra feed for the oxen would cost $3.
- As the oxen would be kept busy ploughing, Chief Ode would get $6.25 less each year for contract carting with them.

(a) Should Chief Ode apply for the loan?
(b) Will any extra profit cover his annual loan payments?

EXERCISE 4.7

Mr Chiphwanya is thinking of providing a 10-tonne capacity concrete maize storage bin costing $350 and expected to last for 10 years.

Budget this proposal if he sells his maize now to the local marketing board depot at $3.85/91 kg bag in bags which cost him 45¢ each, straight after harvest. If he stores maize for nine months after harvest, his less wise neighbours would collect from him, in their own containers at 7¢/kg. The bin would have to be cleaned before filling each year at a cost of $7 and fumigation of the grain would cost $5/tonne/year. Transport to the local depot costs $3 per tonne. Storage losses are expected to amount to 5 per cent.

EXERCISE 4.8

Mr Magagula's most profitable enterprise is cotton of which he grows 8 ha. It has a gross margin of $170/ha and costs $28/ha to weed with casual labour. The remaining 20 ha of his farm is planted to maize which has a gross margin of $84/ha. Shortage of casual hand labour for weeding limits the cotton enterprise to 8 ha so the proposal is to increase this area by using a herbicide to avoid the labour 'bottleneck' at weeding.

Use of a herbicide would enable cotton to be increased to 14 ha and would reduce the cost of casual weeding labour to $7/ha. The herbicide would cost $11.20/ha and a power take-off pump, tank and boom costing $630 would have to be bought. This equipment is expected to last for four years and to have running costs, including the tractor, of $42 a year.

Should Mr Magagula invest in this equipment in order to increase his cotton area?

EXERCISE 4.9

You are now growing 100 ha of cotton at a gross margin of $625/ha. You are thinking of changing to sugar cane which has a gross margin of $750/ha. The equipment used for cotton would all be used for cane, but, in addition, one more tractor, two trailers and a mobile crane would be needed.

A summary of these machinery costs is:

	Tractor	Each trailer	Crane
	$	$	$
Capital	5625	3125	16250
Depreciation (years)	5	10	5
Annual repairs	810	500	2500
Annual fuel/oil	625	–	940

Calculate the expected change in trading profit and the likely return on the extra investment.

EXERCISE 4.10

You are in charge of pest control in a 40 ha citrus orchard. You now use an old, labour-intensive high-volume sprayer and you are considering replacing it with a new mist blower. Given the following information, assess the financial effect of trading-in the sprayer for a mist blower.

	New mist blower	Existing high-volume sprayer
Capital cost	$6000	–
trade-in value	–	$275
Expected life	5 years	2 years
Trees/ha	395	395
Ha/day	4	1.25
Chemicals/tree/year	5	5
	$1.43	$1.76
Labour needed		
Drivers at $1.20/day	1	1
Assistants at 60¢/day	2	15
Tractor running costs/day	$13.20	$13.20

EXERCISE 4.11

As manager of a citrus estate with an increasingly severe labour shortage and rising wage rates you must raise the productivity of the labour force still available. One possibility is to halve the density of the trees to permit greater mechanisation of operations, especially spraying.

The present layout and density of trees in the 40 ha grove make it hard for tractors to move between the rows. You wish therefore to assess the financial effect, over five years, of removing half the trees to allow tractors to work between the trees.

The following data are available:

Tree density Unthinned — 395/ha
Thinned   — 197.5/ha

Expected yields (18 kg 'lugs'/tree/year)

Year	1	2	3	4	5
Unthinned –	12.0	11.5	11.0	10.5	10.0
Thinned –	14.0	15.0	16.0	17.0	17.0

Income

Sixty-five per cent of fruit is packed for export in 15 kg cartons at the on-farm price of $1.50/carton.

Twenty per cent of fruit is sold locally in 10 kg netting bags for 30¢ each.

Fifteen per cent of fruit is crushed for juice. Each tonne of fruit produces $20.63 worth of juice.

Variable costs
Annual spray chemicals

Unthinned   Five applications each of 230 litres/tree at 0.12¢/litre
Thinned   Five applications each of 90 litres/tree at 0.12¢/litre

Labour for spraying

Unthinned	Driver and 15 labourers output — 1.5 ha/day
Thinned	Driver and 2 labourers output — 4.0 ha/day
Wages	Driver — $2.30/day; labourers — $1.75/day

Annual fertiliser cost
This is related to expected yield as follows:
  12.5¢/tree for the first 5 lugs/tree/year
  9.4¢/tree for the next 5 lugs/tree/year
  6.2¢/tree for the next 5 lugs/tree/year
  3.1¢/tree for the next 5 lugs/tree/year

Annual irrigation cost	19¢/tree/year
Annual cultivations	$95/ha
Annual pruning	12.5¢/tree
Annual harvesting cost	4.4¢/lug
Tree removal cost	$1.90/tree

EXERCISE 4.12

Mr Yesufu has about 1100 hours of work each year for a tractor. His present tractor is getting old and starting to incur heavy repair bills. He could buy a new tractor, likely to last for four years, for $6600 and get a $1100 trade-in on the old one. Given the following cost estimates, would it be financially worthwhile trading-in the old tractor for a new one?

	Old tractor	New tractor
Annual overheads	$	$
Depreciation	880	$( )
Interest	1100	1320
Insurance	165	275
Road licence	44	44
TOTAL	2189	( )
Operating costs/hour	$	$
Fuel and lubricants	0.58	0.53
Repairs and maintenance	1.03	0.47
Labour	0.15	0.15
TOTAL	1.76	1.15

EXERCISE 4.13

Mr Kamanga has an old combine harvester that he uses on 60 ha wheat and 30 ha soya beans each year. His average costs over the last three years, compared with those of a new machine, are shown below.

	Present machine	New machine
	$	$
Annual overheads		
Depreciation	2000	4500
Interest	1200	2100
Insurance	200	400
Licence	200	200
TOTAL	3600	7200
	$	$
Operating costs/ha		
Fuel and lubricants	17.20	14.90
Repairs and main-tenance	25.10	10.20
Labour	13.20	15.20
TOTAL	55.50	40.30

Would it pay the farmer to replace his combine with a new machine now? Should he do so?

EXERCISE 4.14

An estate estimates the following costs for one lorry type (new price $8200):

Year	1	2	3	4	5	6
Running costs	1250	1425	1575	1900	2200	2750
Resale value	6600	5250	4200	3400	2875	2600

At what age should the lorry be replaced (according to the holding cost concept)?

EXERCISE 4.15

A new tractor is bought for your farm at $5600. Its running costs are expected to be as follows on the basis of past experience with this model of tractor.

(i) Fuel and routine maintenance — 90¢ an hour
(ii) Tyres — 35¢ an hour
(iii) Repairs — $110 at 1600 hours
$1900 at 3200 hours
$450 at 4800 hours
$1340 at 5600 hours
$670 at 6400 hours
$1790 at 7200 hours
$1570 at 8000 hours
$2130 at 8400 hours
$2800 at 9200 hours
$3140 at 9600 hours

At 4000 hours you are considering 'trading-in' the tractor for a new one. The trade-in price offered by the dealer is $1680. Is this acceptable?

EXERCISE 4.16

As an estate manager, you are responsible for selecting the range of tractors and implements that would enable 150 ha to be prepared each year most cheaply within 120 days. Ten working hours are available each day.

Implements	Purchase price	Depreciation	Running cost/hour	Working speed	Width	Field efficiency	Smallest tractor possible
	$	%	$	(km/h)	(m)	%	
Ripper	2750	20	0.55	5	3.0	75	A
Ripper, single tine	275	15	0.12	6	0.9	80	E
Plough (6 furrows)	3300	25	1.10	5	2.0	70	C
Plough (single furrow)	825	20	0.45	5	0.75	80	E
Discs heavy (16 × 90 cm)	3300	25	1.10	6	2.5	75	A
Discs heavy (20 × 60 cm)	3575	25	1.10	8	4.0	75	A
Discs light (double offset)	825	20	0.55	6	2.4	80	E
Ridger and fertiliser:							
Type 1	2200	28	0.55	7	3.0	60	C
Type 2	2200	28	0.38	7	1.5	60	E

The following two alternative land preparation methods are possible:

(i) Rip 40% of the area, plough, two passes with light, double-offset discs, ridge and fertilise.
(ii) Rip the whole area, disc once with heavy discs, followed by two light discings, ridging and fertilising.

Using the data shown, select the cheapest, adequate range of equipment and calculate the cost per hectare of land preparation assuming that the equipment will be used only for this job.

Tractor types	Purchase price	Annual depreciation (straight line)	Running cost/hour
	$	%	$
A	24750	15	4.40
B	11600	20	3.30
C	9900	20	2.20
D	8800	25	3.30
E	4950	25	1.40

5 Financial control

MANAGEMENT INFORMATION AND CONTROL SYSTEMS

Many say that there are few opportunities for managerial control in farming. Individual farmers have little influence on the markets in which they buy and sell. They produce in an environment that is fully exposed to the variations of the seasons, the weather and disease. Each can seriously affect the biological processes of farming and give rise to price and yield variations more than in other industries. Nevertheless, there are opportunities for farmers to make decisions controlling their production, financing, staffing and marketing while executing a budgeted plan.

Control is that element of management that provides for feedback of information as the plan is executed, so the essence of business control is adequate records. The collection of valid information is an essential part of the management task and there is a need to seek efficient ways of doing this. In business literature, control is usually described as the procedure of preparing a plan or budget, regularly comparing actual results with it, to identify variations quickly, and taking remedial action whenever these are serious. This may mean changing the plan to take full advantage of changing external conditions. Control thus consists of checking whether or not performance conforms to the plan adopted, the instructions issued and principles set. The control process by the manager thus consists of:

- explaining differences between actual and budgeted flows and balances to date
- assessing the need to depart from or return to the present plan and
- taking suitable action

Financial control consists of the procedures set up in a firm to ensure that its assets and properly protected and fully used, that its transactions are correctly recorded and that its financial position is correctly presented at suitable intervals. The essential features of good financial control are as follows:

- Management is given sufficient, prompt and accurate information, in the form of financial statements, budgets, statistics, progress reports, etc., so that progress can be checked properly and problem areas and opportunities are found quickly.
- Accurate annual accounts are presented promptly at the end of each financial year.
- The theft of assets, and other types of fraud, are effectively discouraged and, if they occur, are soon detected.

The term 'management information' should be interpreted widely to include financial accounts besides statistical data. Clearly, no business can be run successfully unless management receives promptly all the information needed for proper planning and control. It is less readily apparent, but almost as important, that no manager should be loaded with too much detail; production of irrelevant material wastes valuable time and diverts attention from more important matters. Detailed, relevant information should be available in a suitable form soon enough for deviations from plan to be found and for remedial action to be taken. Decisions and actions will be neither efficient nor helpful if they are made on faulty, insufficient or badly interpreted information.

Management information may be divided into broad groups relating to:

- future probabilities and expectations, policies and objectives; plans, projects and programmes
- present results and performances
- past results, performance reports and general background data
- past, present and future combined in the form of trend reports, graphs or charts

Information is a valuable resource and knowledge is power. To run a business effectively a manager must make decisions; otherwise the business will run the manager. Wherever possible, decisions should be based on facts, though opinions may also be relevant. However, information is seldom complete so decisions are mostly made on imperfect knowledge. Some decisions must be made with insufficient facts because time is short or because it would cost too much to get more facts. Nevertheless, the more managers' decisions are based on reliable information, rather than inspired guesses, the greater the chance of better decisions being made.

Factual information is of use only if compared with some concept of what is satisfactory or not. A common standard of comparison is with the same time period in a previous season but there are other types of comparison. In more advanced, alternative management information systems, comparison is with a pre-decided target of performance; a budget or standard.

Management information for tropical smallholder farming is needed to describe present farm systems and to estimate inputs and outputs, costs and returns; for planning changes to present systems and for measuring the effects of changes after they are made. The sources of these data are usually external, for example, local experience, published material, surveys, censuses, technical experiments, case studies and unit farms.

It is just as important for a foreman to have information showing the affairs of his section as for the board of directors to know the profit or loss of the whole business. Indeed, if every supervisor could ensure that his section ran according to plan, top management should have no need to worry about overall results.

Most commercial farmers already produce annual financial statements. If it takes over a fortnight to prepare a rough draft of these accounts after the end of a year, they should find ways of speeding the process. The old method of waiting eight to ten months after the year end for the auditors to come and do the job is no longer adequate. If monthly or quarterly accounts are prepared for management purposes, so much the better.

Accounts and statistics prepared as management information must be available as soon after the events to which they refer as possible; any delay in managers being able to correct adverse trends is a loss to the business. The frequency of comparisons and the time lag between recording and analysing information are critical in any control system. A month is probably the best frequency for recording and accounting for financial control in farming and the frequency should never be less than quarterly.

To avoid abuse and failure, control devices should be carefully planned at the start and should be regularly thoroughly reviewed. After such a review, they can be changed or removed if desirable.

The main purposes of an accounting system are to:

- Record the transactions that occur during the various phases of the business cycle — financial accounting.
- Provide information that will help to plan and implement the action needed at each stage — management accounting.

Industrial office managers have long known the high cost of running separate systems for producing financial accounts and other management information. Financial accounting concerns the way in which funds are used and the preparation of financial statements based on historical data. Whereas accounting is not the whole of control, it is still true that accounts provide most of the data, their advantage being that dissimilar facts are all shown in money terms. Management accounting is a business accounting practice for providing information to managers to help in policy-making and in the day-to-day running of the firm. Whist financial accounting stresses the analysis of transactions, debtor and creditor classification and the preparation of financial statements and other legal needs, management accounting is a system of standards, orders, records and reports concerned with finding and separating problem areas and with the diagnosing emerging trends.

The essential difference between accounts prepared for financial and general management purposes is that the former accounts for only profit realised whilst the latter measure the potential profit of the output achieved. However, this distinction between financial and management accounting is so inexact that the two sets of information overlap. The possibility of integrating accounting processes has become more practicable with the development of computer-aided farm management accountancy systems and services.

In most businesses the development and integration of management information systems evolves over time through predictable stages.

- First most stress is upon the financial accounts.
- Later various management information techniques are adopted separately but with growing sophistication.
- When the need is seen, a third stage occurs in which attempts are made to integrate the information systems with each other and with the financial accounts.
- In the last stage, the availability of computer services enables complete integration of all the systems.

Each stage may take several years and there is often some overlapping. Consequently, decades may pass before the pattern of management information is developed into one integrated whole during which time the needs to be served may have changed. The future trend, however, is clear; a movement towards the capture of basic data to be processed to produce all needed control information. In furthering this trend, every opportunity should be taken to eliminate duplication of work by integrating separate but related systems.

Budgetary control

Managers make plans for the future and watch the outcome closely to try to achieve the desired results. This is the subject of budgetary planning and control.

The complexity of modern business and speed of change imposes an increasing need to plan future policy. Having planned a year's programme and obtained the necessary finance for it, most farmers are happy to forget bookwork until the annual balance sheet is prepared by an accountant. This is a shame as having, one hopes, produced a potentially profitable plan, it is sensible to monitor progress during the year to ensure that things go as well as possible financially.

Until the 1960s, controlling the business was a neglected aspect of farm management. However, an ideal plan is no use unless it is executed successfully; but even this is not enough. Planning has little use without control and control is ineffective unless it measures performance for comparing with plans. It is essential to make regular checks to see how the plans are progressing and whether the targets are being met. We need control systems to enable us to make any necessary small changes promptly so as to achieve or exceed our budgeted aims. The purpose of budgeting control is *action*.

Much of a farmer's success as a manager depends on how effectively he applies the management cycle of planning, executing and controlling so that business operations are forecast, checked and adjusted as necessary. Budgetary control, probably the most well-known type of management control, stimulates this and therefore greatly fosters efficient management.

Executives vary in their approval and acceptance of budgeting procedure and the results produced. These are in growing use, however. Budgeting is familiar to all of us who must balance our income and expenditure. However, it is only during the last 50 years that this vital concept of cost control has been tried and developed in industry and commerce and even more recently in farming. Most large farm businesses, and many small ones, now use budgets of some kind as management tools.

A budget is a predetermined statement of management policy for a given time and it provides a standard for comparing with the results achieved. It is a planning tool while the executives of a firm are producing it. The plan must be based on observable events within a determined time framework and state, in both money and physical terms, the work to be done and resources to be used, in that time.

Budgets are based on forecasts. Whereas these are only opinions, a budget does reflect management's *intentions*. Budgets thus remain useful as plans are effected as they enable progress to be assessed. They are thus also control tools.

A budget usually covers an area of responsibility of one person, such as a cattle manager, or group of persons, so that his performance is measurable at the end of the budget period. As the only people who can fulfil budgets, which are plans, are the managers responsible for budgeted programmes, it follows that budgets should be prepared by or together with those responsible for achieving the planning results. As wide participation is advisable for developing understanding and loyal acceptable of plans, it is good for departmental and section heads to help prepare their own budgets. In this way they know their goals and this tends to increase precision and confidence. Higher management may adjust a junior's proposed budget but, nevertheless, responsibility is still delegated without loss of control. Participation in initial planning helps in controlling junior managers, as they must plan and organise reasonably well. They will try to estimate their needs precisely for, if too high, their budgets will be severely pruned; if too low this will be noticed also and top management will wonder why they are not planning more ambitious programmes.

Budgets necessitate expressing policy in money and quantitative terms. This needs careful thought and examination by managers of the costs and income within their control and should have a good effect on them. He knows that, if he exceeds his budget, he will need a good reason for requesting more funds. He also knows that he could be criticised for poor planning if the need for more funds reasonably could have been foreseen. If he does not spend the funds, he may be blamed for poor planning and for not getting the best results from the resources allocated to him.

Farmers trying to apply budgetary control should be honest with themselves. It is one thing to draw an optimistic picture for the bank to help to get a loan but quite different to forecast estimated costs as a basis for management control.

Two kinds of targets or standards are possible — ideal standards of perfection and practicable standards. Ideal standards do not allow for waste, technical or economic inefficiency or idle time and too high a standard will frighten and discourage staff. Targets should be realistic; for psychological reasons they should be attainable, though production targets should err on the high side to stimulate effort.

It is worth spending time on budgeting and study of previous years' financial statements should indicate the likely level of various costs. Items such as seed, fertiliser, feed and so on will be related directly to the cropped area or scale of activity. Other costs, like repairs, fuel and labour, may be less closely related but previous years' costs should be a useful guide.

Differences between actual and planned performance, called variances, need explaining. This demands managerial effort and may enable remedial measures to be taken or the plan to be revised. Only by budgeting a reasonable forecast of results and then critically examining and comparing actual performance with budgeted results can control be exercised. Clearly, keeping of both

detailed physical records and financial accounts is essential.

Although budgetary control, as usually practised by farm managers, concentrates on 'year end' comparisons, this has only limited use for control. For effective control, frequent comparison is essential. It is pointless to worry about lack of cash when it is too late in the season to cut costs anywhere. Again, it does not help to find at the year's end that twice the poultry feed was used than was planned. To correct any adverse variance the manager should detect it soon after it starts. This approach changes the budget from a static planning tool, which is examined and recalculated at the end of each trading years, to a dynamic control tool that is used throughout the year. Cash flows and physical data, based on the budget, are forecast on a short-term basis, the length of which depends on the features of the farm enterprises and the whole business.

Rapid-turnover, intensive enterprises, like pigs and poultry, are likely to need at least monthly reports detailing important input and output factors like feed cost and offtake. Variances between these items and their forecast values give an early warning that the enterprise is not performing as planned. The frequency of feed-back and its form should relate to the special needs of any enterprise. There is a danger with computer-based reporting systems, because masses of data can be produced fairly cheaply, that so much data may be reported that information needed for a particular decision may be obscured.

Any adverse deviations, or variances, from planned performance should be remedied immediately if they are controllable by management. However, they call into question not only actual performance but also the soundness of the budget. It may be necessary to adjust the plans themselves if changing conditions make this seem desirable. Without remedial action, budgetary control, as such, has not been applied.

Budgetary control, then, is the setting of *goals* or *targets* of performance and the *comparison* of actual results with these goals on a *timely* enough basis for *remedial* action to be taken.

The basic principles of budgetary control are:

- make plans to act as both short- and long-term performance goals
- record actual results
- compare the results achieved with those planned
- calculate the variances, and analyse the reasons for them

- act immediately, if necessary, to either correct the variances or change the plan

Below, briefly, are the main aims of budgetary control:

- To combine the ideas of all levels of management in preparing the budgets.
- To co-ordinate all business activities. (Budgetary control needs a policy and is impossible without one.)
- To centralise control.
- To decentralise responsibility on to each manager or assistant concerned.

Individual budgets for each manager and their comparison against actual performance enables managers to see their own contribution towards the results of the business.

- To help guide management decisions when unforeseeable conditions affect the budget.

Experience in farm management shows clearly that sometimes situations go wrong too fast for other techniques to correct them.

- To plan and control capital outlays in the most profitable way.
- To plan and control income and spending to achieve maximum profits.
- To ensure that sufficient working capital is available to efficiently run the business.

This is especially important for farmers using mainly borrowed funds. Care is needed that spending through the year is phased closely to the supply of credit.

- To give a target against which actual results can be compared.
- To show management what action is likely to correct an adverse situation.

Farmers can accurately find weak areas in their management and take quick, effective remedial action.

Budgetary control applied like this combines planning, budgeting and control together into a single system that enables farmers and their staff to see the value of recording; relevant record-keeping becomes accepted as an essential part of farm business management.

There is no single best way to set up and use budgetary control, even in the smallest business. Each business should be carefully examined to find what information is essential and what controls it could provide. Also, the skills, background and peculiarities of the managers should be considered because it is useless to obtain control information if management does not act on it.

The design problem is finding methods that could be used with little or no help by any intelligent person. Before control is possible, it is necessary to systematically collect, record, analyse and summarise phsical and and financial data but it is not enough merely to keep records; they must be used sensibly and regularly. The system should be designed to make comparisons and draw the maximum response to them from managers.

When a loan is applied for, it is usually necessary to submit a cash-flow budget setting out the season's expected pattern of costs and returns. This provides a basis for budgetary control. Detail that generally is left out of an annual budget (whose main purpose is to indicate profit or loss from the business) becomes essential for effective budgetary control. Having decided the likely annual level of each type of cost, it is necessary to break this down into months. For some items paid regularly, like bond interest or salaries, this is easy and the split can be fairly accurate. For other items, like repairs, fuel and chemicals, less accuracy is possible and sometimes one must simply divide the total expected cost evenly over the months in which it is expected to fall. When this breakdown is done, a forecast of costs throughout the season is produced like that shown in table 5.1.

Budgets should show physical quantities of inputs and outputs as well as expected costs and returns. No target expressed in terms of the result alone, without some ideas of how that result is likely to be produced, can provide a really satisfactory check on the success of the business.

The two columns for each month are for entering estimated and actual performance and costs. There are two ways of making the entries – either when a transaction is agreed or when payment is made or received. However, for the most effective cost control, unlike cash-flow budgeting, costs are best recorded as they are incurred rather than when they are settled. If spending on one particular item gets out of hand but is not due for payment for six months, little opportunity for control exists if this excess spending goes unnoticed because records are based on payment dates.

If there are important variances between actual and forecast items, there are three possible courses of action. Either the plan may be changed to make it more realistic, remedial action may be taken to bring performance up to the planned level or, if performance is better than planned or the variance is too small to justify management intervention, no action need be taken. The choice of which course of action to take depends on the individual case but this type of control allows farmers to detect weaknesses in their management and to base future plans on more accurate knowledge of likely performance.

Few farmers are likely to set up this sort of control if it needs too much work. As budgetary control has no value for income tax accounts, high accuracy is not needed. To save time, cents may be ignored and costs rounded to the nearest, say, $20. As adding is so unpleasant and errors so often occur, the more noughts there are to add up the better.

A fault of many budgetary control systems is that they are imposed on section managers after seemingly secret discussions between top management and the accountants. Because neither of these really know the problems at field level, the budgets are often unrealistic. This does not mean that budgetary control must be set up this way; it merely often happens. The other main weakness is that budgets concentrate on money values, on items measurable in money terms. Many aspects of business, such as innovation in products or techniques, good labour relations, knowhow, etc., are hard to value.

VARIANCES

Comparison of actual results with a previously made budget will usually show that the business has done better or worse than expected. This could mean that the budget estimates were wrong; that the organisation has been more, or less, efficient than expected or that changes have occured in outside conditions that could not reasonably have been foreseen when the budget was drafted.

Variation from a budgeted cost or income tells us nothing but only suggests that further enquiries may be worthwhile. Indeed close accordance between budget and result could hide unsatisfactory results; external circumstances could have changed in such a way that results should have been better than planned; or there may have been both bad budgeting and poor results, one balancing the other.

Variance is the difference between actual performance and planned, budgeted or standard performance — especially a difference between actual cost and budgeted or standard cost. 'Variance' in this context means 'disparity' or 'divergence'. It does not have its technical, statis-

Table 5.1   Budgeted and Actual Costs — July 1986 to June 1987

	July		August		September		October		etc.
Item:	Est.	Act.	Est.	Act.	Est.	Act.	Est.	Act.	
	$	$	$	$	$	$	$	$	
Fertiliser	3750	3910	—	—	—	—	—	—	
Seed	15	15	—	—	—	—	800	845	
Biocides	130	149	—	—	—	—	300	—	
Assistant's salary	300	300	300	300	2950	3100	300	300	
Labour wages	700	668	650	588	700	722	800	845	
Labour rations, etc.	100	122	100	80	100	133	100	96	
Interest and loan repayments	—	—	—	—	3600	3600	—	—	
Fuel	550	240	430	668	800	642	800	695	
Power and coal	80	64	80	67	80	69	80	68	
Car repairs	55	—	55	147	55	—	55	—	
Tractor repairs	270	—	270	—	270	670	270	133	
Machinery costs	160	135	160	360	160	—	160	—	
Tobacco barn repairs	—	—	550	497	—	—	—	26	
New equipment	—	267	—	—	—	—	3500	3103	
Transport	270	240	270	374	—	187	—	—	
Packing materials	—	—	—	—	—	—	—	—	
Sundries	400	470	400	374	400	428	400	486	
Private drawings	275	304	275	240	275	272	275	379	
Totals	7055	6874	3540	3695	9390	9823	7890	7576	
Difference: Overspent				$155		$433			
Underspent		$181						$314	

ical meaning. If actual results are better than planned a *favourable* variance exists; if they are below standard an *adverse* variance exists.

Drawing attention to variances is perhaps the most valuable help a budgetary control system makes to managerial 'control by exception'. It simplifies comparison of the important items to the exclusion of the trivial. The only items reported are those where actual performance is seriously out of line with budget. Variances highlight those situations where actual results are not as planned, whether better or worse. Concentrating on variances enables managers to give most time and attention to the important items. Variances help to pinpoint responsibilities so that management can see where the blame or credit for them lies. The important point is that the reason for any variance should be found, explained and, if necessary, remedied.

For each cost item, there is a total variance known as 'cost variance'. This is a combination of 'price variance' and 'volume variance'. For example, 'materials cost variance' consists of price sub-variance and use sub-variance. Similarly, for sales there is a total variance known as 'value variance'. This also consists of price variance and volume variance. Some accounting systems include labour variance, comprised of wage rate variance and labour efficiency variance. This could be useful in labour-intensive farming enterprises.

An adverse material use variance shows that extra material cost may be due not to a higher price but to inefficient use of inputs. In this case responsibility can be placed on the supervisor in charge of the particular job where the inefficiency occurred. It may be found that the variance was caused by, say, inefficient handling, waste, theft, buying poor quality materials, incorrect machine settings or poorly trained or unsupervised staff.

Let us look at an example of material use variance:

Estimate: 500 kg fertiliser/ha at $6/50 kg bag = $60/ha
Actual: 600 kg fertiliser/ha at $6/50 kg bag = $72/ha

This is an adverse variance of $12 and is illustrated in figure 5.1.

Adverse material price variance could be caused by poor bargaining power, lack of discounts, inflation, etc.

The following example is shown in figure 5.2.

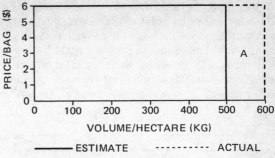

ESTIMATE ---------- ACTUAL

Area A is the adverse variance — 100 kg at $6/50 kg = $12

Figure 5.1   Material Use Variance

Estimate: 500 kg fertiliser/ha at $6/50 kg bag = $60/ha
Actual: 500 kg fertiliser/ha at $5/50 kg bag = $50/ha

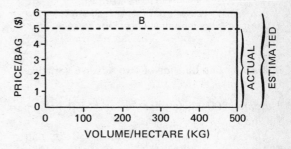

Figure 5.2   Material Price Variance

This is a favourable variance of $10, shown as area B. The size of the variance is 500 kg at $1/50 kg = $10. It is favourable as the actual result falls inside the estimated one.

Figure 5.3 shows the effect of combining the above adverse use variance and favourable price variance.

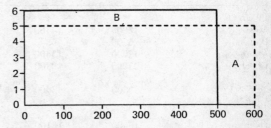

Figure 5.3   Combination of Favourable and Adverse Variances

Estimate: 500 kg fertiliser/ha at $6/50 kg bag = $60/ha
Actual: 600 kg fertiliser/ha at $5/50 kg bag = $60/ha

In this case, there is no material cost variance as the use and price sub-variances exactly balance each other so areas A and B are equal ($10).

If both material use and material price variances are adverse, there is a problem in deciding the separate effect of each sub-variance on material cost variance. This is shown in figure 5.4.

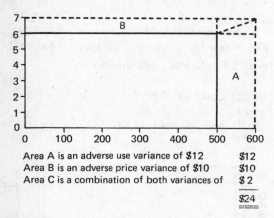

Area A is an adverse use variance of $12 $12
Area B is an adverse price variance of $10 $10
Area C is a combination of both variances of $ 2
                                                   ⎯⎯
                                                   $24

Figure 5.4   Combination of Two Adverse Variances

Estimate: 500 kg fertiliser/ha at $6/50 kg bag = $60/ha
Actual: 600 kg fertiliser/ha at $7/50 kg bag = $84/ha

Accountants have a standard convention for dealing with this common area. They always use estimated prices for measuring use or volume variance in money terms. Price variance, therefore, must be measured in terms of actual quan-tities used or produced. In the example above, therefore, they could split the effect of two sub-variances as follows:

Use variance   = 100 kg/ha at $6/50 kg = $12
Price variance = 600 kg/ha at $1/50 kg = $12
                                          ⎯⎯
                                          $24

despite the fact that a split of $13 and $11 respec-tively would be more precise.

VARIANCE ANALYSIS

Basic variances, though useful, do not meet all management needs. They measure differences without explaining their causes. Table 5.2 shows that if 'profit' is less than planned, variances show which particular items of income and cost are responsible but it does not show which aspects of management are faulty and therefore need cor-recting. The variances are measured but not explained. From table 5.2, however, it is clear that low income from the milk and cotton enter-prises and high feed costs are the main reasons for not reaching the planned net farm income.

Variance analysis is the process of explaining the cause or causes of a variance. This is necess-ary if variances are important so that managers can find out which aspects of management were most responsible for each variance. Knowing the facts, management can pinpoint responsibility for deviations and decide what can and cannot be done to remedy them. Immediate action may be

Table 5.2   Comparison of Actual and Planned Whole Farm Annual Results

	Budget	Actual results	Variances
Income	$	$	$
Milk	18000	14460	− 3540
Porkers	6500	6840	+ 340
Maize	12000	13400	+ 1400
Cotton	23000	18300	− 4700
Totals	59500	53000	− 6500
Costs	$	$	$
Feed	16500	18100	+ 1600
Fertilisers	3600	3200	− 400
Wages	12000	12200	+ 200
Power and machinery	9500	8300	− 1200
Other	3400	2900	− 500
Total	45000	44700	− 300
Net farm income	14500	8300	− 6200

possible to reduce adverse variances or to make favourable ones permanent. If, for example, machinery repairs cost twice as much as expected, investigation is needed. Are the drivers careless? Could you use a cheaper repairer? Are you trading-in soon enough?

Farmers need to judge how much detail it is worth calculating. This depends on the size of the variance and how easy it is to get information. It is not worth wasting hours analysing minor changes in income or cost items. However, variances of more than, say, 20 per cent in large items are usually worth analysing. Many variances, of course, are easily justified; for instance, fuel costs may be irregular and not monthly. Others, such as higher labour or repair costs, should be closely examined.

Budgetary control means more than assessing annual performance and planning future changes. These are types of farming where frequent checks are both possible and necessary because the total effect of small volume, cost or price changes in several items each month can be disastrous. The so-called 'factory enterprises' with a continuous, controllable flow of feed inputs, like poultry and pigs, are a good example of this. However, applied at the right time of year and at the right intervals, regular performance checks help in many other enterprises.

Variances may reflect bad budgeting, poor performance or new conditions that should be considered. There are many technical and commercial reasons why a particular forecast is not attained in practice. Explanation of differences needs further breakdown of each important cost or income variance. All adverse variances occur because of one or more of the following going wrong:

	Cost	Income
Plan:	Size of enterprise (e.g., hectares, head)	Size of enterprise
Volume:	Physical inputs/ hectare or head	Yield/hectare or head
Value:	Cost/unit of input	Price/unit of output

The budget may, however, have been too optimistic. This is common in the early years of budgeting: especially if few records are available or if a major policy change has occurred, such as introduction of a new enterprise, owner or manager. By comparing budgets with results, accuracy of estimation and the art of decision-making are likely to improve.

Some variances will be caused by factors within the manager's control; others not. For example, an adverse cost variance may be caused by a general price rise in inputs and, therefore, be unavoidable. If this is so, it is still necessary to find this cause so that it is allowed for in any remedial action taken in other areas. Naturally, a manager should be answered only for variances that are within his control.

Let us look in more detail at the annual variances shown in table 5.2 and see how these might be analysed. The main adverse variances were in milk income, cotton income and feed cost. Tables 5.3–5.5 give detailed breakdowns of these adverse variances.

Table 5.3 Analysis of Milk Income Variance

Cause of variance	Plan (no. cows)	Volume (litres/ cow/year)	Price (cents/ litre)	$
Budget	22	3900	21	18018
Actual	20	3805	19	14459
Variance	−2	−95	−2	−3559
Share of variance	−$1446[a]	−$397[b]	−$1716[c]	

[a] 2 cows producing 3805 litres at 19¢/lb
[b] 22 cows each losing 95 litres at 19¢/lb
[c] 2c/litre lost on 22 × 3900 litres.

Table 5.4 Analysis of Cotton Income Variance

Cause of variance	Plan (ha)	Volume (kg/ha)	Price (cents/kg)	$
Budget	27	2130	40	23004
Actual	27	1653	41	18299
Variance	—	−477	+1	−4705
Share of variance	—	−$5280[a]	+$575[b]	

[a] 477 kg/ha lost over 27 ha at 41 ¢/kg
[b] 1c/kg gained on 27 × 2130 kg

This detailed analysis of the important variances now serves as a guide to what remedial action is needed. The adverse variance in milk income is caused mainly by the drop in price. This could be unavoidable if the market price has fallen. However, if further investigation shows that the main cause is a fall in milk quality, possible remedial measures should be examined. Action is also indicated either to increase the average number of cows in milk throughout the

Table 5.5   Analysis of Dairy Enterprise Feed Cost Variance

Cause of variance	Plan (no. cows)	Volume (tonnes/ cow)	Cost ($/tonne)	$
Budget	22	1.5	375	12375
Actual	20	1.8	392	14112
Variance	−2	+0.5	+17	+1737
Share of variance	−$1411[a]	+$2587[b]	+$561[c]	

[a] 2 cows eating 1.8 tonnes at $392/t
[b] 22 cows eating an extra 0.3 tonne each at $392/t
[c] 22 cows eating 1.5 tonnes each at an extra $17/t

year or to revise the plan.

The adverse cotton income variance is wholly caused by low yield per hectare. The reason should be found by careful review of cultural practices. Is pest control adequate? What about time of planting? Did the crop suffer fron drought or waterlogging? If either of the latter is common in the area, what about a supplementary irrigation scheme or drainage work?

In the case of diary feed cost variance, the quantity of concentrate fed to each cow should be examined before writing angry letters about the greater cost. If it is a home-mixed ration, are you sure all ingredients are going into the cows and not the labour force? Are your measuring scoops accurate?

FINANCIAL ACCOUNTING

Commercial farming involves many transactions and keeping books of account is merely a systematic way of recording these. It need not be difficult and it is an essential part of business management. Financial information, together with physical records, for example, labour cost and man-days worked, are essential for efficient farm business management.

Farm businesses are the same as any others in that over a given period they use resources and convert them into a product or products for sale. The resources all have a cost and the product has a value. One of a firm's measures of success or otherwise is the efficiency with which it converts its resources into saleable products. To measure this efficiency accounts or financial statements must be regularly produced.

Many farm decisions are technical ones but most of them have eventual financial effects. It

follows that one of the most useful tools available to any business manager seeking optimum use of all resources is a well-presented set of accounts.

Whilst knowledge of the ways in which accounts and statements are used by managers does not need detailed knowledge of the methods used to prepare them, it is necessary to understand the main terms used.

Books of account are records of business transactions. They need not be complex and no farmer should be deterred from keeping them because it is too difficult. Information given in the books usually is summarised yearly into a set of accounts or financial statements in the form of a profit and loss account for the previous year and a balance sheet — showing the overall capital position of the business at the end of the year.

Accounting systems should be designed to provide information, at least cost, efficiently, quickly and accurately. Thue system should also offer some protection to the business by revealing any theft or fraud.

Financial or stewardship accounting provides a control by showing how managers are using the funds provided. Financial accounting also provides information on the financial position existing between the firm and outside parties, thus ensuring that, at suitable times, the correct amounts of cash are paid or received.

Company law in some countries requires regular submission of accounts to shareholders, the registrar of companies and the tax authorities.

Financial accounts are historical. They are overall summaries of what happened in the past and they are often produced too long after the events to which they relate to be of much help to managers in promptly controlling their business affairs. This does not mean that they cannot help managers. They depict the profit or loss and financial position of the business resulting from all the efforts of managers to plan and control the firm's affairs. Trends can be seen and many useful control ratios developed and used but, to control business functions effectively, managers need detailed, relevant information, prepared and presented to them quickly and in a suitable form.

There is a consistant logic about the way the accountant works, but it is obscured from many managers by a jargon that seems sometimes to stop them from getting close enough to relevant figures.

Much work has recently been aimed at defining the basic ideas on which accounting theory is built. It may seem strange that accountants did their practical job for so long without eemingly needing a clear, agreed definition of concepts. The reason is that acounting theory developed from practice; it is, in essence, an applied discipline. Accounting concepts are those that are accepted widely rather than being proven truths. The main concepts are:

   the business entity concept
   the duality concept
   the historical cost concept
   the realisation concept
   the accrual concept

Through the business entity concept, the business is regarded as quite distinct from its owners. Accounts, therefore, report events from the viewpoint of the firm itself and refer to the owner only so far as to declare his interest in it. Though owners' equity' is not a liability in the usual sense, the owner is treated as an outsider in financial statements, with the business seen as being entrusted with looking after his property.

The duality concept relates to the fact that any financial transaction automatically affects the business in two ways. Some asset value will rise; another asset value will fall or a liability item rise including the 'liability' of profits earned and owed to the owner(s).

The historical concept concerns asset valuation, which is normally based on original cost and does not reflect current values. Depreciation of fixed assets, in this concept, is seen as charging the cost against income over successive time periods based on the estimated 'life' of the asset and not as an attempt to calculate the falling market value of the asset. 'Book values' on balance sheet, therefore, are simply the balance of historical costs that must still be recovered in future accounting periods as depreciation. It has been until recently assumed that sale of fixed assets is not usually intended and, therefore, there is no reason why balance sheets should reflect market values. However, as mentioned in chapter 4, it is now accepted that the historical cost concept is inadequate in times of inflation or deflation and steps have been taken by governments and accounting bodies to introduce supplementary statements to be published with the historical cost and to show the effects of inflation on both the profit and loss account and the balance sheet if the replacement costs or 'values to the business' were substituted instead. With current assets the effect of using the historical cost concept is often modified as quick realisation is the sole reason for holding such assets.

If it seems that an asset's net realisation value is likely to be less than book value, the former will always be chosen in accordance with the conservative or prudence convention.

The realisation concept concerns the recognition of income; just when can income and the profit element in it be said to have been earned and be recorded as such? Usually such recognition is given when value is transferred and something else of value is received in return (payment or the right to receive payment). The 'something received in return', however, must satisfy the requirements that it is both reasonably certain and capable of objective monetary valuation. This usually means that recognition of realisation occurs at time of sale, when either cash or a debtor (reasonably certain to pay a fixed sum) is received.

The accrual concept rules the matching of costs against income. It demands, first, that costs so charged are based not on payment during the period but on the basis of liabilities incurred and, secondly, that incomes so matched are not based on cash received but on income earned. The latter, of course, links with the realisation concept already described.

Besides these accounting concepts, there are other basic ideas or accounting conventions governing the preparation of accounts. These are:

- materiality
- conservatism
- consistency
- objectivity

An example of applying the convention of materiality is seen, for example, in the purchase of a stapling machine or a set of spanners. Strictly according to the concepts, expense is incurred each time these are used and cost should be spread over the life of each asset. However, this treatment of such small items according to pure theory would incur recording and reporting costs unjustified by the value of the extra precision thus obtained. The convention of materiality, therefore, allows us to modify the cost and accrual concepts and to treat cost as an expense in the one initial period. Deciding what is material and what is not is a matter of reasonable judgement in each case.

When alternative valuations of items are possible the convention of conservatism advises us to adopt the figure most likely to understate, rather than overstate, income, profit and assets; with the reverse as regards expenses, losses and liabilities. The usual view that accountants are a miserable breed of long-faced pessimists probably arises from this convention or rather from the former habit of over-playing it.

These concepts and conventions leave much room for adapting their interpretation to given situations. However, the convention of consistency guards against misuse of this by insisting that the same interpretation is used in successive accounting periods or that, if a changed interpretation would improve the reporting, the fact and effect of such change should be clearly stated.

In making the choices and judgements demanded of them, accountants try to ignore subjective feelings and seek facts based on independent, objective checking; not on unconfirmed opinions.

To become logically acceptable, these accounting concepts require two basic assumptions — those of continuity and stability.

In the absence of information to the contrary, it is assumed that the firm will stay in business indefinitely. This assumption of a 'going concern' gives rise to the adoption of arbitrary time periods and the operation of the accrual concept.

Although it is known that monetary units change in real value, these units are the accountant's only conceivable measures so an assumption of stability is made. Also, it is usually thought that current purchasing power adjustments (at least within the basic accounting system) would arise more problems than they solve.

Financial accounts are used not only for the internal needs of management — to check whether or not performance has been satisfactory and resources converted into saleable products most efficiently. They are also needed to provide information to people outside the business, that is, who are not directly concerned like a manager or sole owner but have various interests in it.

External interests likely to need access to financial statements include the following:

- The shareholders or owners of a company to see how well or badly the firm is run and whether they are likely to receive a fair return on their investment.
- Possible investors to assess whether investment in the business is likely to be profitable or not.
- Creditors to estimate their chance of receiving payment.
- The government tax authorities to assess how much tax is owed by the firm to the state.

Most commercial farmers pay qualified chartered accountants to prepare their annual financial statements for the purpose of negotiating tax liability. Consequently, a small sole trader, or even a small partnership, must seek the advice of an accountant whose task is usually to organise a pile of documents into a profit and loss account and balance sheet, adding a report or opinion as to whether or not the final accounts show 'a true and fair view' of the firm's results for the period and its position at the end of that period (the period usually being a financial year). To prepare these statements, accountants need:

- details of payments and receipts during the accounting year
- list of debtors and creditors at the beginning and end of the year
- valuations of assets at the beginning and end of the year

Many farmers, conscious of their own clerical limitations, do the least possible paperwork but pay an accountant to sort things out, prepare the annual statements and liaise with the tax inspectors on their behalf. Most firms need the services of an accountant to keep them aware of the likely effects on the business of financial legislation, decisions on investment, claiming or carrying over tax allowances, machinery replacement, personal assurance and many other matters that few farmers can absorb in their limited time.

By keeping systematic books of account and records themselves, farmers reduce the accountant's work and consequently his professional charges as most accountants' fees are proportional to time spent. This means that farmers with poor records pay more than those with good ones.

Detailed recording of all transactions to give information for both management accounts, such as gross margin accounts, and financial accounts is not too troublesome and has become less so with the growth of both farm secretarial services and, in many countries, computer-based management accounting services.

In a typical management accounting service farmers must send in details of their payments and receipts on forms provided each month. They also gradually complete cropping and livestock record sheets and submit them at the year end with their valuations. In return, the service, using an electronic date-processing system, provides a cash flow showing the financial position to date every two months with a detailed set of management accounts showing the position of each enterprise; also a profit and loss account including overhead costs, at the year's end. A farmer's financial data are often shown besides data from a group of similar farms for comparison and a report is provided detailing strengths and weaknesses of the business to identify likely profitable changes.

Over the last 50 years, record-keeping systems have been developed in many countries specifically for farmers. Most farmers, however, still record only the minimum information needed for taxation purposes. Few farmers see the accounts as anything more than documents produced under duress with the aim of reducing tax liability. Accounts, however, should enable business efficiency to be assessed and give information needed for planning. This deficiency has been somewhat reduced by the availability of standard data that can be used as a planning guide. However, the value of such data is clearly limited by the lack of precision that standards provide in specifying key input: output relationships on particular farms.

Many small farmers' records consist of an ever-growing file of bank statements, invoices, credit notes, statements, etc. They battle with their paperwork as and when time permits but the tax authorities insist that recognisable statements of profit and loss are submitted each year. One of the most important reasons why farmers do not keep records that would help them is that they do not know exactly what is needed. Some excuse themselves from paying proper attention to this side of management by stressing the 'historical' aspect of recording. They argue that the information relates to what is past whereas a manager must always look ahead; not waste time on post-mortems or in dreaming of what might have been.

By spending a few minutes a day and, perhaps, three or four hours a month on accounts in the office, farmers could save themselves hundreds or thousands of dollars — far more profitable than the same time spend directly on production — yet most farmers still consider office work a waste of time. Most farmers fail to see a direct cash return from it but a well-organised filing system, even if only in cardboard boxes, where anything can be traced at once, and possibly a calculting machine, can save much valuable time. Farming is moving away fast from the time when calculations on the back of a cigarette box were adequate. It is as important to allocate time to record and account keeping as it is to spend time inspecting the cattle or arranging daily work schedules. The office should be the centre of the business — a workshop for decisions. If you have insufficient time to keep and interpret records, the business is being run inefficient because one of your resources (management) is overworked and is probably preventing efficient use of others (land, labour, capital and time).

Accurate record keeping by individual farmers is essential for effective farm business analysis and subsequent planning. Recording, of course, incurs costs besides providing returns. These costs may consist either of fees for professional help or the opportunity cost of the time of the manager, farmer or family member, which could have been spent on other farm work or on leisure. Diminishing returns applies to the money and time spent on recording just as it does on any other variable resource and a time comes when more records, detail or accuracy are not useful enough to warrant the extra cost of obtaining them.

Farmers could throw all invoices and other vouchers into a box for the accountant to sort out once a year, but if this 'method' is adopted they should not complain if his charges are high. Few financial accountants, even in farming areas, have much technical farming knowledge. Their analysis of the farm business will be in accountants' and tax men's terms: the statements finally returned to the farmer will reveal little information useful for making husbandry policy decisions.

The ideal therefore seems to be to do all routine accounting on the farm. As soon as the annual accounts are ready, copies should be sent to the accountant for his comments and advice. If the accountant is to deal with the tax authorities, all supporting documents must also be sent, so that he can check and certify the accounts. If either very high or very low profits are likely, it may be worth submitting budgeted annual accounts some six to eight weeks before the end of the financial year in case the accountant can advise speedy action to improve the position. Managers are well advised to consult an accountant whenever any large investment is considered or when large reserves have accured. There should thus be a partnership between manager/farmer, secretary and accountant.

There are several reasons why keeping a good set of financial accounts for his business amply repays a farmer's time. Farm accounts form the basis for a continuous process of analysis, planning and budgetary control *throughout the year*.

The value of keeping detailed financial accounts throughout the year are as follows.

- By classifying and recording all costs and sales not only can regular accounting reports be prepared to show a farmer how much profit he has made and how one year's profit compares with another, but they also enable him to find out, at any time, the exact financial state of his business. Knowledge of the profit or loss and the overall position enables farmers to adjust their personal drawings and expenditure. Regular accounts also draw attention to all sums owing to or by them, thus providing a spur to improve debt collection and cash flow.

- Capital invested in the farm is determined so the return on capital can be compared with that in alternative investments.
- Reliable financial accounts are needed for submitting to the tax authorities each year as tax liability is based upon profits made. Basically, tax authorities require simple financial accounts as the complex adjustments made for tax assessment are dealt with, not in the accounts, but by separate tax computations.

Simple financial accounts are, however, no substitute for detailed management accounts because, although what they reveal may be interesting, what they conceal is vital. Separate management accounts are needed, to assess the efficiency of the business, both as a whole and by enterprises, and to plan improvements.

Management accounts should help managers make the best use of all available resources. They should be specially designed to aid decision-making. Whenever possible, for management analysis, seasonal variations should be reduced by considering at least three years' accounts before praising or faulting the results.

Management accounts, when used together with comparative analysis and accurate budgeting, should enable managers to gauge whether or not performance has been adequate and whether the firm's efficiency in changing resources into saleable products could be raised. Record-keeping should be a continuous, regular task. Analysis of records and accounts should show the strengths and weaknesses both of the farming system and of individual enterprises and indicate where returns need improving or costs need reducing. Reliable records enable farmers to produce meaningful budgets for proposed changes and this favours them when competing for available credit. Budgets based on a farmer's own records should be much more accurate than those based on standard figures. Management records and accounts form the basis for planning because the first step towards improvement is to find out the weaknesses.

The reason for the great differences between management and financial accounts is that accountants work within rigid tax laws but, nevertheless, still try to minimise tax liability. Management accounts are prepared mainly to show which enterprises are profitable and which are not. Financial and management accounts are prepared for quite different purposes. Different conventions are used for calculating some costs. Some income and expenses may be included in financial accounts that are irrelevant for management accounts and therefore are omitted. Furthermore, management accounts are often prepared to a different year end from financial ones. Basically, there is no reason why the same year end should not be used, although there are situations where a particular year end may be best for taxation purposes but does not suit the overall farming situation, or, perhaps, individual enterprises. Nevertheless, the difference between the two accounts are normally explainable and both accounts are basically correct.

One of the main differences between financial and management accounts concerns valuations. Financial accountants apply the wise commercial concept, accepted for tax purposes, that account is taken only of profits that have been realised by actual sales, so they value assets based on cost, that is 'cost of production' because they do not account for the profit on unsold crops, etc. (The converse of this commercial concept is that if market value is known to be *less* than 'cost of production', then the lower figure is used as it is wise to account for known losses.) However, as management accounts are concerned with the *potential profit* of the *production achieved*, we must establish the total selling value of the year's production and to do this we must account for all unsold crops, cattle, etc., at valuation time at their *selling value* and not merely at the costs of production.

Examples of different valuations often used in financial and management accounts include farm produce used to feed the labour force or livestock, bought inputs in hand and livestock valuations.

Financial accounts usually ignore crops grown and fed within the year and may value, say, maize in hand far below its market value. Management accounts fully allow for all home-grown feeds used and value feed in hand at its market value less selling costs. Consider the case of management accounts that valued all maize in store in $120/tonne sale value. This was all right but the next year's accounts showed the same maize being fed to cattle at $90/tonne. Not surprisingly the financial profit did not look as good as the management results.

Management accounts are designed to include all the expenses for each enterprise and its total output. For crops like potatoes, this may involve either omitting or adding to expenditure and income shown in a set of financial accounts. Few crops of potatoes, onions or other vegetables are valued realistically in financial accounts.

Livestock valuation is an important area where differences arise. Standard values a head used in accounts for tax purposes, are usually far less than the market values used in management accounts. When many livestock move on or off the farm a valuation differences of, say, $150 a head can vastly alter the profit shown.

Farmers are notorious for not paying their bills promptly and many suppliers, especially contractors, are slow to submit statements. Whereas a management accountant would automatically exclude transport and contract charges relating to earlier years — and would question the farmer if these were missing for crops grown and sold during the year — a financial accountant seldom does.

Other differences between management and financial accounts occur but they are probably far less important than those mentioned.

The extent of the differences between financial and management accounts, as presented, stresses the danger of using financial accounts for uses for which they are not designed. It is easy to draw wrong management conclusions from them and to take the wrong planning decisions.

Whether accounts are produced for taxation or management purposes or both, in every case there must be reasonably accurate data and consistent analysis. Farmers should ensure that their separate financial and management accounts are compared, agreed and made compatible. Anyone suggesting that there cannot be (or even should not be) a relationship between farm management and farm financial accounts is not only doing the industry much damage by encouraging continuance of unacceptable accounting standards; he is also revealing a dangerous level of ignorance. There is a relationship, easily understood, that is vitally important to managers.

I suggest that most farmers should keep a cash book or similar purchase analysis record in a form useful to both their accountant and a management consultancy service. This would provide up-to-date cost informaiton on the farm at all times. Also, as the farmer would have much more direct

concern in record keeping, it would help him make a more meaningful interpretation of the resulting figures when assessing advice on either taxation or management matters. This information is, anyway, essential for budgetary control which has become ever more necessary. Clearly, however, the farmer/manager or one of their wives would need some book-keeping and recording skill.

The common lack of understanding of the relation between management and financial accounts is only now starting to lift and at last we are hearing about 'integrated accounts'. Possibly the farming industry is about to mature in the accounting sense and the opportunity now exists for healthy growth of integrated accounting services. There are still too many farm businesses in which the accounting system needs adapting to become a more sophisticated and effective management tool.

Financial accounts must comply with legal requirements. It should be remembered, however, that accounts were developed mainly to provide management information. Separating financial accounting methods from those producing management information merely duplicates work. There is no reason why accounting methods cannot be adapted so that both taxation and management accounts can be prepared from the same set of farm records and accounts. Indeed, it is essential that they should be to avoid unnecessary clerical work.

Although integrated accounting can certainly be applied to many farming situations, especially small to medium sole traders and partnerships, there remain many situations, such as companies, where it is convenient or necessary to extract a less detailed profit and loss account; but, even so, this can still be produced directly from integrated accounts.

Surprisingly, an integrated profit and loss account, like that shown in table 5.6, can be prepared for *any* farm using little more than the informaiton needed to prepare any reasonable financial statement. More information is needed to prepare separate enterprise gross margins. Even so, the essential basis is a correct analysis of all transactions.

RECORDING TRANSACTIONS BY CASH ANALYSIS

Farmers familiar with the work of accountants and book-keepers know that most of them use the

Table 5.6   Integrated Farm Profit and Loss Account

	$	$
Gross margins		
Cropping enterprises	25000	80000
Liverstock enterprises	3000	22000
Less forage costs		
Whole farm gross margin		102000
Common costs		
Labour (including own if applicable)	9000	
Power and machinery	13000	
Property (including notional rent)	9000	
Administration	8000	39000
		63000
Sundry farm income		
(Grazing fees, rents)		1200
Management and investment income		64200
Deduct Interest charges	4000	
Management salaries paid	8000	52200
Add Own labour		500
Net farm income		52700
Add Notional rent		6000
		58700
Valuation variation		
Opening variance } between management	{ 30000	
Closing variance } and taxation valuations	{ 33000	
Deduct increase/Add decrease		3000
Farm profit		55700
Add Other income (e.g., profit from farm store)		2200
		57900
Deduct Other expenditure (e.g., provision of		4200
farm school)		
Financial profit		53700

double-entry system. For most farmers, however, double entry would need too many accounts, most of little value for management use. If onfarm recording of payments and receipts is done, then only a simple analysis book is needed.

The main aim is to show the effect of financial transactions on the bank current account and to provide most of the basic information needed for banking, taxation and management.

Farmers who wish to efficiently record their financial transactions need some sort of 'office' — even if this is only a separate shelf in the living room — where all financial accounts are arranged in an orderly, accessible way. Neatness is sensible laziness; it is the untidy person who works himself to a halt.

The following steps are necessary before annual financial statements can be prepared at the year's end:

AT THE BEGINNING OF THE YEAR

Step 1.   List all farm assets and value them (opening valuation).

Step 2.   List all farm debts owed by you (creditors) and to you (debtors).

DURING THE YEAR

Step 3.   Keep a full record of all receipts and payments set out so that each item is recorded under a suitable heading. A cash analysis book is recommended for this.

AT THE END OF THE YEAR

Step 4. List all farm assets and value them. (Closing valuation and opening valuation for the next year.)
Step 5. List all farm debts owed by you or to you.

To produce annual financial statements a full record of the value of each item bought and sold by the business through the year (Step 3) is needed. Much thought and skill has gone into planning recording systems suitable for use on most farms. The result usually has been some types of analysed cash book as produced in many countries by co-operation between farmers' organisations, accountants and government departments. Many similar layouts are available commercially, pre-printed or with blank headings, in book or loose-leaf form.

The cash analysis system records all transactions through the bank, with analysis columns for classifying details under suitable headings. This is the master record of all receipts and payments made and is simple, easy to check and gives a clear, accurate running analysis of transactions to date during the year with little work.

The cash analysis book itself may have space for the annual statements — opening and closing valuations, creditors and debtors, profit and loss account and balance sheet. Some account books are strongly bound and big enough to last several years. However, if the accounts are audited yearly, book-keeping can be delayed until the cash book is returned by the accountant. Therefore, it is usually best to start a new book each year or to keep two books used in alternate years.

Farmers or managers wishing to keep their own accounts as recommended need the following basic books and equipment:

- cash analysis account book
- valuations book
- wages book
- petty cash book
- petty cash box or safe
- duplicate order book
- cheque-book counterfoils and bank paying-in slips
- files

The valuations book is used to list debts owing to you or by you at the start and end of each financial year. It also contains records of the book value of capital assets like buildings, fencing, conservation works, machinery, liverstock and stocks in hand.

A wages book is needed if more than one or two people are hired. Wages are usually paid in cash and the book should record information such as individual earnings, advances, loan repayments, pay-as-you-earn tax deductions, etc. As pay often requires many additions, corrections and entries, it is clearer to transfer only monthly totals to the cash analysis book As described in chapter 7, this wages book is easily combined with the keeping of physical labour-use records by enterprise or task. It is thus best to record cash paid to regular and casual labourers separately.

Recordings of financial transactions is simplified if all payments are made by cheque rather than cash and all cash payments are banked. These then become single entries in the cash analysis book. It is impossible, however, to avoid cash transactions entirely, especially for small day-to-day payments. Use of cash is best limited to wages and petty cash items, like stationery, stamps, tips, etc., that are too small to warrant a cheque. The less cash is used, the more accurate the accounts should be. Those small cash transactions that do occur should be noted in the petty cash book and analysed totals are then transferred montly to the main cash analysis book.

A petty cash book is merely a small pocket notebook in which to separately record cash payments and receipts that are both small in amount and frequent. The type of petty cash book now almost universal is ruled into analysis columns. Items are classified and posted as monthly totals to the relevant column in the main cash analysis book. This relieves the main cash book of unnecessary detail.

Unless some kind of petty cash system is adopted, bank reconciliation becomes impracticable as this depends on the assumption that all payments are made by cheque. Obviously, it is impracticable to draw cheques for small payments as this would need much unnecessary clerical work and delay in getting the cash needed for these payments. Small cash payments can add up to a large sum over the year. If they are ignored and omitted from the accounts as business expenses, taxable profit would rise.

An 'imprest' system is recommended, as this provides a guide on each reimbursement date as to whether the level of expenditure is normal. A lump sum of, say $100 is drawn by cheque. This is the 'float' and is placed in a petty cash box with the petty cash book. When all but, say, $20 is spent, a further sum for petty cash is drawn by cheque to return the float to $100. Suppose that, at the month's end, payments totalled $64.62, after checking the balance, which should be $35.38, a cheque is drawn for $64.62 and the cash placed in the box.

The imprest system of handling petty cash necessitates the following:

- The float is a fixed sum estimated to cover petty costs for a fixed time — normally a month or a week.
- Further sums of cash are added at these intervals. If replenishment is needed before the normal time has passed, abnormal expenditure is immediately highlighted and can be checked.
- The sum added must exactly equal the sum spent during the previous period.

Cheques drawn for petty cash are entered as 'petty cash' in the details column of the expenses sheet in the cash analysis book and also entered in the relevant analysis column(s). This sum is entered as a receipt, and payments as expenses, in the petty cash book.

It is often easiest to pay small business expenses with private cash and take the appropriate sums, as recorded in the petty cash book, from the petty cash box on return to the farm. It is important to use the cash box only for business uses.

Small business cash receipts are best kept aside from the petty cash float and should be paid regularly into the bank intact. This is then entered on the receipts sheet of the analysis book with 'cash' in the detail column and the sum allocated to the relevant analysis column(s).

Farmers should always order goods and services, or confirm a verbal order, by duplicate order book with carbon or pressure-sensitive paper, in writing; asking suppliers to quote the order number and invoice them separately for each order. It is always useful to keep delivery notes or your own 'goods received' book to check with the suppliers' invoices when received.

Cheque counterfoils and paying-in slips list most receipts and payments that pass through the bank account and they are the main sources from which the cash analysis book is kept. Save bank advice slips recording those items like bills paid and bank commission, that do not appear in the cheque book.

Files, preferably box files, are needed for storing each year's receipts, invoices, statements and letters. No large payment is accepted by the tax authorities, if queried, unless proof is available.

The form in which records of all financial transactions are kept should satisfy the following principles.

- Amounts must be accurate and the layout must enable easy cross-checking.
- The system must ensure nothing is omitted.
- Analysis should provide useful management information.
- Full value of goods and services must be shown, even for part exchange when the cash sum is only the price difference.
- Final totals should, if possible, be suitable for transfer to the annual statements.
- It must be possible to reconcile the cash book with bank statements.
- Provision is needed to integrate the cash book with the wages book, petty cash book and, possibly, with separate capital, loan or reserve accounts.
- If a family farm is run with a single bank account, it is essential to be able to separate business affairs from private ones.

The cash analysis book should be filled in and totals checked regularly through the year so a full record is available of all transactions through the business. This works smoothly if the farmer or manager remembers the following:

- Pay all cash and cheques received into the bank.
- Make all payments by cheques or by cash drawn from the petty cash box which is replenished by drawing cheques for cash.
- Note details of all transactions on the cheque counterfoil or paying-in slip duplicate.
- Keep all invoices and fix them to the relevant monthly statements. (This is essential, in any case, for tax reasons.)

There are two alternative layouts for a cash analysis book: either receipts on the left-hand sheets and payments on the right-hand ones; or receipts are entered in one part of the book and payments in another. The latter provides more analysis columns and is thus more suitable for many farms.

Two essentials for satisfactory use of the cash analysis system are choosing the most suitable analysis column headings for your own particular farming system and adhering rigidly to these headings when making entries. There are many ready-printed column cash analysis books on the market but few have the best column headings for farm management use.

Anyone starting a new system or improving an existing one should choose column heading carefully. The basic needs are that entries should be simple and easy to make and that the account should yield as much useful information as possible without further processing. There is much room for choice and, therefore great need for understanding both farming and business management in making the best choices.

Each column of the cash analysis book usually provides one item for the profit and loss account. Therefore, the more columns used, the more detail can be given in the final statements. However, if too many columns are used the cash book becomes large and awkward. The main need is to be able to identify the variable costs of each enterprise to calculate their gross margins.

The number of headings chosen needs careful planning. On the one hand, to avoid, as far as possible, later need for breaking down individual columns, there should be a separate heading for each main item. Obviously, however, the total number of columns available is limited, unless the analysis continues over more than two facing pages, which should be avoided. Separate columns for items, even important ones needing few entries during the year, wastes space.

The *number* of items rather than the size should determine whether or not a separate column is justified. Vegetable sales, for example, may have weekly totals and merit a column, even if only a small enterprise. Conversely, big capital items could share one column as they are easy to separate later if necessary.

You often have to group items together into one column and use a code to aid analysis at the end of the period. It usually helps to group dissimilar items unlikely to come from the same supplier. Different colours can help; alternatively, there could be a code of single symbols to differentiate items.

On arable farms, you could have a payment analysis column for each main crop, and items like fertilisers could then be directly allocated to the consuming enterprises. However, this is not usually done so sufficient details should be given to enable it to be done later for management analysis. Capital items are usually kept separate as they need to be depreciated.

Payment headings are similar for most farms but receipts vary widely with the farming system. Any unsuitable headings should be omitted, for example, livestock purchases and feed on farms with no liverstock; this enables some grouped items to be split.

Figures 5.5 and 5.6 show the general layout of a cash analysis book.

However limited the analysis may be, so many transactions must be recorded, even on a medium-sized farm, that much delay can occur in producing final accounts unless recording is kept up to date. This ensures that when a profit and loss account and balance sheet is needed, no delay is caused by having to enter many unrecorded transactions.

Before starting to keep one of these simple, single-entry cash analysis books, an important decision is necessary. Should entries be made when a purchase or sale is effected or only when payments are made or received? As keeping debtors' and creditors' ledgers is usually unwarranted and makes extra work, items are usually entered only when payment is made or income obtained — not when the transaction is agreed to or when goods are received or dispatched on credit. Entries are made from either bank statements or cheque-book counterfoils and paying-in slips related to the period. At the year's end adjustments must be made for funds owing to or by the business. Keeping all unpaid accounts in a separate file makes it easy to total creditors and debtors at any time. The accounts are refiled by date or by firm when payment is made or received.

The cash analysis book should be completed weekly or at least monthly. Timing depends on individual choice. On small farms a monthly half-day might be enough to deal with the dozen or score of invoices. Even on large farms there may be few more invoices though the amounts are larger. The following recording intervals are suggested.

Daily — Mark labour tickets and record petty cash transactions while you still remember them.

Weekly — Write up the wages book and send cheques received to the bank.

Monthly — Issue cheques for payments due, pay wages, record details of the month's transactions in the cash analysis book from the cheque counterfoils and paying-in slips and check the balance against the bank statement.

The first entry in the cash analysis book is the bank balance at the start of the year — a credit balance on current account at the head of the 'total received' column for the first month of the financial year or a bank overdraft in the 'total paid' column. This first entry is the only one that is not repeated in the analysis columns.

NOVEMBER 1986

Date	Details	cheque no.	Total paid	STAFF	REPAIRS AND SPARES	FUEL AND LUBRICANTS	SEED FERTILISER CHEMICALS	LIVESTOCK PURCHASES (STOCKFEED)	OVERHEAD	CAPITAL DEVELOP- MENTS	SUNDRIES	PRIVATE DRAWINGS
	Brought forward from Oct.		745·69									
2	Petty cash	076	44·62		13·43	7·64					23·55	
4	Advance Relay tractor tyres	077	140·64		140·64							
7	1½ tonnes urea	078	210·00				210·00					
8	20 feeders @ $90	079	1,800·00					1,800·00				
8	Hail insurance (tobacco)	080	123·70						123·70			
10	Farmers' Co-op. 1 tonne balancer meal + tobacco chemicals. (c)	—	—				74·00	100·00				
25	Wages	081	268·72	268·72								
30	Private Drawings	S.O.	250·00									250·00
	TOTALS FOR NOVEMBER (carried forward to Dec.).		3,856·37	268·72	154·07	7·64	284·00	1,900·00	123·70	—	23·55	250·00

$3,011·68

($3,011·68 + 745·69 − 174 = $3,853·37)

Figure 5.5   Payments. Year Ending 30.9.87

NOVEMBER 1986

Date	Details	Slip	Total received	Coffee	Burley Tobacco	Cassava	Maize	Cattle Finishing	Other Livestock	Capital Sales	Sundries	Private Income
5	Brought forward from Oct.	D.C.	376·84									
5	Supplementary payment	D.C.	167·93		167·93							
8	Benin Products. 1·6 tonnes	23	96·00			96·00						
10	Co-op. 20 goats @ $15 (c)	24	126·00						(G)300·00			
19	Fertilizer distributor	25	125·00							125·00		
24	19 fat cattle @ $220	26	4,180·00					4,180·00				
27	Cash. 15 bags maize;		60·00				60·00		(F)12·50			
30	25											
30	Rent for grazing	28	50·00								50·00	
30	Share dividend paid in	28	162·31									162·31
30	Monthly egg revenue	D.C.	78·16						(F)78·16			
	TOTALS FOR NOVEMBER		5,434·74	—	167·93	96·00	60·00	4,180·00	390·66	125·00	50·00	162·31

$5,231·90

($5,231·90 + 376·84 − 300 + 126 = $5434·74)

Cash at bank should be $5,434·74 − $3,853·37 = $1,581·37. Checked 3/12 and OK.

Figure 5.6 Receipts. Year Ending 30.9.87

Table 5.7   Common Payments and Receipts Headings in Cash Analysis Books

Common payments column headings	*Possible grouping codes*
Staff	Regular labour payments and rations (R)
	Casual labour payments and rations (C)
	Managerial salaries and bonuses (M)
Repairs and spares	Tractors (T)
	Lorries (L)
	Other vehicles (V)
	Buildings (B)
	Fencing (F)
Power	Fuel and lubricants (F)
	Electricity (E)
	Coal (C)
Seed, fertiliser and biocides and water	Seed (S)
	Fertiliser (F)
	Biocides (B)
	Water (W)
Marketing	Transport and railages (T)
	Packing materials (P)
Liverstock purchases	Beef (B)
	Diary (D)
	Pigs (P)
	Poultry (F)
Stockfeeds	Beef (B)
	Dairy (D)
	Pigs (P)
	Poultry (F)
Capital Development	Land purchase, buildings, fencing (P)
	Machinery (M)
Loan repayments and Interest	Mortgage (M)
	Interest (I)
Overheads	Bank charges (B)
	Rent, licences and road rates (R)
	Accounting fees and insurances (A)
	Office expenses, e.g., postage (O)
Sundries	
Private drawings	

Common receipts column headings	*Possible grouping codes*
Grain crops	Maize (M)
	Small grains (S)
	Rice (R)
Tobacco	Flue-cured (F)
	Air-cured (A)
	Sun-cured (S)
Other crops	Groundnuts (G)
	Cotton (C)
	Grams (G)
	Vegetables (V)
Cattle	Beef (B)
	Dairy products (D)
Other livestock	Poultry (F)
	Eggs (E)
	Pigs (P)
	Goats (G)
Capital sales	
Sundries	Contract work (C)
	Grazing rents (R)
	Fruit (F)
Private income	

Enter all bank transactions during the year — once in the total received or total paid column and once in the appropriate analysis column(s) to classify the details. The total columns must *always* record the transaction as it affects the bank account because these columns are used for bank reconciliation. The analysis columns refer to what has happened on the farm itself, having separated all private transactions into separate columns.

The last three numbers of each cheque are written in the cheque number column on the payments pages. Abbreviaitons may be used when no cheque is involved. For example, SO for standing order. For bank charges a blank or a dash may be used. On receipts pages this column is used for the paying-in book serial number.

All columns should be totalled at each month end. Sum the analysis column totals together and then add the balance, at the start of the month and check with the sum of the total paid or total received column. The monthly closing balance should then be checked against the bank statement. These become the opening balances for the next month.

As a farm grows and becomes more complex, the need arises for a way to find the relative success of enterprises within it besides the total profit. Unless this is done, it is possible for an enterprise to make a loss that is not readily apparent when its trading results are merged with those of the whole business. Consequently, it helps to give physical details whenever practicable. For example, numbers of livestock bought and sold, weights of crops and volume of milk sold; also quantities of feeds bought. The cash analysis book is then fully used to record transactions, not only sums of money. Although it is, in effect, an index to both the files of paid invoices and of payments received, as much physical detail as practicable helps in quick calculation of enterprise gross margins.

The sum of the analysis columns should usually equal the sum of the total column less any opening balance. This should be the same when added either across or down the page. This monthly check is important so that any errors are found soon enough for them to be corrected. A second check on the accuracy of the cash analysis book is monthly reconciliation with the bank statement to ensure that nothing is omitted, such as bank charges or dairy board cheques paid directly into the bank account. The difference between the total received and total paid columns should equal the bank balance.

A bank statement is a copy of the customer's ledger account as shown in the banker's books. To reveal fraud or errors, the customer should regularly compare his bank statement with his cash book. The term 'bank reconciliation' or 'bank agreement' refers to these regular checks between the cash book of the business and the statements sent by the bank.

It is essential to check in detail each item in both the cash book and the bank statement. When you are satisfied that, for example, a cheque for $50 appears in both the cash book and the bank statement, you tick each entry — one in the cash book and the other in the bank statement. After doing this carefully, you will probably find some items without ticks against them. Some will be in the cash book and others may be on the bank statement. If, as suggested, all receipts and payments are passed through the bank account, there should be few discrepancies.

The next job is to handle the unticked items according to their nature. Time lags can occur as, when you draw a cheque to settle a debt, time passes before the bank receives it. During this time the bank balance will differ from the cash book balance. Cheques in transit are called unpresented cheques. Other causes of difference include bank charges, standing orders for subscriptions and other regular payments and dishonoured cheques paid in.

Differences are usually due to one or more of the following:

- paid in cheques not yet credited
- cheques drawn but not yet presented for payment
- bank charges for ledger fees, cheque books or interest on an overdraft
- interest credited to a deposit account
- errors in either bank statement or cash book

In a bank reconciliation statement, the last three items must first be dealt with by correcting errors and entering bank interest or charges in the cash book. A statement can then be prepared by taking the balance as shown in the statement and *adding* to it items paid in and not yet credited and *deducting* standing orders and cheques drawn but not yet presented for payment. (If the bank balance is an overdraft, for 'add' and 'deduct' read 'deduct' and 'add'.)

To summarise, if the bank balance is in credit/overdraft:

- add/deduct all receipts in the cash book that are not on the statement

- deduct/add from the bank balance on the statement all payments in the cash book that are not on the statement
- correct any errors on the analysis sheet

Reconciliation Statement, Manda Farm, 30 June 1987

	$	$
Balance as per statement		1351.51
Add Cheques paid to bank but not yet cleared:		
A.M. Phiri	109.25	
R. de Kock	83.33	192.58
		1544.09
Less Unpresented cheques issued:		
M. Magagula	208.51	
Farmers' Co-op.	141.48	349.99
Balance as per cash book		1194.10

Complications like private transactions, discounts on composite payments for several different items, payments on account and contra-accounts sometimes confuse people and should be avoided if possible.

Many business have only one bank account and draw funds from it for private expenses, like school fees and living costs, besides business ones. These can also be payments made into the account that are nothing to do with the business; for example, share dividends, legacies, income from a farm store, etc. It is most important to record these non-farm transactions apart from the farm ones. To minimise recording complications, funds for private expenses should be kept separate from the petty cash receipts and expenses. However, if there is only one bank account, it is best to record all private and non-farm items in separate payments and receipts columns as this enables totals to be reconciled against the bank statement. If the business is large, it is best to have a separate bank account and cheque book for private use.

In most small farm business the distinction between business and private activities is blurred. For example, a telephone bill may include private as well as business calls and the owner or manager is best able to divide it suitably between the private and farm analysis columns. Similarly, part of the capital and running costs of the car may be charged legally to the business. The proportions of such shared costs that may be included in the business accounts are agreed by the tax authorities with the farmer or his accountant.

It is usually best to record expenses net of discounts. If there are several cost items in a bill, which therefore need allocating to more than one column, the discount should be shared between items in proportion to their gross costs. Problems sometimes arise in dealing with discounts that are credited only after the initial account is paid. For example, farmers buying goods from a co-operative may first pay the full charge. The discount is not calculated until the next month and is then either paid to the farmer or deducted from the statement for goods bought during that month. Although it is generally better to record the net cost of all purchases, in such cases the discount should be treated as a sundry receipt and the full cost shown for the goods bought.

If payments on account must be made, because some payment is needed but the financial position precludes settling the whole account, it is better to choose certain items from the total account and pay those, rather than pay a round sum. Subsequent checking is then easier.

Composite payments or receipts must often be split into more than one analysis heading, for example, a cheque made out to a supplier for both stockfeed and fencing wire. Payments to the butcher may have to be split into 'private' for the house and 'staff — (R)' for rations. A farmer or manager himself can best do this each month and it need not be precise to the last cent.

The last complication that can arise in entering the cash analysis book concerns contra-accounts. 'Contra' means 'against' and in this context it refers to the practice of setting payments and receipts against each other with only the difference in value being received or paid in money and and thus showing on the bank statement. If only the net receipt or payment is entered in the cash analysis book, much information needed to produce good management accounts is lost. Unless the gross figures are fully recorded both farm input and output will be understated and may well make any later inter-farm comparisons unsound. These combined transactions should be treated as separate items and should be entered in the cash book as such, with the symbol (C) to mark all hidden transactions. In this way all transactions are available for managerial analysis and the difference between the totals of the receipts and payments columns can still be reconciled with the monthly bank statement.

Figures 5.5 and 5.6 show a typical contra-transaction. The farmer sold goats to the Farmer's Co-op. for $300 and bought feed and tobacco chemicals for $174 during the month. These transactions were settled by receipt of a cheque for $126 which was paid into the bank and thus appears in the receipts total column. The payments total column is left blank but *all* transactions appear in the analysis columns.

Typical examples of contra-transactions include the trading-in of an old machine for a new one, sales of livestock by auction with the return being net of auctioneer's commission, grain receipts less charges for grain bags or transport, sugar cane receipts less seed cane and marketing levy charges, etc.

ADJUSTMENT OF CASH ANALYSIS BOOK FOR CREDITORS AND DEBTORS

If all transactions were in cash and settled before the year end and no valuation changes occurred over the year, the difference between payments and receipts would be the profit. However, this is not so and adjustments and additions to the cash analysis book are needed to calculate profit.

Detail of all payments and receipts during the financial year is not enough to prepare the yearly profit and loss account and balance sheet. Lists of creditors, debtors and asset valuations are needed for both the start and end of the year.

The receipts and payments entered in the cash analysis book are those already settled by cash or cheque. Purchases and sales on credit are not entered in the cash book when the goods are bought or sold; only when funds are paid or received. These items do not, therefore, show in the cash book. On valuation day there can be goods that have been received on the farm and goods that have been dispatched for which there is no entry in the cash book because the accounts are not settled.

At the financial year end, a list is needed of those who own money (debtors) and those to whom it is owed (creditors), together with the amounts and the items for which funds are owed or owing. These sums relate to sales and purchases made during the financial year and they must therefore be included in that year's financial statements. Similarly, sums owed or owing at the start of the year will probably have been paid during the year and must be excluded because they refer to sales and purchases made last financial year.

During the annual valuation of assets, lists of business creditors and debtors, with the items and amounts, should be prepared. You can check on creditors only by thorough analysis. The first step is to list all delivery notes and packing notes, representing goods received, for which there are

no invoices; also all invoices still unpaid at the year's end. This should give a full list of current creditors.

Valuing debtors on valuation day can be difficult if farmers neglect to record all goods leaving the farm. This also needs thorough analysis.

It is convenient to list creditors and debtors in the valuation book. Instead of doing this once a year, I suggest that lists be kept throughout the year, as shown in figure 5.7. As debts are settled they should be crossed off and transferred to the payments or receipts page of the cash book. Items left on valuation day provide a full list of creditors and debtors. When these items are paid after the year end, they should be entered as 'debtor' or 'creditor' in the detail column of the cash analysis book and be crossed off the list of debtors or creditors.

Neither private debtors nor business ones unlikely to be paid should be transferred to the cash analysis book. Doubtful debts unlikely to be received should be disclosed to the tax authorities. Bad debts should be written off by entering them in the profit and loss account as an expense and by omitting them from both the closing debtor adjustment to the cash analysis book and the trade debtors item on the closing balance sheet. Any written-off business debt that is paid later should be entered as a receipt in the cash analysis book.

To close the cash book for the financial year, the amounts are totalled in each column and the analysis columns are adjusted for creditors and debtors to reflect only the current year's trading and not transactions relating to the previous or next year's accounts. They are not added to the totals columns as this would prevent bank reconciliation. After the columns have been ruled off and checked, debtors are listed below the appropriate receipts columns and creditors below the relevant payments ones. Receipts, adjusted for debtors at the beginning and end of the year, are called 'income'; payments, adjusted similarly, are called 'expenditure'. It is the sums of income and expenditure that are transferred to the profit and loss account.

Figure 5.8 shows how the cash analysis book is closed at the end of the financial year. The opening creditors and debtors for one year are the same as the closing ones for the last year.

Instead of adjusting the analysis columns in the cash analysis book, opening and closing creditors and debtors may be included directly in the profit and loss account. This, however, hides much useful management information.

CREDITORS   1986–87

Date	Creditors	Details of Debt	Amount Owed
3/10	Cesco	Repair to crawler	$157.20
8/10	Industrial Gases	Oxygen and acetylene	84.15
16/10	Mr Avila	Hire of lorry to cart bricks	45.00
17/10	Belize Building Co.	Improvements. Coffee pulpery	397.62
etc.			

DEBTORS   1986–7

Date	Debtor	Details of Debt	Amount Owing
16/10	Mr Belafonte	Contract ploughing	$375.00
21/11	Grain & Milling	242 bags rice	1452.60
25/11	Mr Patel	Bananas	357.40
etc.			

Figure 5.7   Continuous Record of Creditors and Debtors

We have now listed all payments and receipts during the year, adjusted for creditors and debtors. All that still is needed to work out the profit for the year is a list of both opening and closing asset valuations and the value of produce consumed on the farm.

Valuations

Assets are valued when book-keeping is first begun and at the end of each year afterwards; the closing valuation of one year serving as the opening valuation of the next. Keeping accounts without valuations is pointless as even the fullest record of transactions is worthless unless the farmer knows the value of his stocks at the start and end of each financial year.

Valuation, or 'stocktaking', means valuing the business assets. We are considering annual valuations for a 'going concern', an important accounting concept that means any business that will, and can financially stay in business. An outgoing/incoming valuation is quite different as it relates to an actual sale/purchase of assets — the end of one business and the start of another — and requires professional valuers.

Income and expenditure do not fully show the value of production and inputs used during the year, even after adjustment for creditors and debtors. With inputs this is because of changes in the amount and value of materials in store at the end of the financial year compared with the start. On the output side it is because of changes in the amount and value of unsold (or unconsumed) products on the farm at the start and end of the year.

If the total value of inputs in store at the end of the year exceeds that at the start, then the value of inputs used during the year is less than the expenditure on them and *vice versa*. If the total value of products in hand is more at the end than at the start, then the value of output has exceeded the income and *vice versa*. Allowance for such changes must therefore be made to calculate the farm profit or loss or enterprise gross margins. Furthermore, asset values must be included in the balance sheet to calcuate the capital position and net worth of the business.

Valuations are thus made for two main uses: one, to find the value of assets for entry in the balance sheet as at the date to which the accounts are drawn up; the other, to find the cost of stock used during the period concerned. Stocks in his sense covers stocks of inputs, both home-produced or bought, stocks of products awaiting sale and work in progress. The profit and loss account should be charged with only the material used for the production of the period, whatever the value of purchases.

PAYMENTS

Date	Details	cheque no.	Total paid	Staff	Repairs and Spares	Fuel and Lubricant	Seed Fertiliser Chemicals	Livestock Purchase (Stockfeed)	Overheads	Capital Developments	Sundries	Private Drawings
	TOTAL ANNUAL PAYMENTS	—	42,289·96	6,643·76	8,326·55	6,661·24	4,995·98	5,210·00	3,400·00	4,200·00	587·43	3,200·00
	PLUS: CLOSING CREDITORS			78·43	743·24	361·34	151·43	97·68	246·94	150·00	42·50	
	LESS: OPENING CREDITORS			64·82	831·27	570·23	64·07	453·84	168·32	210·00	40·15	
	ANNUAL EXPENDITURE			5,657·37	8,238·52	6,452·35	5,083·34	4,963·84	3,478·62	4,140·00	539·78	

RECEIPTS

Date	Details	Slip	Total received	Coffee	Burley Tobacco	Cassava	Maize	Cattle Finishing	Other Livestock	Capital Sales	Sundries	Private Income
	TOTAL ANNUAL RECEIPTS	—	60,696·43	19,246·08	12,830·72	6,415·36	9,623·08	6,236·24	4,463·72	976·32	471·93	
	PLUS CLOSING DEBTORS			—	1,283·07	151·43	74·63	—	360·72	—	89·40	
	LESS OPENING DEBTORS			—	1,041·73	—	126·42	426·31	143·86	125·00	63·00	
	ANNUAL INCOME			19,246·08	13,072·06	6,566·79	9,571·29	5,809·93	4,680·58	851·32	508·33	

Figure 5.8 Payments. Year Ending 30.9.87

Table 5.8   A Simplified Profit and Loss Account

Expenditure		Income	
	$		$
Stock at the start of the year — 10 cows at $150	1500	Milk sales	2400
Feed and other costs	1700	Stock at the end of the year — 12 cows at $150	1800
Two cows bought at $150	300		
Profit	700		
	4200		4200

Table 5.8 shows that there has been a change in the asset values and the profit of $700 consists of $400 in cash ($2400 − 1700 − 300) plus $300 worth of cows bought during the year. Therefore, although profit is usually *stated* in terms of cash it is often represented, at least party, by a change in the value of assets. A large profit does not always mean that more cash is available. Profit is the excess of income (receipts adjusted for opening and closing debtors) over expenditure (payments adjusted for opening and closing creditors) *adjusted for changes in valuation*.

Table 5.9 shows that expenditure can exceed income and yet a profit still be made. Unless assets are valued at the start and end of the financial year, it is impossible to calculate annual profit.

Table 5.9   A Simplified Profit and Loss Account

Expenditure		Income	
	$		$
Opening valuation	1400	Income	700
Expenditure	1600	Closing valuation	2400
Profit	100		
	3100		3100

In most farm businesses the financial year starts and finishes on the anniversary of starting the business. However, farmers can deliberately agree a different financial year with the tax authorities. Farmers should usually choose a valuation date, kept to every year, at a time when most crops have been sold or used and the stocks in hand are at a low level. This minimises the task of valuation and the scope for error in estimating the value of stocks. It also allows accurate calculation of crop outputs when the annual accounts are completed. For most farming systems a suitable valuation time is between harvest and planting when there is little seed, fertiliser or growing crops in the ground.

The advice to choose a valuation date when stocks are at their lowest level is certainly wise for management accounts but it is not always the best policy for financial accounts.

An accurate inventory (list) of crops, livestock and 'deadstock' is essential. Deadstock consists of assets other than growing crops and animals; for example, buildings, machinery and inputs and products in store. An inventory involves physical measurement and counting of assets (stocktaking) and valuing them. Physical measurement is usually easier than proper valuation and so will not be described in detail.

On his valuation date, a farmer should walk around his farm and list all his assets in detail, under suitable headings in his valuation book. Amounts of items and the values should be recorded in full. While doing this, he should note any items for which he has not yet paid or for which he has already been paid.

The first annual valuation is a big job on a large farm but it need not be repeated in future years in such great detail because, once the initial list of assets, such as land, house, buildings, improvements and machinery has been made, only changes to the inventory and depreciation are needed for succeeding valuations. What needs measuring is change more than total figures.

The following groups of items need listing:

- Fixed and part-fixed capital assets, for example, land and house, buildings, improvements, plant and machinery.
- Purchased stores, for example, fertilisers, seeds and chemicals.
- Harvested produce on hand, for example, maize and cowpeas — whether for sale or feeding.
- Fodder.
- Growing crops.
- Livestock.

After listing all his assets, the farmer should either value them himself or pass the lists to a valuer or his accountant for valuation. If a professional valuer is used, he needs to know your policy, as does your accountant — not forgetting, if you are a limited company, your shareholders, as the basis for annual valuations should be stated in the published accounts.

Capital stock is the total amount of physical capital existing at any one time in the firm. It is hard to measure this satisfactory in practice. Clearly, it is impossible to aggregate the physical amounts of different types of capital but this raises the problem of defining a useful money measure. Strictly, the value of the current capital stock is the present value of the stream of income that it will generate in the future but it is impossible to measure this precisely. Valuation is mainly a process of estimation as the current value of most assets cannot be known until they are sold.

Alternative principles are available, according to the use for which valuations are needed. Once the basis is decided it should be the same for both opening and closing valuations. If values are updated, as for breeding livestock for example, then both the opening and closing valuations should be calculated using the revised unit values in estimating annual profit for management purposes. The basis of valuation used in any data presented should be clearly stated and should be consistent throughout the period of any series of figures.

To show this, assume a farmer has 20 tonnes of unshelled maize at the start of the year and 25 tonnes at the end, both to be ground into corn-and-cob meal for stall-fed cattle finishing. One valuer considers that the maize is worth \$80/tonne.

	\$
Opening valuation 20 tonnes at \$80	1600
Closing valuation 25 tonnes at \$80	2000
Rise in value of stocks	400

Another valuer thinks the maize is worth \$100/tonne.

	\$
Opening valuation 20 tonnes at \$100	2000
Closing valuation 25 tonnes at \$100	2500
Rise in value of stocks	500

Although the two estimates differ, profit in the second case is only \$100 more than in the first. However, if the maize cobs were valued at \$80/tonne at the start and \$100/tonne at the end, the effect would be as follows:

	\$
Opening valuation 20 tonnes at \$80	1600
Closing valuation 25 tonnes at \$100	2500
Rise in value of stocks	900

In this case the profit would rise by \$900 and be very misleading. It is important to be as accu-rate as possible and to use conservative rather than optimistic values. However, it is even more important to use the same basis for the valuation at both the start and end of the year.

The nature of the asset and the reason for which it is held determine the method of valuation. Assets held mainly for sale are easily measured in terms of their contribution to the business whereas fixed and part-fixed capital assets are bought and sold less often so a less direct valuation method must be used. The latter type of asset contributes indirectly to the production of income whereas the former contributes directly by sale.

Stocks are often valued at cost price or net realisable value, whichever is least. This ensures that value and profits are not overstated and provision is automatically made when the net realisable value is found to be below cost and is therefore used as the basis of valuation. No 'paper profits' accrue and losses resulting from falling markets are absorbed straight away. Usually, therefore, valuations are based on cost of production or market value less selling costs — whichever is least.

The 'true' value of assets, which must be known for management decisions and for estimates of the state of the business by a creditor, may be hard to fix. At any rate, these 'true' values are, for several reasons, likely to differ from the assets values given on the balance sheet. Reasons for different valuations for the same assets in management and financial accounts include the use of standard values for trading livestock, and valuing of land at historic cost in the latter. Table 5.10 shows a typical variation between asset valuations in management and financial accounts.

Some items bought or used during the year will be in use for more than one year. These are fixed and part-fixed capital assets like buildings and machinery. These are usually shown at cost less depreciation. The theory is that the cost of each asset should be written off against those profits that it helps to earn, which is usually the period of its expected useful life. This means that profits are evenly calculated, as far as the historical cost convention is concerned, but it does not always give an accurate balance sheet value; especially in times of acute inflation.

Sometimes replacement cost less depreciation may better determine value. Long-lived assets like buildings may be valued more realistically this way especially if large changes in price have occurred. Such a method will guard against undervaluation but not against overvaluation.

All fixed and part-fixed capital assets should be listed with date of purchase and details of any sales. All high-valued assets should be listed separately. The initial valuation is adjusted for later valuation by adding purchases and deducting sales and depreciation.

Much of a farm's assets normally consist of land and improvements bought of the steady rise in land values, these assets rise in value many times over. Yet they are usually written down in the financial accounts or kept at original cost which may in no way reflect market value. For instance, table 5.10 shows that a ranch bought several years ago for $20,000 has stayed in the financial accounts at this sum whereas the present conservative market value, with buildings and improvements, is $60,000.

Arguments can be presented in favour of valuing at purchase price as much as of changing it. Changing valuations could distort annual operating results. However, although the land and farmhouse normally stay at original cost in financial accounts, they are often revalued in practice at market value in management accounts. Market value is normally calculated as replacement cost less depreciation in respect of old or dilapidated buildings.

Farm buildings are usually valued at cost less depreciation to date in financial accounts. The book values are often unrealistically low as many buildings are written off for tax purposes when they are still of use and valuable. In management accounts, farm buildings built recently may be valued at cost less depreciation. However, those built a long time ago should be valued at replacement cost less depreciation. The rate of depreciation will depend on the type of building and its use but few of today's farm buildings, except the farmhouse, will still be worth anything in 20 years' time. The replacement cost used should be the cost of a new building to replace the present one for its existing rather than its original use.

Machinery and equipment is usually valued at original cost less accumulated depreciation to the date of valuation. The depreciation charge allowed as an annual business expense for tax purposes is calculated by using standard depreciation rates according to the type of asset. Actual depreciation is not known until the asset is sold or scrapped at which time the depreciated book value can be reconciled with the actual realisation by the use of a scrapping allowance or loss or profit on realisation.

For both management and financial accounts, it is probably satisfactory to depreciate all farm machinery at 20 per cent a year on reducing balance.

A large share of total assets will, on many farms, consist of stocks of bought goods in hand, work in progress, such as growing crops, and products in store awaiting sale or consumption. Consequently, misvaluing these assets can seriously affect the correctness of the profit shown in the accounts. It also invalidates any calculations of return on capital employed.

Stocks of bought supplies in hand usually include fuel, fertiliser, seeds, chemicals, oil and lubricants, packing materials, building materials, machinery spares and other stores. Earned or bought marketing quotas may also be included. These are usually valued at cost or net realisable value, whichever is least. Cost should be net of quantity discounts, but not cash discounts, and include transport to the farm if significant. Value is reduced by deductions for damaged or deteriorated goods and, exceptionally, it may be necessary to use realisable values, for example for machinery spares no longer needed that can be sold only at a low price. If the valuation of grow-

Table 5.10　Differences in Asset Valuation between Financial and Management Accounts

Financial Accounts			Management Accounts	
		$		$
Land	Original cost	20000	Land and improvements (including buildings and	
Buildings	Cost less depreciation	1000	fencing)	
Improvements	Cost less depreciation	100	Market value	60000
Plant, equipment and vehicles			Plant, equipment and vehicles	
Cost less depreciation		6000	Market value	9000
Farm stores in hand Cost		4000	Farm stores in hand Cost	4000
Livestock Standard values		13100	Livestock Market values	67100
Sundry debtors		200	Sundry debtors	200
		44400		140300

Table 5.11   Machinery Depreciation Calculation for Year ending 30 September 1987

Category	Book value 1.10.76	Add purchases	Less sales	Sub-total	Less Depreciation		Book value 30.9.87
					Rate	Amount	
	$	$	$	$	$	$	$
Lorry	4100	—	—	4100	20	820	3280
Pick-up	4200	—	—	4200	20	840	3360
Tractors	5600	—	900	4700	20	940	3760
New tractors	—	6200	—	6200	40	2480	3720
Implements	13200	—	400	12800	20	2560	10240
New implements	—	2100	—	2100	40	840	1260
Totals	27100	8300	1300	34100	—	8480	25620

ing crops is ignored or kept constant for management purposes, then the valuation of stocks in hand must include any seeds, fertilisers and perhaps chemicals that have already been used on growing crops for which it is not intended to calculate individual valuations. These values are often shown in the balance sheet as a separate item called 'crop expenditure carried forward'.

The rule that stock should be priced at cost or market price, whichever is least, is observed wherever possible. But which costs are to be used, as the costs of identical items of stock often change during the year? Shall we use the latest prices, the prices at which each item was bought or take an average? The three main methods are:

First in, first out (FIFO)

This is the most common method. Assets are valued at the price at which each was bought and it is assumed that whenever an item is used it was the first to be bought. The earliest stock prices are charged as production expenses and stocks are valued at the latest prices.

Last in, first out (LIFO)

This method of costing is used by firms that carry many items of stock of the same kind bought at different times and prices. When an item is used, it is assumed to have been bought last. The latest prices are changed as production expenses and stocks are valued at the earliest prices.

In a time of rising prices the FIFO method will give a larger profit than LIFO, the position being reversed if prices fall.

Replacement Cost

This is likely to increase in influence as an important element in current cost accounting.

The valuation of crops, whether harvested or still on the land, presents more problems than any other kind of stock and this is the one where it is common to find the market value wrongly used.

When finished products are available for sale, the market price will usually exceed cost of production. The latter figure is better for valuation; otherwise the profit included in the market price will be anticipated. Cost of production is somewhere between the original cost of the seed, fertiliser and chemicals at one extreme and the full selling price at the other but it is not usually known exactly. Consequently, it is conventional to value finished products in store awaiting sale at market price less costs still to be incurred for processing, storage, transport and marketing; that is, the *net* selling price provides a realistic measure of the current value of these assets to the business. In practice a value of market price less 10–25 per cent is often taken. This allows for deterioration and risk as well as further storage, transport and marketing costs to be incurred.

Home-grown fodder in store, like hay and silage, are best valued at 'standard' costs of production per tonne. These may be based on the estimated average variable cost of production on the particular farm or upon published reports.

In sub-arid areas, without irrigation, where crops do not occupy the land all the year, a financial year end is chosen so that there are no crops in the land on valuation date. If there are crops in the land, they represent work in progress and, strictly speaking, should be valued as costs will have been incurred for seed, fertiliser, chemicals and cultivations. However, the situation should not change much from year to year provided there is no great change in farming system.

If the crop is ready for harvesting but still in the land, it should preferably be valued like saleable crops in store, less estimated harvest costs. Otherwise growing crops should be valued at estimated variable costs up to the valuation date. Perennial crops like fruit trees should be valued at a depreciating rate as they near the end of their production life. Timber should be valued at current value standing, but care should be taken to ignore the resulting profit arising from this item when calculating the taxable profit as it is more in the nature of a capital gain that should be taxed only on realisation.

For valuation purposes, livestock may be divided into two groups: trading stock and productive stock. Whereas trading stock is valued at cost of production, or 75–80 per cent of current market value whichever is least, productive stock is usually valued at fixed prices called standard values.

Productive livestock, also sometimes called fixed or breeding stock, are those animals kept for breeding or for their produce in terms of young, milk, eggs, wool, etc. They are kept on the farm mainly to produce items for sale. The main intention is not to fatten them and sell the animals themselves either to other farmers or for slaughter. They may thus be regarded as capital assets. Under this heading come all livestock not intended mainly for sale, for example, cows kept to produce milk or calves for rearing, bulls, draught animals and all breeding stock. Productive livestock must be sold or consumed when old or otherwise unsatisfactory but this is an incidental receipt and not the main aim for which they are kept. Any profit from sale of productive stock should be regarded as a capital gain, but it is for the tax authority to decide whether or not it regards it as such.

Productive livestock are medium-term investments and their valuation should be isolated from market price changes. They are therefore often valued at fixed, standard book values.

Standard values are fixed at cost of production or market value, whichever is least, for each class of productive livestock — bulls, oxen, cows, heifers, steers, calvers, etc. If you do not have reliable cost of production figures for home-bred stock, use 75–80 per cent of current market value. Once a standard value for a class of livestock is fixed, in agreement with the tax authorities, it is kept in succeeding years to prevent taxable paper profits due to market price changes. Once accepted, these standard values may not be changed without the tax inspector's permission.

On ranches where standard values were set many years ago, these are now totally unrealistic and seriously undervalue the herds so permission of the tax authorities should be sought to change them. Unfortunately, a rise in standard values produces a paper profit that may be liable for tax, unless otherwise negotiated, so this should be done only in a year when a lower than average profit is expected. Highvalue purchased bulls or pedigree stock are usually kept on the books at cost until they are sold or die.

Trading livestock is all livestock kept mainly for sale. This includes steers for sale fat or as stores, pigs or sheep for fattening, table poultry and any livestock reared for sale. Sows, for example, are productive stock but all the litters are trading stock except those kept for breeding.

Trading stock may be valued at cost of production, if known, or at, say, 75–80 per cent of reasonable average market value — ignoring any unusual market trends. The 75–80 per cent method is a good general working rule, saving time and trouble for both taxpayer and tax authorities. However, if a farmer can show fairly convincingly from his accounts that the cost price is less than the normal 75–80 per cent, he is entitled to use his actual costs. Trading livestock, likely to be sold in the coming financial year, may be revalued each year.

It may seem from the above account that valuation is complicated. In practice, it is easier than might be supposed. This description of the methods used by valuers has been given so that, whether farmers hire professional help or undertake their own valuations, they will understand the principles and effects of the methods used.

Now that we have adjusted payments and receipts from the cash analysis book for creditors and debtors respectively and have opening and closing valuations, the profit and loss account for the financial year can be produced.

Profit and loss accounts (operating statements)

At the end of the financial year, the manager or his accountant uses information from the valuations book, lists of opening and closing creditors and debtors and the cash analysis book to obtain the financial results of the year's trading. These results are known as financial statements and consist of a profit and loss account and a balance sheet. These are highly important because the main financial aims of running a business are to run it profitably and to finance its activities so that it stays solvent.

A profit and loss account provides a comparison of the values created in the period with he value consumed in their creation, according o the accounting conventions, such as the depreciation and valuation of stocks, adopted for he business. A profit and loss account is a profitability' statement showing the profit, oss or change in wealth resulting from business activity over a fixed time period. Finding out the profit or loss made by a business is one of the most important results of financial accounting so explanation of the procedure used to determine it s important, therefore. A balance sheet is a solvency' statement recording the net worth on a given date. Solvency is the ability to pay debts on ime and, therefore, it is of great interest to anyone considering lending funds to the business.

Just as a balance sheet may be regarded as a still picture at a given time, so the profit and loss account is a moving picture summarising a farm's business activities over, usually, the full financial year. It can, however, compare expenses against income for any time span. A yearly profit and loss account deals with all financial transactions of a trading or revenue (rather than a capital) nature between the business and the outside world during the year and considers what has been brought forward from the previous year and what is passed over to the next year in terms of both income and expenditure. It takes account not only of money paid out and received but also of debts owing to and by the business and the value of some of the business assets at the start and end of the year.

The term 'profit and loss account' is perhaps unfortunate as no statement can show both a profit *and* a loss. Other terms often are income and expenditure account, revenue account, income statement, profit statements and operating statement.

A profit and loss account can be used to find the profit or loss either for the whole farm or for an individual enterprise. The former figure is needed as a basis on which to assess the tax payable by the business; income tax if the farmer is trading on his own or in partnership, corporation tax if the business has been incorporated as a company.

In most countries a profit and loss account by law must give a 'true and fair view' of the profit or loss of the business and disclose certain items like provision for depreciation and interest on loans and overdrafts.

Often the annual financial statements are regarded as merely a costly, wasteful necessity to satisfy the taxman. If such an attitude is taken, then the opportunity to understand the financial affairs of the farm is missed. By summarising the year's expenditure and the sources of income, a profit and loss account is a valuable source of management information and should be studied carefully by anyone concerned with investment as, together with the balance sheet, it shows the return of the business.

A detailed profit and loss account shows the business income and expenditure, opening and closing stocks, or changes in their valuation, and the net profit or loss for an accounting period which is usually a year. It shows what has been produced and what has been used in its production and also the final surplus available either for distribution to the owners or allocation to reserves. If the cost of resources or inputs used is less than the value of the saleable product or output, the farm has made a profit. If the cost of inputs exceeds the value of production, the farm has made a loss.

Net profit is the most important and interesting figure in the profit and loss account as this must be enough to cover both living expenses and loan repayments. It is the return on capital employed, provides capital for the replacement of machinery and equipment and may even have to provide for payment of tax. A profit adds to and a loss reduces capital in the business but profit differs from cash income. Non-cash 'incomes', like unsold crops or livestock, raise profit but do not add to cash available. On the other hand, non-cash expenses, like depreciation, reduce profit but not cash in hand.

The profit or loss given on an annual profit and loss account is net. It is impossible to see exactly when it was made during the year or, if more than one product is produced and the profit and loss account is not detailed, just how much of the profit came from each product.

In outline a profit and loss account is set out as shown in table 5.8. Having totalled the opening valuation with the true current business expenses and the true business income with the closing valuation, it is unlikely that the two totals will not differ. If the former exceeds the latter, the balance is a net loss and will appear on the right hand side as such.

Actual profit and loss accounts usually give more detail than that shown in table 5.8. Items are grouped, with sub-totals, under rational headings suited to the farming system so that any trends in a series of annual accounts are easily seen. Last year's figures are also usually given for comparison.

Figure 5.9 shows a typical annual farm profit

and loss account. However, this traditional horizontal layout is not at all sacred and it is now increasingly common to present the statement vertically as shown in figure 5.10.

If the payments and receipts column totals from the cash analysis book for some reason have not been converted to annual expense and income totals by adjustment for creditors and debtors, then this must be done on the profit and loss account itself. This means adding closing creditors and deducting opening ones from the expenditure side. Closing debtors are then added and opening debtors deducted from the income side.

It is unnecessary to give valuations fully on the profit and loss account, unlike the balance sheet, as only changes in valuation affect profit. Any fall in valuation is treated as an expense and any rise as an income item.

Not all payments are treated as 'expenses' for charging to the profit and loss account. For example, all personal transactions are omitted; also tax payments on last year's profit, fixed assets

or other capital investments (unless both opening and closing valuations are shown fully, as in figure 5.9) and loan repayments. Last year's income tax is omitted as the profit and loss account is used to assess this year's tax and you cannot get tax relief on last year's tax. Just as when money is borrowed it is not income and is not taxable, so repayments thereof are not expenses and cannot be offset against tax.

Expenses are not always payments. The 'expense' of depreciation, for example, may exceed in a given year any payments made for new capital investment. Conversely, some farm 'income', such as produce used in the house, is not a cash receipt.

Although tax is an appropriation of profit and not an expense for the purpose of calculating taxable profit, an estimate of tax payable on the profit of a period is often made, as in figure 5.10, so that sufficient funds are retained to meet this debt when it falls due.

As it is impossible to tell from a profit and loss account whether or not the profit was earned

S. & J. Akbanie Ltd

1985			1986	1985			1986
$	*Opening valuation 30.9.85*	$	$	$	*Income (Receipts)*	$	
2530	Buildings and fencing	2300		15176	Coffee	19246	
14375	Vehicles, plant and equipment	12500		10074	Burley tobacco	13072	
760	Productive livestock	700		—	Cassava	6567	
2270	Trading livestock	2370		9341	Maize	9572	
38755	Coffee in land	37600		5214	Cattle finishing	5810	
6450	Cassava in land	5000		173	Other livestock	4681	
1176	Stocks in hand	1216	61686	—	Capital sales	851	
				437	Sundries	508	60307
	Expenses (Payments)				*Non-cash receipts (Benefits in kind)*		
5512	Salaries and wages	5657		640	Rental value of house	640	
6850	Repairs and spares	8239		470	Produce consumed in house	530	
254	Seed	297		483	Private use of car & telephone	516	1686
3478	Fertiliser	3896			*Closing valuation 30.9.86*		
755	Chemicals	890		2300	Building and fencing	2070	
1378	Livestock purchases	1655		12500	Vehicles, plant and equipments	13500	
3170	Stockfeed	3309		700	Productive livestock	760	
1080	Capital expenses	4140		2370	Trading livestock	1650	
5173	Fuel and lubricants	6452		37600	Coffee in land	32800	
860	Interest	870		5000	Cassava in land	—	
1836	General overheads	2609		1216	Stocks in hand	897	51677
481	Sundries	540	38554				
6551	*Net profit before tax*		13430				
103694			113670	103694			113670

Figure 5.9   Outdated Format — Profit and Loss Account for the Year Ended 30.9.86

S. & J. Akbanie Ltd

	1986		1985	
	$	$	$	$
Income (Receipts)				
Livestock	10491		5387	
Crops	48457		34591	
Miscellaneous	3045		2030	
Valuation appreciation	—	61993	—	42008
Expenses (Payments)				
Materials	28348		22158	
Labour	3357		3512	
Marketing	1109		536	
Financial	1870			
Administration	10009		2761	
Valuation depreciation		48630	4630	35457
Net profit before tax		13430		6551
Taxation based on profit for the year		2898		840
Net profit after tax		10532		5711

Figure 5.10 Modern Format — Profit and Loss Account for the Year Ended 30.9.86

evenly over the whole period, managers often request more frequent than yearly presentation. Such statements are now often produced quarterly or even monthly. This demand for frequency stems from an enquiring manager's mind and is not required by law.

Net profit, after tax, is the surplus available for distribution to the firm's owners, as income, and as a source of revenue and funds for new investment. A profit may be used to:

- distribute to the owners as dividends or shares of profit in cash or kind; in advance or in arrears and/or
- increase working capital investment and/or
- increase fixed assets investment and/or
- increase investment outside the business

A loss is a deduction in capital investment in the enterprise and as such has the effect of:

- reducing working capital invested and/or
- reducing fixed asset investment and/or
- reducing outside investment

These summaries of how profits may be used and the likely effects of making a loss should be understood fully by any manager with control of business investment.

Sources and application of funds statements

A sources and applications (allocation, disposition, uses or flow) of funds statement usually shows how funds raised by a business, through both trading and non-trading activities, like an increase in share captial, have been used.

Money cannot simply vanish in a business. However, when shown a profit and loss account that reveals a profit of, say $18,000, many farmers say that, because they have nothing in the bank, they cannot have made a profit of more than a few dollars. It is true that the profit shown on a profit and loss account does not necessarily represent money in the bank; often it represents money already spent or 'tied up' in non-cash items. However, if the year's profit is not reflected by the cash available, investigation will show where it has gone.

A sources and applications of funds statement improves understanding of the capital position of the business. It compares the way in which funds were raised with the way in which they were used. It can show clearly if the profit has been used for personal purposes, for taxation or for a change in the owner's equity in the farm. It usefully links the profit and loss account and the balance sheet and should aid understanding of the often confusing position when an accountant's assessment of a good profit goes together with greater bank borrowing.

A statement of sources and application of funds essentially takes the declared profit (or loss) on the account, makes allowances for non-cash items (valuation changes, depreciation, notional items) and for non-trading account items (debtor/creditor changes, personal receipts, tax, drawings, grants, loans, repayments, net capital expenditure, etc.) so as to establish the cash position for the year.

Both sources and uses of funds may be divided into three classes — trading, capital and private. Trading sources shown in the profit and loss account need adjustments for non-cash changes such as depreciation of assets. Likewise, receipts, like home use of farm produce and private use of the car, should also be omitted. Decreases in valuations and debtors and increases in credi-

tors, in that they respectively represent a release of or further access to cash, are also added to the funds generated by the farm's operations. Capital sources are receipts from the sale of land, machinery and fixed equipment, grants and any new loan capital. Private sources include cash from outside the business and private receipts.

With regard to the uses of these funds, trading uses show the cash used to raise valuations, pay creditors or extend credit to others. Capital uses are represented by the purchase of land, build-

Table 5.12   A Typical Sources and Application of Funds Statement

Sources of funds	Year ended 30 Sept. 1986
Trading	$
Trading profit or loss before tax	13430
Add Depreciation of machinery	1250
Depreciation of buildings and fixtures	230
Decrease in valuation of:	
Livestock	720
Harvested crops	—
Growing crops	9800
Bought stores	319
Decrease in debtors	—
Increase in creditors	—
Deduct Notional receipts	
Produce consumed (Benefits in kind)	530
Private use of car, telephone and house	1156
Capital	
Add Receipts — Disposals of land and buildings	—
— Disposals of machinery and fixed assets	—
New loans and mortgages	—
Private	
Add Cash introduced	—
Private receipts	—
Total disposable funds	27435
Application of funds	
Trading (Net current assets)	
Increase in valuation of:	
Livestock	60
Harvested crops	—
Growing crops	—
Bought stores	—
Increase in debtors	—
Decrease in creditors	—
Capital	
Purchase and improvement of land and buildings	—
Improvements to land and buildings	3470
Purchase of machinery and equipment	2250
Loan and mortgage repayments	10000
Private	
Private drawings:	
Family living	6500
Transfer to other investments	2650
Total allocated funds	24930
Proof	
BALANCE: Cash/Bank decrease	—
Cash/Bank increase	2505
TOTAL (should equal Total Disposable Funds)	27435

ngs, machinery and equipment and by property improvements and loan repayments. Private uses include personal drawings and income tax payments.

If disposable funds exceed the sum of the uses, the balance will show as an increase in cash in hand, at the bank or both. If, however, the sources exceed the applications, the reverse will occur.

After making a profit of $13,430 and certainly not living riotously for the year on drawings of $6500, this farmer thinks that he is better off than 12 months before by only a $2505 rise in his bank balance. However, the capital account on his balance sheet will look healthier because of the partial repayment of a long-term loan and the higher value of livestock and equipment in hand; although this is little consolation to most farmers.

Balance sheets

A balance sheet is merely a specialised statement at a stated date, of a firm's net assets, or liabilities, and how those assets have been funded. It distinguishes between:

- The amounts *owed by* the enterprise (liabilities)
- The things *owned by* and amounts *owed to* the enterprise (assets).

Balance sheets can be unintelligibly boring — or revealing documents giving much information on the business. Farmers often regard them as the former and bankers as the latter.

Managers have two main financial tasks important to accountants; to run the business profitably and secondly, to finance its activities so as to keep it solvent. In the short run, these two aims can conflict.

In contrast to a profit and loss account, which reflects the success or otherwise of a business over a period of time, a balance sheet is a statement of affairs at one point in time from which the financial state of the firm can be diagnosed. It would be impracticable to draw up a new balance sheet after each transaction so they are normally 'struck' at the end of each financial year. Some of the information already used in compiling the profit and loss account, such as provision for outstanding liabilities, is used again in making up the balance sheet. Extra details are mainly of capital assets and long-term liabilities. A balance sheet is a statement of a firm's wealth on a given date whereas a profit and loss account records the transactions that change that wealth over a time period.

A balance sheet's basic features are the sources of capital of the business on the one hand and its investment in assets on the other. It summarises what is owed by the business (its debts or liabilities) and what it uses (its possessions or assets). It is a list of the assets and liabilities of a business, grouped methodically to show the financial state of the firm on a given date. Land, for example, is an asset, as are growing crops and fixed equipment. An overdraft, however, is a liability; along with shareholders' investments, hire purchase debts and creditors.

The assets part of the balance sheet shows the use of funds; that is how they are spread in various forms through the business. The liabilities part shows the sources of these funds; that is whether they were obtained by borrowing, indebting the firm to outside interests or were supplied by the owners. A balance sheet thus shows two sides of the same picture. The assets part classifies capital according to its form and purpose; the liabilities part classifies it by its source.

Typical balance sheets show the following:

(i) The capital of the business, either supplied by the owners (share capital in a company) or by long-term lenders of funds (loan capital).

(ii) The accumulated undistributed profits or losses.

(iii) The fixed assets (those the business has bought to hold and earn income and not for resale).

(iv) The current assets (those assets the business uses or produces and intends to add value, sell at a profit and turn into cash).

(v) The current liabilities (those liabilities, like trade creditors, that the business has incurred and must pay before the next annual balance sheet is struck).

Items (i) and (ii) are often grouped under the heading 'capital employed' and items (iii) to (v) as 'employment of capital'.

A farm balance sheet should show:

- the financial structure of the business in a clear, concise statement giving any interested person a summary of the state of the whole business,
- the owner's net worth at a stated date,
- in a partnership the division of that net worth between the partners,
- the present financial liquidity or solvency of the business at a given date; that is, whether it can pay its cash debts from liquid assets.

These uses of balance sheets are most important, both for internal use in the business and for interested outsider and it is a fault of much current farm business analysis that balance sheets are often ignored.

Business objectives include more than profit maximisation so stress on net profit, ignoring net worth, may distort a farmer's real financial state. Farm financial accounting methods often deal inadequately with the capital assets of the business. This has occurred partly owing to failure to study solvency as well as profitability. Farm businesses are often highly profitable but insolvent when growing fast. Over the long term their profits usually have enabled them to put their financial affairs right and to stay in business but there is always a danger of their becoming bankrupt and being wound up.

Both land banks and commercial banks like, and may insist on, seeing recent audited balance sheets when a farmer applies for a loan. For a bank or other lender, a series of balance sheets, coupled with profit and loss accounts, is a handy way to find where the business is, where it has come from, where it may be going and hence how much may be lent safely to it.

Balance sheets reflect the firm's past and may not always be a reliable guide to its future. They are usually several months old when prepared. Nevertheless, study of a series of balance sheets of the same business, prepared in the same way, can reveal much about its overall health and direction. A true view of a firm's finances can be seen only by considering trends of events as shown by a series of balance sheets, the future prospects of the firm and the intentions of management. Study of a series of annual balance sheets will show whether the capital position is improving or not — whether net worth is going up or down, relative changes between debtors and creditors, the amount of debt related to the owner's capital and whether liquidity is improving or not.

The assets of a business consist of items of value used by that business, together with any claims it has on items of value held by others; in other words, debts owed to the business. Assets consist of property (the land, buildings, cattle, money, etc.) and also 'advantages' to which a money value can be fixed. Apart from debtors, assets are what a business has obtained in terms of physical ownership.

The ease with which the right to money can be changed into cash are called the 'liquidity' of the asset and both assets and liabilities are listed in balance sheets according to this concept of liquidity. Assets are usually classified as being either of a 'current' or 'fixed' nature according to whether they are held to aid production (fixed), for use in (stocks) or an end result (debtors) of production. The last two are usually held for only a short time and must be converted into cash as soon as possible so that the firm can meet its liabilities.

Current (or floating) assets represent circulating or gross working capital — always on the move. They are fairly easily realised and tend to rise and fall directly with trading activities during the year. Current assets are used and renewed during the production and selling cycle. They are usually turned quickly into cash and include debtors, growing crops, livestock for sale and materials in hand — all of which turn-over to produce a regular cash flow. Payments in advance are usually grouped with debtors for convenience. Current assets are obtained with the definite intention of conversion to cash, so whether an asset is current or fixed depends not on its nature but on the purpose for which it was acquired. Current assets are usually split into liquid and physical working assets.

Liquid (or quick) assets are those assets either in money or in a form easily turned into money. They are realised with little or no delay and consist of cash in hand or at the bank and 'near-cash' assets like debtors i.e., current assets other than stocks.

Debtors, including prepayments, are the only current assets not represented by physical ownership. When a debtor pays, debtors will be reduced and cash increased but total liquid assets stays the same. An example of a prepayment is an insurance premium paid during the financial year, whose benefit extends after the end of that year. If, say, an annual premium covers a further six months after the financial statements are drawn up, half the payment will appear in the profit and loss account as a running cost and the rest will show on the balance sheet as a debtor.

Physical working assets are temporary assets usually meant to be turned into cash in a year or less. They include livestock (other than stock kept mainly for breeding), harvested and growing crops and deadstock (other than machinery), for example, materials in store. Working assets may be finished products awaiting sale, products still being finished or assets that will be used up in a single production cycle. They are the stocks of the business or the assets that are brought or provided with a view to resale in the course of

ormal trading. To summarise, liquid assets plus
hysical working assets are current assets; and
urrent assets plus fixed assets are total business
ssets.

Fixed or 'capital' assets are obtained to help
arn income for a long time; not to earn profit by
e-sale. They are held as part of the permanent
quipment of the business and selling them
vould affect its productive capacity and hence its
tability and future profitability. Fixed assets rep-
esent an investment over several production
ycles and cannot be sold without replacement
y another similar asset, or a change in man-
gerial policy, or production would fall.

Fixed assets include the farmland and build-
ngs, if freehold, plant, machinery, breeding
vestock and draught animals; also vehicles and
ffice equipment. They form a high proportion of
otal assets in commercial faring, especially with
wner-occupation, when compared with other
ndustries. Both the cost and accumulated
epreciation on fixed assets are usually shown in
ne accounts together with their 'net book values'.
his is suitable for the legal and accounting uses
f balance sheets but 'cost less depreciation' or
et book value can mislead in decision-making
vithin the business unless the assets have been
evalued recently. During inflation, the money
alue of some fixed assets may rise over time.
lthough balances sheets do so, it is confusing to
ombine fixed assets at old prices with current
ssets at today's prices. This problem is now
ecognised by accountants and measures have
een taken to suggest how current value financial
tatements should be produced.

The classification of assets into current or fixed
epends upon the type of business. It is not
ecided on by the nature of the asset itself but on
vhat it is used for. For example, a tractor is a
ixed asset for farmers but a current asset for
ealers. Again, land to an owner-occupier farmer
s a fixed asset but to a property dealer it is a
urrent asset — something held for resale when
he opportunity arises.

The expression 'financial position', which is
owadays applied to balance sheets, is rather
ague and has various meanings. It was probably
rought into accounting language to remove any
dea that balance sheets are intended to show the
urrent value of the business or its assets,
specially in times of inflation. Balance sheets
ontain some fact and much opinion. Many
ariables like stocks, creditors and overdrafts may
ave changed greatly in a short time. Some asset
alues recorded may vary widely from their

current values, whether the latter are interpreted
as the sum these assets would fetch if sold on the
open market or as the cost of replacing them at
current prices.

Farming balance sheets, especially those of
owner-occupiers, can also mislead because
assets, especially land, are often shown at cost
rather than market value. Moreover, because the
balance sheet is meant to describe the position at
a given time, it may not give a true view of the
average capital employed on the farm over a full
year or production cycle. The fact that balance
sheets give little idea of what assets are worth in
real terms is exploited by take-over bidders who
try to find out the real value of their victim
company's assets. The activities of take-over
bidders have made most boards aware of the
need to show current values.

The amount of assets changes owing to bringing new capital into the business, its trading
activities or to changes in value of individual
assets. A particular asset can rise in value because
of market trends. For instance, land or investments on the stock exchange may rise in value —
they may also fall! Bulls rise in value from birth to
full productive activity, as do other classes of livestock, and then drop in value with age. Other
than land, precious stones and 'works of art',
assets must eventually depreciate in value over
their expected life. This is a loss involved in running a business and is therefore charged as an
expense on the profit and loss account in the form
of depreciation. This also helps to prevent the
value of assets on the balance sheet being overstated.

In a 'normal' balance sheet, prepared by a
bookkeeper or accountant for tax purposes, certain fixed assets, such as land, are shown at their
historic cost, but most are shown at 'cost less
depreciaiton'. The rates of depreciation are sometimes those allowed for tax purposes (that is,
'wear and tear' and special initial allowances),
but more often are calculated to write off the cost
of the asset over its expected useful life.

Liabilities are the claims of outsiders on the
business; they are the claims against the business
by the various suppliers of funds to it. They consist of funds owed by the business that must
eventually be repaid to its creditors or lenders.
Total liabilities consist of current liabilities,
deferred or long-term liabilities and the owner's
net worth.

Current or short-term liabilities are debts that
may have to be paid in the short term — usually
within a year. They include trade creditors, bank

overdrafts and short-term loans, tax payable within the coming year and, in the case of companies, interest payable on loan capital (debentures), dividends due to any preference shares and any dividends declared and due for payment to ordinary shareholders.

Bank overdrafts are usually regarded as 'at call' and are therefore included in current liabilities. However, if as is usual, the bank allows the overdraft to run up to a certain limit or if there is a special long-term arrangement, part at least of the overdraft facility may be regarded as a long- or medium-term loan and therefore be grouped with deferred liabilities.

Large commercial and industrial firms, hiring professional accountants, calculate their tax debts and show them with other current liabilities on the balance sheet. Small farmers can rarely do this precisely but it is still worth including an estimated provision for tax. It could be argued that the tax is not payable until the year after the completed trading year but the eventual tax bill would have to be paid — effectively out of the net worth — were the business to be wound up. Consequently, a more realistic net worth figure is produced by allowing for tax before it falls due.

Deferred, fixed or long-term liabilities are those that do not have to be repaid in the next financial year but are likely to remain for a longer time. These include long-term loans, like remaining mortgages, medium-term loans (usually on fixed interest terms for a fixed time) and the owner's own net worth. The fixed liabilities of a company include its debenture capital, the capital subscribed by shareholders and its reserves. Mortgages, debenture stock and loan capital are borrowed funds that may or may not have a fixed repayment date. Issued share capital is the money put up by shareholders. As it is permanent capital, there is no repayment date and it cannot be removed quickly.

The *net capital employed* in a business, its net assets, is found by deducting current liabilities from total assets. This figure is often used as the denominator in calculating the rate of return on capital. It shows how the available long-term finance has been used. Confusion is often caused by a belief that buying fixed assets raises the capital employed. This is true only if funds for buying these assets come from outside the business. In all other cases, the purchase price would come from existing cash resources so that the net capital employed in the business would not change.

Another term often used is *working capital* or net current assets. The difference between thi and net capital employed, or net assets, is the book value of the fixed assets because working capital is current assets less current liabilities. Working capital represents the volume of fund available for the day-to-day running of the business. If current liabilities exceed current assets the firm is insolvent and may be forced out o business. (Working capital as defined above i sometimes called net working capital when gross working capital means the total value of curren assets.)

There are several balance sheet layouts and the method used mainly depends on current fashior and personal taste. Always, however, the top o the sheet gives the name of the person or business and the date. Moreover, the balance sheet shoulc *always* balance. Total assets must equal tota liabilities. If assets grow, then either liabilities, ne worth or both must rise equally.

To fully understand the cash-flow path, it helps to depict the balance sheet of a business as shown in the diagram.

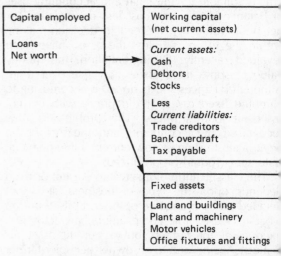

The capital employed provides funds for buying fixed assets and for working capital. From the diagram it is seen that this long-term finance enables the business to make necessary investments in both fixed and current assets, although some of the latter are financed in the short term by the current liabilities.

The date of the balance sheet is important because the state of the assets and liabilities changes continually and the picture refers to the close of business on only one day of the year whereas the financial position of any firm may change greatly within a few months because of external changes in, say, the supply of credit, demand for products or in the cost of inputs needed. Some unscrupulous people have been known to manipulate their assets at this date in order to present a favourable picture to shareholders. The next day the position could be very different! Comparative figures for the last year should be shown on any balance sheet.

In the traditional 'old-fashioned' balance sheet layout, assets are shown on the right and liabilities on the left. This division of the balance sheet into two separate lists is probably familiar even to those with only a slight knowledge of the document.

Figure 5.11 outlines this traditional layout (although the headings of 'assets' and 'liabilities' are usually deemed unnecessary) of the balance sheet of a sole trader on a freehold farm.

A variation to the traditional layout is to deduct current liabilities from current assets on the right-hand side, the result being shown as net current assets or working capital.

Traditionally, assets and liabilities are arranged in an assumed order of either increasing or decreasing liquidity and urgency of repayment. It is recommended to use decreasing order of liquidity. The most liquid asset (cash) and the most liquid liability (money at call) are shown at the top, moving down to the least liquid (land and long-term loans) at the bottom.

It has become ever more common in the last 10 years or so to present balance sheets in a vertical or narrative form with current assets less current liabilities at the top. Fixed assets are then added to net current assets to give total net assets. This alternative layout stresses that the main components of the balance sheet of any business, however large, are its capital structure and net assets. What is changed is the way in which items are shown — current liabilities being deducted from both sides of the balance sheet. They are now shown as a deduction from current assets rather than as a separate item on the left-hand side of the balance sheet. The whole balance sheet is then put in the form of a column where total assets less current liabilities equals the owner's equity (net worth) plus long- and medium-term loans. As shown in figure 5.12, the bottom half of the new layout shows where the

1985	Liabilities (Sources)	1986 $		Assets (Uses)	1986 $
	Current liabilities			*Current assets*	
3177	Creditors — trade, hire purchase	2900	47	Cash in hand	83
332	— employees	254	1577	Cash at bank	4046
1163	Provisions — unpaid tax	2898	3286	Debtors	1050
—	Bank overdraft	—	843	Prepayments — licences & insurances	560
				Products on hand	
	Short-term loans	—		— Crops	
			42600	— Trading livestock	32800
			2370	Stocks on hand	1650
			1216		897
4672		6052	51939		41086
	Deferred liabilities:			*Fixed assets:*	
—	Medium-term loans	--	247	Office equipment	261
	Long-term loans		700	Breeding livestock/oxen	760
18512	— outstanding mortgage	9716	7500	Machinery & equipment	9000
			5000	Vehicles	4500
			2300	Buildings	2070
8512		9716	100000	Freehold land at cost	100000
144502	Net worth (owner's a/c)	141909	115747		116591
167686	Capital	157677	167686	At cost	157677

We have prepared the accounts set out above and on pages 6 to 12 from the books and vouchers supplied to use which in our opinion have been properly kept. We obtained the information and explanations we required. In our opinion the accounts give a true and fair statement of the farm's affairs as at 30.9.86 and its profit for the year ended on that date.

Signed: Straight, Narrow and Partners, Chartered Accountants.

Figure 5.11   Balance Sheet as at 30 September 1986. Kunaka Farm

funds came from. Figure 5.12 represents a typical balance sheet for a company using the vertical layout.

Those funds employed in a business that belong to the owner(s) of that business are called *net worth*, net capital, capital owned or *owner's equity*. This is a balancing item calculated by deducting outside claims on a firm's assets from its total assets, that is, total assets less long-term debts and current liabilities. Net worth is the owner's stake in the business and, in theory, is the value of assets that would remain to the owners if the business were liquidated. Should liabilities to outside interests exceed total assets, the difference is a net capital deficit and the business is insolvent. The net worth of a business is the balance sheet value of assets left to the owner(s) of the business after all outside claims against those assets have been paid and thus gives some idea of the firm's solvency.

In a company the ordinary shares and reserves are the equity or ownership interest in it.

Net worth is of special use if compared with the earning it provides. The proportion of earnings to net worth should be watched to ensure that the 'return' is reasonable. Study of a firm's balance sheets for three or four years will show whether the capital position is improving or not. If profits are used to expand the business, net worth will rise; if they are all spent on living expenses, it will not change. If private drawings exceed profits, net worth will fall and the farmer is living on his capital. In a well-run business, net worth should rise steadily each year.

Some people are puzzled that, on a traditional balance sheet, net worth appears as a liability. The reason for this is that the owner's equity *is* a liability of the business — so there is no inconsistency. The business is a separate entity from its owner(s). Net worth is what the business owes to the owner(s) or shareholders.

If external liabilities exceed the realisable value of all assets then the business is insolvent and should be wound up unless a rescue operation can be mounted. If net worth is negative it is called a capital deficit. This is shown as an asset on a traditional balance sheet being, in effect, what the owner(s) would owe the business on liquidation.

There are deficiencies in the net worth figure shown on balance sheets. Total assets less outside liabilities is unlikely to equal the saleable value of the business. This is mainly because, accounts record depreciated original costs instead of saleable values. A true estimate of net worth requires

revaluation of assets. Nevertheless, although the net worth shown on balance sheets seldom represents values in terms of current money, at least it forms a basis for annual comparison — provided each year's accounts are made up in the same way.

Changes in annual net worth on balance sheets are from profits (or losses) earned by trading and drawings from the business, for example, distribution of dividends to shareholders or private drawings for personal use. This is because net worth consists of the original investment plus profits earned and held as reserves within the business. Changes in net worth result from one or more of the following:

- trading profits or losses
- drawings, whether of cash, *goods or services*, made by the owner(s) for private use
- bringing new funds, other than loans, into the business
- withdrawal of funds from the business for tax payment or off-farm investment

These items connect one balance sheet to the next.

As a business grows, a high share of profits made is usually kept in it, to provide capital for further growth. These retained profits belong to the owners or shareholders and either appear as revenue reserves on the balance sheet or some may be used to increase the issued capital by issuing more ordinary shares to existing shareholders — a bonus issue. Revenue reserves are profits earned from operating and available for distribution that arose from trading in earlier years. Capital reserves, however, are profits that arose from a rise in capital assets and they cannot be distributed as dividends unless they have been sold.

In non-company farms, one balance sheet may seem to bear little relation to the last one, even after studying the year's activities summarised by the profit and loss account. The reason is that personal transactions are not recorded on both the balance sheet and the profit and loss account.

Reconciliation of the non-company balance sheet with the profit and loss account is done in a 'capital' or 'proprietors' account, either on the balance sheet or separately.

The transactions excluded from the profit and loss account include those concerning private cash and farm goods or services used privately (benefits in kind) and shown as income on the profit and loss account but for which no cash was received. Private drawings reduce the firm's bank

NIGER FARMING ESTATES LTD
BALANCE SHEET AS AT 31.12.86

31.12.85		$
	Current assets	
631	Cash in hand	572
8971	Bank balance and deposits.	9850
763	Debtors and prepayments	1194
30750	Growing crops	32700
9560	Trading livestock	8650
23700	Stocks in hand	25600
74375	Total current assets (1)	78566
	Current liabilities	
1053	Creditors	1263
8000	Dividend recommended	9500
15000	Company tax provision	16500
10654	Short term bank loan	12197
34707	Total current liabilities (2)	39460
39668	Net current assets: (3 = 1 − 2)	39106
	Fixed assets	
11430	Breeding livestock	12530
61560	Plant, machinery & vehicles	56480
163340	Land and buildings	168480
236330	Total fixed assets (4)	237490
	Total net assets (3 + 4)	
	These net assets are financed by (capital employed):	
	Loans	
9499	Medium-term bank loan	9999
82535	Long-term Land Bank loan (mortgage)	62200
20000	Debentures	20000
	Shareholders' funds (net capital)	
60000	Revenues reserve (retained profits)	80000
3964	Capital reserve	4397
100000	100,000 ordinary shares at $1 each	100000
275998	*Total funds employed*	276596

Figure 5.12   A Typical Balance Sheet for a Company Farm

balance without helping the business; private income paid into the business account has the reverse effect. Also the tax paid during the year is not shown in the body of the profit and loss account. Entry on a capital account of personal drawings is especially useful as a position of the owner(s) living above his means will show first on the balance sheet and not on the profit and loss account.

An example of the capital account section of a balance sheet for a farm business follows:

	$
Opening net worth (Closing net worth of the last balance sheet)	144562
Plus	
Trading net profit	13430
Private income paid into an account shared with the farm	—
Less Private drawings from the farm account	(11499)
Income credited in profit and loss account for private use of farm produce, car, etc.	(1686)
Tax paid on last year's profit.	(2898)
Equals	
Closing net worth	141909

The closing net worth calculated in this way must equal excess of assets over outside liabilities.

Whatever change occurs in an asset or liability, a balancing change must occur on the same or on the other side of a traditional balance sheet. This can be understood by thinking of a pair of scales. If a weight is added to one side, then an equal weight must either be added to the other or withdrawn from the same side to keep the balance. An example of a balancing change on the same side is selling cattle for cash. The asset 'cattle' falls but the asset 'cash' rises by an equal amount. Again, if you inherit money and use some of it to repay a bank overdraft, the liability 'bank overdraft' disappears but net worth rises equally.

An example of a balancing change on both sides is buying a tractor on credit. The asset 'tractor' is added but an equal liability — either 'creditors' or, maybe, 'medium-term loan' is added on the other side. Payment of this liability would reduce assets (cash) and liabilities (creditors or medium-term loan) equally.

Balance sheet interpretation

It needs practice to read a balance sheet properly. However, this skill is aided, first, by setting out the relevant figures in such a way as to help make intelligent judgements easy; secondly, by using suitable ways of valuing assets and thirdly by wise use of certain balance sheet ratios. Several tests used in industry are sometimes used in farming — especially in large companies and partnerships. These tests are mostly based on the balance sheet and on net profit.

Apart from the profitability of the business, as shown by its profit and loss accounts over a few years, other features vital to its strength and ability to survive are *stability*, *liquidity* and *flexibility*. A past record of *growth* also helps to increase credit-worthiness.

It is not enough to make a profit — the capital position must be sound so that the business stays solvent and can pay its debts when they arise. It is only through full understanding of balance sheets that the financial security of the business can be assessed.

Of all the business statements used, the balance sheet is the least understood, easiest to misinterpret or manipulate and yet remains the most widely used document for making business decisions. Accountants tend to surround it with an air of mystery and stock-exchange jargon which may explain its frequent misuse. One should not take balance sheets on their face value. The lawful manipulation of accounts to make a balance sheet show a company or other business in the most favourable way, and so attract investments or raise the shares' market value, is called 'window dressing'.

To understand the significance of the figures in the annual financial statements, ratios are normally used because absolute figures can mislead when assessing business performance. The relationships between components of the profit and loss account and of the balance sheet and those between and within the main classes of assets and liabilities enable useful financial ratios to be prepared. This is called ratio analysis. A financial ratio is a ratio of two quantities, both measured in money terms. Various ratios can help to assess investment and trading performance, credit-worthiness and to find reasons for an inefficiency. We should, however, follow these simple rules:

- decide what information is needed from the accounts
- find the relevant figures
- calculate the ratio
- assess its meaning and what action, if any, is needed

Many financial and operating, or profit and loss account, ratios can be obtained from financial statements. These help to show the strengths and failings of the business. The main ones are as follows:

Short-term position	Long-term position
Tests of liquidity (availability of funds)	Net capital ratio
— Current ratio	Percent equity
— Liquidity ratio	Fixed assets: Net worth ratio
	Fixed assets: Total assets ratio
Working capital: fixed assets ratio	Capital gearing ratio
	Net profit: Capital employed ratio

Four general rules govern the use of financial ratios:

- calculation should be rapid
- calculations take time so only those ratios of direct use should be compiled
- ratios should be presented in the most suitable way for the firm
- ratios do not give financial control but identify areas worth examining.

Ratio analysis helps forming first opinions, pointing the way to useful areas for examination. It is of special use to shareholders, financial managers, creditors and lending institutions. Comparison can be made within the business over time and is called trend analysis or with other businesses within the industry and is called comparative analysis or inter-firm comparison. The right way to judge performance is to look at a set of ratios and results together and draw an overall opinion.

Firm conclusions should not be drawn from a few separate ratios. The trend in them over time always gives a better assessment than those for one year.

Care is needed to avoid forming conclusions, by ratio analysis, that later prove wrong. Accounting ratios are only a guide and no standards can be laid down for them as they will vary with type and size of business and the timing of the financial year. They must always be used jointly with other comparable information. Ratios should be related to one another before interpretation. If, for example, sales rise by 6 per cent in a year and profits by 3 per cent, the apparently satisfactory position is proved otherwise if this was achieved by injecting more capital.

Before computing any ratios, the existing asset valuations shown on the balance sheets being used should be examined and revised if they are thought to poorly reflect current realisable values.

Whether or not the firm can pay its bills is important and several control ratios are useful for finding the current financial position. One of the most important is the *current, or working capital, ratio*. This ratio is compiled by taking from the balance sheet the total of current assets and expressing them as a ratio of the total current liabilities. For example, if current assets and current liabilities are $6000 and $4000 respectively, then the current ratio is 1.5:1.

Examining the ratio of current assets to current liabilities indicates the short-term solvency of the business and a firm should normally have more of the former than the latter. In other words, current ratio should exceed 1:1 — otherwise there is no working capital.

Current assets become cash quickly, or are already such. As the decision on how to spend the cash is almost completely flexible, they give managers more opportunity to exploit economic and technical changes to their advantage. Flexibility is becoming more important with the growing rate of technical change. It is increased by a higher net cash flow and by higher provisions for depreciation. Only easily realised investments, tax reserve certificates and short-term debtors should be included.

The current ratio indicates the short-term trading stability and solvency of a business and the extent to which current assets are being financed from long-term sources. Changes in this ratio are often useful indicators of solvency and overtrading.

A current ratio of at least 2:1 is usually thought satisfactory. This is because it is safe to assume that if current assets must be turned quickly into cash, they will fetch at least half their book value. As stocks are valued already at the lower of cost or market price, this may be thought too pessimistic and the ratio is often well below 2:1 in farming if a bank overdraft is not regarded as at least a medium-term liability, but rather as the current liability that it really is. The more liquid the current assets, the lower the ratio may safely fall. It should, however, usually exceed 1.5:1. Current ratios much over 2:1 may reflect poor credit control, underused cash or excessive stocks. Idle cash earns nothing!

Current ratios in farming will vary with the type of business and the season. No attempt is made to suggest exact rates at which farmers should aim as the 'safe' level for any given ratio will vary with the type of business, the structure of the main classes of assets and liabilities and the personal qualities and position of the farmer.

Narrow current ratios can cause problems if large bills fall due at the wrong time. A low ratio may indicate a lack of working capital. Indeed, if current liabilities exceed current assets, the business is in short-term danger, even if its long-term position is basically sound, because it risks having to sell fixed assets to meet short-term claims, thereby reducing its production potential. Continuous narrow current ratios are embarrassing and often indicate a need for new credit or other financial measures to relieve managers from trouble. The higher the ratio, the safer is the firm in the short run because it is more likely to survive unforeseen demands from creditors or the bank manager by being able to obtain the necessary funds quickly.

Liquidity ratio indicates the *immediate* solvency of the business; as against its *ultimate* solvency as measured by the net current ratio. It indicates the availability of cash or assets that can quickly be turned into cash to cover money at call and is a more precise measure of a firm's creditworthiness than the current ratio. It relates the short-term obligations of the business to the funds likely to be available to meet them.

It is normally considered that the liquidity ratio should be at least 1:1, indicating that the liquid assets should at least equal the current liabilities to enable them to be paid. This would avoid either the forced sale of physical working assets or, worse, sale of fixed assets. Too high a ratio may indicate favourable trading but conversely it may indicate that funds could be used better.

The term 'overtrading' describes the position when there is a shortage of cash or, more exactly, when liquid assets fall seriously short of current liabilities. A business in this position might need to obtain further loan funds to reduce current liabilities or to increase liquid assets. Contrary to what might be expected, the liquidity ratio may fall in prosperous times as more activity may lead to larger stocks and more debtors but less cash. The liquidity ratio gives a better idea than the current ratio of the amount by which we would be squeezed if we had to pay creditors quickly.

Farmers should normally try to ensure that their short-term financial position is not endangered by too narrow current and liquid ratios. Other things being equal, higher ratios indicate greater financial strength and ability to meet a crisis. However, too much liquidity can be harmful as there is usually a conflict between the liquidity of assets and their profitability. Stocks in hand may be highly liquid but continuing to hold them is likely to incur costs and hence reduce profits. Cash is the most liquid asset but earns nothing until it is invested. The liquidity ratio is likely to vary in the short run.

The ratio of debtors to creditors is also a gauge of overtrading and solvency, the danger sign being a steady fall in this ratio. This can be seen by a rise in bank borrowing. Overtrading is a common problem in small, growing firms. There is a temptation to use any extra cash in the bank account to buy fixed assets and a stage may then be reached where there are insufficient funds to pay creditors, resulting in a sudden cash crisis.

The management of *working capital* is vital to all firms and no firm deserves to survive if it neglects this. Working capital must cover current assets, like stocks of inputs and work in progress. It is reduced by current liabilities, like the amount of credit obtained from suppliers.

Working capital, or net current assets, is current assets *less* current liabilities. It should not be confused with liquid capital which is that part of the working capital that can be turned quickly into cash, such as debtors, quickly realisable investments and bank balances. Working capital is a vague term for that part of the business capital that is always circulating and its availability shows the short-term solvency of the business.

If working capital is positive, part of the current assets are financed by fixed liabilities — a generally healthy state of affairs. If it is negative, the business is in an unhappy position as it has insufficient funds or other liquid assets available to cover its current liabilities.

The amount of working capital needed varies with the type of busiess and its commercial practices; for example, on whether its products are sold on credit or for cash. Farm enterprises with a short production cycle, such as eggs, vegetables, porker or milk production, need little working capital because sales income comes in regularly. (The initial capital outlay in fixed assets may, however, be heavy.) Enterprises with a long production cycle, such as beef rearing or plantation crops, need much working capital as costs are constantly incurred but there is a long delay before any receipts arise.

The ratio of *working capital: fixed assets* explores further the adequacy of the working capital balance. It can reveal a situation of capital being invested in fixed assets at the expense of working capital or alternatively it could show that fixed assets are being run down to provide funds for current production. The trend in this ratio shows how the balance between achieving adequate liquidity and sufficient fixed assets is being kept.

If the ratio is high, this may indicate that fixed assets are being reduced, and future production endangered, so as to obtain adequate working capital. If it is low, this may indicate weakness in that capital is being invested in fixed assets at the expense of working capital. Thus the business may be unable to pay its current production costs. This is yet another sign of overtrading.

Another ratio sometimes calculated is that between *net worth and creditors*. This ratio indicates the stability of the firm; its ability to deal with variations in profits and still stay in business.

This ratio shows a whether the business is financed much by short-term capital borrowing that is liable, both in theory and practice, to have to be repaid quickly. As recent experience has shown some farmers, this can be embarrassing. Just as a forced sale always puts the seller in a weak position, so forced refinancing of loans always puts the borrower in a weak position. A high share of ownership interest gives stability to the business.

A more common ratio is the *net capital ratio* which is calculated by dividing the total assets of the business by its total liabilities less the owner's capital (or net worth). It is the ratio of total assets to outside liabilities, for instance.

$$\text{Net capital ratio} = \frac{\text{Total assets (\$25000)}}{\substack{\text{Total liabilities (\$25000)} - \\ \text{owner's capital (\$10000)}}}$$
$$= 1.7{:}1$$

This ratio relates to the long-term health and safety of the business and it often indicates this better than the actual net worth figure. It also gives a good idea of the farmer's chance of raising more external funds. Net capital ratio makes it easy to estimate by how much the value of assets would have to fall before that value is exceeded by the outside liabilities. In the example above, therefore, the value of assets would have to fall by 40 per cent before that value was equalled by the outside liabilities. The main value of the net capital ratio lies in estimating the likely effect of any future changes on the financial structure of the business.

A question often asked is, 'What is a safe net capital ratio?' No general answer exists because a safe ratio depends on the type of farm and the current level of uncertainly in the business. If crop failures or wide price or cost variations are common, a higher ratio is needed for safety than in more stable times. Maybe the best answer is to make different assumptions about the value of farm assets and calculate the effects on the net capital ratio. This means studying general economic conditions and the trend of product prices. Many bank managers expect a ratio of about 2:1.

Another important financial ratio is *percent equity* which is closely related to the net capital ratio. This is just the owner's equity or net worth expressed as a percentage of total assets. It is the share of the capital that belongs to the owner(s) of the business and it thus measures both the stability of the firm and its ability to borrow funds for growth. The rest of the capital, (creditors and loans), is provided externally.

The higher the percent equity, the safer the business. Lenders like to see the owner(s) providing at least half of the capital — and nearer two-thirds for stability. The higher the percentage, the less vulnerable the business is to an unforeseen drop in the value of its assets or to occasional trading losses. In theory, percent equity represents the owner(s)' stake in the business. In practice, this may be far from true unless assets are revalued before calculating percent equity.

The relationship between *fixed assets and net worth*, usually expressed as a percentage, also interests both lenders and owners. Wise financial management suggests that they should be fairly close. In farming, the financing of the land as well as most of the plant, machinery and breeding livestock may well require the owner to obtain loan capital.

It is important to match the sources of capital with the types of assets possessed. The total value of fixed assets should always be less than the long-term outside liabilities plus net worth, otherwise some fixed assets are being financed by short-term borrowing. This is dangerous and could lead to forced sale of some of the fixed assets. This basic point is imparted by such maxims as: 'Never borrow short term to invest long term.' This matching of the type of borrowing to the type of assets can be seen by analysing a series of balance sheets. It is wise to know whether increases in fixed assets have been matched by increases in retained profits and deferred liabilities or, dangerously, by a rise in current liabilities.

A high fixed assets:net worth ratio usually implies a low current ratio and inadequate liquid resources. Conversely, a low ratio may indicate a top-heavy structure with inadequate earning power.

The flexibility aspect of the business is shown partly by the current ratio but also by the balance of the different types of asset. Just as too high a proportion of 'fixed' costs in total costs is considered a sign of weakness in management, likely to lead to poor profits, so is too high a proportion of fixed assets to total assets. The analogy of an over-bodied car might be drawn. Such a car, which has a poor power:weight ratio, usually has poor acceleration and is easily passed by its competitors on the road.

The ratio of *fixed assets to total assets*, usually expressed as a percentage, shows the share of capital, tied up in equipment, buildings, land, etc., that cannot be turned quickly into cash. It is also a guide to flexibility.

This ratio will vary greatly between farms and especially between owned and rented farms. If too high a percentage of assets are fixed, there may be insufficient working capital to use them profitably. In this case, either loans should be obtained on the security of part of the fixed assets to increase current and medium-term assets of, if this is impossible or for some reason undesirable, the possibility of selling some of the fixed assets should be considered. This is the main idea behind the sale and leaseback of land.

Another important relation bearing on the stability of a business, and its attraction to a would-be lender, is the ratio between long- and medium-term, fixed interest term borrowing in the form of loan capital (deferred liabilities) and net worth or ownership interest. This ratio is called the *gearing* of the capital. The importance of gearing to the financial stability of the business derives from the fact that interest payable on loan capital is a first charge on profits and only the balance is a return on the owner(s)' own capital.

I define gearing as the ratio of long-term, fixed interest loan capital, including preference shares and debentures, to owner's interest, equity or net worth, including ordinary share capital and reserves. (Unfortunately, this ratio is often expressed the other way round as the ratio of owner's equity to loan capital.)

Gearing shows the extent to which a business is working on borrowed capital and indicates its vulnerability to profit changes.

Firms with a high proportion of long-term, fixed interest debt in relation to owner's equity are highly geared. As shown in table 5.13, high gearing increases the real return on owned capital if the profit on total capital exceeds the real interest charge but, if the profit rate falls below this, high gearing becomes a burden and sharply reduces the return on owned capital. This is known as the principle of increasing risk. The higher the gearing, the greater the chance of getting a high return on one's own capital but

also the greater the chance of going bankrupt. High gearing should therefore be practised only when consistently good profits seem almost certain. Table 4.4 shows that where profits are high in a highly geared firm, this benefits ordinary shareholders as they get both capital growth and good dividends. If profits fall, they may get no dividend since fixed interest capital takes all available profits. Highly geared businesses gain much from inflation as the fixed interest they pay on borrowed funds drops each year in real terms.

If all long-term funds are provided by the owner(s), the business has no gearing. It should be realised, however, that no or low gearing may mean that profitable investment opportunities must be missed. It is usually considered acceptable for a business to have a gearing of 1:2 — in other words to borrow up to one-third of its long-term capital on fixed interest terms, as in the case of Farm B.

The percentage of *net profit to capital employed*, or return on funds, is widely used to assess business profitability. This is only

Table 5.13   Effect of Gearing and Profitibility on Return on Owner's Equity

	Farm A (no gearing)	Farm B (medium geared)	Farm C (high geared)
	$	$	$
Gearing	0:1	1:2	2:1
Loan capital	0	10000	20000
Owners equity at start of year	30000	20000	10000
Total long-term finance	30000	30000	30000
Good profit year:			
Profit before tax and interest (30%)	9000	9000	9000
Interest at 13%	0	1300	2600
Tax at 30%*	2700	2310	1920
Profit on owner's equity	6300	5390	4480
Rate of return on owner's equity	21%	27%	44.8%
Poor profit year:			
Profit before interest (5%)	1500	1500	1500
Interest at 13%	0	1300	2600
Tax at 30%*	450	60	0
Profit on owner's equity	1050	140	(1100)
Rate of return on owner's equity	3.50%	0.70%	(11%)

(Figures in brackets are negatives)

* Tax is paid on profit *after* interest

common-sense. When we invest money in a venture, be it a business, stocks, shares or tangibles, we look to the annual reward that such an investment will earn us. If the yield is unsatisfactory, we may be tempted to make a change in hope of a greater future return. A high, or a trend towards a high, rate of return is desirable.

This percentage of net profit to capital used is called either the return on capital employed or the primary ratio. It is somewhat vague as there are many ways of defining both profit and capital employed and so many problems involved in each definition that it is doubtful if it has much value.

Net profit is usually defined as profit *before* tax or interest charges. Capital employed may mean total assets. This allows for the fact that when goods are supplied on credit, the suppliers are, in fact, supporting the buyer financially during the period of credit. However, it is generally taken to mean net assets — total assets less current liabilities.

The importance to the owner of knowing the amount of capital employed is that it enables him to measure the degree of success or otherwise of his farming operations. The larger the profit made in relation to capital employed, the greater the success. The converse is obviously true in that smaller profits made with the same employed capital reduce this measure of success.

From the viewpoint of the owner(s), however, the ultimate measure of success is the relation between profit *after* deduction of tax and interest on borrowed funds to net worth or shareholder's funds. This shows the return made on the investment of the owner(s) and, with profitable trading and high gearing, it is usually much greater than if calculated on total or net assets.

An informative statement, already described, is one called either a *sources and application of funds* or, more shortly, a flow of funds statement. It is also called a cash and fund flow analysis.

The ratios described can greatly help managers to understand the financial state of their business and point the way to better control of profits. The profit and loss account offers much scope for more detailed ratio analysis and, if managers decide what they wish to control, it is easy to develop relevant ratios and use them to assess the position.

Study of the financial information relating to a business and presented in various forms, such as profit and loss accounts, balance sheets, flow of funds statements and so on, can reveal the trends in the main elements of the financial structure of the business. Such tests are useful to anyone examining the business from the outside; for example, a sleeping partner or a possible buyer. He should note if any ratios are worsening each year as this might indicate that the firm is heading for trouble.

Any interpretation of trends must be adjusted by knowledge of the personal and technical situation of the business if it is truly to show whether the business is gaining or losing financial strength. Such strength and ability to meet misfortunes and exploit opportunities make the business attractive to lenders on whom farmers are likely to rely more in future. It is of only slightly less interest to those farmers who are mainly self-financing. All farmers should know whether they have this strength and how to show it.

Ratios are often presented to management as a series of tables that can be hard to read and interpret for all but accountants. The simplest presentation is graphical so that trends can be seen quickly and easily.

Tables 5.14 and 5.15 show some typical balance sheet ratios for British owner-occupied and tenanted farms respectively.

Ratio analysis of farm balance sheets can be used as successfully in farming as in industrial businesses. However, the rule-of-thumb guides accepted by experience in industry may not always apply directly to farming. Consequently, balance sheet ratios must always be used with much care.

One big difference is the high proportion of total assets made up by land on owner-occupied farms. Although current ratios are often higher than seem necessary, liquidity ratios can be low. This, however, probably reflects the fact that short-term lenders to agriculture will usually wait until the end of the production cycle before requiring repayment.

The ratios described have little or no meaning unless assets are realistically valued, even though such valuations can be only rough estimates. Land is usually valued at historic cost, livestock at low historic standard values and other fixed assets are often depreciated rapidly for tax purposes in financial accounts — or even written-off while still of value. Realistic market values should be used for management decisions and for estimates of the state of a business by creditors or potential buyers.

Another point is that financial ratios should be looked at in relation to each other, so as to give an overall view of the financial structure of the business. Further, it should be remembered that the balance sheet relates to one point in time; on a farm, a month or two earlier or later and the position may look very different. Finally, even more important than the *level* of ratios are the *trends* in those levels over time.

PHYSICAL RECORDS

This section may be the most unpopular one in the book to many farmers and managers. Any record keeping is widely disliked by most farmers but they should realise without records the chances of getting regular, satisfactory profits are much reduced. Up-to-date records kept methodically throughout the year are essential for good management. Consequently, farm management advisors try hard to persuade farmers to keep more useful records.

To produce accounts and financial statements, as described previously, it is necessary to record certain financial and physical details. For managers to make correct decisions, they must also have other detailed information about the business. This is specially important when management, rather than purely financial, accounts are wanted. Physical records are concerned with the source and disposal of materials and services.

The advice given on physical record keeping is not designed only for those farmers who keep their own records but also for large farming companies with full-time office staff. Few of them now keep the best records for management use.

Improved management must come from inside the business; it cannot be imposed from outside. If a farmer knows that he has an economic prob-

lem, he must provide records on which a so diagnosis can be based. The remedy is o obvious so no outside advice is needed. We not take all our aches to a doctor!

Physical and financial records are c plementary. Financial accounts show income expenditure and physical records relate these units of output and input. Together they meas the efficiency of use of both physical financial resources. Without physical records, use cannot be made of financial ones. example, financial accounts show how m labour and machinery cost in total. The only to split these costs accurately between enterpri is by having labour- and tractor-use records.

Information has a cost and the more in mation collected, the higher it will be. Rec keeping takes time, usually the farmer's o time, and this is one of his scarcest resources the information is not used, the money and t spent collecting it is wasted. For example, i possible to use much costly time collecting breaking down data on tractor running costs allocating it to individual enterprises. However the information is not used profitably, both c and time will have been wasted.

Record keeping has for too long been c sidered a refinement that big farmers can p with if it amuses them but for which ordin farmers are too busy. Most farmers probably g less time and care to clerical work than any ot businessmen, although record keeping brings i focus various aspects of the business and remo much guessing when planning improveme Records are valuable aids to good manageme Consultants cannot be of much use to farm who keep few if any records. An hour or

Table 5.14 A Simplified Balance Sheet for an Owner-occupier

Liabilities		%	%	Assets		%	%
Current:	Creditors	2		Current:	Liquid	4	
	Overdraft	2			Working	9	
		—	4			—	13
Deferred:	Loans	6		Fixed:	Breeding livestock	7	
	Net worth	90			Machinery and equipment	5	
					Land and buildings	75	
		—	96			—	87
			100				100
Current ratio		3.2:1		Net worth: total assets of			90 %
Liquidity ratio		1:1		Gearing			1:15
Working capital		+ 9 %[a]		Working capital: fixed assets			1:9.
Net capital ratio		10:1					

[a] Normally expressed as a lump sum.

Table 5.15   A Simplified Balance Sheet for a Tenanted Farm

Liabilities		%	%	Assets		%	%
Current:	Creditors	8		Current:	Liquid	10	
	Overdraft	8			Working	34	
		—	16				44
Deferred:	Loans	4		Fixed:	Breeding livestock	30	
	Net worth	80	84		Machinery and equipment	26	
						—	56
			100				100
Current ratio		2.8:1		Net worth: total assets			80 %
Liquidity ratio		0.6:1		Gearing			1:20
Working capital		+ 28 %		Working capital: fixed assets			1:2
Net capital ratio		5:1					

regularly spent compiling and analysing records can be far more profitable than staying awake to supervise night ploughing — probably unnecessary anyway if sufficient thought had been given to planning tractor use.

Record keeping may seem to be slow and hard work but, as the aim is to make farmers aware of their input and output levels, the time spent is small compared to the benefits gained. Any records kept should enable, potentially at least, farm profit to be raised enough to warrant the time and effort. Clearly, the benefits gained will depend on the quality and accuracy of the records.

Farming is a highly competitive and diverse business needing frequent decisions. Records supply the facts upon which to base these decisions.

Most farmers keep few records and tend to live too closely to the daily problems of broken-down tractors, bad weather and labour problems to study them closely. They may achieve some beautiful 'trees in the forest', like high yields, reliable tractors, a stable and efficient labour force and perfect farm roads, but the forest as a whole may be in bad shape. The business may not pay well despite some perfect aspects.

As the prices of both products and inputs often change, physical records are often more important than financial ones. This is especially true if data are needed for planning.

The advantages of keeping good physical (and financial) records are as follows:

- Those farmers who use records to produce budgets are best able to compete for credit.
- Properly analysed records provide facts on which to base decisions. Decisions should never be taken in isolation as they are all related.

- Farmers with records can plan to help ensure the best use of their available resources.
- Records enable comparisons of yields and inputs to be made between seasons, lands (fields) and livestock. On most farms some lands need different treatment. The size of these differences is seldom realised unless a record of the husbandry methods and their results is kept for several seasons. This information should lead to a definite policy of soil management and of feeding and thus remove much of the guesswork on which fertiliser and feed levels are often based.
- Records provide data that farmers can compare with published 'standards' for comparable enterprises in a similar physical and economic environment.

To summarise, physical records are kept to:

- provide some of the necessary information from which detailed financial information can be produced regularly
- check and control physical performance
- guide future decisions
- provide the basic data needed for planning

Records are essential control tools because the main need for effective production control is accurate information. Timely information is provided on the effectiveness of men and machines and the availability of inputs. If, for example, too much feed is being used, it is important to detect this adverse variance quickly so that, if it is controllable, remedial action is taken *before* big losses occur. Sensible informed decisions needed for such control can be made only when facts are both available and used.

Effective planning depends on the quality and quantity of data available to planners who must therefore set up reliable and timely information sources. Without adequate records there is no basis for planning except memory, guessing, instinct or opinion — all very unreliable as none lead to informed plans because these would be based on unreliable information and could not reasonably predict the future. We must rely on information collected in the past to guide us on what is likely in the future. This is more reliable than guessing or luck.

Skill is needed in collecting and organising physical information but even more skill and experience are needed to interpret the facts and draw right conclusions about their relevance to a given problem.

Record keeping is not itself productive, so it is pointless to record anything unless the record can be profitably used to make informed decisions.

What records to keep

Recording systems usually fall into two extremes. First, those that provide much information about the business, often collected at great effort and cost, but little used. Secondly, those that provide a minimum of rather inaccurate information used to produce financial statements for tax purposes. The best systems lie between these extremes.

There is a vast range of possible physical records and it is unlikely that any farmer will keep none at all. Most managers seem to be fairly good at keeping records of output, like yield per hectare or milk per cow, but they are often poor at recording inputs, like feed used by various types of livestock. The lesson needs to be learnt that all records that can help in running the business efficiently should be kept.

Most advisers agree that the most suitable planning data are those obtained from a farm's own past records. These reflect the particular conditions of an individual farm and the quality of its management and staff. Unfortunately, such data are never perfect. Besides possible recording errors, situations and staff change. Changes occur in techniques and the prices of both inputs and products; the composition of the labour force may change; the farmer or manager's experience will grow and his aims and motivation may change. Nevertheless, data from the farm itself still provides the best available guide to the future, even though such information may need 'normalising', (that is, adjusting to expectations for a 'normal' future year), because of unusual past or changing conditions and, in any case, expected future prices must be applied to the physical data collected.

We keep records to help us farm; we do not farm to keep records! Unlike financial accounts, which are fairly standardised, the physical records needed on farms vary widely with the enterprise and production methods used. For example, no reasonable farmer would expect to develop a successful pig breeding unit without knowing many facts about each of his sows and boars. The choice of type of records and how they are laid out depends not only on what enterprises exist on the farm but also on their intensity, on the preferences of those who are going to *use* the records and on the accounting system used. Only the least records considered essential are described in this book.

Physical records should include the following:

- Tractor and fuel use records by enterprise.
- Stock control records, for example, fuel, fertiliser, feed and spares held in store, received and issued.
- Labour records classifying its use by type of labour, time of the year and enterprise.
- Crop records including areas, yields, planting and harvesting dates, inputs used, including irrigation water, and rotations.
- Livestock records including a monthly record of numbers, births, deaths, sales, farm slaughterings and purchases; and breeding records.

Labour records will be described under staff control (chapter 7) and crop and livestock records under production control (chapter 9).

It is impossible to gauge the need for information exactly but many keen managers tend to overestimate the amount needed. There is no point in keeping a record unless it fully satisfies an essential need. Farm consultants are all familiar with the large farm office, its walls covered with graphs and records, on farms that rarely make a profit. Records not used for making decisions, as is often true of rainfall records, may be interesting for discussion in the bar but are of no value. Records can become a fetish and get so complex that they mislead instead of aiding logical decision-making.

Good farm records should satisfy the following criteria:

- they should serve a definite purpose and be used
- they should be easy for the recorder to fill in accurately
- they should be up to date so that any action needed is taken as soon as possible
- the information should be of maximum use to both field managers and accountants

There is often a clash on large estates between the needs of the accountant and the field manager both for information and for its presentation. Sometimes an accountant's presentation is such that a manager cannot use it effectively. For example, information on tractor use in hours each year is usually built into accounts in such a way that it becomes a single figure of no use to a manager making decisions on replacement or the tractor-hour needs of each enterprise.

I suggest that farmers and managers should decide what physical records are worth keeping by asking themselves the following three questions:

- What decisions do I have to make?
- What information is needed to make these decisions as rational as possible?
- What records will give me this information as easily and reliably as possible?

Individual record form design vary between countries and businesses depending on conditions and the needs of the managers and accountants concerned. For this reason, no particular set of forms is recommended. The important point is that any existing system should be modified, if necessary, so that it meets the criteria given and questions posed.

Until recently, the criterion in designing farm records was that they should be suitable for tax purposes. Although records designed with this main aim must often be used for management purposes, because there are no others, this should not hide the fact that they are usually unsuitable. This is not the only case of farmers getting their priorities wrong but it is one of the most important. It cannot be overstressed that records designed mainly for tax purposes will rarely satisfy management needs but that those designed mainly for management needs will satisfy tax ones.

Most physical records are kept by the farmer, his wife, manager or clerk, as only they have access to the data needed. However, in many countries computer-based, electronic data processing and farm-recording services are available to produce regular management accounts from raw data. Computer systems, if available, can do much to reduce the hard, repetitive work needed to analyse business detail and therefore have much more potential than merely being a quick way to obtain tax records. The crucial element for success is to provide timely decision-making information; computerised systems will be little used unless they show problem areas, suggest the nature of the trouble and indicate immediate remedial action needed.

To have information available when needed, to ensure that as many daily and yearly decisions as possible are good, we must collect it and store it so that we can find it again. There is only one way of collecting information and that is by *measurement*; there is only one way of storing it and that is by *recording*.

METHODS OF MEASUREMENT

Accuracy is often hard to obtain but no comparison of husbandry methods is valid unless it is based on correct quantities of inputs and outputs.

Scales should be used often to check liveweight gains. Small farmers could buy and share the use of a scale through co-operative syndicates — or even use weighbands. Semi-direct methods of assessing weights fairly accurately include check-weighing a few sacks or bales. This is especially important if home-grown, bulk-stored grain is used as stockfeed. If grain or fodder crops fed to livestock are not weighed, the arable as well as the livestock results can be hidden. Indirect methods must sometimes be used to give at least a rough idea of the weight of materials like silage, hay or bulk grain. The capacity of a clamp, pit or silo should be calculated and multiplied by density to give a fairly correct weight, for example, a cubic metre of bulk, shelled maize grain weighs about 700 kg. The capacities of storage bins, trailers and concentrate feed measures should also be known. A scoop shaped like a beer glass, with a small surface area, gives more consistent measure than one shaped like a dish.

Tractor meters are often merely revolution counters. If the engine normally works at 1500 r.p.m., then one tractor meter 'hour' is 90,000 revolutions whether this is achieved in 55 minutes ploughing or two-and-a-half hours cultivating. Meter hours may be used for maintenance purposes but clock hours are used for recording the time taken by a job.

Specialist instruments, like maximum and minimum thermometers, grain-moisture meters, hydrometers and hygrometers may be needed for some measurements. Other measurements are obtained by sampling, for example, soil tests, germination tests and feed analyses. Other technical measurements are sent to us automatically, for example, tobacco-grading details, weight and length of baconers.

The net areas of lands and of the various classes of arable and grazing land should be surveyed accurately. Unplantable bunds, termite mounds, roads and outcrops should be excluded.

Measurements should be as exact as necessary. It may be important that one strain of bacon pigs is consistently 5 mm longer than another. Widely accepted units of measurement should be used — not 'debbies', tins, baskets, carts, bushels or donums.

It is pointless to measure things unless we know what we are measuring. Separate animals and batches of materials must be identifiable. Grain in sacks can be identified by using different coloured twines for stitching the tops — with a note of the colour code on the store wall. Grain in numbered bins can be identified on a blackboard on the wall. Cattle can be identified by freeze branding, ear punches, ear tags, neckbands, etc.; pigs by ear notching and poultry by leg bands.

METHODS OF RECORDING

Most records, especially on livestock farms, are made on paper with a pencil or ballpoint pen. Wet paper disintegrates. Large papers easily tear and small ones are easily lost. Mud and thumbmarks are often clearer than the figures. It is best to make temporary records on the spot and to transfer them to permanent records in the calm of a clean office. (In some cases, like milk recording, permanent records may have to be made on the spot.) Ideas for temporary records include the following:

- Coloured discs or tokens for different kinds of feed. Suitable tokens are put into a box whenever a batch of feed is mixed.
- Use of blackboard and chalk or a pegboard with matches or golf tees.
- Using a fibre pen on transparent plastic over a form mounted on a board on a wall.

Machinery cost and use records

Tractor and machinery inputs costs add up to about 15 per cent of total production costs on commercial arable farms in many parts of the tropics. This proportion naturally varies with the efficiency of machine use, the degree of mechanisation and the main farm enterprises. It is greater on, say, highly mechanised maize farms than on more labour-intensive tobacco farms. Nevertheless, machinery costs are substantial and inefficient tractor use in particular can seriously affect costs and profits. Although most of the tropics are less mechanised than temperate coun-

tries, rising wages and output are increasing t[he] need for careful and able tractor and machi[nery] use.

Machinery records are tiresome to keep, nee[d]ing daily entries, and they give figures that [an] experienced farmer might be able to estima[te] fairly well. However, if accurate records are kep[t] experience often shows that farmer's ideas [of] work rates are often too optimistic. They tend [to] forget the times when things went badly an[d] estimate 'normal' performance without regard f[or] breakdowns, bad weather or other delays.

Records of machinery (and labour) use are us[e]ful because they:

- enable job performance to be compared wi[th] other farms or with 'standards'
- provide planning figures
- provide a check on whether work is progres[s]ing as planned

Their main use may be in showing the proportio[n] of tractor time spent on non-productive uses, lik[e] maintenance of roads and fences, rather than [in] revealing the actual hours or litres of fuel used [in] specific enterprises.

Tractor and equipment logs should be ke[pt] either in a notebook or on a ruled blackboard [in] the workshop with a line for each main machine[.] This helps to ensure regular maintenance accord[-]ing to hour-meter readings and to check on th[e] extent of repairs and maintenance. The latte[r] gives reliable facts to help judge when to replac[e] a tractor or machine. However, if a roug[h] analysis from accounting statements shows tha[t] machinery costs are too high, there is a stron[g] case for more detailed analysis of them.

Where detailed tractor records are kept the[y] can be used to allocate tractor and field machin[-] ery costs to enterprises.

The main machinery costs are:

- fuel, oil and grease and electricity, approx[.] 25%
- repairs, maintenance and operating labou[r] approx. 25%
- depreciation and interest approx. 50%

The first two groups of costs are variable and ca[n] be allocated to enterprises if detailed records ar[e] kept.

Ideally, the repair and maintenance cost o[f] each main machine on the farm should be sep[-] arated so that a history develops that can be use[d] for gauging not only when to replace the machin[e] but also whether to replace it with a similar o[r] different type. The main routine services shoul[d] be recorded to prevent omissions and this can raise trade-in values. Unfortunately, it is often hard to allocate the cost of spare parts from dealers' statements; nevertheless the effort is worthwhile.

Depreciation and interest are 'fixed' costs however, much work is done. To spread these fixed costs it is important, for example, to keep a tractor working usefully for at least 1000 hours each year. However, it is normally thought that a second tractor is needed when annual total tractor need exceeds 800 hours. This and further tractors should each cope with a further 1200 hours.

Notebooks should be kept to record the costs of tractors, other farm vehicles, stationary engines, like irrigation pumps, and the number of power units used by electric motors. The cost of farm labour used for maintenance should be included.

Choice of the physical unit in which tractor work is best recorded depends on practical convenience and the availability of meters. The three main units are actual time used on the job, meter hours and litres of fuel used.

Using actual hours of work is not a good way to allocate costs to enterprises — although convenient. An hour's ploughing will cost more in wear and tear on a tractor than an 'hour's' light carting. A farmer's estimated hours worked does not allow accurately for stoppages. Nevertheless, tractor timesheets should be kept for planning the time needed to do jobs as described in chapter 4.

If meters are available on each tractor, this is much more precise but does not relate closely to fuel use. One tractor when ploughing can use 6 litres/meter hour whereas when disc harrowing it can use 3.5 litres/meter hour; nor is this unit similar between light and heavy tractors.

I suggest that tractor and field machinery running costs are best assigned to enterprises in proportion to the fuel used in each. This is favoured because fuel use varies with load, gear, speed and time and is thus closely related to work done, even though light work may be done at high r.p.m. with high meter hour readings. Fuel use tends to be high when working with costly machines (sub-soilers, ploughs and disc equipment) whereas the opposite occurs when using equipment that is cheap to run (trailers, spike harrows and mills).

Fuel itself represents up to 40 per cent of total tractor running costs. Checking fuel use also deters waste and petty theft which occurs alarmingly on some farms. In one area of Zimbabwe, annual diesel use fell by a mean of 2700 litres on those farms that began recording fuel use. Fuel records indicate engine condition and facilitate timely servicing and oil changes. A sharp rise in fuel use usually means that the engine needs tuning; it may indicate a leaking fuel pipe. Light and medium tractors cost similar amounts for each litre of fuel used, though they differ greatly in hourly cost.

There are several ways to record tractor fuel use, the following are some of the most common:

- Fuel gauges on the tractor. This is often broken and soon becomes inaccurate.
- Fuel gauge on bulk storage tanks. This is a good but costly method.
- Gravity level. This is a vertical clear plastic tube connected outside a fuel tank. In time the plastic dulls and the level is hard to read.
- A dipstick welded to the fuel tank filler cap. The tractor is parked on level ground with a full tank, five litres are drained out at a time and the dipstick is marked each time so that it shows how *empty* the fuel tank is at any one time. In this way, by looking at the dipstick after finishing a job, one can see how much fuel was used on that job.

Fuel use can be noted by the driver at any time and be entered onto a daily record by the farmer, manager, assistant or clerk. If a tractor has done several jobs during the day, an estimated fuel allocation is made to each enterprise. A monthly summary is made (as in labour use recording). The details of fuel use records will vary between farms but a permanent routine should always be set up.

Table 5.16 Estimated Average Fuel Use in Litres/ Hectare/Year for some Tropical Crops

Flue-cured tobacco	190 litres/ha	Groundnuts	65 litres/ha
Burley tobacco	180	Soya beans	55
Cotton	110	Sunflowers	55
Maize	100	Sorghum	55
Irrigated wheat	85		

A useful daily fuel use record is shown in figure 5.13. The codes indicate enterprises, maintenance and development work to suit a particular farm. For example, on the third of the month Tractor A and the combine harvester were working on sunflowers and Tractor B was being serviced. Figure 5.14 shows a monthly summary of tractor fuel use by enterprise or job.

At the end of each year the financial accounts and the physical records of non-specialised tractor *and implement* running costs are added together to enable those running costs to be allocated logically to separate enterprises. (All the running costs of specialised equipment, like a sugar-cane loader, are allocated to the specific enterprise.)

Let us say a manager's year's end figures are as follows:

Total running costs of non-specialised tractors and implements for year ending 30.9.86

	$
Fuel, oil and grease	4050
Repairs and maintenance	6075
Total running costs	10125

Litres of fuel used − 22500 litres

Then, running costs/ litres fuel used $= \dfrac{10125 \times 100}{22500} = 45¢$

Now, if say, 100 litres/ha fuel were used on maize and 100 ha were grown then the non-specialised machinery running costs assigned to the maize enterprise should be $(45 × 100 × 100)/100 = $4500.

Table 5.16 indicates common annual fuel use per hectare for some tropical, mechanised estate crops.

Few farmers overcultivate, far more plant crops in too poor a tilth for maximum germination. Thus, a notably heavy use of tractor hours or fuel does not necessarily merit criticism. Sometimes higher than average machinery costs go with higher profits. Knowing the number of hours or litres for an enterprise, therefore, may be more misleading than helpful. Of greater importance is

knowing the cost of running a tractor or lorry and records should be designed for this main aim. This means that a record is needed of the litres of fuel used by each vehicle.

Most tractors will hold about 70 litres of fuel in their tanks and use about 10 litres an hour when pulling hard at 2 km/h and use about 5500 litres a year. Typical figures of hourly fuel use of small — medium tractors are given in table 5.17. Where consumptions below about 1.8 l/h are recorded, this indicates that much of the tractor's life is spent idling.

Table 5.17   Average Fuel Use by Operation

Operation	Fuel consumption in litres/hour
Rotary cultivating	4.5
Ploughing, cultivating, disc harrowing	3.6
Combining	3.2
Harrowing, rolling, ridging, planting and belt work	2.7
Hoeing, light harrowing, fertiliser distribution, mowing hay and silage, baling	2.3
Spraying	1.8

Large tractors, used mainly for heavy work, tend on average to be used at a higher percentage of their potential power than smaller ones. Thus, typical average fuel use is 2.2 l/h for a 30 kW tractor, 4.5 l/h for a 50 kW, 9 l/h for a 70 kW and 11.2 l/h for a 75 kW one. Maximum consumptions are about four-fold for the small tractors and about double for the larger ones.

There are three degrees of detail in which tractor fuel may be recorded:

		Tractor A				Tractor B			Combine Harvester		
Date	Code	Litres fuel	Litres oil	Driver no.	Code	Litres fuel	Litres oil	Driver no.	Code	Litres fuel	Litres oil	Driver no.
1	MG	15	—	1	MG	20	—	2	SU	32	—	3
2	MG	13	—	1	MG	18	—	2	SU	24	—	3
3	SU	16	1	2	MT	—	4½	—	SU	27	3	3
4	TB	13	—	1	MG	10	—	2	SU	30	—	3
5	TB	12	1	1	TB	22	1	2	SU	25	—	3
etc.												
29	CT	14	—	1	TB	17	—	2	MT	—	—	—
30	CT	12	—	1	TB	15	1	3	MT	—	4	—
31	CT	13	1	3	CT	21	—	1	—	—	—	—
Totals	—	352	12	—	—	424	27	—	—	450	18	—

Figure 5.13   Daily Record of Machine Fuel Use — October 1986

Enterprise	Code	1	2	3	4	5	6	7	8	31	Total litres used	Litres used B/F	Cumulative total and C/F
Maize grain	MG	35	31	—	10	—	—	12	20	16	201	374	575
Citrus	FR	32	24	43	30	25	—	—	—	—	154	297	451
Burley tobacco	TB	—	—	—	13	34	27	29	30	—	307	563	870
Cotton	CT	—	—	—	—	30	24	22	25	34	264	512	776
Maintenance	MT	—	—	—	—	—	—	12	—	15	64	119	183
Development work	DV	—	—	10	—	8	—	—	7	—	36	73	109

Figure 5.14   Monthly Summary of Tractor Fuel Use — December 1986

Whole farm use

The total tractor litres used during the season are recorded from the financial accounts. A farmer can thus calculate his average variable cost per litre of running a tractor.

N.B. The variable cost excludes depreciation.

Comparison can be made with other farmers' costs per litre. From this the farmer will get an idea of the efficiency of his tractor use.

Single enterprise

Operational

The total litres are recorded for each operation. Farmers can detect any weakness or inefficiency.

Preventive maintenance of machinery

Preventive maintenance is work done to prevent failure. It should not be neglected, even by the non-mechanically minded. Enthusiasm is as catching as apathy so, before expecting too much from his operators, a farmer should look critically at his own, and perhaps his assistant's approach to routine maintenance — more so because he will often have relatively untrained operators.

Maintenance can include both inspection and planned repair or replacement and also breakdown maintenance. The available maintenance strategies are all variants on one or more of the following:

- Run until failure and then repair or replace.
- Preventive maintenance based on some fixed period, e.g. a pump after 500 running hours.
- Preventive maintenance based on the condition of the plant, e.g., tyres when the remaining tread is down to 1mm.

A complex piece of plant may need different routines for different components and for varying

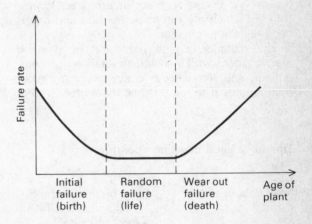

Figure 5.15   The Three Stages of Failure of a Plant

stages in its life cycle. Three different stages can be distinguished with different types and rates of failure.

Planned preventive maintenance helps to minimise the problems of breakdown maintenance. It recognises that each item, like moving parts, bearing surfaces and parts subjected to sudden pressure or vibration, has built-in weaknesses. A system of planned maintenance will increase the preventive maintenance done — especially when the work load is heaviest. It should always be done either at the end of the day or early in the morning before going to the lands.

There is a wide range of efficiency in machinery use and many farm managers neglect even to read instruction books and prepare preventive maintenance and servicing schedules. Good farm management, however, must include good machinery management to reduce unit costs.

With growing efficiency and mechanisation, the need to use plant and equipment close to its full capacity becomes ever more important. It is pointless to organise labour to a high degree of productivity unless plant and equipment allows the full use of such labour. This challenge usually has not been met. Machines are bought to reduce labour needs without realising that they need skilled labour and this situation, unless checked, can raise total costs.

The present level of maintenance accepted by many farmers is merely repairing breakdowns. This negative attitude does nothing to prevent breakdowns. It inevitably results in high labour costs, lost production through unavoidable breakdowns and excessive costs incurred on spares, some of which are run to destruction and others need replacing too soon.

These things can be prevented by planned maintenance, which anticipates failure of components and stimulates effective action to avoid such failures. The result is that more time is spent preventing breakdowns than in repairing them. If a machine is properly lubricated, routine adjustments made, all the nuts tightened and worn parts replaced *before* failure, it can break down only through an accident.

Preventive maintenance reduces damage, wear and cost. Timely care and attention extends the useful life of a machine, reduces fuel costs and improves both safety and efficiency while reducing the number and severity of breakdowns. It can also lead to better working conditions for the operator. Machinery repair and maintenance costs can be reduced by up to 50 per cent by regular and correct maintenance. An engine tune-up could reduce fuel use by 15 per cent and raise maximum power by over 10 per cent. On a 75 kW tractor this could mean 15–18 litres saved each day and a bonus of 7–8 extra kilowatts of power.

The main cause of premature failure of farm machines is delaying things until tomorrow. The costly results of fairly small 'oversights' by an operator are well known but worth repeating.

- A major engine overhaul for a tractor after only 55 hours work?
  Ridiculous! But it happened owing to excessive engine wear caused by too little oil in the air cleaner.
- A crawler track pin bush worn down rapidly in about a sixth of its usual life due to over-tightening of the tracks but, also, this bush was not turned. Turning would have extended its life by over 40 per cent.
- A hydraulic pump vane ring worn out in about 100 hours when its usual life should have been about 5000 hours. The extra wear was caused by dirty oil.

Many farmers will put an untrained tractor driver on a machine and send him off to the land with blind faith. The tractor, now disappearing fast down the road, may be the most costly single movable asset on the farm.

The real key to success of any maintenance scheme is the effectiveness of the training given to those who use it. Tractors and equipment are becoming ever more complex and the standard of operator *must* improve to keep pace. To raise the standard of maintenance and enable a meaningful incentive scheme to be introduced, it is often thought best to allocate the same driver to each tractor every day.

The system to be described is as fool-proof as can be, considering human failure, and *can*, if the initial training and setting up of the system is

Tips for a good machine operator

DO	DON'T
Read your operator's manual.	Jump on the machine, start it and take off with a load and a cold engine.
Perform daily maintenance.	Start the engine and turn it off before it has warmed up.
Check the machine for damage and potential failure before starting it.	Jam the transmission into gear.
Run the engine until it is warm before putting it into gear.	Snap, jump or 'ride' the clutch.
Let a hot engine run a few minutes without load so that it cools before shutting it off.	Spin the tyres.
Keep nuts and bolts tight.	Overload the capacity of the machine.
Watch the instrument panel.	Let the engine idle for long periods.
Keep the machine properly serviced.	Store the machine without properly preparing it for storage.
See a serviceman for major repairs.	Neglect periodic maintenance on schedule.
Operate at safe speeds.	Remove safety shields.

done with enthusiasm, become almost self-running and ultimately demand only a small but important share of a manager's time in direct supervision. This system of tractor and vehicle maintenance falls naturally into three groups — daily, weekly and periodic routines.

DAILY ROUTINE

This consists of a fixed number of points to be checked. The following suggestions could be taught and explained to tractor drivers and, if necessary, be displayed pictorially in the workshop:

- air cleaner oil
- engine oil
- transmission oil
- radiator water
- battery water level
- some grease points
- tyre pressures
- fuel
- report anything wrong like cuts on tyres or a sudden drop in oil level

A tractor driver should need no direct supervision during this daily check and needs only three to ten minutes for the whole routine. Most tractor drivers can remember these nine points fairly easily and will attend to them carefully *if they are explained*. If the farmer occasionally stops a tractor and checks, for example, the air cleaner or battery this will help to keep the driver alert and show the farmer's enthusiasm for proper maintenance.

WEEKLY ROUTINE

This consists of a written checklist, like the one shown in figure 5.16, that can be designed for any make of tractor or vehicle. It would be unfair to expect drivers to remember everything needing weekly attention without a list.

Owing to the importance of this weekly service, supervision should be available; if not the farmer himself, then his assistant, workshop foreman or a particularly responsible and capable driver. This person's duty is to remind any illiterate driver of the points to be checked and also answer any queries that arise.

An attempt is made to avoid the tendency with checklists for items to be neglected, by requiring measurements to be taken (see clutch, brake, tyre pressure) whether adjustment is needed or not. The other important features of this checklist are that it not only forms part of a permanent record of attention (each one is filed) but the operators' names are recorded. They soon learn from experience that they will be held responsible for later breakdowns caused by poor servicing.

PERIODIC ROUTINE

This covers items like oil changes and filter changes that do not often arise, for example, every 150, 600 or 1200 hours. The timing of these operations is usually noted on a blackboard on the workshop wall with tractors listed and, opposite each one, the hours at which engine oil and transmission oil, etc., must be changed.

For plant and equipment used only seasonally, end of season maintenance is important before storage; as follows:

- If easily done, dismantle working parts and clean.
- Check for wear, missing parts and breakages.
- Make repairs and obtain any spares needed.
- Remove any corrosion, by heat, chemical or mechanical means, and take preventive measures like painting.
- Reassemble and lubricate
- Leave belts slack, support tyres, take special measures on engines and batteries until equipment is needed.
- Store under cover.

All too often a tractor tyre's life is shortened by the operator failing to give it the necessary timely attention. As tractor tyres run at low speeds over fairly soft surfaces, they are usually in service longer than car or truck tyres and are therefore exposed longer to the dangers of poor maintenance. Tyre operating costs are minimised if regular inspection, correct air pressures and protection from accidental injury are practised. Regular inspection ensures early detection of cuts and other damage.

Stock control

The function of stocks is to provide an economic linking of supply and demand for stocks of materials, spares and finished goods.

Sometimes called inventory control or material control, the purpose of stock control is to optimise the level of stocks held by regularly arranging receipts and issues to ensure that stock balances are adequate for current consumption, with due regard to economy in all costs. Stock control requires:

- assessing items to hold in stock
- deciding how much of each to hold
- regulating inputs into store

Stocks provide a pool from which needed inputs can be taken continually and allow the accumulation of products for sale. Every business holds three main types of stock: inputs from suppliers, work in progress and products awaiting dispatch or collection. A problem in any business is the size of the necessary investment in plant, equipment and stocks before realising any return.

Stocks tend to fluctuate and are harder to contro than plant or equipment.

Holding stocks always costs money so stoc levels must be kept as low as possible withou preventing efficient running of the business Critical examination of stocks to see if variou methods of stock control are applicable usuall results in savings. The two main problems are t fix a reasonable minimum figure and to assess th reliability of suppliers.

The aims of stock control are as follows:

- To provide information to managers on stocks for example, to advise them on slow-movin stocks.
- To ensure that too much stock is not kept, tha too much working capital is not tied up i

---

WEEKLY TRACTOR SERVICE

Tractor: Date:

Hours: Serviced by:

TICK

1.   Air cleaner:
        Wash bowl and gauze . . . .
        Check stack pipe . . . .
        Renew oil (Type:) . . . .
        Check hose clamps . . . .
2.   Brakes — Free play . . . . cm . . . .
3.   Clutch — Free play . . . . cm . . . .
4.   Fan belt tension correct . . . .
5.   Fuel sediment bowl cleaned out . . . .
6.   Engine oil level correct . . . .
7.   Injection pump oil level correct . . . .
8.   Diff. and gearbox oil level correct . . . .
9.   Final drive oil level correct (Fiat) . . . .
10. PTO input and output oil levels correct (Fiat) . . . .
11. Hydraulic oil level correct (Type:) . . . .
12. Water level correct . . . .
13. Greased all grease points . . . . There are . . . . grease nipples
        damaged or missing at the following points: . . . .
14. Battery water level correct . . . .
        Terminals tight and clean . . . .
        Frame bolts tight . . . .
15. Tyre pressures when checked . . . . front . . . . rear . . . .
        Corrected to . . . . front . . . . rear . . . .
16. Check front wheel bearing play . . . .
17. All nuts and bolts checked . . . .
18. Tractor washed . . . .
19. Remove and clean magnetic filter plug (Nuffields only) . . . .
20. The following points were noticed: . . . .
        . . . .
        . . . .
        . . . .

Figure 5.16   Example of a Weekly Tractor Service Checklist for a Farm with Two Makes of Tractor

stocks and to minimise storage costs. The aim is to hold the optimum level, all things considered, not the minimum level one can get away with.

- To provide just sufficient stocks to prevent production stopping if suppliers do not deliver on time. Materials and parts must be there when wanted. There is thus a conflict between the need to provide good service and to economise in stock holdings. An adequate compromise is needed.
- To help in standardisation and variety reduction, thus reducing paperwork.
- To allocate storage space for in-process and finished goods, like cured tobacco in bulk before grading and after baling.
- To record and store all goods safely and in good condition.
- To relate materials use to available finance.
- To plan input supplies on long-term contracts against production programmes.
- To benefit from seasonal price changes and quantity discounts.

Physical measures are needed to protect stocks. Locked, wired-in enclosures are needed for 'attractive' goods liable to theft, for example, sacks, paraffin and rations. The following protective steps are recommended:

Cement and fertiliser	— off the floor on dunnage.
Timber	— sawn timber parted by scantlings for air circulation.
Tyres	— away from heat and upright on treads in racks to avoid weight damage to the internal fabric.
Grain	— protected against pests and moisture.
Machinery and equipment	— well-greased and wrapped in plastic when not in use.

Goods often used or needed quickly should be stocked. Readily obtained goods with infrequent demand need not be stocked. Holding spares for plant and machinery depends on the technical importance of a part and its expected delivery time. Some standby spares, like generators or alternators for vehicles, must be stocked more to insure against loss of time than to allow for any known rate of use.

Holding too few stocks means frequent ordering and handling of goods and there are dangers of items running out, lost production and profit.

Holding too much stagnant stock ties up capital that could have been used elsewhere. It costs money to hold stocks; in terms of storage space, equipment, insurance, deterioration, time, interest or opportunity cost and obsolescence, besides the direct cost of the stocks. The annual cost of holding stock is usually between 15 and 40 per cent of the value of the stock. It is unwise to hold large stocks of items that often change in design. A balance is needed between having too little or too much stock.

It is hard to obtain reliable information on stockholding costs but a general rule is to order large quantities of low unit cost items and small quantities of high cost ones. However, this may not apply if attractive quantity discounts are available for large orders or if a price rise is expected soon.

Pareto's Principle states that in any series of elements to be controlled, a selected small fraction of numbers of elements usually accounts for a large fraction in terms of effect. This is also called the concept of the 'vital few and the trivial many' and it has wide significance and application in management. Its simplicity and universality make it valuable when looking for ways of raising efficiency, reducing unnecessary work, etc. Awareness of this rule is to be aware of a simple way of finding those few items that matter most. Thus 85–90 per cent of the total value of stocks held is probably represented by only 25 per cent of the items. It is the numbers of these items that must be optimised most urgently by *selective* stock control.

Table 5.18 Stocks ranked in Order of Total Value of each Item

Part no.	Total value	Cumulative value	Rank
	$	$	
A 164	14260	—	1
A 784	14000	28260	2
:	:	:	:
D: 78	23	65786	633

Use of Pareto's Principle in stock control simply means ranking all the items stocked according to total and cumulative value and finding a more precise rule for your own business; for example, tight control of the most valuable 25 per cent of items and looser control of the rest. This concept is shown in table 5.18 and figure 5.17.

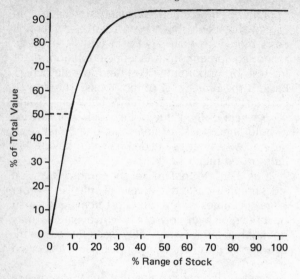

Figure 5.17   Illustration of Pareto Principle in Stock Control by Lorenz Curve

The systematic approach to cost reduction considers the main costs from two angles: the amount bought and the price paid. Some managers seem too ready to accept that neither of these can be reduced without affecting quality.

The best value for money is not always the same as low cost. It can be obtained in several ways:

- By getting a better article at the same cost.
- By getting the same article cheaper.
- By paying more for a much better quality article. This is justified and gives better value if, in use, full advantage can be taken of the extra quality, for example, by using less of it or if it results in less labour, less maintenance or more output. Use of urea instead of sulphate of ammonia could come into this category.
- By paying much less for a lower quality article if the quality is adequate for the purpose. This is a case of over-specification — a useful and usually overlooked way of cost reduction.

Purchases should be made to best advantage. They fall into four groups:

Regular orders

For example, fertilisers, fuel and stockfeeds. Ask several firms for their terms for the order for a year. Either negotiate a discount for cash or arrange *in advance* to withhold payment during those months when working capital is likely to be scarce. Examine the economics of early delivery or quantity discounts.

Capital purchases

Such as machinery, ask several dealers to quo in writing for the new machine and, if relevar for their trade-in price on the old one. Accept th best tender and notify unsuccessful tenderers th you were given better terms elsewhere. They w get the message and try to be more competitiv next time. Tendering for your needs helps establish a reputation for fair albeit tough tradin It avoids the position of the last person you spe to on the telephone extracting from you the be offer to date and then undercutting it.

Normally, for all important items, bids a requested from several suppliers before the initi order is placed. Repeat orders are subjected the same process at intervals — perhaps yearl Usually, two or three bids are requested, with th larger number applying to the purchased items highest turnover. The enquiry should be made o a printed form, prominently marked as 'Enquiry 'Quotation Request', etc., to avoid it being mi taken for an order.

Each potential supplier should be aske specifically to provide such information as:

- The price at which goods can be supplied the delivery point including any quantity trade discounts.
- Terms of payment.
- Confirmation of delivery date.

Unavoidable but irregular costs

For example, repairs and veterinary bills. Pa immediately, or within the allowed time, an request the standard discount. If you suspec overcharging, *say so* and transfer your custor elsewhere.

Unplanned, impulse purchases

See salesmen only by appointment and arrang either that you will contact them when needed o that they should contact you only if they hav something new to sell.

The functions of purchasing are as follows:

- Assessing of alternative suppliers and using small but effective number of them for th more important inputs.
- Buying at competitive prices; giving du regard to total long-term cost rather than t spot price at any one time.
- Rating suppliers for delivery, service an quality, besides price.
- Maintaining adequate stock levels.

The method of buying should suit the nature of the demand for the item and existing market conditions. A *spot order* is a single order for goods needed immediately. This method applies to emergency measures and to goods so seldom used that they are not stocked. It may or may not be repeated and a price is negotiated.

A *contract order* is a contract entered into whereby the supplier agrees to deliver a quantity of goods over a medium or long time period. Delivery orders are specified as to time and quantity later but a price is negotiated for the whole contract at the start. contract buying is suitable for materials needed in quantity over a long time, like coal and fuel. If the contract calls for deferred delivery, it is possible to take advantage of prices in force when the contract is placed, while spreading delivery according to estimated future needs. The buyer assures himself of deliveries without undue stock investment.

Speculative buying occurs, without close reference to the needs of the production plan, if a buyer purchases excess stocks when prices are low as he expects them to soon rise again.

On large commercial estates, problems often arise in deciding where to off-load goods. Sometimes it is necessary to have more than one off-loading point to ease later handling. To avoid problems in these cases, precise instructions should be given to both staff and suppliers.

With regard to records, invoicing is a check on goods sold; ordering and its allied procedures are a check on goods received. The simplest way to reduce mistakes is to use a triplicate order book. When goods are ordered, the details are entered on an order form, the top copy of which is given to the supplier. The second copy is kept in the desk tidy or in a bulldog clip and is a reminder of what has not been delivered. Critical items can thus be checked easily and are less likely to be forgotten.

All deliveries should be accompanied by a delivery note which should go promptly to the farm office if not taken there by the person delivering the goods. Once the invoices are to hand, they are checked against the relevant delivery note(s) which are then attached to the invoices to await receipt of the statement. When payment is made, the copy invoice is destroyed and the third copy in the order book marked with the date paid and the amount. The sum of all unpaid invoices at any time is the sum owed to sundry creditors. Services, like telephone, water and electricity, should be entered on an order when paid so that a full record is kept in the order books.

Another advantage of using an order book is that it can enable a farmer to trade with wholesalers who will do this more readily if given an official-looking order. This could lead to large cost savings.

A discount or rebate is a drop in the price of goods or services allowed for some reason like early payment, early delivery or bulk purchases. Discounts allowed for the last two reasons should exceed any extra storage costs.

Storing products for which demand is seasonal, like fertiliser and some machinery, is costly for the manufacturer or dealer. He must provide or hire a store and supply or borrow the capital tied up in stocks. It is therefore in his interest to give farmers an incentive to buy out of season. However, farmers then incur these costs and, although storage space may cost little or nothing, the capital certainly will — either in interest or opportunity cost.

Most companies offer discounts for bulk purchases. They can do so because economies arise in dealing with large orders; the same vehicle may be needed to deliver the load, whatever its size, and the clerical work for each customer changes little. Benefit from any quantity discounts, provided the extra amount to be bought to qualify for such a discount does not cause excessive stock holdings.

A natural development of co-operative selling of products is co-operative buying of inputs. Quantity discounts are obtained by the ability of co-operative societies or buying groups to buy in bulk and the common problem of farmers having to buy retail and sell wholesale is overcome. Certainly, no one who has negotiated for feeds, fertilisers or traded-in old machinery can doubt the value of organised buying; even those farmers who do not belong to groups benefit from the competitive selling that buying groups can stimulate.

Provisioning means determining needs in advance and deciding both when to order and how much. It usually involves fixing minimum stock levels. These are danger levels and depend on the severity of a runout on production, the rate of use and the 'lead time' between order and receipt of goods.

One method widely used for small stock items in constant, uniform use, is very simple. The stock items are placed in two bins and, when one is empty, another order is placed for the particular item and those from the second bin are used. This two-bin, or fixed order quantity system avoids keeping detailed records of stock and saves much time. The second bin holds a quantity equal to the 'buffer stock' plus expected demand during the lead time.

There are three costs involved in holding stocks — the purchase price itself, costs connected with purchasing and the costs of storing the stocks. Purchasing costs include those of obtaining quotes, preparing and sending out purchase orders, receiving goods, moving them and accounting costs — all net of any discounts.

Assume that a material is used at a constant rate thoughout the year; for example 97 units a week. Total annual use is therefore 5040 units. This could be bought in several ways — 5040 once, 2520 twice, 1680 three times a year, etc. On the one hand, buying a large quantity infrequently minimises purchasing cost but results in a large average stock-holding cost. Conversely, small frequent purchases results in high buying costs and low stock-holding costs. This is shown in figure 5.18.

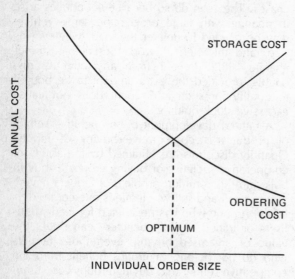

Figure 5.18   Effect of Ordering Frequency on Total Annual Cost of Keeping Stocks

To examine this more closely, let us assume that the unit purchase price of the item is $6, the cost of placing and servicing an order is $85 and the yearly cost of holding stocks is 20 per cent of their purchase price.

If only one order is placed for 5040 units, then on average there will be 2520 units held for the whole year and the annual storage cost of the material will be:

$$\$(0.5 \times 5040 \times 6 \times 20)/100 = \$3024.$$

The annual ordering cost will be $85 and the purchase price of the material will be:

$$\$5040 \times 6 = \$30240.$$

As the purchase price is assumed to be constant each year, however many orders are placed let us ignore it and concentrate on the sum of the storage and ordering costs — the acquisition cost. The variation of acquisition cost with the number of orders placed is seen in table 5.19. This shows that the best policy is to order 840 units every two months.

It is necessary to consider the time between placing an order and receiving the goods — the replacement or lead time — and also the safe level of buffer stock below which stocks must never can be allowed to fall before replenishment is made. That is, one should determine the reordering level. The main technique for deciding how many units to order, the 'optimum re-order quantity' and when, consists essentially of variations on the simple equation:

$$Q = \sqrt{\frac{2\,PY}{1\,C}}$$

where Q = optimum re-order (economic purchasing) quantity

P = the cost of placing one order

Y = the annual rate of use.

1 = annual stockholding cost as a decimal of their cost

C = unit cost of the input.

Thus, in the example above:

$$Q = \sqrt{\frac{2 \times 85 \times 5040}{0.2 \times 6}} = 845$$

This simple equation is a useful guide to stock-holding decisions under conditions of certainty when demand for an item and the lead time for obtaining it are known. However, when demand and lead times are uncertain the two main approaches are to hold lot size constant and vary the time between placing orders and, conversely, to hold the time between placing orders constant but vary the lot size. These methods satisfice rather than optimise the decision.

There are problems in obtaining figures for the optimum re-order quantity formula, especially in times of inflation. However, with so much of a firm's capital tied up in stock, it is important to avoid holding too much stock.

Figure 5.19 shows the stock cycle. Average stockholding is the buffer stock plus half the re-order quantity.

Table 5.19   Variation of Acquisition Cost with the Annual Number of Orders

No. of orders placed/year	Quantity/order	Mean annual stock	Annual storage cost	Annual ordering cost	Acquisition cost
			$	$	$
1.	5040	2520	3024	85	3109
2.	2520	1260	1512	170	1682
3.	1680	840	1008	340	1096
4.	1260	630	756	340	1096
5.	1008	504	600	425	1025
6.	840	420	504	510	1014
7.	720	360	432	595	1027
8.	630	315	378	680	1058
9.	560	280	336	765	1101
10.	504	252	302	850	1152
11.	458	229	275	935	1210
12.	420	210	252	1020	1272

Stock control records

In general, these records are less useful in farming than labour and tractor-use records. Fairly precise cost allocation of fertilisers to different enterprises can usually be made from memory, especially if a list of the quantities and costs of each kind of fertiliser is available and similar considerations might tell against the keeping of fuel records. Nevertheless, a simple store keeping system can help on small farms as well as on large ones.

Stocks should usually be kept to a minimum provided there is no danger of them being absent when needed and that likely price rises have been considered.

Stock records are documents recording each day full details of individual receipts, issues and balances of stocks in hand. As daily entries are shown, this system of records is known as 'perpetual inventory'. Entries should include only what is essential for efficient control.

Replacing stocks takes time after ordering and is cheaper if done as a routine rather than when unavoidable — by special transport, telegrams, telephone calls or a personal visit. A proper system should ensure that materials are in hand when needed and thus cut out journeys made just to get a few routine items like plough discs. If it is 60 km to the local stockist, the vehicle cost at, say, 20¢/km would be $24. Add the cost of the driver and you have a costly way of obtaining spare parts. A more economical practice is to order ahead for collection during normal business trips.

A good stock control system ensures that just sufficient spares are held and that the least capital is tied up. It also allows seasonal needs to be foreseen accurately, as they are based on records rather than memory. This could enable bulk orders to be made at favourable discounts.

The purpose of stock records is to show what is in hand without needing a physical count, to decide how much to order and when and to supply information for the annual valuation (stock-taking). Information about stocks is needed in terms of value. Stock values are needed yearly to calculate the profit made each year and prepare a balance sheet. Another advantage of stock recording is that if a machine is sold, the spares

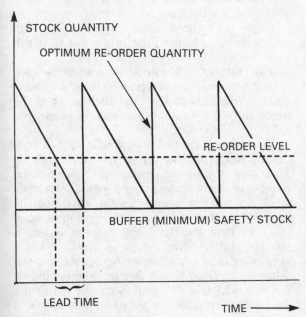

Figure 5.19   The Stock Cycle

for it can be sold together with it.

Records of materials, like feeds, grain, fertilisers, seeds, fuel and pesticides, not only help to ensure timely re-ordering but they also tend to reduce both theft and waste. An optimum re-order level should be set for all stocks. A simple system of stock record cards, bin allocation and coding of spare parts virtually removes the problem of a farm losing money because men and machines are idle owing to absence of an essential spare part.

Stock records, which may be in the form of cards or loose-leaf ledger sheets, are the basis of stock control. Well-designed cards kept together in suitable filing boxes or cabinets, properly grouped with guide cards, are usually adequate. These cards may show only quantity, quantity and unit price or quantity, unit price, value of each transaction and value of the balance.

Farmers should record all deliveries into the store and all drawings to sale, individual lands, labour rations or to livestock enterprises. There may, for example, be no predetermined breakdown of the use of, say 5 tonnes of cottonseed. Its cost can be shared between individual livestock enterprises only from use records.

Other methods of stock recording, besides individual cards for each item of stock, include loose-leaf books, and proprietary visible index systems — with or without coloured signals to give extra information, such as, below minimum buffer stock, no movement for a year, etc.

Although cards may be kept in a box in alphabetical order, there is a risk that the card may be forgotten if someone is in a hurry whereas a card fixed to a bin should act as a reminder to make an entry.

Bin cards are records fixed to physical stock in bins, racks, etc. They are marked each time items are bought, ordered or issued and can either be made or bought ready printed. They should be fixed to the bin front so they can be seen easily and be written on. An advantage of a bin card is that each time an item is removed the remaining balance can be counted and checked against the figure on the card. In this way a constant stock check is kept and any differences appear at once.

Figure 5.20 shows the sort of bin card that can greatly aid efficient stock control.

As parts are taken from the bin the entries on the card will in time give a history from which the maximum number of stock can be calculated whilst taking into account seasonal peak needs. In the same way minimum stock is obtained by finding how long it takes to get a new supply; the minimum being the number needed during this lead time. It is essential to check the cards as a routine — with preferably one person responsible for this.

If a varied collection of stocks is held, it is usual to divide them into 'groups', arranged numerically and alphabetically. For example, suppose group no. A is tractor spares. A separate card would be made out for each size and class and each item given a part number. The item classified as part no. 3 in group no. 5 would then be coded as 5/3. With this system all goods with the same characteristics are kept on adjoining cards and a 'master' card, kept at the front of the tray or drawer, will show totals of each group at any time.

The storeroom should be used for this purpose alone. It should be dry and lockable. Some thought should be given to weight distribution, keeping heavy items as low and near to the outside walls as possible.

The bins are easily built from wood and make a 'bad weather' job for farm staff. It is worth placing a small board across the front of the bins that, besides preventing parts falling out, provides a place for fixing the bin cards.

A storekeeping rule is to place similar items in different parts of the store so that they will be selected by the system rather than be 'recognised' and selected by error.

Given adequate storage for materials, stock control is little trouble. For each item in stock a section of floor space, a bin or a pigeon hole is allotted and a bin card is provided.

Figure 5.21 shows a typical bin system with bin cards fixed to the front of the bin to which they refer.

The problem of finding an item when the card is on the bin can be overcome by using a location code for the bins plus an alphabetical list of the items and their location; for example, filter, oil, MF 175....A3, etc.

Visual stock control is a simple way of controlling stock by quantity in the store itself by separating in different bins or tying together the number of items representing the minimum stock level. An order is made when this fixed quantity has been reached. For low value items in constant use, like nuts and bolts, this two-bin system is suitable and involves no paper work. A small container, such as a kilogram milk-power tin or a 5-litre oil can may be selected to represent the re-order level. This is filled with the particular material which is re-ordered when only the tinful is left.

A.

Stock:	Fertiliser compound
Description:	20:20:0 + 4 % S in 50 kg plastic pockets
Suppliers:	Kano Producers' Co-operative
Price/unit:	$12
Minimum stock:	—

1985–86	Remarks	In	Out	Balance
1/10	On hand	—	—	64
20/10	Maize	—	37	27
10/11	Pastures	—	11	16
15/11	Cowpeas	—	7	9
27/11	Garden	—	2	7
12/12	Vegetables	—	3	4
1/7	Received	64	—	68

B.

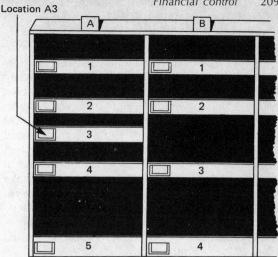

Figure 5.21 A Simple Bin System

ORDER FROM: Olympus Garage, Juba										FITS: MF 135, 165, 175 PRICE: $3		
1985/86	Ref.	Order	In	Out	Bal.	1985/86	Ref.		Order	In	Out	Bal.
29/8	O/No. 256	6			0	22/12	Tractor 4				1	5
2/9	Ac./No. 1378		6		6							
3/9	Tractors 3 and 5			2	4							
17/10	Tractor 1			1	3							
27/10	Tractor 2			1	2							
28/10	O/No. 272	4										
4/11	Ac/No. 1492		4		6							
DESCRIPTION: Oil Filter		PART NO (OR CODE) 5/3					LOCATION A3			MAX. 6		MIN. 2

Figure 5.20   Two Typical Bin Cards

COST CONTROL

Modern cost contrl means control of all costs by frequently comparing actual expenditue with standards or budgets, sot that adverse variances are found and remedied quickly.

Systematic attention is needed to achieve the potential cost reduction possible with almost any product. Managers should decide beforehand what their cost should be. Then, if these estimated costs seem too high, action can be taken before they eat into the profit. These cost estimates are then used as standards. This is the essence of cost control: setting cost targets; ensuring that people know how to achieve them; checking actual performance; finding the causes of any variances and taking remedial action.

All managers try to keep costs down but many do not understand the basic differences between *cost control* and *cost reduction* and neither of these has anything to do with those unplanned economy drives that often do more harm than good. In cost reduction, the aim is to reduce the target cost by any suitable means, for example, using cheaper inputs of equal quality, better methods or better equipment. Attempting to reduce costs irrespective of the effect on output would be irresponsible but, nevertheless, a rational examination of farm costs can make a significant contribution to the improvement of business finances.

Many firms have cost control systems but few have effective control over costs. The reasons are deeper than the common faults of over-complex paper work, out-dated and inaccurate information, etc. The trouble in practice is that most cost control systems tell the manager only what went wrong yesterday. They do not help him avoid tomorrow's problems. The only way to prevent excess costs is for a manager to take action in advance. No analysis of adverse variations can undo what has already happened.

Top management must ensure that there is personal accountability for both labour and material costs at the operational level.

The work of cost control may be confined to those items where variances fall outside acceptable tolerances and remedial action is needed. This system of 'control by exception' frees senior staff from having to examine all costs in detail.

The main areas needing detailed cost control are:

- labour costs
- insurance
- taxation
- material costs — preventive maintenance
  — stock control
  — buying policies
  — timing and size of orders

Some managers regard material costing as too hard or as a waste of time. If the raw inputs were gold coins they would take some notice but many items are regarded as of little value. However, it makes little sense for a firm to try reducing labour costs without giving materials similar attention. Well-organised materials cost control, as previously described, is essential.

Lack of cost control causes mismanagement of productive resources and lower productivity. It arises from incorrect decisions on methods and materials, wrong priorities and misdirected production plans.

Labour cost control

In most tropical countries, farming is labour-intensive and labour is one of the most important resources used to achieve the target output. Commercial farmers throughout the tropics usually hire regular labour and this can make up as much as half total production costs on commercial arable farms. However, even smallholders often hire labour, for example, coffee farmers in Kenya and Tanzania and cocoa growers in West Africa. A study of labour cost is therefore one of the most rewarding that a farmer or manager can make because it is within his control whereas most other production costs, like materials and selling costs, are mainly fixed by technical advice or set charges. They are thus less controllable. Sometimes it is easier to reduce labour and machinery costs than to raise output.

Farmers in temperate (*mainly* in the sense of climate) countries often envy the low wages that will attract labourers in most of the tropics, although low prevailing wage rates sometimes make it hard to recruit farm labour. They rarely realise that the productivity of our labour is also usually low so we are not as lucky as they think as labour cost per unit of output can be high. More important though, farmers outside the tropics rarely realise how much time is spent supervising unskilled labour. Thus one of our greatest assets, abundant and *seemingly* cheap labour, may be a liability if we must neglect our other managerial duties. In practice, many farm managers are mainly labour supervisors.

It is a manager's job to find out what is retarding labour productivity and to try to remove any blocks. Some things are harder to correct than others but usually, much improvement in our methods of labour management is possible.

Farmers often think that a higher wage is needed to raise labour productivity. Few seem to have looked for ways of raising living standards; an approach that would often be more cost-effective. For example, labourers must often walk up to 6 km to their work place or to work for seven to eight hours before having their first meal. Arguments that tropical labourers often traditionally eat only two meals a day are easily refuted by the fact that they do not work traditionally for long periods as they must when employed.

Efficiency and economy of labour use depends upon sound supervision, careful choice of methods, choice of the correct type of labour, and so on; but there is nothing here that a manager's technical knowledge and experience should not enable him to handle.

The only satisfactory way to estimate labour needs and establish a basis for later control is to compile a labour budget from the output budget by using labour standards found for each enterprise. Labour standards are best fixed by work study but managerial estimates are better than nothing.

Labour costs can vary for unavoidable reasons, like government legislation, or for avoidable reasons, like waiting for tools and orders. Excess labour costs can arise for the following reasons:

- Waiting time paid to workers when they are unable to work.
- Low productivity if only a guaranteed minimum wage is paid.
- Poor control of overtime.
- Use of the wrong type of labour.
- Unused capacity.
- Excessive piece-work payments, or excess time spent owing to faulty materials or equipment.
- Poor methods. Many jobs involve excessive labour cost and wasted effort. Each time something is picked up, moved and put down by a person, cost but no value is added. Method improvement is nearly always possible.

Labour costs can be reduced by better training and by reducing labour turnover.

Workmen's compensation protects employers from claims arising through injury to, or death of, employees resulting from accidents at work. Employers may be forced by law to have such insurance.

INSURANCE

Insurance is a contract by which, for a certain specified payment, called a premium, the insurer contracts to pay the policyholder a specified sum of money on the occurrence of a specific event. The premium varies with the type of risk and with the amount insured. The system works on the basis of spreading the risk so that sufficient premium income is received to enable the insurer to pay all agreed claims and still have a profit.

If trading conditions become adverse, as they invariably do at times, there is rightly a tendency to cut costs as much as possible. It is easy to reduce costs merely by under-insuring, presumably on the assumption that 'it will never happen to me'. This course has often proved disastrous, especially if the property damaged or destroyed consists of all the 'worldly goods' of the farmer in a situation involving the use of an average clause in the policy. 'Average', in insurance, means that the insured is taking part of the risk himself as, for example, stock valued at $20,000 being totally destroyed by fire but insured for only $12,000. The insurance company will accept a claim for only $7200 (12,000 × 12,000/20,000) as it is considered that the insured personally accepted 40 per cent of the risk. The main policies are those for life, theft, fire, hail, employers liability and motor vehicles.

Householder's comprehensive insurance provides cover against risks to the house and its contents and covers such risks as fire, theft and storm damage.

Motor vehicles may be insured against claims by third parties. This is usually essential before they can be taxed and used on public roads. This does not cover the insured for personal injury or damage to his vehicle. Alternatively, they may be insured for third party, fire and theft or comprehensively. The latter covers injury to the insured and damage to his vehicle.

Insurance is a complex subject and advice should always be sought from a reputable insurance broker who charges his clients nothing for advice. Brokers obtain their income as commission from the insurance company to whom they introduce your business.

TAXATION

Although tax evasion is deliberate criminal fraud, tax avoidance is legal. Avoidance means knowing the nature of income or company tax and finding ways to reduce it through full use of available allowances or one of the many complex financial schemes devised by clever tax accountants. It is a legitimate arrangement of affairs to minimise tax liability legally. Nobody owes duty to pay more than the law demands; it is only evasion of the law that can cause trouble.

Unfortunately, few farmers understand tax laws well enough to ensure that they minimise their tax payments. They have little knowledge of tax abatements or ways by which the taxable profit or loss may be legally adjusted to minimise total tax liability over the years.

Taxes vary between countries and over time so it is possible only to give a general outline here. It is one of the jobs of accountants to give detailed, local advice on tax matters. However, avoidance of unnecessary taxation still depends partly on a broad understanding of the subject by managers responsible for decision-making.

Farmers, whether working as individuals or companies, are generally taxed in the same way as businessmen in other industries — paying income tax or corporation tax respectively. Certain extra specific capital allowances are, however, often allowed to farmers. Income tax is the most important of the taxes levied by governments and it is usually progressive; as shown in table 2.4. With progressive taxes, higher income groups pay not only more income tax than lower income groups but proportionately more. Corporation tax is usually paid at a constant rate.

Any decision to invest in capital equipment can be affected by the tax system, especially by the rate of tax, the timing of its collection and the allowances given by the tax authorities. Taxation aspects can be decisive when considering investing in a new enterprise or farm or in extra plant or machinery.

It pays farmers who are making good pre-tax profits to write off their capital assets as quickly as they are allowed and so claim the maximum tax relief while building up a good set of equipment. The tax rate paid by a farmer influences the purchase of capital assets. The higher his tax rate, the less the net cost to him of, say, buying machinery. If farmers carry out their own repairs, it pays to do the most costly repair projects in high profit years if possible.

It is always wise to keep a rough check on the share of the farm car's annual distance travelled on purely private use. For most farmers an arbitrary allowance for the reasonable cost of the business use of the car is agreed between accountant and tax assessor in the lack of any specific evidence. Those few farmers who do not keep a record of this sort usually obtain a rise in the cost chargeable as a business expense. It is surprising what a small share of the annual distance is private for most farmers.

Other expenses that farmers may usually charge as a business expense include:

- repairs to plant, vehicles and farm buildings
- a share of the costs of a telephone used for business
- loose tools unlikely to last more than a year
- bolts, nails, sacks, veterinary medicines and medicines for the labour force
- selling charges, railages, levies, etc.
- rent
- sometimes the costs of eliminating bilharzia and liver flukes

Investment allowances are sometimes granted if a government wishes to encourage certain types of investment in the national interest. This is often the case with *new* farm improvements, fencing, plant and machinery (including tractors but excluding motor vehices) that you have created or bought during the year and brought into use for farming operations. If an investment allowance is, say, 15 per cent on top of special initial and wear-and-tear allowances, a farmer is granted a total deduction equal to 15 per cent of the cost of any asset that qualifies to be set against profits.

Special initial allowances, applicable only in the year of purchase and calculated on cost, are often available for the erection of farm improvements, such as staff housing, and the purchase of farm machinery and buildings. This allowance, which is optional, is often misunderstood and incorrectly interpreted by taxpayers. It is true that it gives an advantage to a taxpayer in the year that he obtains can asset and brings it into use and therefore encourages capital development. However, it is not true that the state is helping the purchase of the asset and, if there is a loss in the year when the asset is bought, this allowance merely arises the loss. Typical initial allowances vary from a tenth to a third of cost, thus raising the share of cost that can be written off in the first year.

Annual wear-and-tear allowances on machinery, implements, buildings, vehicles and tractors are calculated on a diminishing value after deducting any special initial allowance. As described previously, they may be calculated by the straight line or reducing balance methods. In the past, 'standard' rates of wear and tear allowances were designed to be roughly in proportion to the rate at which a particular asset is expected to depreciate. Nowadays accelerated depreciation is often allowed as government policy.

When a capital asset that has received initial and annual allowance is disposed of by sale or destruction and the amount realised on the sale or insurance recovery is less than its book value, the excess of the written-down value over the sum received is allowed as a balancing allowance against the firm's taxable income. If the sum realised on disposal, destrucion or scrapping of a capital asset that has received capital allowances in previous years exceeds its book value, the excess is added back as taxable income; a balancing charge.

Table 5.20 shows that the actual allowances to be claimed from capital investments do not vary if initial allowances are claimed or not. There may be times when there is an advantage in claiming initial allowances if new assets are bought in a high profit year but this will reduce tax allowances in later years.

Table 5.20 Example of Capital Allowances

	No initial allowance		Initial allowance of 20 %	
Assumptions:	A sugar cane harvester is bought for $36000 and is sold in its third year of use for $25000. The annual wear and tear allowance is 20% on reducing balance.			
	Allowances	Book value	Allowances	Book value
Year 1.	$	$	$	$
New machine bought		36000		36000
Initial allowance			7200	
Annual wear and tear	7200	28800	7200	21600
Year 2.				
Annual wear and tear	5760	23040	4320	17280
Total allowance given	12960		18720	

	$	$
Sale price of harvester	25000	25000
Book value	23040	17280
Balancing charge	1960	7720
Total allowances given	12960	18720
Less balancing charge	1960	7720
Net total of allowances	11000	11000

GLOSSARY

Annual accounts financial statements showing the state of affairs at a stated date and the result of operations during the previous accounting period.

Capital reserves funds in a company that belong to its shareholders but may not be distributed or are not, in the director's opinion, available for distribution.

Deadstock a farmer's assets (capital) other than growing crops and animals, for example, buildings, machinery, stocks and stored products.

Insolvency a person or organisation is insolvent if he or it cannot pay debts when they are due. Insolvency differs from bankruptcy or liquidation of a company — though this may follow from it. A rich man can be insolvent if his assets cannot be realised when he needs cash.

Market value a term, often used in published accounts, for the amount an asset would fetch if sold in a completely free market. Distinction should always be made between value in present use and value in alternative use. The latter may be much greater, for example, if farm land could be profitably developed for housing.

Net assets a term often used in published accounts for total assets less current liabilities. The significance of this total can be somewhat doubtful, depending on whether it is considered equal to capital employed — as is often the case.

Paying-in slip a document to be completed when making a bank deposit, showing how much has been deposited in different kinds of cash and in cheques.

Payment in advance an accounting term for money paid for goods or services not yet received and that cannot usually be recovered. For example, rent paid for a period ending after the date to which the accounts are made up.

Revenue reserves a general heading for all reserves other than capital reserves. They are funds available for distribution but for various policy reasons not yet distributed.

Standing order an instruction to a bank to pay a certain fixed sum at stated intervals; for example, rent or hire purchase instalments.

Trade creditor a term for a person or business to whom money is owed in the course of trade.

RECOMMENDED READING
Books

Anderson, R.G., *Corporate Planning & Control* (Macdonald & Evans, 1975).

Baily, P.J.N. and Farmer, D., *Materials Management Handbook* (Gower, 1982).

Barnard, C.S. and Nix J.S., *Farm Planning and Control*, 2nd ed. (C.U.P., 1979).

Batty, J., *Management Accounting* (Macdonald & Evans, 1975).

Boland, R.G.A. and Feathers, J.A., *Budgetary Control* (English University Press, 1970).

Brech, E.F.L. (ed.), *The Principles & Practice of Management*, 3rd ed. (Longman, 1975).

Brown, J.L. and Howard, L.R., *Principles & Practice of Management Accounting*, 3rd ed. (Macdonald & Evans, 1975).

Castle, E.N., Becker, M.H., and Smith, F.J., *Farm Business Management*, 2nd ed. (Macmillan, 1972).

Coy, D.V., *Accounting and Finance for Managers in Tropical Agriculture* (Longman, 1982).

Garbutt, D., MaCallum, C. and Rennie, E.D., *Guide to Accounting Foundations* (Pitman, 1980).

Hammill, A.E., *Simplified Financial Statements* (Institute of Chartered Accountants, 1979).

Hosken, M., *The Farm Office*, 3rd ed. (Farming Press, 1982).

Koontz, H., and O'Donnell, C., *Principles of Management*, 5th ed. (McGraw-Hill, 1972).

Leeds, C.A. and Stainton, R.S., *Management & Business Studies* (Macdonald & Evans, 1974).

Lockyer, K.G., *Production Control in Practice* (Pitman, 1975).

Magee, J.O., *Basic Accounting*, (Macdonald & Evans, 1973).

Hutchinson, H.H., and Dyer, L.S., *Interpretation of a Balance Sheet*, 5th ed., (Institute of Bankers, 1979).

Morrison, A., *Storage & Control of Stock*, 2nd ed. (Pitman, 1974).

Oxford, R., *Be Your Own Accountant* (Nelson, 1975).

Stanier, W., *Maintenance Aspects of Terotechnology*, (HMSO, 1975).

Sturrock, F., *Farm Accounting & Management*, 7th ed. (Pitman, 1982).

Taylor, A.H. and Shearing, H., *Financial & Cost Accounting for Management*, 6th ed. (Macdonald & Evans, 1974).

Vause, R. and Woodward, N., *Finance for Managers* (Macmillan, 1981).

Walker, J.W., *Making Sense of Finance and Accounts in Business* (The Bodley Head, 1977).

Warren, M.F., *Financial Management for Farmers* (Hutchinson, 1982).

Watts, B.K.R., *Business and Financial Management*, 2nd ed. (Macdonald & Evans, 1975).

Westwick, C.A., *How to use Management Ratios* (Gower Press, 1979).

Periodicals

Barnard, C.S., 'Data in Agriculture; a Review', *J.agric.econ.* (1975) 26, 3.

Davies, G.D. & Dunford, W.J., 'Analysing the Financial Stability of the Farm', *Farm Economic Notes no. 25*, Agric. Econ. Unit, University of Exeter (1979)

Gamble, B., 'Money Matters: The Role of the Accountant', *Farm Management* (1983), 5, 2.

Hall, W.K., 'Better Maintenance is on the lands', *Power Fmg* (May, 1979)

— 'Chart a Smooth Course for Workshop Routine', *Power Fmg* (June, 1979)

Howe, S., 'Tractors and Power Systems — A Problem of Investment', *Power Eng.* (March, 1981)

Kerr, H.W.T., 'Measuring Capital in the Farm Business', *Farm Management* (1975), 2, 11.

Theophilus, T.W.D., 'A New Look at the Balance Sheet', *Farm Management* (1982) 4, 9.

EXERCISE 5.1

Your annual tobacco income varies adversely from forecast income by $6728.
Variances are as follows:

> *Budget:* 60 ha producing 1800 kg/ha and fetching 88¢/kg
>
> *Actual:* 56 ha producing 1900 kg/ha and fetching 83¢/kg

(a) How much does each variance contribute to total income variance?

(b) What possible remedial action do you suggest?

EXERCISE 5.2

The following states the affairs of Mr Ndhlovu's farm during October and November 1986. Set the information out as if entering it into a cash analysis book, including details of produce consumed by Mr Ndhlovu and his family. Finally, calculate his farm profit for these two months, ignoring any valuation changes that may have occurred.

On 2 October he sold two cows to Mr Kanyemba, one for $48 and the other for $50. Four days later he sold six bags of maize at $4 each. On 10 October he sold 10 eggs at 5.5¢ each. On 15 October Mr Ndhlovu bought four hoes costing 62¢ each. On 23 October he sold 15 eggs at the same price as before. On 25 October he bought 50 kg of layer's mash for his hens for $7.50. On 30 October he sold two 50 kg bags of beans at $4.20 each.

On 31 October he paid casual labourers $30 for doing maintenance work on his house and tobacco barns. About a quarter of their time had been spent on his house.

Mr Ndhlovu bought a lock and an axe, which cost 40¢ and $2.10 respectively, on 4 November. On 5 November he sold 20 eggs at 5.75¢ each. On 7 November he paid $2.75 to hire two oxen and a cart. On 10 November he paid $15 for two bags of fertiliser. Ten days later he sold some bamboo for $8.30. On 3 November he took $18 out of the farm cash box for his own use and bought a maize production guide for 45¢.

Mr Ndhlovu and his family consumed seven baskets of maize during October and November. These cylindrical baskets are 0.5 m in diameter and 0.6 m high. He estimates that a cubic metre contained 300 kg of maize on the cob. Shelling percentage is about 70 and maize could have been sold for $5 90 kg bag. Ndhlovu also used 12 eggs in October and 9 eggs in November that he could have sold for 5.5¢ and 5.75¢ each respectively. On 11 October a goat was consumed that could have been sold for $6.50. Finally, he estimates that he and his family consumed vegetables worth $1.20 during the two months.

EXERCISE 5.3

Calculate the profit made by Penga-Penga Estates Ltd for the 1985–86 financial year from the information given in tables 5.3.1 and 5.3.2.

Table 5.3.1   Valuations

Valuation as at 1.10.85

	$		$
Land (at cost)	30000	Vehicles	5500
Buildings	8000	Office equipment	250
Fencing	600	Stocks in hand	2750
Plant and machinery	17500	Cattle	34000
Debtors	600	Creditors	1200

Valuation at 30.9.86

Stocks in hand	2500
Cattle	36000
Debtors	1400
Creditors	1400

(Depreciate fixed and part-fixed assets, other than land, as follows:

	%		%
Buildings	8	Fencing	5
Plant and machinery	15	Vehicles	20
Office equipment	20		

Table 5.3.2   Cash Analysis — Annual Totals

Payments		Receipts	
	$		$
Seed and fertiliser	9000	Tobacco	27000
Biocides	1500	Maize	13000
Wages	5500	Beef	7000
Repairs and		Values of produce	
maintenance	2300	used in house	630
Feedstuffs	4500		
Overheads	2700		
Tractor (bought on			
31.3.86)	7500		
Private drawings	7200		

EXERCISE 5.4

From the following figures relating to a financial year for the Monaragala Coffee Producers' Co-operative, prepare a profit and loss account for the year ended 31 December 1986.

	$
Stocks (1.1.86)	8559
Trade debtors (1.1.86)	3788
Trade creditors (1.1.86)	6895
Receipts	51134
Payments	41761
Discount received	1117
Cash sales	16117
Stocks (31.12.86)	9307
Trade debtors (31.12.86)	4084
Trade creditors (31.12.86)	7492

EXERCISE 5.5

At the end of the 1985/86 financial year you have $3520 worth of cattle, $2530 worth of pigs, $77 worth of poultry, $5225 worth of crops and $396 worth of stockfeeds and fertilisers.

A local butcher owes you $198 for pigs, you owe Mr Dhlamini $44 for tractor hire and your bank overdraft totals $564.

Prepare a balance sheet to show your net capital worth.

(Assume that $1000 worth of pigs and $2000 worth of cattle are breedings stock.)

EXERCISE 5.6

Mr Hamid's financial year ends on 30 September each year.

The balance sheet as at 30 September 1986, with last year's figures for comparison, was prepared by his accountant as shown below.

When Mr Hamid compared the two sets of figures, he expressed some doubt as to the correctness of the $9554 trading profit for 1985–86. He pointed out that he had started the year with $2248 in the bank but at the end of the year he owed $2726, a net *fall* of $4974, and this in spite of the fact that he had paid in an extra $3513. He maintained that if he had really made a profit of $9554 his bank balance would have risen considerably.

(a)   Prepare a statement showing Mr Hamid how the net worth of his business has risen during the year, showing in detail what caused this.

(b)   Compute the amount of Mr Hamid's working capital at the start and end of the year and prepare a statement reconciling these two amounts by showing the sources of any rise in working capital and how any falls have occurred.

	1984–85	1985–86		1984–85	1985–86
	$	$		$	$
Opening capital	22400	23042	Fixed assets at cost	16298	25009
Further cash introduced	—	3513	*Less* Total depreciation		
Annual profit	7306	9554.	provided to date	1686	3878
	29786	36109		14612	21131
Less					
Drawings	6744	9274	Stock	7306	6365
Net worth	23042	26835	Debtors	3934	5100
			Bank	2248	—
Bank overdraft	—	2726			
Creditors	5059	3035			
	28100	32596		28100	32596

EXERCISE 5.7

Profit and Loss Account for the Year Ended 30.9.87 Mr Jere

Income	$	$
Sales	123.00	
Valuation appreciation &		
Livestock ($186.96 − 162.36)	24.60	
Home consumption	14.76	162.36
Expenses		
Purchases	12.30	
Valuation depreciation —		
Deadstock ($202.95 − 157.03)	45.92	58.22
Net profit before tax		104.14

This is a simplified trading account as it excludes creditors and debtors.

Now, from Mr Jere's records, you find that on 30 September 1986 he had a bank overdraft of $30 but his neighbour, Mr Chiphwanya, owed him $21 for tobacco. He also had cash at home of $17.80 but he owed $2.15 tax.

However, at the end of the year, on 30 September 1987, his bank overdraft had been cleared but now he owed Mr Munthali $6.15 and Mr Chiphwanya still owed him $4. In his cash box there was $2.50 and during the year his personal drawings were $126.

Prepare a modified profit and loss account to include creditors and debtors. Then prepare:
(a) an opening balance sheet as at 30 September 1986;
(b) a closing balance sheet as at 30 September 1987;
(c) finally prove the accounts.

EXERCISE 5.8

Mr Yesufu is a vegetable trader who does not keep full books of account or have a bank account.

On 1 January 1986, he had the following assets and liabilities:

	$		$
Creditors	385	12 % loan from Mr Edje	330
Cash	165	(Interest payable on	
		31.12.86)	
Pick-up truck	572	Value of vegetable	
Cash register	55	stocks	270

During the financial year Mr Yesufu kept details of his transactions and provides you with the following information:

	$
Payments to suppliers	1815
Rotten vegetables returned to suppliers	88
Owing to suppliers on 31.12.86	462
Total annual cash receipts	2475
Cash in hand 31.12.86	176
Value of vegetable stocks 31.12.86	358

He estimates the running expenses of his pick-up truck at $11 a week and other sundry expenses at $6 a week. Cash not otherwise accounted for should be regarded as private drawings.

Prepare:
(a) a summarised cash account for the year;
(b) his profit and loss account for the year (depreciating the truck at 20 %);
(c) his balance sheet as at 31 December 1986.

6 Work planning

PERSONNEL MANAGEMENT

Personnel management is that part of management concerned with people and their relationships within a business. It aims to bring together and develop into an effective organisation the people who make up a business and, considering the well-being of each person and working group, to enable them to contribute to its success. It is that part of management concerned with the policies, procedures and practices governing the recruitment, selection, training, promotion, payment and working conditions of the people employed in a business.

Good personnel management gets the best out of people at work, at an economic level, and ensures that the social system of the business does not prevent this. A personnel manager must try to fit the right person into the right job, help to fit employees for any higher post of which they may be capable and plan ahead the firm's manpower needs. The main aim of the personnel function is to raise the effectiveness of staff by substituting co-operation in the common taks for the suspicion and hostility that have so long characterised relations between employers and employed; put more simply, getting the best out of employees by winning and keeping their wholehearted co-operation. This may be done by:

- scientific selection, training and development of workers
- developing a spirit of co-operation between workers and managers
- division of labour between workers and managers so that each does only the work for which he is most suited
- finding the best way to do jobs by systematic analysis of all methods and tools used

One of the main lessons of the last 55 years has been that managing the people who are a living part of the business is just as important as proper care of the other inputs used in production. No manager can think only of organisation structures or job descriptions; he has also to think of how James and Samuel will get on if James is given a vacant job. When someone is being recruited from outside, the selectors will often ask themselves, 'Is he likely to be accepted by his colleagues?' In the last resort a business will be as effective as the people who manage and work in it. The capabilities and co-operative activities of people represent the most important resources available to business.

The best managers usually treat staff as individuals and choose them carefully after checking their backgrounds. They fit people into the most suitable jobs where their inherent disabilities are least disadvantageous; for example, women on light work or work needing manual skill and juveniles on unskilled work. Useful special skills and the characteristics of individuals are recognised and developed. Their potentialities are then used to best advantage by giving basic or even specialised training to help them to develop those skills. For example, tractor drivers can be sent on machinery courses and assistants be encouraged to enter ploughing competitions. This raises their usefulness, self-respect and status.

The importance of the human element to a firm's efficiency and productivity cannot be overstressed. The prosperity of a business depends on the efforts of the people staffing it. If each employee is suited to his job, works efficiently and enthusiastically and actively promotes the interests of the business, the latter will prosper as much as it is allowed by external conditions. Several studies have shown that differences in productive efficiency between firms in the same industry, and even between sections within one firm, are not due only to variations in working methods or the technical organisation of work. Differences in productivity are often due to differences in how people are managed.

The personnel function is inherent in management and, therefore, is practised by all managers and supervisors in a business. It is thus part of the job of anyone who manages others besides being the job of specialists called personnel managers or personnel officers. 'Line' authority has immediate responsibility for personnel management, for treating people and organising them to make best use of their abilities, thus reaching maximum efficiency for themselves and the firm.

TYPES OF FARM WORKER

Family workers usually outnumber hired workers on peasant holdings in the tropics. As it is often the only type of labour used, family size often limits the scale of farming undertaken. Family labour is becoming less plentiful owing to the decline in polygamy and the rise in schooling. This labour source is often insufficient for the optimum farming system and limits further production growth. Moreover, farmers are not always efficient managers of family labour.

Communal labour is often important in areas with communal land tenure. Several families help each other in rotation, the host usually providing food and beer.

Hired labour is predominant on fully commercial farms and is provided by regular, contract or casual workers.

Regular labour falls into two groups: managers and assistants, like foremen, and labourers. Managers and assistants usually get lower salaries than their counter-parts in industry or commerce but they also usually get free housing, other perquisites and often a share of the farm profit. Regular labourers on commecial farms are mainly unskilled or part-skilled and come from subsistence farming areas — often from other countries. If married, they tend to leave their families back home and are thus somewhat unstable, tending to stay on a farm for only a season or so. In 1970, commercial farmers in Zimbabwe employed about 166,000 indigenous labourers and about 123,000 from neighbouring countries. However, only 30 per cent of the total labour force had been on the same farm for more

than two years and these people came mainly from squatter families. A growing number of peasant farmers employ one or more regular labourers throughout the year.

Contract labour is used mainly for short periods for capital development projects like tree stumping, brick-making, building and bunding.

Casual or temporary labourers are used to cope with seasonal work peaks, like cotton reaping or tobacco desuckering. They may be transported to estate from neighbouring subsistence farming areas when work is slack there. Another source of supply is the wives and children of regular farm workers. When wives are used to carrying out light work like gleaning after harvest, hand weeding or hand threshing, they seem to prefer to work in the cool of the early morning and in the later afternoon and evening. This means that jobs are done at the most comfortable time of the day and the hot mid-day is left free for cooking, washing and generally tending the family. By working these hours, output can rise greatly and the women are pleased that their farm work does not interfere with their domestic duties.

On some commercial estates, primary schools are provided to help stabilise the regular labour force. Government may supply and pay the teacher and ensure that educational standards are adequate. The farmer usually provides the building and often books and food. The children, often attracted from beyond the estate, may be available for part-time casual work during busy periods.

LABOUR PROBLEMS IN TROPICAL FARMING

Skill in farming usually means ability to do many jobs rather than specialisation. This skill grows with experience but few labourers stay long enough to gain sufficient experience to merit satisfying salaries. Few farmers pay sufficient attention to training, although the more skilled labourers become, the more they are worth paying.

The greatest loss due to unskilled labour takes the form of reduced managerial time. Farmers tend to spend so much time supervising that too little time remains for true managerial work like decision-making and planning. Few of us realise the high cost of managerial time spent checking what labourers are doing. A manager standing behind his gang when he should be planning or investigating new techniques might easily find that his opportunity cost is $100 to $200 an hour at some times of the year.

Growth of the exchange economy has created new job opportunities. Thus male family heads may migrate to towns, mines or plantations to seek paid work. The loss of these potential innovators from the smallholders' sector reduces the quality of management on their holdings. There is a marked preference for work in towns despite severe urban unemployment in most tropical countries. Both from the viewpoint of individual families and in the national interest it is desirable that as many young men as possible find work in farming.

Future labour availability can be ensured only if a reliable and dependable pool of local labour is created. This is possible only if there is a desire or willingness of people to work on farms. Such an attitude is often absent as potential labourers feel that both working conditions and pay on farms are poor. People usually assign a relatively low status to farm workers and the image of the work and workers is not good. Even the prospect of long periods of unemployment and hardship is not enough to divert the urban unemployed to the rural areas where they are needed.

Although costing more per man-day, better quality labour would probably cut labour cost per unit of output because of its higher efficiency.

Excessively labour turnover is a source of inefficiency and can be very costly. High turnover can adversely affect the morale of the rest of the staff — especially the supervisory staff. Every time someone leaves the skill he gained during his employment is lost and there is a possible loss of output if equipment is unmanned until he is replaced, or higher overtime costs; extra costs are incurred in training the new employee and extra work organisation and supervision may be needed as a result of an employee leaving. For effective use of manpower, a stable, motivated workforce is necessary but, in most of the tropics, turnover and absenteeism are high.

Labour turnover = (annual no. of leavers/ average no. of employees) × 100.

Labour turnover may be due to many causes that can be divided into:

Internal causes	External causes
Wages — amount, time and method of payment	Transport
	Poor housing or schooling facilities
Unsuitable work	Poor health
Dislike of job	Domestic troubles
Inability to perform job	Marriage
Breach of discipline	Leaving district
Poor working conditions	Retirement
Poor working relationships	
Lack of suitable incentives	
Wish to improve position	
Wish to gain more experience	

If jobs are hard to get, workers will stay on the farm and any dissatisfaction they feel will find other outlets.

An exit interview to seek the real causes of leaving may show how these can be removed and reveal a need to reappraise certain personnel policies and procedures or reveal inadequate supervision, thus reducing the rate of labour turnover.

A growing number of young students from colleges and universities are placed straight into supervisory jobs. Older labourers resent being supervised by young newcomers to the industry. However, training with a good experienced supervisor and 'personality' soon enables them to handle large labour gangs successfully.

Although low pay and social status may be the main hindrance in obtaining qualified staff, other changes are needed if farmwork is to become competitive with other jobs.

SUPPLY OF LABOUR

A large problem on many commercial farms is recruiting and keeping sufficient qualified farm staff right down from managerial to manual jobs. However, good working conditions lessen recruitment problems. Staffing involves manning and keeping manned the posts provided for the job to be done.

Unfortunately, there is little or no organised labour selection for farm jobs. This undoubtedly causes much waste of time and money. It seems only logical when recruiting manpower to project its future costs much as returns would be estimated for a proposed capital investment. A $6000-a-year manager employed for 20 years is an investment of about $150,000 on a net present value basis (see chapter 11), allowing for pay rises and other staff costs. Consider the weeks or months of staff analysis and the hours of board time often spent on a $150,000 capital proposal; then compare it with the hasty appraisal of likely costs before establishing or filling new posts.

A manager or foreman should be held fully accountable for the work of his staff only if he has participated in selecting them. The first choice should be made by the person who will be directly in charge of the new employee. Similarly, provided there are reasonable safeguards against victimisation, any manager or foreman should have the right to remove a worker from his section. Whether this includes dismissal should be a top management decision.

Many farmers hire staff with no knowledge of their skills and little of their past. This often results in misplacement, higher costs, poor labour relations, discontent, loss or damage of materials and equipment and many other problems. By comparison, large firms in industries like mining use ways of testing workers before hiring them. Staff selection by means of adaptability tests would help to reduce turnover. Cheap tests are available that can be supervised by non-specialist staff. Results of analyses of these tests have proved highly accurate and people are placed in jobs most suited to their abilities which, in turn, reduces the training time needed to make them fully productive. Many tests developed for the South African mines are now used daily in Asian countries like Indonesia.

Even a practised interviewer is unlikely to select the best person for a job unless he knows exactly what he is seeking. This requires job analysis which includes both a job description and a personnel specification. The latter might require special aptitudes; for example, tobacco graders should always be checked for colour blindness.

Nobody should accept a post as a farm manager or assistant without a formal, written contract, however friendly or well-acquainted with the employer he is.

The contract should cover the following basic points:

- Name of employer, employee and farm.
- Timing of contract.
- Duties of employee.
- Monthly salary. When due and how often reviewed.
- Bonus. When due, how calculated and provision of a certificate by the employer's accountant.

- Pension rights, if any.
- Entitlement to housing and perquisites, for example, vehicle, fuel, farm produce and rations for servants.
- Responsibility and control of operations during absence of employer or manager.
- Leave entitlement.
- Sickness and injury benefits.
- Secrecy about employer's affairs.
- Maintenance of records relating to duties.
- Safety of employer's property.
- Grievance procedure.
- Default. Employer's right to dismiss for breach of terms or misconduct and provision for arbitration.

In peasant smallholdings much time is spent collecting water and fuel, processing food, marketing, on social obligations and in illness. The amount of labour available for farm work on these family holdings depends upon how many family members work on the holding and how long each member is prepared to work. If non-family labour is also used, the labour supply will depend also on the availability of hired labour when needed, the amount of cash, goods or land the farmer can offer the employee and how long hired labourers are willing to work.

LABOUR MEASUREMENT

Physical labour use may be expressed as man-days, standard man-days, gang work-days, hours or minutes. These may be quoted per head of livestock, per unit of area or by time period. Another concept used to indicate labour availability is man-units, which equate the labour inputs of women, children and casual labour to adult, male, heads of household.

Defining how much labour is available is somewhat arbitrary, as it depends on the age at which children are expected to work on the farm, on whether women and old men are included, on how many hours they are able and willing to work and their rate of working. The amount of labour available and the amount used may differ if there is unemployment or underemployment of labour, either permanent or seasonal.

For many farm management purposes, variations in the rate of working are ignored and labour inputs are measured in man-hours. The labour input in man-hours is the product of the number of men employed and the average hours worked. It is therefore assumed that the labour of one man for 60 hours is equivalent to that of two for 30 hours, three for 20, four for 15 or five for 12 hours. This assumption is slightly unrealistic but adequate for most purposes provided it is realised that there is a time limit on most tasks so that more than one man may be needed to finish a job on time.

A man-day is the work done by one man in a working day. This will clearly depend both on the rate of working and the length of the working day. There is no valid reason for assuming that standards in other parts of the world and on commercial farms in the tropics, such as a 40-hour week, should apply to tropical peasants. Ultimately, the total hours cultivators work is fixed by the hours of daylight or by the need for rest.

The length of day worked in tropical peasant farming varies with the type of worker, and the level of mechanisation. However, it is generally much less than eight hours except during occasional times of peak labour demand. Thus a typical wage labourer elsewhere might work about eight hours a day 250 days each year — a total of 2000 hours a year — but surveys in Malawi indicate that the average time spent by adults on *field work* is only about two hours a day. Various surveys in West Africa indicate the following average times spent on *farming* activities.

	Average hours/year	Average days/year	Average hours/day
Men	1070	169	6.3
Women	650	82	7.9
Children	240	81	3.0

In their non-agricultural time, subsistence farmers must find time to build and repair their houses, perform many religious and civic duties and do many other tasks that commercial farmers have done for them. Wives are involved in trading and housework and children in schooling.

For comparisons to be made between different types of labour, it is necessary to express days and hours in terms of a common denominator like man-days and man-hours. To assign male adult equivalents to different sex and age groups, two assumptions are usually made. First, that physical labour productivity show initially a positive correlation and then a negative one with age and, secondly, that the physical labour productivity of women is less than that of men. Little information on the relative rates of working is available but the following conversion scale is typical:

Labour class	Age (years)	Man-units	Equivalent man-days/ month
Small child	Less than 7	0.0	0
Large child	7–14	0.4	10
Male adult	15–64	1.0	25
Female adult	15–64	0.8	20
Male adult	65 or more	0.5	12.5
Female adult	65 or more	0.5	12.5

If we assume that the average month has 25 working days to allow for rest days, sickness and traditional duties, then equivalent man-days/ month are as shown.

These conversion coefficients have severe limitations. The performance differential between men and women narrows as jobs become lighter, suggesting that physical strength strongly influences performance. The relative working rates of men, women and children vary from one tasks to another. When a woman works at half the speed of a man on one job and twice as fast on another, fixing her equivalent value on the basis of the first will grossly under-estimate family labour capacity on the second. Assumption of a constant performance differential between men, women and children, which is implied by using a single coefficient over the whole season, is unrealistic as relative performance varies according to energy requirements. Certain tasks occur at busy times of the year and the relative rates of working on these critical jobs, found by work study, should be used as man-unit equivalents.

The amount of farmwork a self-employed person is willing to do depends on:

- the amount of work needed to obtain his subsistence needs
- the potential gain from doing extra work
- the physical health and diet of the person
- climate
- presence of markets
- other job opportunities and
- his motivation.

SEASONAL DEMAND FOR LABOUR

We all know that different crop enterprises need different amounts of labour. Cotton and tobacco need more labour than groundnuts, which in turn need more labour than maize. With annual crops in the sub-arid tropics, the task of raising labour productivity over the whole year is complicated by seasonality in field operations and the existence of peak labour periods. It is these peak periods that usually determine the labour and machinery needed on the farm throughout the year. For the family work force and regular hired workers, the supply of labour is relatively fixed throughout the year. It is therefore probable that farmers will either have less labour than they want during work peaks, more than they want at slack times or both.

Busy periods or 'work peaks' alternate with slack periods or 'work troughs' within each enterprise, apart from much livestock work. However, the busiest months for one crop are not always the busiest for another. We must therefore know not only the total annual labour needed but also the labour needed each month, half month or even each 10-day period for critical operations.

In practice, there are large variations in labour used per hectare owing to different levels of yield and mechanisation. Also, the seasonal demand for labour can be adjusted for those operations for which timeliness is to critical, like tobacco grading or ground-nut shelling. In peasant farming, the traditional division of labour is breaking down, with women performing more tasks that were done only by men in the past. They are saving time by making more use of mills for preparing food but their growing awareness of nutrition, hygiene and homecraft reduces the time available for farmwork. Thus there is always likely to be differences in labour used between farms, between areas and over time.

Surveys have been carried out on actual average labour use on some crops. These figures should be used only as a guide as different people work at different rates and crop labour demand varies with yield. Table 6.1 refers to largely mechanised, high-yielding crops in Zimbabwe and table 6.2 shows labour use on lower yielding, hand-cultivated crops in Malawi and Sri Lanka.

Work peaks occur because critical jobs like planting, weeding and reaping are closely related to the seasons and must be finished in a limited time. Delays usually cause loss of yield so the labour needed to finish the job is compressed into a peak period. At work peaks, more labour would greatly raise total output, either because of more timely completion of the job or because a larger area could be handled. Other tasks, especially repair and maintenance work, allow greater flexibility of timing. However, some seasonal unemployment or under-employment is almost inevitable in farming.

It is wise to break down labour use by enterprise and/or operations at all times as a routine so that average use over several years can be calculated to allow for seasonal variations. However, some managers may select one or more enterprises or particular tasks each year for careful study. This procedure, continued for several years, can amass a valuable fund of facts for management decisions.

Figure 6.1 shows some of the enterprise/labour use data given in tables 6.1 and 6.2 plotted as labour profiles. This shows graphically the seasonal labour use for each enterprise and the effect of different levels of yield and mechanisation.

By combining all planned enterprises into a whole farm labour profile, using hatching or colouring to show individual enterprise, an indication is given of when labour supply and management problems are likely to occur by revealing peaks and troughs. Whole farm labour profiles are also useful in showing the effect of changes in the scale of enterprises, introducing new ones or removing existing ones on seasonal labour demand. Labour profiles can also aid in choosing enterprise combinations to minimise peaks and create a more even demand for labour over the year. Much calculation is involved in preparing a whole farm labour profile but the result is usually instructive and the effort spent is worthwhile.

Consider a mechanised, commercial farm growing 30 ha of flue-cured tobacco and 60 ha of maize. Table 6.3 shows the farmer's own average labour use obtained from his physical records over the past three seasons and figure 6.2 shows the labour profile for the combined crops.

If the farmer has 65 regular labourers, each of whom works 25 days a month, then his regular labour supply is 1625 man-days each month. The profile immediately shows that this is too low in the peak tobacco months of January to March. Knowing this, however, is half the battle and the farmer or manager can take action, to be described later, to remedy this situation.

While they are useful guides for labour planning, profiles have their weaknesses. Actual labour supply will vary with sickness and, more important, the time available for field work will vary with the weather. Profiles are not specific enough about the actual jobs or operations that must be done and they ignore the concept of gang size. The importance of the team or gang in labour organisation cannot be overstressed.

WORK PLANNING

Most farmers plan in their minds the work to be done on a daily basis. Always plan today the

Table 6.1   Average Labour Use for Estate Crops in Zimbabwe (Man-days/ha)

	Flue-cured tobacco	Burley tobacco	Cotton	Groundnuts	Maize	Soya beans	Ration beans	Sorghum	Sunflowers
Sep.	14	5	1	—	7	—	—	—	—
Oct.	29	18	3	—	4	—	—	—	—
Nov.	38	46	6	5	8	—	—	—	—
Dec.	23	8	4	3	8	2	—	—	3
Jan.	59	31	12	2	2	3	5	1	2
Feb.	84	78	6	—	1	—	5	1	—
Mar.	98	39	5	1	—	—	—	—	2
Apr.	29	8	37	3	5	8	3	8	13
May	21	115	64	32	7	1	10	7	7
June	15	86	35	47	6	35	10	1	1
July	15	—	2	32	4	—	—	2	—
Aug.	12	10	3	—	5	—	—	—	—
Totals	437[a]	444	178	125	57	49	33	29	28

[a] Owing to the level of mechanisation in the USA, the equivalent is about 29 man-days/ha there.

Table 6.2   Average Labour Use for Hand-Cultivated Crops in Malawi (Man-days/ha)

	Fire-cured tobacco	Cotton	Groundnuts	Rice	Maize	Cassava
Sep.	8	6 (15)	5 (21)	9 (20)	7	5 (20)
Oct.	29	17 (32)	14 (25)	21 (39)	15	6 (40)
Nov.	40	13 (16)	21 (18)	11 (20)	12	9 (9)
Dec.	29	24 (15)	26 (8)	22 (21)	22	6 (8)
Jan.	27	31 (22)	18 (34)	14 (34)	19	9 (7)
Feb.	24	24 (47)	18 (44)	7 (12)	9	9 (6)
Mar.	40	21 (24)	21	12	5	6 (10)
Apr.	59	43	33	2	12	5 (10)
May	43	48	31	40	11	5 (10)
June	25	42	20	38	9	5
July	18	32	13	16	5	5
Aug.	—	10	11	10	3	2
Totals	342	311 (171)	231 (150)	202 (146)	129	72 (120)

(Figures in brackets refer to rainfed crops in Sri Lanka. Note the huge differences and thus the need to use local data.)

Table 6.3   Expected Monthly Labour Demand in Man-days/Month

	Tobacco (30 ha)	Maize (60 ha)	Total
Sep.	420	300	720
Oct.	810	360	1170
Nov.	1200	420	1620
Dec.	630	480	1110
Jan.	1650	180	1830
Feb.	2460	60	2520
Mar.	2940	60	3000
Apr.	810	360	1170
May	660	420	1080
June	390	420	810
July	420	240	660
Aug.	270	270	540

work for tomorrow and always tell everyone each evening at the latest where he must report and with what tools in the morning. Most farm jobs are best begun early so that most of the work is done before it gets too hot. Nothing lowers the morals of a gang more than to have to wait, as precious time passes, for instructions, equipment or transport. If possible, it is best to give each man only one job a day because, understandably, workers like to see the end of the task.

Daily planning is fairly simple but for long-range planning the complexity of farmwork is such that mental planning is inadequate. Formal planning is unpopular with farmers because it is a mental rather than a physical task and, as such, does not seem like work. Nevertheless, I will outline a method of labour planning that could be used by farmers, perhaps with the aid of a friend or adviser. The process should lead to deep thought about all farm jobs and reveal faults in

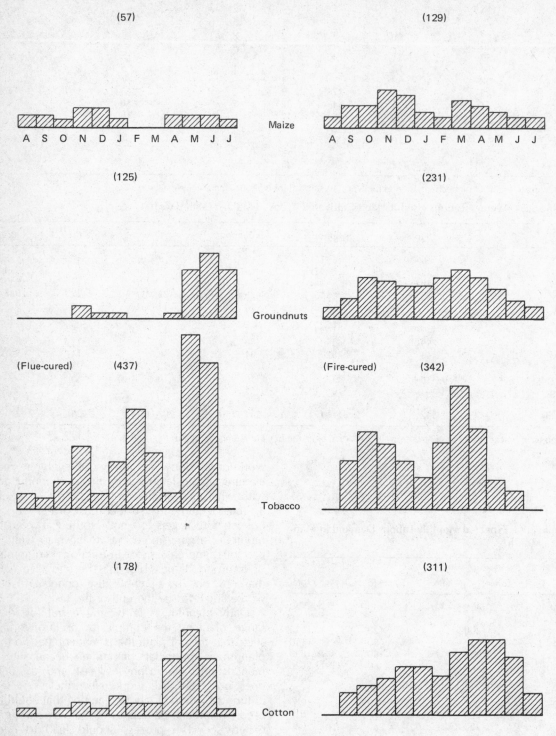

Figure 6.1   Estimated Labour Use: Man-days/ha

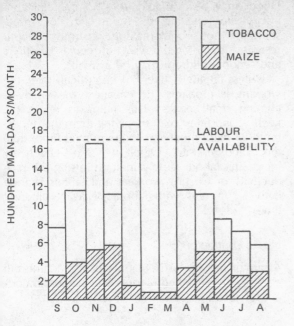

Figure 6.2   A Whole Farm Labour Profile

- has sufficient labour to carry out good husbandry practices on all his enterprises.
- does not have much excess labour. However, if excess labour is unavoidable, we should advise him how to use it effectively.

The advantages of work planning can be summarised as follows:

- Peak and slack periods of labour demand become clear and can be allowed for in planning future production.
- As the relevant information is more detailed than in a labour profile, ways of removing or reducing peaks are more easily seen.
- Planning enables farmers to organise their operations and adjust their methods to raise the efficiency of labour use.
- There is little doubt that large savings can arise from well planned labour (and tractor) use.
- It enables farmers to plan the recruitment and training of new staff. Training new staff is sometime a long and costly job, so adequate warning of staff needs is essential.

Some jobs can be done by a gang. In these cases, labour needs are often expressed in gang work-days which represent the amount of work or area handled by the available gang in a working day.

The information needed for seasonal work planning is as follows:

- the size of each enterprise,
- a list of jobs to be done in each enterprise and the method to be used,
- the number of workers (gang size) needed for each job (some work can be done only if handled by a gang of a minimum size and this, rather than average monthly labour demand, must be allowed for),
- the rate of work for each job — usually in ha/day,
- the times within which each job must be started and finished,
- the likely number of available days when work is possible within each time period; allowing for bad weather, national holidays, etc.,
- the type of workers needed — regular/casual, male/female/juvenile, etc.

Collecting and compiling this information, as in table 4.31, is a hard task. It can be eased by using published data on work rates but data collected on the farm itself are more relevant. Space should be left on the calendar of operations to note actual results as this will facilitate making next year's plan.

records and information, which can then be remedied. The final result should ensure more effective use of labour.

The first step in a systematic study of labour use is to find the timing of peak work loads and those jobs that cause them.

As far as possible we should equate the seasonal supply of and demand for labour. However, the climate dictates that there will be uneven demand for labour for field work and, as described, different enterprises need varying amounts of labour at different times. Thus, we must accept that we are unlikely to obtain a perfect fit between labour supply and demand.

It is wrong to level out labour need *too* much over the year. Besides the obvious need to allow time for holidays, trouble is likely if labourers have to work flat out all the year. Most people are willing to work flat out for long hours in bursts of activity for several weeks at a time but it is too much to expect them to keep this pace continually, even on commercial farms, without increased labour turnover. Besides this, some flexibility should be allowed to help us cope with bad weather.

When helping a farmer plan his farming system, we should try to ensure from a labour standpoint that he

For each crop in turn, the operations are listed, as in tractor use planning, together with the method to be used. The expected output for each job is recorded, using the farmer's own estimate of what he should achieve. The timing of each operation is also recorded, usually in relation to other operations, but sometimes with calendar dates for critical operations like planting.

When planning, the farmer may not have decided what methods he will use; in fact he may expect some guidance from the plan. Alternative methods can then be recorded and compared in terms of their use of labour (and tractor) time.

The plan is drawn up on graph paper, as shown in figure 6.2 with the horizontal scale representing months. The vertical scale shows the number of labourers, with regular labourers above the base line and casual ones below. Dotted lines help to show the amount of labour available. Foreman, herders and domestic workers are excluded from the labour plan; so are tractor drivers as their work is shown on the tractor plan.

The operations in each enterprise for which timeliness is important, such as planting, cultivating and maybe reaping, are plotted first and then adjustments are made to fit in all other jobs when most convenient.

Together with the labour plan it is worth listing any preparatory work for each operation, like maintenance of equipment and buying seed and fertiliser, working back from the target date for the job. This should prevent the common situation of a farmer finding that his equipment is defective when he wants to start a vital job.

Whether forecast labour needs are depicted as a labour profile or a graph showing how operations fit together, period of excess demand (peaks) and excess supply (troughs) are likely to arise. However, if aware of them, a manager can do much to reduce these. A work peak is large amount of work that must be done in a limited time, usually at a certain time of the year. The word 'operations' is used to describe the activities carried out by a business to provide the services to customers that are its basic reason for existing. In a business producing and distributing a physical product, its 'operations' will be all those activities related to production and distribution.

The following are ways of coping with work peaks:

Increase the labour supply

Achieved by more use of casual labour or overtime or by working more efficiently. Contract labour may be available from commercial contractors for specific short-term jobs like tree stumping or brick-making. Family labour reserves, including the wives of regular labourers and school children, may be available.

Unless fatigue is likely to be critical, overtime working is probably the cheapest way of overcoming small peaks. This helps to retain the keenest and fittest man by giving them the opportunity of extra money as they are paid more than the standard rate. Other ways of effectively increasing labour supply include raising the productivity of existing workers and labour sharing between farmers with labour peaks at different times.

Use of contractors

A farmer has the use not only of a machine but also usually of skilled labour. This can be important for jobs like combine harvesting.

Selective mechanisation to reduce labour needs

Here there are two levels of decision. One is whether or not to mechanise a job now done by hand; the other is whether to substitute a new machine for an existing one to do the job quicker, with less men or both.

New technology

Using herbicides or desuckering oils reduces labour demand as does the use of a range of crop cultivars that will mature at different times.

Extend the time during which a job can be done

For example, water-planting tobacco, cotton or maize or dry-planting maize or cotton before the first rains. With *some* operations this may reduce yields but the loss can be less than the saving in combined labour and machinery costs.

Adjusting the enterprise combination

Clearly, a peak can be lowered by growing less of the crop or one of the crops responsible for it. Naturally, any change in the balance of crops should be considered within the context of the financial plan.

Labour troughs can be used for overhead and maintenance work like land clearing, building, fencing and the making of roads and soil and water conservation works. If this is not enough, a

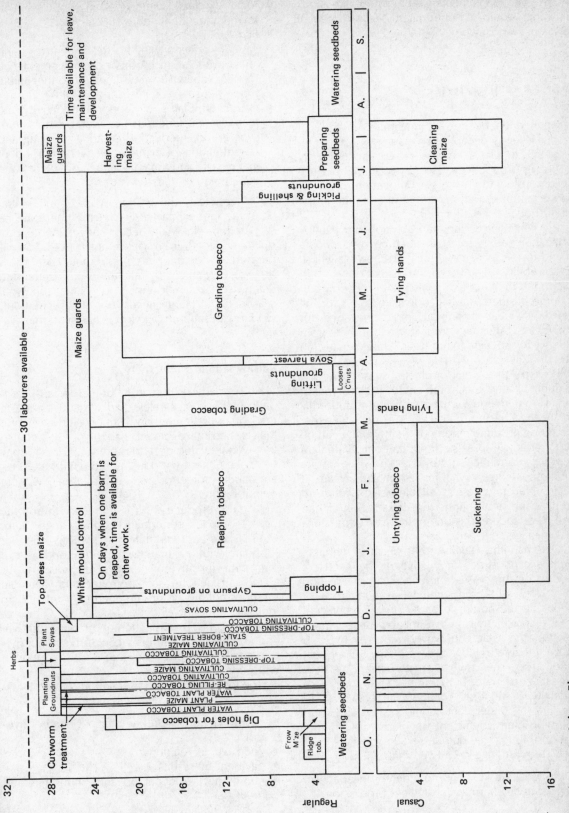

Figure 6.3   Labour Plan

short-term, supplementary enterprise, like stock finishing, could be introduced — provided it raises net farm income.

LABOUR PRODUCTIVITY

As mentioned before, the cheapness of tropical labour is partly deceptive so managers have every incentive to use it efficiently and effectively by raising its productivity. Two aspects of labour use need consideration: the best allocation of labour and developing and using the full potential of the labour force once allocated.

On some farms the staff are not told exactly what their job is so that sometimes two people attend to the same tasks, neither knowing whose responsibility it is, or some task is not done at all, each man thinking that someone else is doing it. In large businesses this can lead to chaos for the firm and great frustration and annoyance for employees. So many problems can arise from failure to specify employees' tasks that job descriptions can only help.

Perhaps the worst problem caused by not defining exactly what a job entails is that it makes it hard to define just what qualifications and skills are needed by the man who is to do it. Neither the hiring of suitable staff nor their training can be carried out with precision; nor can one evaluate a man's performance if no one knows what he should be doing or what his responsibilities are. It is also hard to assess a suitable wage. Failure to prepare job descriptions for all staff, other than unskilled labourers, is one of the worst omissions that management can make.

Overmanning usually arises because management has no exact knowledge of the needs of production processes through lack of method study and work measurement. Calculations of labour needs should be based on a 75 per cent level of productivity. On Zimbabwean commercial tobacco estates, however, a survey in 1960 showed that farm labourers were only 50 per cent productive. Their average productivity is now almost 70 per cent but for other enterprises productivity is still low. For the competitive world we live in 75 per cent should be the aim. Bad planning and inefficient methods are mainly responsible for low output.

A study in Nigeria showed that:

- the local workers had no inherent incapacities or attitudes harmful to efficient production.

- local workers' willingness to work was no lower than that of workers in developed economies,
- careful selection, financial incentives and close observation by higher management will reduce supervisory weaknesses to negligible proportions.
- the determinants of labour productivity are management functions like work organisation, supervision, production control, planning and co-ordination of work, effective provision of incentives and maintenance of plant and equipment.

Nevertheless, the will to work — no matter what the farmer may do — must come from the worker himself. Many factors affect labour productivity, apart from yields and prices, if it is measured as output per unit of labour cost. These include investment in machinery and buildings, enterprise size, field size and shape, soil type, seasonality of labour needs, use of casual labour, work organisation, wage levels and man management.

Raising labour productivity

The use made of human labour in the tropics is often ineffective, careless and wasteful owing to inadequate planning and supervision of work, that is, bad management. Labour productivity — the average value of output per man-day — may be raised either by producing more with the present labour force or by saving labour without output falling.

To raise productivity it is necessary to find the causes of poor performance and change *them*. Causes, like poor ventilation or lighting in indoor jobs, are within a manager's control and he can offer either positive or negative rewards.

Job satisfaction results from good management but does not always ensure good performance. In industry, scientific evidence has shown a connection between job satisfaction and labour turnover and absenteeism, but it is inconclusive about performance. It is not certain that a satisfied worker will perform well.

The following help to raise labour productivity:

- reducing work peaks
- raising yields per unit area or per head of livestock
- good labour relations
- payment by results
- improving the layout of lands, buildings and roads

- selective mechanisation
- adopting labour-saving work methods

If there is sufficient land and capital to extend cultivation, the real constraint is not the total supply of labour but that at peak periods. It this 'bottleneck' can be eased or removed, some increase in cultivated area will be possible. Peak labour needs can be reduced by changing the jobs done, extending the time over which each job is spread, by working faster or more efficiently or by changing the farm system.

Greater productivity is achieved only by raising the working rate on key tasks during work peaks. These include daily routines like fetching water, feeding livestock, milking cows and tapping rubber trees. Reducing the peak labour needs for jobs that depress output reveals new peaks, one of which then becomes the new constraint on output.

Much can be done to raise labour productivity by improving the layout of lands, buildings and roads. In smallholder farming, holdings are often fragmented and much time is spent walking between plots. Whenever possible, farmers should try to exchanged plots so that each one's plots are consolidated. Some farmers have their homes a long way from their plot(s). This results in wasted time, walking to and fro. Farmers should be encouraged to build their homes near their plot(s).

Labour needs per hectare fall as the area rises. The shape of lands is also important; rectangular lands up to about 450 m long need less labour per hectare than square ones as less time is wasted in turning machines which can still be filled at the headlands. Storage buildings should be sited near where the stored materials are to be used and gravity should be used where possible.

Labour is often limiting on smallholdings so a rise in output needs either more labour or the substitution of capital for labour to cope with labour peaks. However, hired labour can be a hindrance to raising profit unless it is used efficiently. Selective mechanisation and efficient hand tools, like swan-neck hoes and wheelbarrows, to reduce labour demand, boost morale and reduce drudgery should be continually sought. Keeping hand tools in good condition can improve labour efficiency. Much time is saved if carts are used for carrying loads instead of people carrying them on their heads.

Labour-saving work methods reduce labour use per unit of output by making work easier, simpler and less tiring. Method study does not substitute for specialist knowledge and experience but it helps these to be put to best advantage. Rational methods and work measurement could probably reduce labour costs by 15 per cent and this would have a large effect on annual profit.

Work study refers to those techniques, especially method study and work measurement, that are used to examine all aspects of human work and that lead systematically to scrutiny of anything that affects the efficiency of the situation being studied so as to improve it. Method study and work measurement are normally used together as work measurement provides the means for testing better ways of doing work found by method study. Each compliments the other and makes it more effective.

Work study has reduced production costs in industry by a huge extent. It offers similar savings to farmers. There is nothing mysterious or hard about work study. It is simply organised application of commonsense to planning farmwork and the turning of casual work methods into orderly routines.

We cannot master all the detailed techniques that work study specialists use. However, we should know what work study is, what it can do on a farm and how it is best applied there. We will not get the same results as a fully trained specialist but we should get *some* results, possibly better results than we would get spending the same time and energy on other ways of raising labour productivity.

For small family farmers it may be impossible to reduce the labour force as labour saving would merely cause unemployment. The only alternative is to raise output as a high output per worker is related to a high output per hectare and high profits. If population density is high and the area of land limits production, growth can be achieved only by intensifying and raising output per hectare but, if there is surplus land, productivity can be raised by extending the cultivated area.

Labour needs per hectare rise by only about 0.5 per cent for each per cent rise in yield per hectare. The labour used to produce each unit of output therefore drops at high yields because fixed labour needs are spread over more output.

Good labour relations are necessary for efficient working — see chapter 7. Labourers should be selected carefully and treated fairly. You will often see advertisements for managers stating, 'Must be good with labour'. This is important as there is little worse than supervising a large labour force that is resentful, unwilling and continually annoying you by dishonesty, imaginary sickness or rudeness.

Methods of payment will be described more fully later. The principle, however, is to encourage greater labour productivity by relating pay to the amount and quality of output rather than to the time spent. Skilled workers, like tractor drivers and foremen, are usually paid higher basic wages than unskilled workers and piecework or bonus schemes should be used whenever practicable. Incentive schemes should be used to encourage good workers to stay with you.

Work study aims are to help managers to make the best use of resources by more effective use of labour, plant and machines.

Method study provides a way of examining routines, faulty decisions or an inefficient environment. It can suggest better work methods and layouts and turn our attention to the use of labour-saving devices and methods. One of the most important features of method study for a farmer is the attitude of mind that it produces. Once he understands what method study involves, he will examine all farm jobs more critically and he can often make large improvements without even using a wrist watch.

It is sometimes wrongly believed that work study applies only to repetitive jobs and that its scope on farms is therefore limited. However, on farms where peak periods place a heavy strain on the farmer, his men and equipment, a saving of seasonal peak work may be more valuable than a saving in repetitive work.

Method study can be used to greatly reduce the drudgery of much farm work. This is especially useful on family farms without hired labour.

The gains from work study are usually highest when management is already good, the labour position is tight and higher production is planned or occurring. In these circumstances, labour saved can be profitably used on productive work.

Little can be gained by applying work study if vital technical factors are the main cause of poor results. A further limitation of work study is that it is no cure for bad labour relations and indeed, can be damaging if clumsily applied. It can do good only if both managers and workers understand what is being considered and if the workers welcome a study of their methods. Workers may interpret criticisms of their methods as implying criticism of themselves. They may, therefore, resent the results of seemingly successful work study. They may, moreover, not welcome reorganisation if it reduces their total earnings by reducing overtime. A good solution that implies criticism of previous management may be rejected unless tactfully presented.

Method study

Most people work to live, not live to work, so unnecessary work should be eliminated. When the plant, buildings and paths are sited best, the way is clear for one of Man's most interesting activities — method study. Every farm job needs labour, equipment and materials. Many of these jobs are themselves complex and need a complex mixture of labour and machinery. Only by careful study and orderly analysis of each aspect of a job is it possible to minimise its cost. There are few ways of doing jobs that cannot be improved, however much they have been developed and however well they are done now. Wasteful practices are not always obvious — especially to those who are doing a job. Method study is one of the oldest and consistently effective techniques — one can hardly fail to make improvements by its systematic use, savings are possible at little cost.

Managers can practise method study themselves, in basic form, with little or no formal training. Many labour-saving principles are so clear and easily understood that farmers need not undertake systematic work study before introducing a useful change. Even without formal training in method study, progressive managers can ask pertinent questions such as:

- What work (or travel) is unnecessary?
- Are loads large enough and easily handled?
- Are jobs organised so that each new operation can, whenever possible, start where the previous one ends, to avoid time-wasting, picking-up, putting-down, storage or transport?
- If a job consists of a series of operations, is there a suitably sized stockpile of 'work' between each worker to prevent anyone being kept waiting?
- If jobs are done by a team, is the manager ensuring that no one is underworked or overworked — is there a good team balance?

Method study may be defined as:

The systematic recording and critical examination of existing and proposed ways of doing work, as a means of developing and applying easier and more effective methods and reducing costs.

This is the creative branch of work study that is of particular importance and value to famers.

It follows from the definition of method study that it aims basically to do three things:

- Reveal and analyse fully the facts about any situation. The record must be factual and cover what really happens and not what someone says happens or, worse, what someone thinks should be happening.
- Examine the facts critically. The result of his critical examination, being based on facts, will usually lead to a better working method.
- Develop from the examination of the facts the best solution possible in a given situation.

Obviously, with so many different problems to be tackled, a simple, flexible procedure is needed that can be applied to all types of work. The following steps form the basic procedure for method study:

- *Select* the work problem to be studied.
- *Record* all the relevant facts of the present or proposed method.
- *Examine* the facts critically and in sequence.
- *Develop* the most practical, effective and economic method.
- *Train* for the new skills needed.
- *Install* this better method as standard practice.
- *Maintain* this standard practice by regular routine checks.

The endless pressure of routine chores makes habit an especially strong influence in farming. The same old routines cope with repetitive work although sheer muscle power often now has been replaced by technological ingenuity, all that has usually happened is that we have merely acquired a new set of habits.

Method study develops a questioning mind. It pays to question accepted practices, however well established, and consider whether another way of doing a job or change in the organisation might save time, labour, money or all three.

Method study can improve productive efficiency through:

- better basic processes
- better farmstead and workplace layout
- better design of plant and equipment
- better use of materials, equipment and labour
- more efficient materials handling
- better flow of production and processes
- standardising methods
- better safety standards
- better working environment

The effects of applying method study can be dramatic. Many firms have cut costs or greatly raised output. On average, I should be dis-appointed if a method study did not improve efficiency on any one job by about 15 per cent. This could have a marked effect on annual profit.

The first step in method study is to *select* the work to be studied. Selection of fruitful areas to study needs sound judgement. The work to be studied must offer scope for savings in manual effort, the prospect of better machine use and, if possible an opportunity for skilled workers to fully use their skill instead of wasting their energies upon needless tasks. Detailed scrutiny of costs can help in selecting work to study. Obvious starting points where substantial savings are likely:

- Poor use of materials, layout or machine capacity.
- Unnecessary movement of materials.
- Bottlenecks, e.g. operations during periods of peak labour demand.
- Unpleasant, fatiguing and dangerous jobs.

Other pointers to a need for method study in some fields include too much handling and movement; bottlenecks in production or too much work in progress. It is not always the work with obvious faults that offers the most valuable improvements.

A basic principle of method study is that, whatever caused the decision to study a job, the larger activity of which this job is a part is first challenged by the relevant steps of the basic procedure — 'record' and 'examine'. Only if it is decided that the larger activity is valid in the existing situation should the study proceed to greater detail.

Method study need not be costly; indeed it may cost little more than the time of those carrying it out. However, the cost rises with the level of detail and also if specialist knowledge and equipment must be used. Clearly, no study should be more detailed than is justified economically and so the selection of a job must be based on an estimate of its time and cost.

When a job is selected and the detail and extent of the examination decided, systematic analysis of the work is begun. A *record* is made of all the relevant facts relating to the present or proposed method. On an existing job the procedure is to watch it being done through all its phases and to record exactly what each part of the product at every stage. This forms the basis for the critical examination and the success of the whole study depends directly on precise, effective recording.

The usual way of recording facts is to write them down but to describe in writing even a simple process can be very tedious. It is often hard to picture the whole process from a written record. Consequently, symbols are used to combine a shorthand system with a system of analysis. Five basic types of event, into which all work can be classified, are represented by the symbols shown in figure 6.4.

○ Operation — Produces or accomplishes

□ Inspection — Verifies (By checking quality or quantity)

◊ Transport — Moves

▭ Delay — Delays or interferes (Temporary storage)

▽ Storage — Holds or keeps

Figure 6.4   Process Charting Symbols

In this way any process can be represented as a series of symbols, against each of which is recorded the essential information about the event. Such a record of sequence of events is called a *process chart* and enables the main facts of a whole process to be pictured and methodically examined. Charts used to record process sequence, using symbols, are:

* *Outline Process Charts* used to give an overall view of a process or procedure with little recording effort.
* *Flow Process Charts* (1) (men, material, document or multiple activity).

Charts using a time scale are *Multiple Activity Time Charts* (1) (men, materials, equipment).

Charts and diagrams used to record movement:

* *Flow Diagrams* (2) for simple moves.
* *String Diagrams* (2) for complex ones.

(Charts types 1 and 2 are complementary and used together to give a full picture).

Flow process charts, using all five symbols, follow events either as they affect a material or a worker. Man and material should never both be recorded on one chart; confusion is avoided only by strictly separating them and making separate records.

On a flow process chart the distances involved in any movement can be recorded close to the symbol but this is not always sufficient to show the real problem to be tackled. It may be necessary to consider the actual path of movement followed by the subject of the chart and for this purpose a *flow diagram* is drawn. This is a scale drawing of the working area on which the sites of the various events and the actual paths of move-

ment followed by the subject are shown.

A *string diagram* is like a flow diagram with one important exception. Instead of drawing the movement of the subject, coloured threads or wool, held in place by pins, are used. This variation of a flow diagram is used to show the paths of movement of workers, material or equipment within a limited area over a fairly long time, especially when paths of movement are repetitive or too complex to be drawn satisfactorily.

The diagram is a scale drawing of the working area mounted on softboard. Pins are placed in positions where operations or inspections occur and thread is used to show the path of movement followed by the subject when visiting the places where these operations occur. Coloured threads can be used for different materials, components or workers and extra pins are needed where the direction of moves between end points changes.

String diagrams clearly show areas where congested movement occurs and where scrutiny may be needed. The pattern of movement shown also helps to suggest how equipment might be better sited. Templates, cardboard cut-outs of plant to the same scale as the drawing, help in improving the layout as they can be moved conveniently until the best one is found. If you want to measure the distance travelled by the subject, then the amount of thread used can be measured and the distance calculated according to the scale of the plan. String diagrams are easily understood and help when explaining proposals and the reasons for changes.

Flow diagrams often reveal that the siting of gates, doors, feedstores and watering points is poor. A weary pigman was heard to say that the only possible ways the sties could have been sited was by driving 100 sows into a field and building the pens around them when they fell asleep. He knew because he had to carry the feed and water and remove the dung. Such situations are easily avoided or remedied if tackled with patience.

Let us look at a simple example of the use of a flow diagram. Look at the plan of the poultry house in figure 6.5 and notice how the nests are scattered away from the door and the space between feeders.

Now consider that most people collect eggs three times a day. If two minutes are saved on each trip, this equals four or five eight-hour working days a year. In figure 6.6 the nests are near the door and a saving of about 35 km of walking in this pen would be achieved each year. There is no substitute for adequate forethought in siting and internal layout.

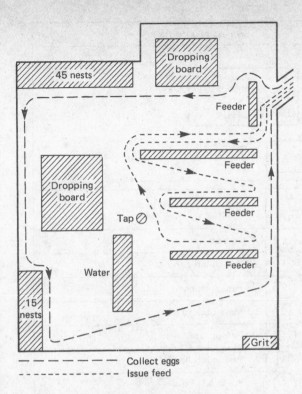

— — — — — — Collect eggs
- - - - - - - - - Issue feed

Figure 6.5   Inefficient Poultry House Layout

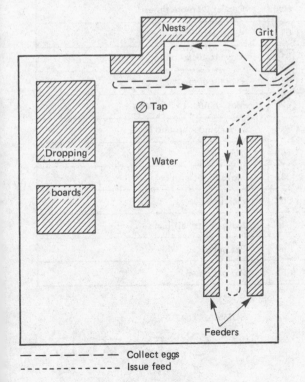

— — — — — — Collect eggs
- - - - - - - - - Issue feed

Figure 6.6   Efficient Poultry House Layout

The techniques so far described are for recording an overall process rather than the details of any particular event. Sometimes it may be profitable to study in detail the work done at a single workplace. If so, interest centres on what the worker does, and especially what he does with his hands. For this purpose a *two-handed process chart* is used, showing the corresponding activities for both hands. Figures 6.7 and 6.8 show the use of this technique to improve the process of tying 'hands' of tobacco after grading.

The third step in the basic procedure is the key to method improvement. Recordig has produced charts of the process showing the relevant details so that these can be assessed readily. All the facts are now critically *examined*.

In every job a positive critical approach can achieve improvement. Such an approach demands freedom from bias and an urge to learn, experiment and above all to think. In examining each part of a job a challenging approach is needed. Glib answers and hasty judgements must be avoided. 'Bright ideas' must not be acted on until all the facts have been examined: only when a better method is being developed should they be considered, and this must not be done until all undesirable features of the existing procedure have been found. Without systematic criticism full improvement will not be made, even if the technical ability needed to examine a process properly exists.

The examination is not carried out casually. There is a well-established, sound and logical system to be followed as shown in figure 6.9. It begins by asking what is achieved and goes on to consider how, when, where and by whom. It ends by deciding what else should be done, by whom and why. The evaluation is completed by costing any changes suggested and assessing the likely gains. All method study results can be costed so that their value can be judged by all concerned in deciding whether to accept them. This is one of the strengths of methods study.

The full questioning procedure will indicate how improvements could be achieved. Often an unnecessary event can be entirely removed. Alternatively, work can possibly be done in a more suitable place or an event form part of an improved sequence or a task be done by a more suitable person. The use of each question in turn ensures that every aspect of the problem and all possible alternatives are fully examined.

When all the events in the process or procedure have been examined, the unnecessary ones rejected and the true interrelationships of the

CHART NO. 1 SHEET NO. 1	WORKPLACE LAYOUT
DRAWING: of grading shed layout.	
OPERATION: tying hands	
LOCATION: Next to check grader	
OPERATOR: Female	
CHARTED BY: R.W.D. DATE: 2/9/86	

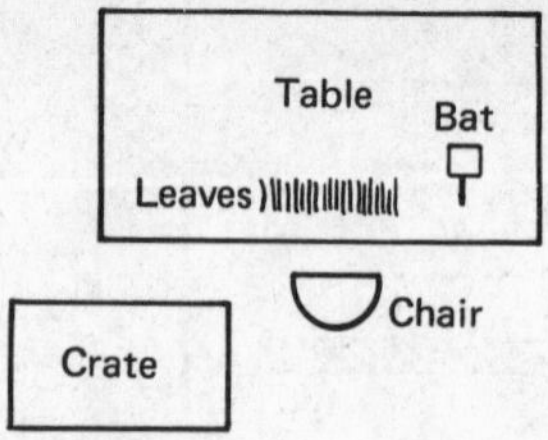

Table   Bat
Leaves)||||||||||||||
Crate   Chair

Left-hand description	Symbols		Right-hand description
	LH	RH	
Idle	1	1	Pick-up leaf
Idle	2	◁1	Pass to left hand
Grasp leaf	1	2	Put in left hand
Hold leaf	▽1	2▷	Move back to pile
Hold leaf (0.011 min)	▽2	3	Pick up leaf
Hold leaf (0.008 min)	▽3	◁3	Pass to left hand
Grasp leaf (0.003 min)	2	4	Put in left hand
Hold leaf (0.008 min)	▽4	4▷	Move back to pile
Repeat 28 more times	Rep.	Rep.	Repeat 28 more times
Change grasp of leaves	31	61	Assist left hand
Hold bunch of leaves	▽89	61▷	Move to pick up binder
Hold bunch of leaves	▽90	62	Pick up binder
Hold bunch of leaves	▽91	◁62	Move binder to left hand
Wrap binder around butts	32	63	Wrap binder around butts
Tuck in	33	64	Tuck in
Hold bunch of leaves	▽92	63▷	Move to pick up bat
Hold bunch of leaves	▽93	65	Pick up bat
Hold bunch of leaves	▽94	◁64	Move bat to left hand
Hold bunch of leaves	▽95	66	Pat butts
Put tied hand down	◁1	65▷	Put bat down

SUMMARY:

◯ 33   ▷ 1   ▽ 95   D 2

SUMMARY:

◯ 66   ▷ 65   ▽ —   D —

Figure 6.7   Two-handed Process Chart

rest revealed, it is possible to pass to the fourth stage of the basic procedure and *develop* realistic, constructive proposals for improvement. These must conform to standards of safety, quality and good practice.

The first step is to build a framework for the improved method by arranging, in correct order, essential 'key' operations in their revised form. Any ancillary work is then combined as efficiently as possible. The results of the systematic

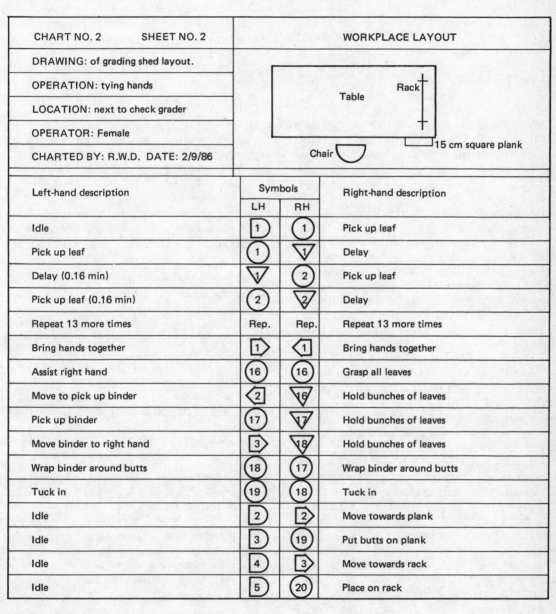

CHART NO. 2 SHEET NO. 2			WORKPLACE LAYOUT	

DRAWING: of grading shed layout.
OPERATION: tying hands
LOCATION: next to check grader
OPERATOR: Female
CHARTED BY: R.W.D. DATE: 2/9/86

Table   Rack   15 cm square plank   Chair

Left-hand description	Symbols LH	Symbols RH	Right-hand description
Idle	1	1	Pick up leaf
Pick up leaf	1	▽1	Delay
Delay (0.16 min)	▽1	2	Pick up leaf
Pick up leaf (0.16 min)	2	▽2	Delay
Repeat 13 more times	Rep.	Rep.	Repeat 13 more times
Bring hands together	▷1	◁1	Bring hands together
Assist right hand	16	16	Grasp all leaves
Move to pick up binder	◁2	▽16	Hold bunches of leaves
Pick up binder	17	▽17	Hold bunches of leaves
Move binder to right hand	▷3	▽18	Hold bunches of leaves
Wrap binder around butts	18	17	Wrap binder around butts
Tuck in	19	18	Tuck in
Idle	2	▷2	Move towards plank
Idle	3	19	Put butts on plank
Idle	4	▷3	Move towards rack
Idle	5	20	Place on rack

SUMMARY:

○ 19   ▷ 3   ▽ 14   D 5

SUMMARY:

○ 20   ▷ 3   ▽ 18   D —

Figure 6.8   Two-handed Process Chart

examination must be used throughout to ensure that work is done by the most suitable people, in the best place and by the simplest means. Compromise is often necessary between the possible ways of improvement. At this stage creative thought, based on knowledge, is invaluable. The 'bright ideas' that were noted but firmly placed asided during the systematic examination can now be seen in their true perspective and, if justified, can be introduced into the method. Parts of the new method may have to be tested; mock-ups may have to be made or equipment improvised. All people, directly or indirectly affected by proposals, should be asked to help and advise.

Finally the improved method should be drawn up in process chart form. This is then given the same close examination as were the charts of the original method. Thus we ensure that the improved method is logical and best, provided existing conditions do not change.

All the work that went into developing a workable, better method can be wasted unless equal care is taken in *installing* it. Much depends on proper preparation beforehand, including, full explanation of what is proposed. If the study was properly done, all concerned with the changes should feel that they are partly responsible for their success. If this feelilng is created, the change can be affected knowing that people are confident in it and will actively support it. It is vital that every detail of a method is carefully prepared before installation and the job must be supervised actively during the initial period after the changes are made.

To *maintain* the new method, the need for regular checks is obvious. This last step is important yet often ignored. Casual changes in method can arise for several reasons. Sometimes they are easily observed; sometimes they consist of a build-up of small changes that cause a gradual drift away from the new method. Regular checks ensure that changes of either kind are noted. If they are useful they can be built into the method; if they lower efficiency they should be removed.

This brief review of method study has been made in terms of the steps in the basic procedure. Any successful manager must have used many of the principles outlined but the techniques described enable the use of old principles in a more ordered way.

One branch of method study is concerned with the details of hand work on jobs in which little or no equipment is used, like cotton reaping and thinning and tobacco grading. Where manual work is done at one place, as in the latter case, the following concepts affect output:

- The worker should sit, if possible, and a comfortable working position with suitable working heights should be provided and maintained.
- Both hands should be used equally and at the same time.
- Balanced, co-ordinated motions (obstacles to the movements must be removed) are the easiest. The hands should be kept close together, preferably within 10 cm, so that both can be controlled by eye without excessive head movement. Continuous circular motions of the hands at a steady rate are better than jerky, irregular ones.
- Hands should not be used for holding things. Holding devices leave the hands free for productive work.
- Materials and tools should be within easy reach.
- The position of the feet is important.
- We should try to change the work so that we use finger motions before wrist motions, elbow motions before shoulder motions and so on — always choosing the least tiring motion.

Figure 6.10 shows how these principles were applied to designing a tobacco grading table at which the grader is seated.

Systematic method study is not needed to detect many of the common inefficiencies on farms and the following points may help managers to see their own weaknesses. Some of them are basic but, nevertheless, often overlooked. Managers should ask themselves whether their own practices are efficient in each of the details mentioned.

The following principles affect men using equipment or working together as a gang:

- Equipment should be fitted to the job.
- The work should be planned to avoid idle time for men or machines.
- A man's output is usually more when he is in a small rather than a large gang.
- All machinery should be able to bear its seasonal loads without breakdowns but spares should be available in case breakdowns occur.

In routine work in the around buildings, the following concepts should be observed:

(1) A building is a means to an end and not an end in itself. The jobs within the around it represent continuing labour demand and production cost. This is where savings can be made usually and yet all too often new buiding may be decided

WHAT IS DONE?
Identify the job. The form of words does not matter.

WHY?
Find out why the job must be done. Is there one main reason? If this reason were changed is it possible that the job would disappear? Or are there several other reasons that must still be satisfied?
Find out what would have to be changed to eliminate or reduce the job. No written record is needed.

Consider the possibilities: what would happen if the job were stopped. What could you do to make elimination acceptable? Would it be easier/cheaper to go on doing the job? Could you make the job partly unnecessary? Can any specification be relaxed? Write down only decisions or ideas for action. Don't record argument.

Do any ideas have any effect on the skill demanded of the worker? On the time needed for the job? On its frequency?

HOW IS IT DONE?
Think about the way the job is done in terms of (a) what is changed by doing the job (b) the principle of the method (c) the tools, aids and equipment used.

Explore reasons mentally as for WHAT above.

Consider the possibilities: eliminate unnecessary parts of the job. Simplify the essentials remaining. Specify what must be done to make the simplification work.
Check whether tools and aids are standard and efficient. Is the man serving a purpose better performed by something mechanical? (E.g., does he act partly as a jig or a source of energy?) What could you mechanise? Now? Next year? Write down workable alternatives.

Would method changes permit more economical manning?

Check on safety provisions.

WHEN IS IT DONE?
Find out the TIME aspect of the job under the heading of:
Sequence
Frequency
Absolute time
Duration

Find out what factors govern the TIME aspect.
Does it form a vital link in a chain of activities? Does it control the timing of some other job?
Is the frequency governed by outside factors or does the content of the job affect it?
Does the job *have* to be done at any special time? Or is it just 'convenient'. What is it convenient for? What is convenience worth?
What factors control duration?

Consider sequence only when related jobs are being studied.
Decide what the vital linkages are.
What is the absolute minimum frequency? What safeguards would you need? Do you really need to increase the frequency? Might this reduce overall costs? When is the most suitable time? How close can you get to his is practice? Is the job such that you can do it for a shorter while without failing to reach the objective? Record changes.

Use Multiple Activity Charts or Arrow Diagrams to integrate jobs.

Devise time changes on the basis of redefined and re-evaluated time constraints.

WHERE IS IT DONE?
This is unlikely to be significant but check on the convenience or advantages of the present site.

Where would you like to do the job?
Could you shorten the job or simplify it by doing it elsewhere?
Is there sufficient attraction in other sites to make a change worthwhile.

Record a confirmed or changed site.

WHO DOES IT?
Find out the number and qualifications of people doing the job. What is thought to govern the present manning arrangements? How sound are the reasons? Are they logical or arbitrary?

Evaluate the known alternatives to decide what is the simplest manning arrangement to perform the job to the REVISED specification. Integrate with the evaluation the changes that must be made in order to make manning change acceptable. Record changes.

Figure 6.9 Guide to the Use of a Critical Examination Sheet

on to secure improvement when better methods in existing buildings could produce equal benefits at less or even no cost.

(2) Work areas should be close together to reduce movement; for example, a feed store in the piggery and tools and supplies as near as possible to where they will be used. Site new buildings or re-site enterprises to reduce or eliminate unnecessary transport and plan stock buildings and yards to reduce work on feeding and littering.

(3) Circular travel eliminates back-tracking.

(4) Full use should be made of gravity. Now that mechanical equipment is available for lifting heavy loads, like front loaders on tractors, use can often be made of gravity even if buildings are on a level site. For example, a bulk-feed hopper above a milking parlour enables feed to drop down an enclosed chute straight into the parlour. Again, grain storage bins, filled by auger, can be erected so that lorries can drive under them for loading by gravity.

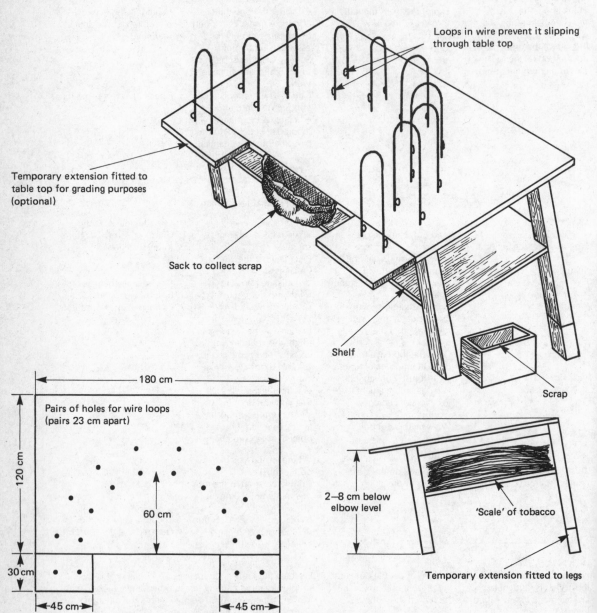

Loops in wire prevent it slipping through table top

Temporary extension fitted to table top for grading purposes (optional)

Sack to collect scrap

Shelf

Scrap

180 cm

Pairs of holes for wire loops (pairs 23 cm apart)

120 cm

60 cm

30 cm

45 cm

45 cm

2–8 cm below elbow level

'Scale' of tobacco

Temporary extension fitted to legs

Figure 6.10   Well-designed Tobacco-grading Table

(5) Transport on wheels is better than handling. Passages should be wide and smooth enough for trolleys or tractors and trailers.

(6) A job should start where the previous one ended.

(7) Full loads reduce the number of trips.

(8) Double-handling should be avoided.

(9) Materials should be kept in bulk whenever practicable.

(10) Materials should flow in one direction. A plant for grinding and mixing stockfeed is an example of an installaton that needs careful planning so that the bought feeds join the home-ground meal at the appropriate place and that the mixed rations are stored near the door from which they will leave.

Materials handling merits more attention. The best system of handling is no handling. Each time a material is handled, that is, lifted, moved and placed down again, cost but not value is added to it. Some handling is unavoidable but much of it is unnecessary. Consequently, a main aim should be to minimise handling. This may well involve changing layouts — if it is justified financially. Alternatively, some materials-handling equipment, such as fork trucks, pallets, wheelbarrows or sackbarrows, may be necessary. The most simple form of 'mover' for unsewn bags consists of wire hooks that pass through the sack and hook onto a pole carried by two workers as shown in figure 6.11.

It is important to use cheap, efficient mechanical devices and hand tools to simplify or cut out work. Examples include ball-value controls for water troughs, cattle grids instead of gates and hand tools of the right design, weight and capacity. Figure 6.12 shows how to make a suitable container for hand distribution of seed or fertiliser, or for reaping certain crops, from an ordinary fertiliser sack. Weight is evenly spread on the worker's shoulders and both hands are free.

Some examples of ways in which method study has helped to develop better methods of doing field work in the tropics follow. Figure 6.13 shows the best method found for hand-reaping maize cobs.

Figure 6.14 shows as improved way of reaping tobacco. Traditionally, tobacco leaves were separately picked by hand by reapers. The reaped leaves were passed to a waiter who in turn carried bundles of leaves to a road or other collecting area. At this point the leaves were either tied in small bundles over a stick or they were placed loose in crates for transporting and later tying at a

Figure 6.11 Pole and Hooks for Moving Bags

central farm depot. Finally the sticks were placed in curing barns. This harvesting method not only wasted much labour; it also caused too much waste and breakage of leaf.

The reaping method now used on most commercial tobacco farms requires the reaper to pick and tie the reaped leaves simultaneously. Furthermore, the stick has usually been eliminated so that now most tobacco is either handled on strings or on specially designed clips. On average, labour use has been halved and wastage is at least 10 per cent less. As a net result, a grower can now hold a smaller, permanent labour force rather than the large, fluctuating one he used to.

Another good example of the use of method study is in cotton reaping. Traditionally, reapers held a reaping container, such as bucket, in one hand and picked the bolls with the other. They picked all open bolls and then a gang later had to grade the seed cotton. Method study has now resulted in a system whereby a reaping pocket is tied around the waist of the reaper. He is therefore free to reap the inside of two adjacent rows with *both* hands and he reaps only one grade at a time. This eliminates any later cleaning or grading. Studies have shown that single-row picking can be as much as 37 per cent slower than picking from the inside of two adjacent rows. Similar results have been applied to coffee reaping. The simple change to reaping coffee cherries into a pocket tied to the reaper instead of into baskets on the ground can raise hourly output for minor, intermediate and main reapings by 46, 79 and 124 per cent, on average, respectively.

Table 6.4   Metres of Cotton Row Thinned by Hand each Hour by a Worker at 75% Productivity

No. of pulls/ 10 metres	Method			
	A	B	C	D
50	375	340	240	160
60	310	280	200	130
70	270	250	170	110
80	230	210	150	100
90	210	190	135	90
100	190	170	120	75

Methods: A. Single row, two-handed thinning of continous or hill-drop planted cotton.
          B. Single row, one-handed thinning of hill-drop planted cotton.
          C. Single row, one-handed thinning of continuous planted cotton.
          D. Single row, one-handed thinning and measuring of continuous planted cotton.

Finally, table 6.4 shows the effect on labour productivity of different methods of hand-thinning cotton. It can be seen that the most output is obtained if the worker is trained to use both hands. Output depends on the number of pulls needed to thin the cotton to the required stand. This is found out by trial. It is not necessary for the worker to use a measuring stick all the time. He can be taught to estimate the needed spacing.

Work measurement

This is the second of the two complementary parts of work study. It is defined by the British Standards Institution as:

The application of techniques designed to establish the time for a qualified worker to

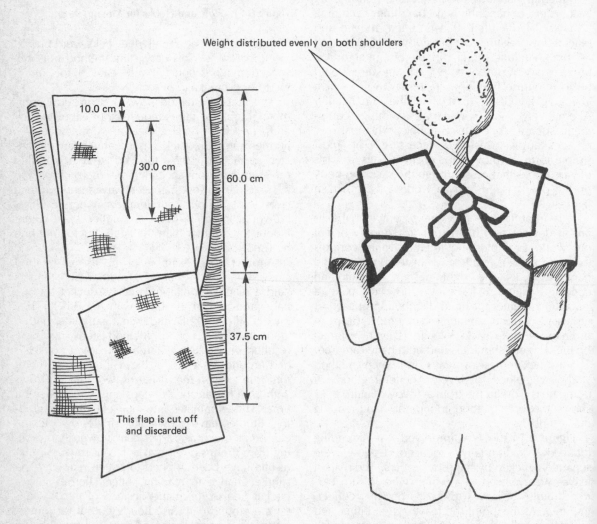

Figure 6.12   An Easily-made Container for Hand-distribution of Materials

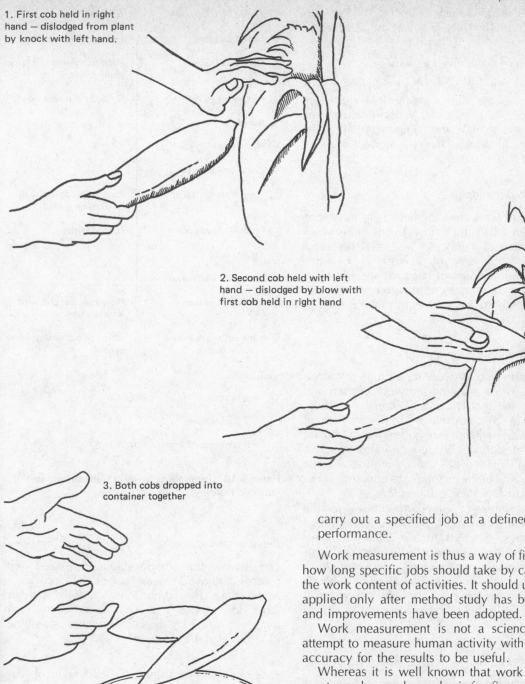

1. First cob held in right hand — dislodged from plant by knock with left hand.

2. Second cob held with left hand — dislodged by blow with first cob held in right hand

3. Both cobs dropped into container together

Figure 6.13   Good Method of Reaping Cobs

carry out a specified job at a defined level of performance.

Work measurement is thus a way of finding out how long specific jobs should take by calculating the work content of activities. It should usually be applied only after method study has been done and improvements have been adopted.

Work measurement is not a science but an attempt to measure human activity with sufficient accuracy for the results to be useful.

Whereas it is well known that work measurement can be used as a basis for financial incentive schemes, it is less often known that it has equally, if not more important, wider uses. Work measurement gives a quantitative basis for comparing alternative methods but its uses are not limited to this. Good planning and control must be based on sound data and, if incentive schemes designed to raise workers' performance are to be fair and rational, there must be a reliable way of measuring that performance.

The uses of work measurement are as follows:

Comparison of alternative methods

Where two alternative methods seem equally advantageous, the one needing least time for completion is more efficient. Work measurement is the best way to make this comparison, although the quality and value of the end product must be considered.

Correct initial manning

The results of work measurement help to ensure that jobs are correctly manned. The number of men allocated to a job can be based on exact knowledge of the amount of work to be done. Analysis of the amount of effective and idle time existing in a job makes it possible to allocate work efficiently so that each man's time is used to best advantage.

Labour and equipment planning

The data provided by work measurement form a reliable basis for planning the work of men and machines so that managers can use these resources to best advantage. Managers cannot plan accurately without standard times. In farming the problem is greater because conditions vary so much. Past experience is usually used as a guide but, especially if new methods are adopted, accurate standard times are better guides.

Accurate ubiassed standard times are the basis for setting task work, piece work and incentive schemes. Attempts to use these techiques without time standards usually lead to problems for both managers and workers. For example, a farmer might give his workers too much tobacco to grade, in which case the scheme will fall down; or too little, so he spends too much on bonuses.

Management control

When a plan has been made a manager needs to know how closely it is followed in practice. Reliable figures on the actual use made of resources are invaluable if the manager is to make sound decisions on how to raise productive efficiency. Accurate records of waiting time and analysis of its causes, plus a knowledge of the performance while work is being done, give a reliable basis for control.

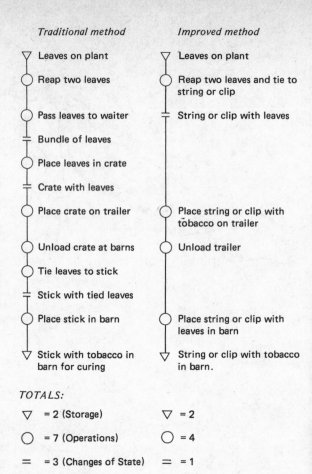

Figure 6.14   Example of Method Comparison in Tobacco Reaping

Reliable labour costing

The indices that are available from well kept records and work measurement data provide a sound basis for labour cost control. Standard times for various jobs can be converted into money terms, using the current wages as a guide, to arrive at standard costs.

A rational basis for sound incentive payment schemes

Incentive schemes, like a bonus paid for exceeding normal output rates, may be based on a firm, factual and fair base. Incentive schemes are only one possible use for work measurement and the techniques are used in the same way whether or not the results are to be used for this purpose.

Production of work measurement figures for

farm use is a specialised job to be undertaken only by trained practitioners with at least several months' practical experience after their training. For example, it needs much experience to make adjustments to allow for working conditions.

The procedure of work measurement consists of four stages:

- *Defining* what the work is — often harder than it seems.
- *Determining* the time in which the work *can* be done, that is, timing the job.
- *Assessing* the performance of the worker being studied: is he working faster or slower than 'normal'? That is, rating the worker's performance.
- *Identifying* the amount of wasted time in a job, that is, using allowances.

The procedure used to find how much labour is needed to do a job is to time how long it takes a qualified man to perform each element of it. It is important to split the job into elements as these elements may be rearranged later between the men. It also helps to improve the accuracy of the timings. Observations should be repeated about 30 times. Average time taken for each element is calculated and these are summed to give the total basis time.

Records of basic times for elements are called synthetic data and allowances are usually added to these, to allow for such things as relaxation time, to provide the time for doing the job at the standard rate of working and for recovering from the effort, that is, that work content. The addition of allowances may be made elements by element (when the work contents of the elements are summed to obtain the work contents of the elements are summed to obtain the work content for the whole job) or the basic times for the elements may be summed and the allowances added job wise, again to give the work content.

The most important allowance in farming is the rest or relaxation allowance. This is an addition to the basic time (usually stated as a percentage) to recover from the physiological and psychological effects of using energy in performing specified work under given conditions and to allow attention to personal needs.

While working, the operator incurs fatigue which is defined as 'a physical and/or mental weariness, real or imaginary, existing in a person and adversely affecting his ability to perform work'. The effects of fatigue can be reduced by rest breaks, during which the body recovers from its exertion, or by reducing the rate of working and thus the use of energy.

Although much is known about the nature and effects of fatigue under extreme conditions, little work has been done on the allowances to be made for recovery from fatigue at usual working levels. To a great extent, therefore, rest allowances are still largely guesswork.

There are both constant and variable rest allowances. Constant allowances consist of a personal allowance to cover needs like washing, drinking and relieving oneself, and a basic fatigue allowance to allow for the energy used even when not working.

Variable allowances are given for certain ergonomic factors that vary between jobs or elements. The main ones are as follows:

A. *Standing*

If possible at a seat should be provided.

B. *Unusual position*

The usual position for working varies throughout the world but in western countries is with the worker standing or sitting and with the work at about waist height.

C. *Use of muscle power*

Someone shovellling a light load must rest up to 12 per cent of the work cycle because of the continuity of the effort alone. Hand sawyers need to rest for about 20 per cent of their time. If heavy weights are lifted, it iş best to provide mechanical aid.

D. *Bad light*

The amount of light needed depends on the type of work. If it is not enough, allowance is made for the strain imposed. If it is too much, consider non-glare wall colourings.

E. *Atmospheric conditions*

When the body works, it produces heat which is lost by sweating. The rate of loss of this heat depends on the temperature and humdity of the air, the ventilation and the local presence of hot machines, walls, etc. If conditions are bad, the rest should be taken away from those conditions for it to be effective. Research in the USA and the UK shows that men work best at about 18°C. Acclimatisation of workers to hot conditions should be considered.

F. *Close attention*

Allowance is made for fatigue if eyestrain occurs, as in chick sexing, removing off-type groundnut kernels passing on a conveyor belts, etc.

Table 6.5   Typical Rest Allowances for Manual Work

		Men	Women
		%	%
Constant allowances			
Personal allowances		5	7
Basic fatigue allowance		4	4
Variable allowances			
A. Standing		2	4
B. Unusual position	— slightly	0	1
	— bending/ crouching stooping	3	4
	— lying, stretching up	7	7
	— confined space	15	15
C. Use of muscle power Force	— 2 kg	0	1
	— 10 kg	3–4	5
	— 25 kg (max.)	14	23
D. Bad light	0 — 5%		
E. Atmospheric conditions			
Reasonable	0 — 5		
Unusually bad	5 — 50		
F. Close attention	0 — 7		
G. Noise level	0 — 5		
H. Mental strain	0 — 8		
I. Monotony	0 — 4		
J. Tedium		0–5	0–2

G. *Noise*

Strain arises from loud, intermittent noises occurring near the workplace, like car horns or slamming doors, or where the work requires listening for changes in sound.

H. *Mental strain*

Fatigue may be caused by prolonged concentration.

I. *Monotony*

Repeated use of certain mental powers at a low level, as in arithmetic and clerical work, can cause strain from monotony.

J. *Tedium*

This is due to repeated use of certain parts of the body and is likely to occur on mass production assembly lines, etc.

Table 6.5 shows typical rest allowances. Unless a large allowance has been made for use of muscle power, because mechanical means cannot be substituted, the total rest allowance should be little over 20 per cent.

The form in which rest allowances are taken varies. In an extreme case, a worker might take no rest breaks but work continuously at a lower rate. Another man might work harder and take fewer but longer breaks. Frequent short pauses are usually more effective than fewer long ones. In farming, the timing of rest breaks can be important at peak labour periods, especially with large gangs. Sometimes the rest breaks for the gang coincide with some other periodic stoppage. This may occur, for example, in water planting when the water container is changed or in land fumigation when the injector guns are filled.

A similar problem with tropical farm workers is that of feeding. It is common for the workers to start work having had no food since the previous evening. Their work efficiency will tend to drop quickly after a few hours and can be raised and kept at a higher level by ensuring that they eat something before arriving in the morning and that some food is provided at work about four hours after work started. By so doing it is possible to achieve the same output at a higher level of efficiency over a shorter time and to cut out the long, wasteful break in the middle of the day.

The definition of work measurement refers to a 'qualified' worker. In this context, a qualified worker is one who has the necessary physical qualities, intelligence and education and has acquired the needed skill and knowledge to perform the job to adequate standards of safety, quantity and quality. He should not, however, be so experienced that he consistently achieves a higher output than most other workers. He should be at ease and working normally. Clearly, it would also be a waste of effort to measure untrained workers.

Work measurement alone will not improve productivity. It is the use of target or standard times, developed from work measurement, and used as a basis for control that raises performance. The end result of work measurement should be the fixing of a standard rate of working that is the average rate at which qualified workers will naturally work at a job if they know and keep to the specified method and are motivated to work.

The following information is usually needed for a work standard:

● The basic time, that is, the time it takes a

worker familiar with the job to do it.

- Allowances to adjust for working conditions and the nature of the job.
- Performance rating according to how hard the man which the standard time apply cannot be overstressed:

Method

This has a great effect on the time for the job. In weeding, for example, whether the worker stands on one side or astride the ridge.

Equipment

This also can greatly affect the time for the job; for example, whether a traditional or swan-neck hoe is used, the weight of the tool, whether the blade is worn and its size.

Aptitude and skill

Usually only trained workers with fair aptitude and skill are studied but this should be stated when standard times are quoted.

Effort

This is usually allowed for by a rating or levelling procedure. Again, only workers putting a fair effort into the job are studied.

Working conditions

Taking the example of cultivation again, such items as soil type, soil moisture and weed density should be defined when quoting standard are times.

If suitably motivated, the average regular African farm labour can work at a level of 75 per cent of standard.

If, for example, a farmer finds on inspecting his 5 ha land that it can be weeded with an average of 140 strokes/10 m and the inter-row spacing is 1 m, then a total of 50,000 m of row has to be cultivated. He expects a 75 per cent level of productivity from his labour force which works eight hours a day. From table 6.8 we see that a worker can weed about 245 m/hour, that is, 1,960 m/day. The total number of man-days needed therefore is (50,000/1960) = 27 days. If the job should be finished in three days, a gang of nine is needed.

When cultivating, some weeds must be removed by hand. However, as the time for hand-pulling hardly varies from that for hoeing,

this element may in practice be regarded as hoeing. For example, if a labourer needs 90 strokes to cultivate 10 m of row and 10 pulls by hand to remove weeds in the row, the element may be taken as 100 strokes/10 m.

WORK ALLOCATION

There are four main ways of allocating and paying for labour. Careful selection of the method used for a particular farm task can help to raise labour productivity. These four methods are time work, task work, piecework and bonus work.

Time work

In this system, sometimes called daily work, employees work for set hours each day, and, apart from any overtime earnings, they receive a fixed daily, weekly or monthly rate of pay for attendance. Time work is probably the most usual way of paying for labour on most tropical farms but labourers prefer task work.

This system is most suitable for those jobs where speed or quality of production cannot be influenced by the energy or skill of the worker or when he has to do many different jobs during the day. It has several disadvantages however. As pay depends on time rather than output, management has less control over output and there is no incentive for workers to put much effort into their jobs. All workers on the same basic rate get the same pay. They merely have to report for work and put in time. According to work study specialists in Zimbabwe, most local farm workers on this system there achieve no more than half their potential output and much supervision is needed.

Task work

Here the worker is given a set task to be done in the day, like hoeing 1000 m of maize row, and when the task is finished and inspected he can go and gets a full day's pay.

Most labourers prefers task work to time work as time off on finishing a task is a strong incentive, its effectiveness depending on the value of leisure to each worker. They will usually work much harder to finish a task quickly. In hot months workers start as early as 4.30 a.m., work and finish as early as 10.00–10.30 a.m. This gives them time to fish, sleep, go to town or relax.

Task work is especially useful in allocating work to casual labourers during times of peak labour demand. If they are hired locally and also have to tend their own land, they can do the latter in the afternoons.

The advantages of task work over time work are that it needs supervision but there is more control over output. This system of allocating and paying for work has, however, the following disadvantages:

- Between 30 and 50 per cent of labour potential is wasted because, to provide the leisure incentive, the tasks set seldom exceed 70 per cent of the worker's potential and are usually much lower. Tasks should be set realistically at 75 per cent of standard.

Table 6.6   Labour Use per Hectare calculated by Work Measurement in Malawi

Maize		Rice		Cotton		Tobacco	
				Hours/year			
Ridging	135	Preparation	458	Ridging	135	Ridging	136
Planting	43	Transplanting	443	Planting	24	Planting	49
Weeding	74	Weeding	342	Weeding	78	Weeding	140
Reaping/450 kg	20	Reaping	370	Reaping	475	Reaping	30
		Threshing	103	Grading	950		

Table 6.7   Cotton Reaping (Double row Reaping into Pocket)

	Level of productivity					
	100%	90%	75%	70%	60%	50%
Bolls/kg (kg a worker can reap each hour — using both hands)						
180	8	7	6	6	5	4
220	7	6	5	5	4	4
260	6	5	4	4	4	3
310	5	4	4	3	3	2
350	4	4	3	3	2	2

(Figures rounded to nearest kg/h)

- The best workers do the same work as the poor ones for whom the task is so often set. They often finish their task in four of five hours so their capabilities are not fully used and their potential earnings are reduced.
- Supervision is necessary until the slowest worker has finished his task — often an hour or more after the others have gone.

There is clearly a need to set tasks at realistic levels and work study standards are available for many jobs. In the early stages, task work seems to work well. However, conditions slowly change over time. Tools or equipment wear out or

Table 6.8   Hand Weeding Cotton Using a Traditional Hoe[a]

No. of strokes/10 m (assessed by trial in land)	Level of productivity					
	100%	90%	75%	70%	60%	50%
	(metres of row/hour)					
50	920	830	690	640	550	460
60	770	690	580	540	460	385
70	660	590	495	460	395	330
80	570	510	430	400	340	285
90	510	460	380	355	305	255
100	460	415	345	320	275	230
110	420	375	315	295	250	210
120	380	340	285	265	225	190
130	355	320	265	250	215	180
140	325	290	245	225	195	160
150	305	275	230	215	180	150
160	285	255	215	200	175	140
170	270	240	200	190	160	135
180	255	230	190	180	150	130
200	230	205	170	160	135	115

[a] A swan-neck hoe, however, is more suitable and can raise output by about 10%.

Table 6.9 Coffee Reaping

Reaping cherries/kg	100%			90%			75%			60%			50%		
	A	B	C	A	B	C	A	B	C	A	B	C	A	B	C
(1. Basket on ground)															
400	7	10.5	14	6	9	12.5	5	8	10.5	4	6	8	3.5	3.5	7
450	6	9	12.5	5	8	11	4.5	6.5	9	3.5	5.5	7.5	3	4.5	6
500	5.5	8	11	4.5	7	9.5	4	6	8	3	4.5	6.5	2.5	4	5.5
550	5	7	10	4.5	6	9	4	5	7.5	3	4	6	2.5	3.5	5
600	4.5	6.5	9	4	5.5	8	3.5	5	6.5	2.5	3.5	5	2	3	4.5
(2. Pocket on reaper)															
400	10	17.5	30	9	15.5	27	7.5	13	22.5	6	10.5	18	5	8.5	15
450	9	16	27.5	8	14	24.5	6.5	12	20.5	5	9.5	16.5	4.5	8	13.5
500	8	14	25	7	12.5	22.5	6	10.5	18.5	4.5	8	15	4	7	12.5
550	7	13	22.5	6	11.5	20	5	9.5	16.5	4	7.5	13.5	3.5	6.5	11
600	6.5	12	20	5.5	10.5	18	5	9	15	3.5	7	12	3	6	10

Level of productivity
(Kg a worker can reap each hour — using both hands)

A. Minor reapings — up to 1 kg cherries/tree
B. Intermediate reapings — between 1 kg and 2 kg cherries/tree
C. Main reapings — over 2 kg cherries/tree

become too few, workers are allowed to develop bad habits or changes occur in the nature of the material handled. Labourers resist any apparent rise in their task and, with worsening conditions and equipment, it becomes ever harder to attain the initial task. Lower tasks force their way in if supervisors take the easy way out and the manager is not in touch with detail closely enough to dispute their judgement.

Despite the fact that many tropical farmers say that leisure is a greater incentive to their workers than money, experience in many industries in Africa shows that properly installed and administered money incentives are preferred by workers. Whenever existing task systems have been partially or wholly replaced by money incentives, productivity rises of from 20 to 100 per cent have usually resulted.

Piecework

This, like bonus systems, is a system of *payment by results* where payment depends on the amount of work done rather than the time worked. Sometimes called contract work, a piecework system can be applied either to separate workers or to gangs. The system is designed to pay workers at a constant rate for each unit of work produced correctly, whatever the number. The amount of work must be measurable and quality standards must be set.

Piecework provides a financial incentive for greater productivity and workers are thus moti-vated to exert themselves if they wish to raise their earnings. Most labourers do not mind working longer hours, even under hard conditions, if the reward compensates them for the extra effort.

Piecework is especially useful if workers are fully integrated into the wage economy, for example, in or near urban areas, because high earnings are possible. The work should be such that the workers, by their own effort and skill, can produce and thus earn more.

Examples of the application of piecework rates are, say:

- 3¢ for each 10 strings of tobacco reaped
- 1¢ for each 10 'hands' of tobacco tied
- 25¢ for each 5 kg of cotton reaped
- 5¢ for each 10 m of planting ridge made

To illustrate a way of calculating piecework rates, let us look at intermediate coffee reapings when there are 500 cherries/kg and reaping pockets are used for an eight-hour day. From table 6.9, at 75 per cent productivity, workers can reap about 84 kg cherries each day. To calculate the piecework rate, relate standard output to wages plus 50 per cent. If wages are $1.28 a day, wages plus 50 per cent is $1.92 and:

Piecework rate = $1.92/84 = 2.3¢/kg
(Contract workers will work for eight or ten hours a day at 90–100 per cent of standard because of self-motivation).

With crops like cotton and fruit that require

several pickings, reduced quantities or size may lessen earning opportunities at certain times in the season. It is usually possible to compensate workers for this. For example, in the USA it is common practice to raise cotton piecework rate for the later pickings in the season.

Piecework is the most widely used form of payment-by-results because of its great advantages of simplicity and fairness. It is easy to understand and workers can calculate their earnings. Consequently, for suitable tasks where fairly standarised conditions exist, it is easy for managers to set up and for an unsophisticated worker to understand. Workers can achieve high earnings and daily output will often rise by about 30 per cent over time work or task work. Furthermore, a manager knows his labour cost per unit of output in advance and the system may lead to better care of equipment to avoid breakdowns and thus loss of earnings.

The manager is responsible for ensuring that workers are not kept waiting for materials. If there are delays, the worker's earnings suffer. It is also the responsibility of managerment to ensure proper machinery maintenance to reduce the risk of breakdown. Both these areas can cause tension between workers and managers.

There are drawbacks to the piecework system and unless workers are motivated by good work conditions and labour relations, the high output may not be maintained. Furthermore, the slowest workers and other workers still paid on a time basis may become discontented. Workers are sometimes reluctant to raise their output because they fear that piece rates may be reduced if output is too high or that they will not receive full payment for their output. It is the manager's job to remove these fears. This is a matter of realistic basic standards and the fairness with which they are applied. An incentive scheme must be adapted to fit the situation on each particular farm.

Standards for the quality of work must be set, otherwise the worker may, in his efforts to do as much as possible, produce a poor standard of work — rocks in cotton reaping bags, etc. When the going is tough, farmers tend to neglect quality in order to get the job done.

A common cause of trouble in cotton reaping is non-payment for dirty cotton because the workers know that it will be sold. Better psychology is to make the reapers clean it first next morning before they are paid. They soon learn that they are losing reaping time and earnings by not reaping to the required quality.

Bonus work

This is really a combination of either time work or task work with piecework. A bonus is paid for output above normal and should be paid not later than the end of the week in which it was earned. To be successful, bonus schemes need intelligent application and a good standard of management.

Bonus schemes reward workers for extra output as they are given a basic task and output above this level earns a bonus. They should be based on standards set by work study and job evaluation and pay more to those who exceed these standards of physical production. They are very effective in maintaining output and quality of work by raising a worker's interest in his work. In fact, output can rise by 50 to 100 per cent over that achieved by time work.

Bonus payments are harder than piecework rates to calculate and for most workers to understand and therefore they need careful explanation. A complex bonus scheme is easily misunderstood and, consequently, may become mistrusted by workers, thereby defeating the purpose for which it was created. Unless a bonus scheme is uniform throughout all sections of a business, there is likely to be discontent among those who are unable to qualify for bonus however hard they work. This discontent will be felt especially when someone is transferred or even 'promoted' from a job in which bonus can be earned to a job for which there is only a fixed basic rate. When incentive bonus schemes are used, they must provide a fair method of rewarding workers for equally increased performance.

Bonuses should be simple to calculate, readily understood by worker, manager and employer, give the worker a fair reward for extra effort but be low enough to raise profit.

Several important points should be considered before starting a bonus scheme.

Organisation must be good so that workers are not penalised or prevented from attaining their bonus. For example, in tobacco production, the row length should not be too long for reaping and the grading-shed layout should be good.

Workers must *not* be allowed leisure as an alternative to bonus work because they may stop work and go home on finishing their basic task. They should be present for set hours. Most workers will finish their basic task in the morning and, rather than be idle in the afternoon, they will start to do bonus work.

Penalties for poor quality work must be set and clearly understood by the workers.

Bonus payments should be frequent, even daily in the early stages of the scheme until the cash incentive is appreciated, and separate from ordinary pay. Weekly, monthly or quarterly payment may be introduced later. However, payment is best as close as possible to the time when the effort was made. Annual bonuses are totally ineffective in raising productivity because of the time lag between the effort and the reward.

The results of incentive schemes usually show clearly that those schemes under which workers are each rewarded for their own efforts are more effective than those under which groups of people share equally a communal bonus. Sometimes, however, several men work as a team; the team being of such a nature that the work of each individual member is interdependent and cannot be measured. In such situations only a group bonus can be calculated. Teams covered collectively by an incentive scheme should be as small as possible and should always consist of members of an interdependent group and not be based purely on geographical location. It has been found in practice that if such a team does not exceed eight, there is little loss of individual incentive. Furthermore, less supervision is needed as workers do not like to see their bonus being endangered by slackers. Members of a bonus group should, if possible, be able to see each other most of the time.

Bonuses are often paid to managers, assistants, foremen and skilled stockmen, who are not usually paid overtime. The bonus itself may be a fixed sum or a share of gross or net income from the whole farm or from a particular enterprise; for example, 10 per cent of milk sales less concentrate costs, over a certain basic amount, in a dairy enterprise.

The level at which the basic task, for which the worker is paid his normal wage, is set depends upon the overall ability of his group but it should be within reach of the poorer workers. Farmers should accept work study standards as a basis. For a normal farm gang the basic task should be usually calculated at the 75 per cent level of productivity. This still enables a worker to raise his earnings if he wishes to do so yet it is a level of output that can be reached by each worker. Farmers should make it clear to their workers that any output over this basic will be rewarded and that if it is continuously exceeded, the basic task will not be raised. This policy may also prevent the common victimisation of those who work harder to raise their earnings.

The amount of bonus paid depends on the levels of the basic pay rate and the basic task. A general rule is for a 33–50 per cent wage increase to be paid for a 50–100 per cent rise in output. A bonus should be just that — not a partial substitute for too low a basic wage.

Let us look at a few examples of bonus schemes. In cotton reaping, for example, one could set a basic task for each reaper at 24 kg a day and pay a bonus of 3¢ for each kilogram over 24 kg if less than 36 kg is reaped; 4¢ for each kilogram over 24 kg if 36 kg or more is reaped.

In coffee reaping, a farmer might set a basic task, at 75 per cent productivity, of 84 kg of cherries each day and pay a bonus of 2.3¢ per kg over basic.

In tobacco reaping, the basic task might be 160 clips for each reaping, the basic daily wage of $2.00. Then the labour cost per clip is 1.25¢ and the farmer might decide to pay 7¢ for every 10 clips over basic. Waiters should be paid a bonus based on the average bonus of the reapers they serve. The trailer loader, recorder and foreman could be given a bonus based on the bonus earnings of the whole gang.

Table 6.10 analyses the cost of wages for 11 sugarcane cutters in Swaziland. The basic task per man-day was to cut four areas, each estimated to yield a tonne of cane. Workers received a basic daily rate of $2.00 and a bonus of 40¢ a tonne for each tonne after the fourth. (Average cost per tonne of the basic task was 50¢.) It can be seen quite clearly that harvest labour cost per tonne fell as:

- a worker's weekly output rose
- a worker's daily output rose
- a worker's total weekly pay rose

This bonus system thus benefited both workers and employer. Good workers earned more and helped to reduce the unit cost of a saleable product.

GLOSSARY

Ergonomics study of the capabilities of people and how work and working conditions affect them.

RECOMMENDED READING
Books

Barber, B., *The Practice of Personnel Management* (Institute of Personnel Management, 1970).

Barnard, C.S. and Nix, J.S., *Farm Planning and Control*, 2nd ed. (C.U.P., 1979).

Beacham, R.H.S., *Pay Systems* (Heinemann, 1979).

Bramam, I., *Practical Manpower Planning* (Institute of Personnel Planning, 1978).

Braithwaite, R. and Schofield, P., *How to Recruit* (British Institute of Management, 1979).

Brech, E.F.L. (ed.), *The Principles and Practice of Management*, 3rd ed. (Longman, 1975).

Currie, R.M., *The Measurement of Work*. (Management Publications, 1972).

Fraser, A.K., and Lugg, G.W., *Work Study in Agriculture* (Land Books, 1962).

Grandjean, E., *Fitting the Task to the Man* (Taylor and Francis, 1980)

Harvey, N., *Farm Work Study* (Farmer and Stockbrocker Publications, 1958).

Johnson, D.T. *Farm Planning in Malawi* (Ministry of Agriculture, Malawi, 1972).

Kanawaty, G., *Introduction to Work Study*, 3rd. ed. (I.L.O., 1979)

Leeds, C.A. and Stainton, R.S., *Management and Business Studies* (Macdonald & Evans, 1974).

Norman, L. and Coote, R.B., *The Farm Business* (Longman, 1971).

Powell, J., *Work Study* (Arrow Books, 1976).

Stewart, R., *The Reality of Organisations* (Pan, 1972).

Thomason, G.A., *A Textbook of Personnel Management*, 4th ed. (Institute of Personnel Management, 1981).

Upton, M., *Farm Management in Africa* (O.U.P., 1973).

Monthly Labour Use Per Hectare in the Peak Months

	Maize	(man-days) Groundnuts	Tobacco	Beef
Nov.	12	30	45	8
Dec.	25	25	35	8
Jan.	22	22	27	8
Feb.	10	20	17	8
Mar.	7	27	35	8
Apr.	15	37	72	8
May	12	30	45	8

Periodicals

Campbell, R.J., 'Planning to Employ a Contractor', *Farm Management* (1982), 4, 12.

Cockayne, A., 'Considerations before Building', *Farm Management* (1982), 4, 5.

Errington, A. and Hastings, M. 'Work Planning in Agriculture', *Farm Management* (1981) 4, 5.

Thompson, H.D., 'Recruiting and Interviewing New Employees', *Farm Management* (1980), 4, 1.

EXERCISE 6.1

Mr Hamid has a farm of 7 ha, 2 ha of which are fenced grazing land unsuitable for cropping. Helped by his local agricultural extension worker, he has calculated his expected enterprise gross margins to be as follows in the average season:

	Gross margins/ha ($)
Hybrid maize	56
Composite maize	45 (Yield: 2,700 kg/ha)
Groundnuts	74
Tobacco	132 (Yield: 1,120 kg/ha)
Beef grazing	59

He has a tobacco quota of 900 kg and insists on not growing tobacco on the same piece of land more than once in four years or groundnuts more than once every three years. He also insists on producing enough composite maize for his family's subsistence needs. Hybrid maize will be sold at harvest because he finds it to be susceptible to storage pests.

Mr Hamid, his second wife and two children work full-time on the farm. His first wife runs a store on the farm. He works 25 days each month. His second wife's work output each month is equivalent to his output in 17 days and each child's output is equivalent to his in 8 days. Assuming the following labour use and grain consumption of 300 kg an adult and 150 kg a child each year, devise a plan to maximise whole farm gross margin.

Table 6.10   Sugar-cane Cutters' Wages — Example of a Bonus System

Employee no.	Av. tonnes cut/ week	Av. days worked/ week	Av. tonnes cut/ day	Av. weekly basic pay	Av. weekly bonus	Av. weekly total pay	Labour cost/ tonne
				$	$	$	$
19	46.0	5.1	9.02	10.20	10.24	20.44	44.4
22	48.4	5.9	8.20	11.80	9.91	21.71	44.9
27	31.2	5.2	6.00	10.40	4.16	14.56	46.7
26	28.0	5.0	5.60	10.00	3.20	13.20	47.1
32	20.4	4.2	4.86	8.40	1.44	9.84	48.2
35	20.0	4.5	4.44	9.00	0.79	9.79	49.0
36	16.4	3.7	4.43	7.40	0.64	8.04	49.0
29	24.8	5.7	4.35	11.40	0.80	12.20	49.2
30	22.4	5.2	4.31	10.40	0.64	11.40	49.3
15	20.0	4.9	4.08	5.80	0.16	9.96	49.8
12	13.6	3.4	4.00	6.80	—	6.80	50.0
MEANS	26.4	4.8	5.50	9.60	2.91	12.51	47.4

7 Staff control

Labour is often the biggest single cost on commercial farms and this, including food or the free use of land, is rising all the time. It is thus important to have the right number of staff and to control them efficiently. A farmer has the right number of staff if it is not possible to:

- reduce the labour force and save more costs than any receipts lost or
- raise the labour force and add more to receipts than in costs

Good workers pay their own wages. A man's daily wage, for example, is about equal to the value of 1–2 kg of cured tobacco or of an extra 2¢ per kilogram over only 100 kg of leaf. A good worker can save that by conscientious work almost every day he handles the crop, if he is well trained, well supervised, fit and satisfied.

WORK PROGRESS CHARTS

Comparing actual against planned performance is useful, especially if action can be taken to remedy adverse variances. Speed and ease are essential for control techniques. A simple method is a work chart, as shown in figure 7.1. This is a graph of the planned and actual time that a job takes. Actual progress is recorded regularly and action can be taken if deviations occur.

A separate chart is needed for each main labour-intensive job. The information needed is the dates when you expect to start and end the job and the relevant area or quantity. The time needed depends — as in the gang-work day chart — on the expected yield, weather and soil conditions and on how the job fits into the overall farm plan. Frequent comparisons of actual against

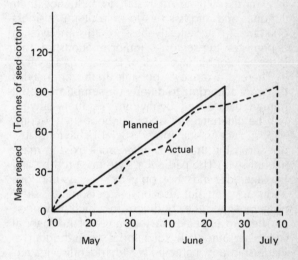

Figure 7.1   A Work Chart for Hand-reaping Cotton

planned or expected results shows immediately not only if the job is behind or ahead of schedule but also by how much.

In figure 7.1, the total has not been completed on time. This could delay land preparation for the next crop or cause a late start on another job. A low performance level may be due to weather changes, delivery quotas or labour problems. It may require hiring outside agencies, like contractors, to ensure completion on time.

A work progress chart should pose the following questions:

- Which is more reasonable, the plan or the actual performance rate, that is, is the planned target attainable under reasonable conditions?
- What are the penalties, that is, loss of profit, for non-achievement?
- Are outside agencies available or can the resources be stretched, for example, by

working 24 hours a day on a shift system? What would the overtime cost?

These are practical problems that managers must solve. Work progress charts merely show where the problems lie.

LEADERSHIP STYLES

Managers usually have juniors and are responsible for their results. Every manager has to behave in such a way towards his or her juniors that they carry out their part of the work that has to be done with the greatest effectiveness. They therefore must motivate their staff to influence their attitudes and efforts so as to help attain business objectives. The effectiveness of this motivation depends on the senior — junior relationship, that is the style of management or leadership adopted.

There is a range of possible managerial styles that vary according to different assumptions about human nature and behaviour. As there always will be differences of opinion about such a complex subject as human and social behaviour, it is not surprising that great differences exist in managerial style. The particular style favoured by any manager depends not only on his assumptions about his staff but also on his personality, skill, judgement and the conditions in which he works.

The two extremes in the range of managerial styles are autocratic (dictatorial or authoritative) leadership and participative (democratic) leadership. The difference hinges on how the manager uses his authority. In the first case, he uses formal authority and says, 'Do this because I am the boss.' In the second, he uses informal authority. He gets work done how, when and by whom he wants it done because his staff *want* to do it that way. If most managers think of the work for which they themselves feel most enthusiasm and commitment, it will be work for which they have had much freedom of choice to influence the task and the way it develops. It follows, therefore, that staff are likely to feel the same way so that participative management is likely to be most effective with them also.

The style of leadership should satisfy the needs of the business, the needs of staff and the manager's own personality. No one approach is best for every manager in a business at all times. No one manager will fall exactly into any one style but will use several to deal with events during his working day. He must learn from experience and study when to be autocratic and when to be more

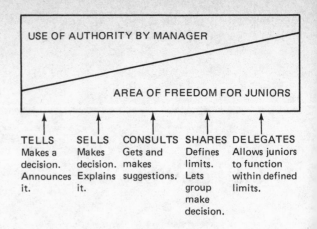

Figure 7.2   Range of Leadership Behaviour

democratic and allow staff to take part in decision-making. He will tend more strongly towards one style, however, owing to his own personality, past experience, working conditions and values.

Figure 7.2 depicts leadership behaviour by comparing how much authority is kept by the manager and how much freedom of decision is left or given to his staff.

Having studied figure 7.2 we should then consider the factors that affect style. These include the following:

Situation

Time available, urgency; complexity, size of group, degree of risk and technology and skill involved; decision dictated from above.

Junior staff

Their experience, skill, knowledge, confidence, commitment and expectations.

Leader himself

His experience, knowledge, confidence and personality. (Does he feel more need for acceptability or for predictability and control?)

Organisational climate

What style is expected and regarded as most suitable by top management — the 'house style'.

The main style factor, however, remains the integrity of the leaders themselves. Juniors' reactions to a leader's style, by various comments, reflect the amount of respect they feel. The following comments were each made to show respect for people who were also successful leaders.

- He is human and treats us as humans.
- He has no favourites and does not bear grudges.
- It is so easy to talk to him. He listens and you can tell he hears.
- He keeps his word and is utterly honest.
- He does not dodge unpleasant issues.
- We always get an answer and, if it is no, we are told why.
- He is fair with praise and criticism; when he criticises, he does not make an enemy of you but gives help.
- He is fair to us as well as to the business.
- He never asks more than you are capable of but helps you develop.
- With an example like that you can't help trying harder.

It is best not to try to copy the style of others but to be oneself. The style should be consistent but flexible. That is, in certain situations the manager can be relied on to act in a certain way. Consistency of style is essential so that junior staff know where they stand. It is no good grinning at a cheeky remark one day and reprimanding someone for a similar remark the next day.

Autocratic management is the easiest form of leadership and the traditional method of direction and control. It is based upon the following assumptions about human behaviour:

- Most staff are lazy and will avoid work if possible.
- Because of this, most people must be driven, directed, threatened with punishment or bribed to make them work hard.
- Most staff prefer to be directed, avoid responsibility, have little ambition and want security above all.

These assumptions lead to autocratic management — authoritarian, dictatorial and hierarchical.

Autocratic managers decide what action to take and impose their decisions on staff without discussion. They try to motivate them by promises of material rewards, like more pay and fringe benefits, and by fear of punishment. They use the power of their office to compel obedience. Autocratic management is based on the idea that the manager is better than his staff and therefore should make decisions alone and direct the staff accordingly.

There are at least two types of autocratic manager. One tells you what he wants but allows you to decide on methods — provided you produce results. The other not only tells you what he wants but also dictates precisely how to do a job.

Autocratic leadership is sometimes best, such as in an emergency, when juniors are disorganised or demoralised or when their past experience leads them to expect firm direction. Formal authority is needed to establish the 'boss image', to lay down administrative procedures and to enforce any discipline.

We can all think of people for whom the assumptions underlying the autocratic management approach are probably right but there is much evidence that only a small minority of people are naturally like this. Many seem to be like it because they are managed as if they were like it. It is a self-fulfilling prophesy — we assume that people dislike work so we firmly control and direct them in their work and, because of this, they grow to dislike their work.

This traditional reward and punishment approach is no longer adequate. In developed countries, it no longer works for manual workers and nowhere can it get the best results from mental workers like supervisors. The main drawbacks of a mainly autocratic style are that it:

- often makes staff apathetic, dissatisfied, hostile and rebellious,
- retards the personal development of staff who are unlikely to accept responsibility for what they have been ordered to do,
- does not use the knowledge of the staff in decision-making and decisions are unlikely to be made in the leader's absence,
- makes juniors too dependent on the boss and suppresses their initiative,
- is costly and often serious if the boss makes a mistake,
- does not build morale or team loyalty,
- is not acceptable to highly educated/trained people.

Participative management stems from entirely different assumptions about staff behaviour, which follow:

- Most people find voluntarily performed work satisfying and as normal as play or rest. They have a psychological need to work.
- People will direct and control themselves towards goals to which they are committed

and respond more to rewards than to threats. Threats and bribes are unnecessary.

- Commitment to goals is related to the rewards that follow their achievement.
- Under proper conditions, most people not only accept but seek responsibility and achievement. Responsible people work best when they are left to get on with the job.
- Imagination, ingenuity and creativity are widespread.
- Most people have potential that is rarely used fully at work.

In participative management, the experience and 'know-how' of the boss and his staff is pooled by the sharing of decisions. As business problems become more complex, this is becoming ever more useful. The boss puts the case for and against a course of action and throws the subject open for democratic discussion. He may and often must set limits within which staff opinion carries equal weight about what is to be done. If junior staff expect this approach and both they and the boss are experienced in it, staff obtain more job satisfaction than if they have been persuaded or forced. By helping to design and agree targets, staff will be more committed to achieving them and apathy and hostility are less likely. They will accept more responsibility for what will be done, take initiative in the leader's absence and identify more with the business. Furthermore, they are more likely to develop and be more ready to move into higher posts as a result of their experience and involvement in decision-making. Unless you have developed at least one of your subordinates to the point where he can take over your own job, it is less likely that you yourself can be spared for promotion. Participation of managers at all levels in fixing plans and standards is essential for management by objectives. The objectives that are agreed between senior and subordinate staff should, as far as possible, be seen to arise directly from the organisation's corporate objectives and corporate plan.

Sometimes participative management is referred to as 'permissive' management which implies management abdication. It is *not* a passport to anarchy or an excuse for weak leadership. The boss does not stop being a boss but the emphasis of his job changes from direction and control towards support, encouragement and help — with more reliance on his staff's self-control.

Evidence shows that team decisions are better than those of the average member of the group although less good than the decision of the best

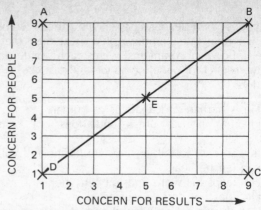

Figure 7.3   The Managerial Grid

member. If the boss is no better than his staff in a certain field, then a decision made in discussion with them is likely to be better than one made without consultation.

Why is it so hard to learn from our own experience and accept the motivational value of participation? Why is it that managers still ask, 'What motivates?' and trainers still run around in circles looking for new ways of telling them? Although participation sounds fine, it is the practical implications that make many managers reluctant to practise it. How much decision-making do they want to delegate, how much do they feel able to delegate and how much backing will they get from the top?

There are drawbacks to participative management in practice. It challenges deep-rooted ideas about the nature of authority and the way leadership should be exercised in organisations. This style of management takes longer than merely dictating orders and needs more skill from both manager and staff — especially self-knowledge and conscious effort to change. If a manager's job is seen by his staff as giving orders and providing structure, delegation may be resisted at first as a sign that the manager is ceasing to manage. Without great care, decision-making can be dominated by one or two vocal members of the group at the cost of the passive majority.

It is hard sometimes for people unused to team work to accept the ideas of others, even in the face of evidence, if these ideas differ from their own. A manager may be reluctant to share decisions because he fears that participation would threaten his job and status. The more he delegates and the less his direct involvement in a decision, the less his 'ownership' of it. (However, what effect would lack of participation be likely to have on his staff's commitment to work? Lastly, participative management means that the boss must be able to take advice and criticism from sources that he may have ignored before. He risks group decisions being reached with which he is less satisfied than his staff and must be prepared for compromise solutions rather than those that he personally thinks best. If he *is* the best member of the group in a certain decision-making area, group decisions will be less good than his own.

Participation will fail if the boss or his staff are immature and in situations where past managerial practice has been autocratic. Participative decision-making must be learned by a work team before the benefits can be gained.

A managerial style between the autocratic and participative ones is the persuasive approach. Here a manager describes the situation and his decision to his staff and then tries to convince them, whatever views they state, that his decision is best. The danger is that the manager's decision may prevail because of his position rather than because it is better. This 'tell and sell' approach often causes the same dissatisfactions, apathy and hostility as the autocratic approach. When, however, there is enough 'give and take' without referring to the manager's authority, this approach may meet the expectations of staff and begin to approach the style of participative management.

To depict various managerial styles, a 'managerial grid' is often described. This is shown in figure 7.3. The horizontal axis represents the manager's concern for results and production and, simultaneously, his respect for himself. The vertical one represents his concern and respect for his staff as people. Each axis is divided into nine.

9.1. A 9.1 manager at position C is an autocrat with most concern for results and least for people. He has a high regard for himself and little for others — 'I'm OK; your're not OK!' He regards people as 'units of production' or 'hands'.

1.1 This manager has little concern for results and low regard both for himself and others. Often called 'the deserter', he contributes little to his job but spends his time doing just enough to keep it and to keep out of trouble. He is the sort of, 'I'm not OK; you're not OK' man who most people thought had retired years ago. He may get promoted to a senior post without substance merely to remove him to a remote office where he can no longer hinder positive action.

1.9. This is the paternalistic, 'be kind to them' manager with a low concern for results but maximum concern for people. He is the 'I'm not OK; you're OK missionary' who goes to great lengths to foster harmony and friendship and gain his staff's affection. These efforts are usually rewarded by irritation, dislike and contempt for him. A manager who becomes so obsessed with the happiness and welfare of the staff that he neglects the achievement of his technical objectives is not helping himself, his organisation or the happy group who may all soon be unemployed.

9.9. This is our democratic or participative manager with maximum concern for both results and people. His attitude is, 'I'm OK; you're OK' and this should result in co-operative and highly motivated teamwork towards achieving business goals.

5.5 This manager shows medium concern for both results and people. He keeps to rules and procedures and tries to produce as much as he can without upsetting people.

It follows that the further one can move along the line from D to B, without departing much from it, the better. However, any basic change in managerial style by a manager or an organisation requires much vision and determination to carry through. No one style is universally ideal. Good leaders change sometimes from the democratic to the autocratic style. Properly managed, the human and technical factors are complementary and should be reconciled by controlling the technical factors *through* the human factors. This ensures that results and efficiency are achieved by enhancing human dignity and satisfaction; not by denying them. Managers should be flexible and use the style suited to each situation.

SUPERVISORY MANAGEMENT

First-line supervisors or foremen are sometimes called *capitaos* or *indunas*. Any time there is a team, someone must be responsible for generally overseeing the job; that is, for supervision. He (or she) controls the activities of others and is responsible for executing management's policies and intentions by leading the group in his charge. His task is to get things done by the people for whom he is responsbile. Supervision is a function of every executive or manager, at any level. First-line supervisors have immediate responsibility for the largest group of staff — farm labourers.

The British Inudstrial Relations Code of Practice defines a supervisor as a 'member of the first line of management responsible for his work group to a higher level of management'. Strictly, a supervisor is anyone in charge of others; those with whom he or she must execute work satisfactorily. This definition is usually confined to those members of first-line management most directly related to the work force at the point of production or service. Supervisors therefore have a vital leadership role.

A supervisor is as much a manager as a director for, however high the standard set by top management, however clear the patterns laid down, the whole process can break down if there is a weak link in the chain. If the concept of a management pyramid is valid, supervisors form the broad and essential base of that pyramid.

The differences between supervisors and higher managers are that the type of person suited to the thought processes involved in planning and control often cannot provide sound supervision. Conversely, the practical man who is an excellent supervisor often cannot grasp the techniques of planning and control. Quiet thinking and the handling of paperwork take back place, in a supervisor's mind, to the need to solve current problems.

First-line supervisors play an important part in the business. As higher management's representative at the point of immediate contact with the workers, and as the workers' first point of contact with the management, the first-line supervisor is a vital link in the two-way flow of communications. To the members of his working group, he is the manager who counts. His group's attitude to the organisation, and the contribution they are willing to make, depends much on how he does his job. High output follows the good supervisor, not the team.

A first-line supervisor is often a key person in the lives of the men who work for him. To the extent that he earns their confidence as a leader, they depend on him for ideas, information, suggestions, approval, guidance and even criticism. They look to him for decisions, timely information, friendly advice and for answers to questions that arise. They expect him to be fair and to use commonsense in working with them.

All instructions to the labour force should be routed through supervisors; they must at least be present when instructions are given so that they know what to look for. As a supervisor is in a key position to influence labour relations, management should ensure that the supervisor:

- is properly selected and trained
- has charge of a work group of a size that he can supervise effectively
- is fully briefed in advance on management's policies as they affect the work group

- is an effective link in the exchange of information and views between higher management and members of his work group

It is unrealistic to expect a supervisor to develop good relations with his men unless those relationships also prevail between himself and higher management. Sometimes workers develop fears, hatreds and frustrations that produce conflict and stress in their jobs. Resentment and anger are then directed towards the supervisor. The attitudes of workers towards their supervisor are affected by general working conditions and by their own experience with particular supervisors. The supervisor is the distributor of rewards and punishments and hence can evoke positive or negative feelings in his staff.

Expectations of higher management can be equally exacting. They give the supervisor the responsibility for communicating orders, policies, information and ideas to the workers. They often also expect him to report workers' opinions and attitudes. A first-line supervisor's role in these communications is vital because he must interpret and translate them into some form of action.

Supervisors are under pressure from both workers and higher management, each with different expectations. This may produce role conflict at times unless higher management recognises the existence of dual loyalties and does not try to gain the sole loyalty of supervisors. It is important for supervisors to be accepted members of both groups.

Perhaps because the supervisor's job is so important, there is a tendency to make it sound impossible. Those impressive lists of the qualities needed in a supervisor are enough to deter anyone from tackling the job. Such lists ignore the facts that:

- people with very different qualities have been successful supervisors,
- each supervisory job is different from every other one,
- particular supervisory jobs can be successfully handled in many different ways,
- the success of the business depends on each supervisor, and his group, contributing positively towards its objectives.

Attention is often directed mainly to short-term production priorities. Although this may have to be accepted, supervisors should realise that their whole job covers not just the technical aspects, which are usually quite clear, but also those human and administrative aspects that are often overlooked. Therefore the basic needs common to all supervisors are as follows:

- Knowledge of what is expected of them; their special functions in the organisation. Workers naturally resent being supervised by anyone who knows less about the job than they themselves. In this situation their morale and sense of responsibility falls.
- Good judgement on how to manage people.

Supervisors need leadership and organising skill, coupled with ability to communicate effectively and to be accepted by the work group, and capacity for growth. A foreman may blame a gang of workers for being lazy and inefficient or for low output when the real cause is his own lack of drive and leadership.

Supervisors are best selected from within the business but it should not be assumed that the 'best' labourer will make the best supervisor. Things to consider in selecting supervisors include a wish to manage, intelligence, analytical skill, ability to communicate and honesty.

SPAN OF CONTROL

A manager's span of control, often called his span of supervision or responsibility, means the number of immediate subordinates with right of direct access to him and who report directly to him. This concept affects the horizontal dimension of an organisation structure.

Some of the most important basic principles of organisation are concerned with the span of control, as the size and structure of a manager's team greatly affects whether his leadership activities will yield results. Clearly, the more people reporting directly to a manager, the harder it is for him to supervise and co-ordinate them. Although many writers are dogmatic about the optimum span of control, there is little agreement about this.

A small span of control enables a manager to treat the members of his group separately and to delegate to them effectively. Small spans tend to be most effective in situations of continuing growth, changes in needs, processes and people; where a consultative and supportive style of management is best. They are also most effective if a manager must spend much time away from his team or if his direct juniors are widely dispersed.

Too small a span of control prevents effective delegation, raises the number of management levels and the level of administrative costs. The manager may retard staff development by trying to do each person's job for him or by exercising too much control over every task. The tall organisational structure resulting from too small a span of control impedes co-operation and clear communications. The more steps a message must pass through in a managerial hierarchy, the greater the risk of distortion or failure. Top managers are too far from the firing line.

Figure 7.4 shows that in a 120 employee firm with a span of control of three there are five levels of management.

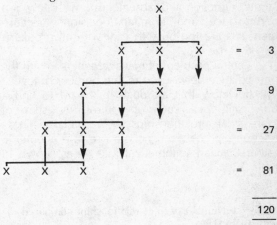

Figure 7.4   A Small Span of Control, Many Levels of Authority — 121 People

Large spans of control mean less management levels, better two-way communication and more need to delegate. It has been suggested that the only way to compel delegation is to give a manager so many staff that he cannot supervise them closely. The fewer levels of management leads to less contacts at which misunderstanding or misinterpretation can arise. Two-way communication is more likely to be fast and timely. Figure 7.5 shows the effect of a large span of control in reducing the number of levels of management by producing a flat organisational structure.

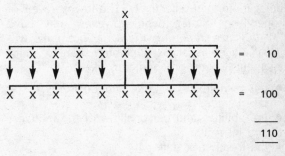

Figure 7.5   A Larger Span of Control, Few Levels of Authority — III People

Large spans of control are feasible in conditions of stability with few changes, when the team is located close enough to the manager for face-to-face communication and when both the manager and his staff are highly competent. However, it can be overdone. If a manger or foreman has too many people directly responsible to him, he cannot give sufficient time to individual needs and juniors have trouble contacting him because he is too busy. The business may suffer high labour turnover and absenteeism with low job performance and unmade decisions because managers are too busy to deal with all problems as they arise.

A central problem of management is to find the best balance between span and number of levels. We do know that although it is hard to find a universally best span of control, the span of managerial relationships is crucial. These relationships between people rise in a geometric rather than an arithmetical rate — as shown in table 7.1.

Table 7.1 Number of Inter-relationships compared with Team Size

Team size	Number of inter-relationships
2 (A,B)	1 (AB)
3 (A,B,C)	3 (AB, AC, BC)
4	6
5	10
6	15
7	21
8	28
9	36
10	45
15	105
20	190

The size of most control groups ranges from five to ten but few managers can effectively control groups of more than about seven people and delegate to them effectively. The lowest levels of supervisory management, however, may have 20–30 men reporting to them, but they will usually have assistants. Optimum spans of management depend on conditions like:

- the complexity of tasks and decisions to be taken
- the abilities and personalities of the manager and his staff
- the location of staff
- the kind of technology used
- the rate at which the business is developing

and changing
- the personality and wishes of the manager's manager

DELEGATION

The commonest complaint from managers is that they cannot find sufficient time to do their jobs, let alone to think. If a business is large, it is often necessary to delegate responsibility and authority for certain areas of it to others — normally juniors. This enables them to make the decisions their senior usually took rather than merely to carry out his detailed orders. It is the main job of management to manage, and that means concentrating on unique actions, detailed policy and planning. Asking other people to do repetitive jobs, and then leaving them to get on with it their way, is one fo the hardest disciplines facing managers.

A manager delegates when he deliberately gives a junior authority to do work that he could have decided to keep and do himself. Juniors are responsible to their senior for the results they achieve on his behalf and he, in turn, is accountable for what his juniors do. Delegation should not be confused with issuing orders or giving instructions to juniors. The essence of delegation is giving authority to a junior — this is more than just passing a job over to be done. Delegation does not free the manager from accountability and is not merely a case of, 'Who can I get to do some of my work?'

Other things being equal, the more authority and the more freedom that can be given to individuals to act within agreed guidelines, the better for all concerned. The hardest job for a delegator is to get out of the way. Keeping out of the way does not mean forever. Regular support is needed. The optimum allocation of jobs in an organisation is all about the comparative advantages that each person has.

Effective delegation is uncommon in farming because it is not easy. It needs patience and skill and should be an inherent part of a manager's style. Delegation clearly involves assigning work from one level of the organisation to another. But how much responsibility and authority should be delegated — if they can be delegated — and how much supervision should be given, are questions usually left unanswered.

Delegation is one of the most important aspects of any manager's job and he often has more freedom of choice about it than about any other part of his responsibilities. For what he chooses to delegate, to whom and when, is almost entirely at his discretion.

The head of a small business does not have to decide what decisions he should delegate as he can make them all himself. He may continue to try to do so as his business grows but its future success will partly depend both on his willingness to delegate and on his ability to decide correctly which decisions he should delegate and which he should retain.

Delegation helps to ensure the best use of executive time because it encourages managers to separate what is urgent from what is important. So much of a manager's time may be spent on immediately urgent jobs that little time is left for more important longer-term ones. By delegating work, they are freed to concentrate on those important tasks needing their personal experience, skill and knowledge; like planning, and forestalling problems. The lower the level at which a task is performed, the lower its cost is likely to be. Cost should be considered, not only in terms of salary but also in terms of opportunity cost — the opportunity that is denied or created for the manager to do other work.

Other advantages of delegation are that it motivates staff and trains them for greater responsibility by enabling them to learn from experience. The decisions that people execute most enthusiastically are those that they took themselves and, even if they made a wrong decision, they will usually, by sheer effort, remedy it. Delegation is the most important single factor in job enrichment. What is routine work to a manager may be a challenge to his junior, besides carrying prestige.

The more that juniors are developed, the more they are likely to be capable of in future. Delegation is one of the most effective tools available to a manager in developing his successor. It can also be especially useful in emergencies such as when the manager is away or there is sudden need for a promotion.

There are several blocks to effective delegation. These include time-consuming coaching in the early stages, the risk of mistakes being made for which the manager is accountable, a secret fear of being dispensable or of being excelled by juniors and a manager's desire to keep doing those jobs he *likes* doing — vocational hobbies. The latter often happens if a manager is especially skilled in an area of his work, especially if he is having trouble with other aspects of his job. This can create a vicious circle so that a person arranges for his activity to consist mainly of work he likes and is good at. The last thing he may want to do is to delegate this so as to concentrate on the more important but, possibly, less pleasant parts of his job.

Few conditions are more frustrating to a person than to be given responsibility for results without authority over methods. Juniors should be positively encouraged to make their own decisions, although they should feel free to seek advice wherever they think it needed. If they gain confidence, but still too often refer questions upwards before acting, they can be discouraged from 'half-taking' decisions if their manager deliberately withholds his own advice and views. Juniors must know the objective of the organisation and they may need training before having a task delegated to them. However, they should be given adequate discretion and not be expected to act exactly the same as the manager had done before.

Delegation reduces the need to supervise staff. However, delegation without control is abdication. Once balanced limits of responsibility are fixed, a control and feedback, or 'warning-light', system of a few key indicators, through which the progress of the individual and the delegated task can be checked, should be set up. Correcting errors always needs tact but the need for it is lessened when the stress is on target setting and when the junior plays a large part in setting the targets. Overcriticism would probably create a preference for punishment-avoidance instead of reward-attainment. Juniors can avoid punishment if they take no risks and do only what they are told. Much reluctance to accept delegated responsibilities can be traced back to this single cause.

In deciding when and what to delegate, the relative strengths and weaknesses of both manager and his juniors should be considered. Delegation usually works best when one or more of the following conditions prevail:

- the parties are physically distant from each other;
- persons in the chain of command are absent for long periods of time;
- workloads are heavy and the main effort is to meet commitments;
- the delegator feels personally secure;
- there is a favourable organisational climate; one that stresses staff development and growth, innovation, creativity and human dignity;
- there is mutual trust between people at all levels in the business.

Deciding what can be delegated and then being determined to do it needs much conscious effort by managers. It is hard to give people the right to be wrong when one will be accountable for any mistakes. Many criteria should be used in deciding what should or should not be delegated and there will always be work areas in which a senior for some reason, including personal whim, will want something done exactly in his way. Usually, however, a task should be delegated by a manager if one or more of his staff could do it:

- better than him
- cheaper than him
- with better timing
- as part of his normal job
- as an aid to his training and development
- to assess suitability for promotion

As mentioned before, all managers must be given sufficient authority to perform the responsibilities delegated to them. This does not, however, free the delegator from accountability to his seniors for his juniors' work. Unless the range of both authority and responsibility is clear, people fear to exceed their authority and may veer in the opposite direction. Let us look at these concepts of authority, responsibility and accountability.

AUTHORITY

People must know the exact limits of their authority if they are to operate freely and ably. Authority is the right to make decisions in the course of performing a responsibility, to require others to accept these decisions and, if necessary, to enforce them. It enables action to be taken so that responsibilities can be successfully met. It is important, especially in routine matters, that power to act should be delegated to those having the needed technical knowledge. Delegation of degrees of authority to staff, according to their responsibility, enables each one to know to whom he reports and who reports to him.

There are three types of authority — positional, charismatic and sapiental authority. **Positional authority** arises from the post that the manager holds in the hierarchy — what the manager is. Naturally, authority does not come automatically with the holding of a title but it is undeniable that a person can be greatly helped by the authority and influence connected to his position. **Charismatic authority** derives from who the manager is — intelligence, general ability and the confidence that a manager inspires. **Sapiental**

authority derives from study and experience — what the manager knows. These personal qualities make him an expert, able to give a sensible, intellectual reason for his opinions or a practical demonstration that his advice is right. Ability to lead and past services are also relevant. If these personal qualities are lacking, then official authority is likely to be a rapidly wasting asset. It is hard to obtain the full co-operation of staff for long if they know that they are merely supporting an incapable leader.

RESPONSIBILITY

This is the obligation of someone to perform a task or set of activities (or see that they are performed) in a way that satisfies criteria laid down by another person or group of people. It is the duty to use delegated authority for the purposes for which it was delegated and also means the duty or duties assigned to a given position. Definitions of responsibility should be clear and positive, especially if jobs overlap, a requirement that implies full understanding of the responsibilities by those who draft the specification.

Responsibility should always be balanced by authority. It is its natural consequence and essential counterpart and, whenever authority is exercised, responsibility arises. Usually, responsibility is feared as much as authority is sought and fear of responsibility paralyses much initiative and destroys many good qualities. A good leader should possess and instil into those around him courage to accept responsibility.

For a person to be able to develop his full potential, to be more creative, responsible and independent, he needs to have personal responsibility and initiative. If his employers prevent this, he can adjust in one of two ways. He can become used to applying only a small amount of his ability to work and seek other compensations; or he can leave.

Inexactly defined responsibilities and terms of reference probably cause more friction between senior and junior staff than anything else. Formal relations can be set out in sufficient detail as job specifications and the general nature of their inter-relations can be shown on an organisation chart.

Naturally, it is unreasonable to hold someone responsible for events caused by factors beyond his control and no change should be made in the scope of responsibility for a job without definite understanding of the change by all concerned.

ACCOUNTABILITY

Although it is possible to delegate both responsibility and the necessary authority, it is impossible to delegate accountability — final responsibility for the success or failure of juniors in performing delegated activities. Consequently, the delegator must maintain control and must always be aware of progress made in relation to the plan so that he can introduce remedial measures if needed. Inability of a junior to perform correctly or to do his work well reflects on the delegator who chose him. This means that the delegator must carefully select those to whom work is delegated. He must be sure that they can carry the degree of responsibility and effort needed for the task.

GOOD LABOUR RELATIONS

Under this heading will be described training and development, discipline, welfare, housing, feeding and communications.

Good labour relations are humanitarian and not costly to apply. They require a basic knowledge of or feeling for human relationships. As managers depend heavily for their success on building and keeping good labour relations, they should understand the response, discipline and behaviour of local workers and make every effort to create an atmosphere in which staff wish to work and develop their skills.

Whereas some farmers have no labour problems and can keep a happy and contented labour force year after year, others, no matter how hard they try and how sincere they may be, are continually plagued by labour problems involving recruitment, retention, poor output and dissatisfaction in various degrees.

You often see advertisements for managers stating 'must be good with labour'. Let us look at the most important aspects of being 'good with labour'.

The oldest and still the best advice on labour relations, especially in the tropics, can be summed up by the four F's: 'Be Firm, Fair and Friendly but not Familiar'. Every effort should be made to develop a common interest between each worker and his colleagues and supervisor; that is, to build a good team spirit.

Without great understanding of your staff and appreciation of their efforts, allied to fair and respectful treatment, you are likely to make little

progress — whether adequate finance for material things is available or not. Healthy relationships between managers and hired workers, with resultant loyalty and strong motivation on the worker's part, need deliberate policies and positive action by the manager and employer. The principles involved are psychological and, as men's psychological make-up and backgrounds vary, methods that appeal to one do not appeal to another. It is useful for managers to study and apply those characteristics of personal treatment to which people tend to react favourably.

The alternative to good labour relationships is high turnover rates, frequent shortage of workers with consequent loss of potential income and dissatisfaction, friction and inefficiency on the job.

The main aim should be to create job satisfaction — to satisfy existing staff by creating working conditions in which they have little or no wish to leave. If this is attained, a desirable stability will have been achieved which, in itself, should greatly help in attracting any replacement. Adequate payment and good physical conditions of work are not by themselves enough to give the job satisfaction that workers need if they are to give of their best. Besides reasonable working hours and competitive wages, constant use of human relations principles is needed to attract and keep employees. This is an area in which many farmers have less experience than they will need in the future. As mentioned before, there is no certainty that a satisfied man will be a high producer, but staff turnover and absenteeism are likely to fall.

More and more informed, enlightened managers realise that individual workers are not motivated solely by financial reward. Job satisfaction, status in the working community and opportunities for self-development are equally important. Motivation is leadership behaviour whereby a person positively influences the behaviour of others without using force. It really means treating staff as humans. Farm workers are not mere hands. They are people with hopes and fears, ambition, pride, humour and anxiety who respond to good manners. We must be able to see things from their view and to realise that they are working to satisfy their needs — not ours. If we can channel their goals in such a way that in achieving them they will help us to achieve ours, then we have succeeded.

The factors that greatly motivate tropical farm workers when seeking employment are the following, but it should be noted that they are not in any order of importance and it is likely to be a combination of all these items that provides the right environment. However, the first three are likely to be more attractive than the rest.

- Earnings.
- Housing (married).
- Schooling for children.
- Shopping facilities.
- Conditions of employment (leave, type of work, hours to be worked, etc.).
- Employer/staff relationship (communication).
- Management practices (systems of payment, bonuses, time off, promotion, leadership, etc.).
- Job satisfaction (does he feel part of the farm team?).
- Job security (including security for his family).

One of the most important aspects in this list is that of communication. If a farmer has trouble obtaining labour, his workers will tell him the reason for this if his communications are good. The main point here is that communication should be two-way between the farmer and his staff. The labour force must know that there is a channel by which they can communicate their ideas, grievances, complaints or whatever to the farmer.

With a large labour force it is hard to have individual communication and farmers might consider encouraging a small group of workers to represent the whole labour force and be the communication point, both upwards and downwards.

The best methods for good communication will vary between farms and each farmer should experiment to find the method most suited to his own situation.

Many aspects of conditions of employment that staff find attractive in other sectors have been offered on farms for some time but there is one main fault — the conditions of service on many farms are not specific. They have never been made fully clear to the staff and workers have evolved their own interpretation of what is needed and this often changes whenever the farmer changes his mind. A better approach that would benefit both employer and staff, is for the farmer to write out his conditions of service, to clarify his own ideas, and then to communicate them clearly to staff as the rule of the farm.

Before reaching specifics, below are some general tips on obtaining good labour relations:

- Have no favourites, and be seen to be fair and impartial.
- Respect workers' beliefs, superstitions, customs, family, tribe and race. To insult them on

these points would make them resentful and 'inner enemies'.

- Admit errors. This is not losing face and a worker will respect his employer's frankness and copy him.
- Let workers present their ideas on how to improve their jobs. They may be able to raise productivity and will be proud of any help they have been able to give. Simplicity of thought can often produce profound observations about simple and complex problems and their solution.
- Encourage workers to feel part of the business and to acquire such status as they are able within it.
- Personally develop the skills expected from workers. Staff work best if they know that the 'boss' *can* do the job himself and that he knows how much should be completed in a given time.

Personnel training and development

There is a definite need for planned training of farm workers to the standard needed. Unfortunately, training tends to be neglected on most tropical farms. Farmers still seem to think that it is the responsibility of their workers to gain the necessary skills to make them more productive. Nothing could be less true. Agriculture must often compete with other industries for labour. As these industries raise labour productivity by training, they can afford better conditions and pay. Only a foolish man would fail to recognise this when he seeks employment. Tropical farm employers should revise their attitudes towards planned training as a continuous training policy seems to be an essential part of farm management.

Most farmers spend time training labourers and higher level staff on the farm but this is usually done during their day-to-day work without setting aside special time. Formal, organised training as carried out in other industries is impossible for most farmers though it might have a place on large estates. Few farmers have the time and there is a vastly wide range of jobs in which workers need training. Successful farm managers, therefore, follow a continuous training policy, probably often subconsciously, and use all available opportunities to train, whether it be for five minutes or half a day, perhaps with one workshop hand or maybe with a whole gang; he can then delegate more unskilled and part-skilled work to staff and spend more of his time on the more demanding aspects of his job.

Human capital is the skills and abilities possessed by people, which enable them to earn income. We can thus regard the income they derive from providing personal service (as opposed to lending money or letting property) as the return on the human capital they possess. We can regard formal or informal training and obtaining these skills as a process of creating human capital, just as the construction of buildings creates physical capital.

If a farmer lucky enough to own a beautiful car let an unskilled person drive it around the farm, he would undoubtedly be thought mad. Yet farmers throughout the tropics do this in effect by letting unskilled workers care for just as costly tractors, equipment and livestock. True, we will replace a man if he does not shape up to expectations over a period of time but by then the damage is done. How much better it would be if prospective workers were given a simple test and remedial training if found wanting.

Training is carried out for one purpose — efficiency. Done effectively, it should lower production costs and raise productivity. If a worker is properly selected and methodically trained, the results are usually worthwhile. Training gives staff a sense of purpose and pride in their work, leading to a happier and more stable labour force. It also reduces the load on management so that managers can spend less time supervising and more time organising and planning.

The education of an adult, even though his formal education is usually slight, differs basically from that of a child in that, with a child, one has to some extent little to work with but with an adult it is often a case of *re*-education. This has both positive and negative aspects. On one hand, the learner already has either some knowledge or skill that should be used positively. On the other, he may have fixed attitudes that are hard to change. If no preliminary attempt is made to find out what useful knowledge or skill the learner already has, which is often much more than many managers think as they do not allow for the strong powers of observation of people who are unaffected by formal schooling, time is wasted showing them what they already know. This can also be demotivating as it is not only irritating to be told what you know but it can also be confusing if you know it in a different way.

As training concerns people and not mechanisms, it cannot be effective in any lasting way without considering the nature of Man. There are many theories about what this is but one fact is certain — he can be developed, wants to develop

and responds to opportunities for development if they are presented correctly. A labour force that is given such opportunities through training will be a motivated one — the strongest factor for improving results. Whereas, owing to their background, there are some aspects of machine operation and maintenance that few local tropical workers naturally adapt to, training can remedy such defects.

Fortunately, most farm workers appreciate training because they see an opportunity for advance and many are genuinely interested in their work and take pride in learning how to do it better.

It is no use training the wrong person, therefore selection is important. Most farmers are skilled at picking the right man for the job but much thought should also be given towards deciding how far the man can go. Clearly the further he can safely go, the more it pays to train him. It is therefore necessary not only to consider his skills, ambitions and interests but also his general education because usually the more educated a person, the more trainable he is; if he is motivated. Usually, younger people are more trainable than their elders but they are more likely to leave so it may therefore be more worthwhile training older, more stable workers who are less likely to leave the farm.

It is a farmer's responsibility to ensure that everyone knows his job. Farmer involvement in training is essential even if the actual on-the-job training is delegated to experienced foremen or operators. The best trainers are those who can guide in such a way that learners can overcome their problems mainly by themselves. Probably few tropical farmers have either the time or the wish to train each new worker and they are unlikely to hire a qualified instructor. One alternative is to divide the staff into groups of 8 to 12. Each group needs a foreman who has been taught training methods.

Training is adopted to assist planned learning. Unfortunately, a training programme cannot be executed casually and few farmers can run a complex training programme. They should be familiar, however, with the best approach to training. This can be summarised as follows:

- The reason for the job must be explained to the worker.
- The trainer should be very familiar with the job to be taught.
- Always train under natural conditions, with the actual equipment if possible.

- Form the correct habits. Good habits can be formed as easily as bad ones. Habit is a valuable aid in raising productivity as it reduces the need for conscious thought. Constant supervision is needed to ensure that recommended methods are used. Correctness in method should precede speed in operation.
- If changing a current method, the worker must be convinced and interested in what he is doing. He should be encouraged to play a part in the development stage if possible; training will then naturally fall into place.

These are four basic steps to teaching someone a new method:

- Ensure that his mind is focused on what is to be learned.
- Detailed demonstration and explanations.
- Practise under supervision. People learn most easily by experience under guidance. 'I hear and I forget; I see and I remember; I do and I understand.'
- Allow the trainee to continue until the required output is attained and immediately correct faults. 'Tell them, show them, watch them, check them.'

Patience is essential in training. A short temper and rough language hardly helps in teaching a new technique. Training is a continuous process needing about 10 per cent explanation, 25 per cent demonstration and 65 per cent practice.

If a worker has not responded as expected to training it may be worth while for the farmer to ask himself the following questions to see what went wrong:

- Was the man responsible, interested and willing? If not, you chose the wrong man!
- Did he understand the instructions? The limitation might be in his understanding or in your communication skill.
- Perhaps you tried to teach him too much at once. If so, start again with shorter sessions and practise in between.
- Did you teach him the right points in the right order? Was every necessary step covered? Do you know the job well enough yourself? Do you know it so well that you perhaps taught him too much detail for him to see the basics?
- Is he physically capable of the job? Does he have the manual skill or strength needed? Is his eyesight good enough?
- Was he taught recently how to do the job? If he has not had much practice on the job since the last season, he may have to be re-taught.

- If a gang were being trained, could they all hear, see and understand the instructions?
- Are labour relations and wages on the farm good? If not, most training will probably be ineffective because motivation and incentive will be low.

Until training facilities are provided on an industry-wide basis, most training will have to be done with existing facilities on the farm. It may, however, be useful for a group of farmers to run an operator training school. If this can be done for machinery, why not for workers? In some countries, tractor driver and workshop training facilities are provided by government and there may be a case for government or commodity associations to set up national training schemes for other jobs like tobacco grading, cotton reaping and stockmanship.

Discipline

It is a basic concept in industry that supervisors are responsible for keeping discipline among the workers they supervise. Discipline, in its true sense, refers to a state of affairs existing in an organisation in which orderly behaviour prevails, according to established rules, and the people in the organisation behave with respect for the needs of the organisation. A well-disciplined unit is one that works at high productivity, while following all necessary rules. Therefore, if a supervisor wants a well-disciplined unit, he must ensure that all his staff are well trained and then motivate them to attain the performance standards desired.

Discipline is, in essence, obedience, diligence, energy, behaviour and outward marks of respect observed according to a system of rules for conduct between the firm and its staff. Whether these agreements have been freely debated or accepted without prior discussion, whether they be written or not, whether they derive from the wishes of the parties to them or from rules and customs, it is these agreements that decide the formalities of discipline.

Discipline is often the backbone, the ultimate act of leadership, therefore handing the power of discipline to a junior could well be more of a problem than a solution.

Most managers feel strongly that discipline is essential for the smooth running of business and that without it no business could prosper. Discipline must exist to a high degree in every organisation. Management authority should be maintained so that staff know that you are going to get what you want from them. This means being resolute that orders will be obeyed, while leaving room for questions and suggestions. It means jumping on any mistake but thinking seriously later to find the reason for it. Good managers stop to ask themselves after every employee's mistake, 'How did I let that happen? How can I stop it happening again?'

Managers sometimes hesitate to demand firm discipline as they fear that it would make them unpopular. However, good supervision and control is always respected. It makes little difference how firm the discipline is; if it is fair and consistent, it will be readily accepted.

All organisations limit the behaviour of their members. We always find rules of conduct prescribing orderly and safe behaviour to protect the safety and welfare of members and enable the organisation to do its work. Rules give predetermined answers to questions that arise repetitively as to what one should or should not do in an organisation. They should be simple, clear and sensible, well publicised and, whenever possible, be positive, rather than prohibitive, and be fairly enforced. At all times everyone should know what can and what cannot be done; what is expected of him and what somebody else will do. Though supervisors and managers must insist that all rules are obeyed, the violation is often too slight to be made an issue of and correction should be good humoured. If someone is late and holds up the whole gang, make him finish the task by himself as the rest leave a few minutes early. If he cuts a small cord of wood, call on him to make up the shortfall and then cut sufficient extra to ensure that he will not do it again. The persistent slacker may often be corrected by good-humouredly making him the laughing stock of the gang. The group itself will often develop and impose sanctions upon the behaviour of members.

It is unwise to ignore mistakes. Rather seek an explanation and, if this is unsatisfactory, give swift, effective correction but never lose your temper — that is regarded as a sign of weakness — and watch the tone of your voice. A quiet instruction to repeat poor work or even a simple mention that it has been noticed and disapproved of is often adequate. In any case, never bear a grudge. Warn, punish if necessary, and forget.

Try to correct people by name rather than give general reprimands to a whole gang. This brings things home most to the men who most need it while, simultaneously, giving the rest a certain self-satisfaction.

The secret of a good reprimand is to make the man feel genuinely ashamed of his faults without belittling him as a man. Never tell a man he is a fool. Ask him *why* he behaved *like* a fool — inferring that he is usually regarded as responsible and that it is disappointing that he should have let the boss down. If he is belittled, he will bear a grudge and may get his own back by idling when he knows it cannot be detected but, if the supervisor shows that he is respected as a man, he will try to live up to this opinion.

The best reprimands take the form of questions. 'Did you do this?' 'Is that what I told you to do?' 'What did I tell you to do?' 'Why didn't you do what I said?' A genuine misunderstanding may emerge owing to poor communication by the farmer and reprimand would then be wrong.

Consistency in disciplinary action is more important than strictness or severity. Penalties should be known in advance and be applied without favouritism. Unfairness hurts more than severity. Any promises, either of a bonus or of hell in technicolour, should always be kept; therefore be short on promises but long on keeping them.

Once a system of participative management is fully effective, there is a change from imposed discipline to self-discipline. The group is zealous for the performance of its members in achieving the group task and group members are keen to conform to group standards. Peer disapproval is stronger than manager disapproval, so the long-term solution is to develop close-knit working groups that are self-supervising.

Discipline can either be positive or negative. Many people ignore the teaching and moulding aspects of discipline, the positive aspects, and tend to think of discipline as punishment or reprimand, which are negative ones. Most people would agree that it is healthier in the long run to stress the positive rather than the negative. The question is, how do we do this effectively?

We can stress the positive by ensuring that workers know what performance standards are desired, training them how to reach these standards and praising them when they perform as they should or encouraging them as they progress towards reaching the standards. Encourage workers when their performance *starts* to improve rather than wait until it is perfect. Positive, preventive discipline begins with private advice and teaching and it is attained and kept through the motivating forces of encouragement and praise.

Unfortunately, even the best supervisor, who has been following the positive discipline approach, will sometimes have a worker who continues to break the rules and may have to be punished or negatively disciplined. The health of the business does not allow, in cases of indiscipline, neglect of certain sanctions able to prevent or minimise their recurrence.

Censure is unpopular among all workers, including managers. People often live in dread of censure, reprimand, ridicule or any form of criticism or disapproval. Used recklessly, unfairly or thoughtlessly, it can be dangerous. A worker who is punished will be unhappy or even angry and angry people often seek ways to lash out destructively.

Constructive use of disciplinary action takes much experience and tact on the manager's part in his choice and degree of punitive sanctions to be used — such as protests, warnings, fines, suspension, demotion or even dismissal. He must try to act so as to minimise anger. Individual people and attendant circumstances should be considered. The following guidelines are suggested:

- Get all the facts and only the facts — ignoring gossip and rumours.
- Ensure that any sanction proposed has been consistently taken against other staff who have committed similar breaches.
- Ensure that the employee knew he was making a mistake.
- Talk with him in private.
- Confine the discussion to the breach itself, keeping personalities out of it and not imply that the worker is a *bad* person.
- Keep calm.
- Encourage the employee to express his views and feelings.
- Get him to agree to the facts.
- Fit the sanction to the facts consistently.
- Bear no grudge, considering the incident closed once a suitable punishment has been given.
- Do not neglect to discipline positively any present or future actions of this employee whenever appropriate.
- Be as friendly as possible.
- Record any action taken.

The ultimte sanction is dismissal. This is unpleasant for all concerned and may lower morale. Whether dismissal results from redundancy, serious misconduct or inefficiency, it should be undertaken only as a last resort and care should be taken to ensure it is justified.

Welfare

This is the part of personnel management concerned with the physical and mental well-being of staff. It is especially important on tropical farms because the staff usually live on the farm, which may be far from facilities that town-dwellers take for granted. As staff become more sophisticated, more attention must be given to social amenities and arrangements for their general well-being. Lack of social amenities, like schools, social centres, transport, medical and sporting facilities, without doubt, deters people from farm work.

Welfare arrangements cover matters like physical working conditions, health, safety and leave. It is good management to provide safe, healthy and, as far as possible, pleasant working conditions for all staff.

A manager should be available to give or secure advice on many staff problems, without becoming too deeply involved in the employees' lives off the job. This means providing an advisory service on personal and domestic matters; giving sympathetic thought to employees' personal problems and never refusing to listen to them when they come to you with domestic problems for advice. Without becoming too familiar, help them in their problem with bureaucracy and show a personal, friendly interest in the welfare of their families.

Leisure amenities, like a soccer pitch and a beer-hall, are important. In the case of the latter, a self-imposed levy on sales may be used by staff to help finance other amenities.

If the farm is in a remote area with few shopping facilities, a well-stocked farm store is appreciated and acts as an incentive to workers to raise their earnings.

Availability of near-by schooling for their children attracts and stabilises the work force.

Land, ploughed and fertilised by the farmer, free for private cultivation by workers or their families is often worthwhile.

Adequate medical facilities and provision of piped water to the staff quarters are important. There should be a daily clinic to provide first-aid and to treat simple ailments, transport to hospital of more serious cases and provision of sick leave and pay.

Employers' liability insurance, or workmen's compensation, is compulsory in most countries to provide compensation to workers injured in the course of their work.

It is usual to extend credit for worthwhile causes, like brideprice or school fees, and possibly loans for, say, half the cost of consumer goods to long-term staff.

Physical working conditions directly affect the health and comfort of staff and need careful attention. They include cleanliness, overcrowding, ventilation, temperature and lighting in farm buildings. Humid and stagnant air and uneven temperatures often cause inefficient work and poor health. Elaborate systems are sometimes needed to rid the air of fumes and dust.

The lighting in many tobacco grading sheds is not bright or uniform enough or of the right type to enable the colours of the cured leaf to be seen. It is usually recommended that, if artificial lighting is used, all natural lighting should be excluded as a combination of both can cause confusion. Recommended lighting is one 120 cm 40 watt power factor corrected industrial tube, with a reflector, mounted 120 cm over each grading table. There should be one tube for each grader. Without a reflector, much light would be wasted upwards.

Proper sanitary facilities should be supplied for both male and female staff; also adequate safety measures. The latter includes fire precautions, shielding of moving machine parts, eye protection and provision of protective clothing when needed. Two aspects of safety are distinguishable insofar as they relate to the well-being of staff. First, specific accidents must be avoided and, secondly, people must be protected from damage or discomfort resulting from long-term exposure to hazards.

Housing

This is an especially important aspect of welfare. Farm life for a paid labourer has so far often been similar to his traditional life, apart from the element of organised work. During leisure hours he and his family live together with others like themselves in a community with a traditionally rural background. He has usually enjoyed the privilege (almost the right) to design and build a home with materials he knows well to a pattern serving his traditional living needs on a site that he chose himself within an allocated labour quarters area.

If he, or the community as a whole, needed more land, he simply took what was available, if the farmer showed no intention of using it himself, for disposing of wastes, gathering fuel, gardening, establishing communications — usually footpaths; for holding religious meetings and burials; for hunting and many legal and other needs and activities.

On many farms today, more intensive land use has caused a shortage of land for use by the labour force and of thatching grass and bush timber for structural uses and fuel. Another feeler shoots out across the fence to still-available sources of these on a neighbouring farm. This leads to conflict as third parties become involved.

The farmer is then forced to provide housing and the era of 'self-domicile' ceases. The apparent need is that of housing but the less obvious need is that of replacing the traditional way of self-helped life with an organised living environment in which the worker works and he and his family are provided for in *all* their domestic needs. The farmer assumes duties as welfare worker, village developer and all-round provider as he is also the employer. To put it bluntly, the situation is rapidly developing in many areas where farm workers and their dependents can no longer live off the diminishing virgin land and must find all their living needs from the employer.

Besides the cost of providing western-type housing for many workers and their relatives, there are human factors that cast doubt on the alleged popularity of attempts to provide this style of housing. A black South African author wrote, 'There is nothing more foreign to the Bantu inclination than a European-style house.' This seems to contradict the many reports of successful labour recruitment from farms where this sort of housing is provided. The solution is found in the many self-built annexes and out-buildings that soon enlarge the original building, the boarding-up of windows and the continued cooking over an open fire even when a stove is provided. It may also be resolved by the fact that the farmer has shown concern for his staff, which is always appreciated. Sociologists, and farm managers, have suggested that the farmer's new-found popularity results from the status symbol he provided.

Considering the deficiencies of both traditional and western-style housing, something intermediate between the two probably would be best. Good quality housing, where a worker can have his wife and family with him, is likely to increase labour stability. Give a woman a house — any woman, any race — and she wants to furnish and decorate it and soon the double-time for Sunday work attracts her husband.

All that is really needed of a house is that it should be fairly warm or cool according to season, reasonably comfortable, draught- and rain-proof. One needs to encourage the birth by self-help of a neat village with space for gardens and animals instead of a crowded quarter.

Feeding

For tropical farm labourers to work best, they must be reasonably healthy and fit. Adequate suitable food is basic to health and fitness. Thus, to achieve high performance levels, attention to the feeding of workers may be necessary. Dieticians agree that the common habit of labourers starting work in the morning, without having eaten for about 10 hours is bad. It is unreasonable to expect labourers to stay efficient for six hours or more if they have had no early morning meal. At least a light, high-energy meal should be eaten before work starts in the morning. The traditional African custom of having only one large meal each day, usually in the evening, is no valid guide to what is best for them when working hard in paid employment.

Many farmers provide their labourers with basic rations as part of their remuneration package.

Those items that can be grown by the workers themselves, need not be issued.

To ensure that the full nutritional needs of the workers are consumed by them, rations must be issued to dependants living with them. Unless this is done, rations will be shared and the workers will go short.

Farmers disagree over the rather paternal practice of issuing rations as part of the wage. Supporters of rationing argue that to pay cash instead of rations is courting disaster because the extra cash will only be gambled or spent wastefully, leaving the farmer to face a regular stream of hungry wives and children. Certainly, when rations are changed for cash, some workers at first adopt a diet of buns and beer, but they soon find this high-life beyond their means and return to more traditional fare.

Workers seem to prefer cash to weekly rations and there are several benefits of this to farmers. For example, easy administration from both a financial and distribution viewpoint, absence of friction in the staff quarters between those who have agreed to pool or share their rations and also removal of a common grievance between casual and regular labourers — usually rationed at different rates.

Usually, if there is a reasonable store nearby where basic foods can be bought, it is probably best to pay all remuneration in cash. A consolidated wage from which workers, who may have a strong leisure preference, must buy their own food, raises the incentive effect of cash bonuses. However, where this situation does not prevail, and especially for single workers, it is often still best to provide weekly rations and maybe also a cook.

Some farmers provide credit on note at a local farm store for food and other essentials. It is administratively simple to recover these debts at the month's end as all credit notes are routed through a clerk for recording.

There is much evidence to show that planned rest periods during the working day can greatly raise productivity. The Zimbabwe Tobacco Association's work study manual states that effective manual output can be raised from 53 per cent to 78 per cent merely by ensuring that workers take breakfast in the morning and by providing a light mid-morning meal.

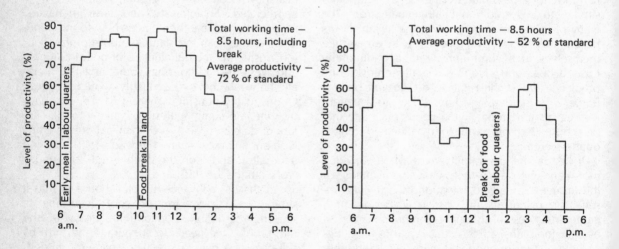

Figure 7.6   The Effect of Meal Times and Feeding on the Productivity of Field Workers

Especially if the work is heavy, the gang should be given a 20–30 minute break for food about three or four hours after work began. This rest break and light meal has been shown to boost output as depicted in figure 7.6. The traditional practice of having a long break around noon, when the workers may have to walk a long way to and from the labour quarters for food, wastes much time and energy.

Providing a suitable meal on the land or other workplace, not more than four hours after the start of the shift, with no long mid-day lunch break is appreciated by both workers and supervisors. Not only is productivity raised, but the gross working day is shortened. This means less supervisory time and an early finish, giving more time for leisure. It is for the farmer to decide whether the food issued on the land or elsewhere should be an extra or should be deducted from the wage or rations issued.

Suitable foods to issue during the mid-morning break includes tea, skim milk, fortified hot soup — based on a soyabean source called textured vegetable protein, a high-protein cocoa mix or partly fermented, weak beer. Any solid food given should not be too bulky but rather a high-protein, high-energy concentrate with sufficient salt to make it tasty.

Good communications

Communication is the process of conveying information from one person to another. It is the exchange of information, ideas, attitudes or emotions between people; the sending of messages that evoke responses. Clearly, this definition does not restrict the concept to words, either spoken or written. It includes non-verbal ways by which meaning is conveyed from one person to another. Even silence can convey meaning and therefore must be considered part of communication. Gestures, facial expressions and other forms of 'body language' also communicate, rightly or wrongly, meaning to others.

Much of every manager's time is spent making or receiving communications. It therefore is not only a skill that every manager must often use; it is also the most neglected. If we accept that the essential job of a manager is to get work done through other people, it follows that his effectiveness as a manager must depend on his ability to communicate with his staff and on their ability to communicate with him.

Good communication depends on clear thinking and has two basic elements — the communi-

cator sends a message that conveys some content and the receiver of the message responds to the content as he understands it. For a reasonable chance of getting a job done well, your staff should understand exactly what is wanted. Messages can be sent by many ways. We usually think in terms of the spoken or written word but it is really much more subtle and both informal communications — the bush telegraph and 'body language' — may be more important. Communication is a tool through which motivation is improved. Used properly, it is a powerful aid in business. Poorly used, it can cause havoc.

Many managerial misunderstandings, frustrations and failures are caused by bad communication; therefore it is important for managers to know when and how to communicate and to whom. Just as a firm may lose business by shortage of stock, so a break in the communication chain can have equally bad effects. Use of accepted channels of communication reduces rumour and misunderstanding. It is through communication that people reach some understanding of each other and influence and are influenced by others; that the manager can overcome the barriers of insecurity, frustration and a feeling of unimportance on the part of staff.

When communicating, you are trying to do four things: to be understood, to be accepted, to get something done and to understand others.

Every business has its informal communications system and, whatever the official system, the 'bush telegraph' is rarely missing. It works fast, and regardless for either the truth or the results of the stories spread. All managers should understand something of this informal method of communication in their own businesses and realise that, though they may be unable to control it, they may be able partly to influence it. It has been suggested that, if the 'bush telegraph' is to help managers, three principles should be observed:

- people should be told about things that may affect them personally;
- people should be told what they want to know rather than what the management wants them to know;
- the information should be given as soon as possible.

Communications may be horizontal, downwards or upwards. Horizontal communication occurs within a management level. It avoids duplication of effort, widens the experience of each manager and helps to co-ordinate departmental relations in fulfilling business policy. A manger's wishes must be conveyed downwards accurately, fully and clearly enough to be wholly understood by his staff. Experiments have shown that information that has travelled down through several layers loses much of its factual content. An average of 20 per cent can be lost through five levels of management. Upward flow of information is specially important but seldom efficient. This partly relates to feedback, like the record of field results and receiving complaints and suggestions from staff. Top management is often told only what the lower levels think it ought to know (or would like to believe) or what hides their own weaknesses.

There are two main purposes of communication. These are:

- to provide knowledge and understanding for efficient work performance
- to provide knowledge and understanding necessary for loyalty and desire to co-operate

Both are needed to achieve maximum productivity. This is shown by the following equation:

The *skill* to work + the *will* to work = best work. Lack of either skill or will always results in low productivity. The will to work emanates from man's wish to succeed and to be recognised. In other words, people's self respect grows when they feel that they belong and are making a useful contribution but suffers when they feel that they are merely another pair of hands. Only through good communication can people be made to feel that they belong and thus gain self-respect.

You can often choose the method of communication and although the best one often depends on circumstances, thinking in terms of people aids the choice. Information should always be sent by the person who knows to the one who needs to know by the shortest most suitable route. The following considerations help in choosing the best communication method.

- Time — is speed vital? Great speed may be costly and of little value.
- The accessibility of the person to be contacted.
- Your knowledge of the person concerned; his personal habits and inclinations.
- The psychological aspects of the situation. The spoken word may be safer than the same thing in writing.
- Will all the possible methods enable the information to be received accurately and, if not, which methods are unsuitable? Consider, for example, the sending of figures. If a telephone were used, might there be misunderstanding and consequent error?
- Is secrecy and recording important? A hand-written postcard is cheaper than a typed letter.
- Is written communication necessary to record the information?
- What will the alternative methods cost?

In both verbal and written communications it is important to express ideas simply — without long words or complex sentences. Published and spoken technical information often leaves much to be desired. Clear, concise orders are necessary to raise production. Many words have several meanings so they must be carefully chosen so that the person receiving the message understands it as intended. Unless the language barrier is broken, much money will be lost by having to repeat work. It is worth ensuring that orders have been understood by having them repeated back to you. However much care and skill a manager spends in 'putting across' his message, he may never know whether it has been properly understood until he follows it up. This can be done by asking questions designed to test the understanding of the receiver and by encouraging him to express his reactions.

Communication is incomplete until there is understanding and acceptance of the message and action results from it. It is a two-way process; there is both an act of sending and one of receiving. Research has suggested ways by which managers and staff can avoid or reduce those factors that reduce the effectiveness of communication.

The answers to the following questions should help in planning effective, suitable communication.

What is the concept, idea or piece of information I want to convey? Is it familiar or new?
Who am I communicating with? What barriers are likely? Are they the right people?
Why do I need to communicate? (To instruct? To inform? To reprimand?)
How can I interest them from the start? How can I best communicate? (Speak? Write? Individually or in groups? Direct or by other means?)
When is the best time? (Working hours or outside them? Next week? Next month?)
Where will it go down best? (At the workplace? In private?)

Oral communications

Oral communication is more than ability to speak a certain language. It is the ability to communicate an idea from your mind to the mind of the person with whom you are speaking. As such it involves the converting of an idea to an image or picture and the translation of this image into words that can be conveyed to the other person. The process is then reversed by the receiver from words to image to idea.

Sender:
　IDEA → IMAGE → WORDS
Receiver:　　　　　　↓
　IDEA ← IMAGE ← WORDS

The directness of the spoken word in face-to-face communication is unsurpassed. It is probably the commonest form of communication and both *what* is said and *how* it is said are highly important. Besides face-to-face communication, both the two-way radio and the telephone are important.

Oral communication has many advantages, like speed, clarity of meaning and maintaining good personal relations. Being a two-way process, it gives communicants an opportunity to check the understanding achieved, to ask questions, check responses and clarify meanings. A few minutes' talk is worth more than pages of writing because warmth and humour can be conveyed. The written word tends to be impersonal and one-way.

Oral communication has the disadvantage that perception of spoken words is likely to be less accurate than that of written ones. There may also be losses along the line and differing responses to the same message. The scope of oral communication is limited to situations where time and the nature of the message permit direct contact.

To be a good oral communicator:

- Know exactly what message to transmit.
- Speak clearly, slowly and deliberately.
- Keep orders simple and give one at a time. Do not wander from one job to the next.
- Accompany orders with sufficient information about the policy behind them.
- Use words that the other person can understand and that have no double meanings. The words used will cause a reaction that will be conditioned by the receiver's ideas and feelings.
- Ensure that the listener got the right message.

Let us look closer at the importance of using simple words. The language used should be that of the person addressed. One cannot overstress how desirable it is for a farm manager to learn the language of his staff, who will often be of a different tribe or race. Communication is hard because, even using the same language, people see different meanings in the same word. We all interpret words by our own experience and no two sets of experience are the same. To Enoch, the word mango implies a delicious, stringless peach mango; to Rashid, it is the stringy, fly-stung fruit, tasting of turpentine, that he used to steal from his neighbour's garden.

Few people have developed the skill to listen carefully, although this greatly aids effective communication by avoiding the loss of important parts of messages. They tend to concentrate, at least mentally, on answering. Let the speaker finish, hear him out and hope that he will listen to you later. A stranger in some countries soon becomes aware of this if he prefers cool water and no ice with his whisky. The custom of splashing ice into a mixture of whisky and tepid water is so strong that stewards react only to the signal 'whisky' and entirely miss the rest of the message. They listen for a particular mental 'set' which distorts reception of the message.

The best way to check that orders have been understood is to insist on them being repeated back. It is amazing how often simple orders are twisted when they are repeated immediately, let alone after a lapse of time. ('A Bell's and water, please.' 'Sorry, sir, we have no Belgian water.') Many local people in the tropics, trained in politeness according to local custom, will not be so rude as to imply that you are not a perfect communicator. Watch their faces to see if they understand. Misunderstandings can cause costly problems.

On large estates, fitting vehicles with two-way radios always aids rapid communication. Telephones also may offer a quick and cheap method. Most telephone calls can be quick. Long, costly calls usually result from the caller not planning what he wants to say in advance. Brief notes should be made beforehand and any necessary papers should be in hand.

Meetings or committees for group discussion, if well conducted for a definite purpose, help if:

- they aid decision-making by bringing together all those whose experience can improve the decision;
- they prevent decisions being taken *too* quickly;
- they help to keep managers aware of the work in other managers' sections;
- decisions made by committees are likely to be more objective than those made by one person;
- they secure the willing co-operation of its members by giving them a say in making the

decisions that they are expected to act upon.

Meetings may have any, perhaps several, of the following aims — to:

- make decisions;
- help the chairman to decide;
- uncover facts;
- produce new ideas;
- pool ideas;
- communicate or,
- make people feel that they belong and are involved.

Barriers to communication

Communication cannot be effective unless you listen and understand as well as tell clearly and honestly. Many things can make communication poor. There may be too much information, too little or it may be of the wrong kind. Managers may be flooded with papers that they have no time to read or even to separate what matters from what does not. They may feel that their time is wasted by too frequent or lengthy meetings. Conversely, they may complain that they never know what is going on; that they are neither told nor consulted about decisions affecting them. It is still worse if they receive the wrong kind of information; figures that are so out of date that they are a poor guide to action or information deliberately selected to give too optimistic a picture of events.

Lack of clarity, too much haste or delay, failure to consider the possible unfamiliarity of technical terms used or lack of sympathy between members of a group can all disrupt the group's communications.

We usually are more aware of barriers to communication between different levels in the hierarchy than of those between sections or departments. The latter can also be important; especially in changing conditions when more inter-departmental communication is needed. At such times it should be made easy for people to talk to each other so that they are not hindered from asking others for information and advice.

Not all communications are worth spreading. An entirely free system of communication in a business would be chaotic. Some of the barriers that inhibit the free flow of information are deliberate. The effectiveness of communication depends more on the wise selection of information to be passed on than on the free flow of all communications.

Although not all barriers of communication indicate trouble, there are many problems in spreading adequate, clear and meaningful information and in developing inter-personal relationships for maximum, effective communication. Among the main barriers are organisational blocks, language blocks and human relations problems. It is necessary to substitute 'gateways' for 'barriers' to communication.

The organisational hierarchy itself affects communication by imposing the idea of 'channels' through which communications must officially travel. In some organisations there is too much stress on keeping to channels. Such rigidity imposes undue hardship on those allowed to contact only their immediate senior and juniors. Limiting people to formal channels makes them feel frustrated and confined. They find it hard to use all their skills and to work at top capacity. Frustration leads to devious ways of overcoming the limitations so as to tap the informal organisation that cuts through and across organisational lines. Although some formalities must be observed to keep activities running smoothly, anyone should be free, with the consent of his senior, to contact whoever can help him with his problems.

Language blocks arise from the fact that messages may get distorted as they pass up or down the hierarchy. Words have meanings that vary with the personalities of both sender and receiver. For example, a foreman may say, 'Tip that drum of herbicide into the sprayer tank', and later find the drum in the tank instead of its contents.

Status differences may hinder the free exchange of information. Status means the feelings of regard, approval, respect and so on that others show towards both a post and its incumbent. It can impede upward communications because a junior knows that his future is influenced by the boss. What his boss thinks and does can affect his future so much that he usually thinks very carefully about what he says to him. Most staff want a happy, pleasant boss so they are reluctant to report their own failings and mistakes to him and hesitate to give him information that will upset him, even if it is not their fault. Workers often fear to tell the truth; they say what they think you want to hear because they fear they will get into trouble even if they are only bearing and not causing bad news. They may prefer a happy, ignorant boss to an informed, upset one and filter the information passed to him. Some drops out; some is greatly changed to achieve a calculated effect on him. If the boss is highly emotional, the juniors' tendency to avoid upsetting him will be exaggerated beyond reason.

Many meetings are inefficient because of some simple physical factor, like the numbers attending, the seating plan, poor ventilation or a connected telephone in the room. Meetings that are too large or small work poorly. The exact number of members should depend on knowledge of the subject for discussion, of the people concerned and on other local factors. Only those who can be expected to make a valid contribution, or who must attend to be informed, should be invited. Any number over twenty is difficult to handle. The best size for most purposes is usually between five and ten. In meetings of this size, people can talk nearly as much as they wish and can all influence each other; yet there is sufficiently varied skill and experience to tackle problems creatively. It is also possible to include each member's ideas in any decisions.

In most meetings, members deal not only with the topics on the agenda but also with irrational 'hidden agenda' items; secret power-struggles, jealousies and anxieties, hidden fears and aggressions and attempts to impress. These emotional factors much affect the way that topics are handled and decisions are made.

Follow-up activities are needed after meetings. A summary should be distributed of all action decisions taken and who is responsible for each. Closely controlled meetings are seen by participants and others as purposeful and effective.

THINK THREE TIMES BEFORE YOU ACT

Think first: How well do I listen?

- Do I reserve judgement until I hear all the facts?
- How many essential facts do I usually remember?
- Do I try to listen from the other person's viewpoint?
- How many questions do I ask to clear vague issues?
- What efforts do I make to check disputed points with other sources?

Think secondly: Why do my communications fail?

- Do I close up when misunderstanding occurs?
- How often do I resist new ideas because they do not fit my way of doing things?
- Do I talk to people but ignore their reactions?

- Do I always insist on having the last word?
- Do I dominate every group discussion?
- Do I use too many words to present a simple idea?

Think thirdly: How do conflicts start?

- Do I often belittle the ideas of others?
- Do I usually order — instead of invite action?
- Am I interested mainly in self-promotion?
- Do I often delay action when a decision is needed?
- Do I expect others to read my mind?
- Do I tend to annoy others?
- Do others often misunderstand me?

EMPLOYEE RECORDS

Wages can account for a high share of costs on commercial farms in the tropics, so good records are essential, but they should be kept to a minimum. Even small improvements in the efficiency of labour use may greatly reduce costs, raise profits or both. Consequently, labour records are useful on any farm unless no workers are hired.

Some of the records to be described must be kept for payment calculations and to satisfy the tax authorities. It is little extra trouble to adapt a normal attendance and payment record into one that breaks work done into either enterprises or operations.

The financial information needed in labour records consists of:

- the basic wage each month, week or 'ticket'
- the days worked over the period
- bonus and overtime payments or any other allowances earned
- tax and other payments deducted
- net wages paid
- any outstanding advances.

The column referring to outstanding advances may not be needed on every farm. Life is simpler if these transactions are minimised.,

In the highly mechanised farming systems of Europe and North America, where the typical units are small family farms, labour costs are often regarded as fixed because they are incurred whether the farm produces anything or not. This also applies to small peasant farms in the tropics. However, on tropical commercial estates labour costs are variable, being dependent on type and number of enterprises, scale of production, degree of mechanisation and efficiency of labour use. Records of the amount of work done on individual enterprises, or even on operations within major enterprises, are useful for both planning and control. Even in Europe and North America, a strong case can be made for recording the number and timing of man-days spent on each enterprise where seasonal labour use varies — as it does with all crops. Records of the use of regular labourers, casual female and casual juvenile labour should be kept separately. The record should be completed daily by the farm clerk or foreman.

A monthly payroll or wages sheet is basically a list of the employees' names, one on each line, with columns for each of the items making up the gross wages and for the various deductions that together fix the net wage due. Preprinted ledgers are readily available from stationers for both daily records and monthly summaries.

A typical layout for a daily labour register is shown in Figure 7.7, and figure 7.8 shows a typical monthly summary.

It takes little more time to use symbols for the type of work done instead of ticks to prove attendance — as shown in figure 7.7. The information needed to build up labour profiles for use in planning is then available. These symbols may take any form, for example,

MC	Maize cultivation	BC	Beef cattle
MH	Maize harvesting	LP	Lucerne planting
MP	Maize planting	WS	Workshop
TP	Tobacco planting	A	Absent
TL	Tobacco land preparation	CR	Cotton reaping

Year   Month:

Name	Sex M F J	Days 1 to 31						Rate/ day or ticket	Bonus/ overtime allow- ances	Gross wage	Tax	Other deduc- tions	Net wages	Signat- ure for receipt	Advances out- standing
		1	2	3	4	...	31								
Enoch	M	BC	BC	BC	BC	...	BC	$1.40	6.00	41.00	8.00	6.50	26.50	X	13.50
John 2	M	TP	TP	TP	A	...	MP	$1.20	0.50	25.70	6.00	–	19.70	X	–
Sam	M	A	Lp	LP	LP	...	W	$1.45	0.70	34.05	8.00	5.30	20.75	X	4.70
Tickey	J	MT	WS	WS	WS	...	MP	$0.65	–	16.90	4.00	–	12.90	Tickey	–

Figure 7.7   A Typical Daily Labour Register

	TOTAL	June	July	Aug.	Sep.	Oct.	Nov.	Dec.	Jan.	Feb.	Mar.	Apr.	May
Tobacco (totals)	5469	344	425	363	471	476	466	333	549	509	524	574	435
Land preparation	570		96	48	95	198	66				6	34	27
Seedbeds	501		39	72	142	130	98	14	2				4
Planting	236						236						
Refilling	36					12		24					
Cultivating	293							132	89	72			
Topping and suckering	321							99	40	98	84		
Reaping and curing	1239							64	418	315	340		
Grading and baling	1813	304	290	243	170						64	102	
Cording and carting wood	297	28			64	103	38				14	408	334
Maintaining tobacco buildings	124					18	16			24	16	18	32
Other	39	12				27						12	38
Maize	322		16		23	18	35	124	52				54
Pastures	76							19			12	17	28
Maintenance (totals)	582	52	27	59	17	12	109	125	32	38	62	26	23
Workshop	303	29	27	36	17	12	24	32	18	38	26	26	18
Roads	220			23			85	93			19		
Other	59	23							14		17		5
Development (totals)	835	213	160	182	93	94							93
Land clearing	391	75	85	44		94							93
Buildings	117	117											
Water supplies	327	21	75	138	93								
Non-farm	402	43	31	39	30	38	30	41	31	29	26	32	31
Absent	1367	96	113	124	108	130	101	123	97	118	142	119	96
TOTAL:	9053	748	772	767	742	768	741	766	761	694	766	768	760

Figure 7.8   Example of Labour Use by Months (Regular Man-days)

Labour-use records kept like this are essential for any careful analysis of labour-use efficiency by enabling comparison of the labour used on various jobs against standards. It enables labour costs to be allocated as variable costs to specific enterprises, if appropriate, and provides valuable planning data for estimating the likely effect of a change in the farming system. It also indicates the labour savings likely to be made by mechanising certain operations.

On large farms note should be made of gang sizes and work rates achieved on critical jobs at the busiest times of the year.

There are three main levels of detail at which labour use may be recorded:

WHOLE FARM

Total man-days are recorded during the year. A farmer knowing his total yearly labour cost, including rations, etc., can compare his labour cost for each man-day with other similar farmers. Although this information is in itself valuable, it is limited.

Regular labour	Farmer A	Farmer B
	$	$
Total annual labour cost	15640	29580
Total annual man-days	12364	19186
Cost per man-day	1.26	1.54

ENTERPRISE

The total days are recorded for each enterprise (tobacco, maize, cattle, etc.) during the season. A farmer who knows he has used 400 man-days per hectare of tobacco can compare his labour-use with other farmers. This is more useful than whole farm comparisons.

Tobacco	Farmer A	Farmer B	Apparent excess
Total days recorded/ha	253	372	119 (47 %)

Farmer B's gross excess cost over Farmer A can be found by multiplying the apparent excess by the area and the average cost of each man-day.

OPERATIONAL

Total man-days are recorded by operation for each main enterprise. A farmer who records in this detail can often see weaknesses in his production methods and thus know in which operations he should become more efficient.

Operation	Farmer A	Farmer B (Man-days)	Apparent excess
Tobacco reaping and barn loading/1000 kg	24.7	98.3	73.6 (298 %)

Clearly, Farmer B's reaping and barn loading is very inefficient in labour use, if the standard of the work done is the same.

REWARD STRUCTURES

The main reward for work done is an adequate wage or salary coupled with fair employment conditions — but these alone are not enough. People also seek other, less clear, rewards, like satisfaction and interest in their work. Over the past 15 years much has been said and written on what makes a man work, what satisfaction he looks for in his work, why he stays in his job, and so on. In fact, a new discipline called *behavioural science* has emerged.

There is no single ideal incentive because incentives vary between societies and between people in the same society. Young men tend to seek challenging posts and opportunity for promotion; older men prefer routine and security to change. Some people value money most; others leisure, achievements, friendly supervision or promotion opportunities. Some may be content just to retain their jobs and avoid any trouble.

If worker performance is below standard, it is necessary to find the causes of this and change *them*.

Wage payments

Earnings are important motivators and few workers mind spells of prolonged heavy work if the monetary rewards are good. The cash payment made to a worker does not truly reflect either the cost of hiring him or the value of the job to him. Besides direct payments, an employer must usually pay a workmen's compensation insurance premium, overhead costs like housing, rations, welfare provisions and, possibly, a pension contribution.

A wage or salary policy should aim to attract, keep and motivate workers at all levels so that the business keeps its best staff and makes best use of their skills. Remuneration should be fair, encourage keenness by rewarding effort, not lead to excessive over-payment and, as far as possible, it should satisfy both employer and staff.

Farm wages are usually paid to regular workers on a monthly or 'ticket' basis for a fixed number of hours each day with a bonus for overtime. There usually is no problem in getting extra hours of work from staff. In fact, if several workers are involved, a fair spread of overtime between them can be a problem. Overtime is usually paid for at 20–100 per cent above the basic rate. If leave preference is strong and there are peak work loads followed by slack times, compensation for overtime may take the form of equivalent time off when the work allows.

It is important to pay workers promptly and regularly whatever the payment system used — this is even more important with casual labourers than with regular ones and weekly payment is best. Workers have regular financial commitments and they prefer a definite payday which should be adhered to if possible. Once the labour force knows that payday is on a fixed day of the week or month, they can plan accordingly.

The mechanics of making up wage packets is eased if the farm staff includes a clerk able to keep wage and credit books, attendance records and all the other records, like fuel use, that so many farmers neglect. A trained clerk should be intelligent and therefore can be used for many other jobs, including supervisory ones. If he can drive a vehicle this raises his value. Never, however, should the clerk who compiles the payroll either fill up the pay packets or hand them out.

Preprinted pay envelopes are available on to which can be written the workers' names and the various amounts of basic pay, overtime pay, tax, other deductions or additions and net pay. Some workers do not like others to see these details and prefer a plain envelope with the details inside. Preprinted slips may be used or adding machine print-out slips may be annotated with what each figure represents. Manager and secretary alike will delight in the streamlining simplicity of a good, microcomputer-based payroll programme. Only those data that change, such as overtime hours or deductions, need to be entered. The machine/programme will not only compute how much is to go into each wage packet but also list the necessary notes and coins to achieve that total.

It is necessary to ensure that pay policy is decided rationally so that all workers are paid in a fair, logical way. Minimum rates may be set by government but actual rates depend, first, on circumstances apart from the employer's will or the employee's worth; like cost of living, abundance or shortage of staff, general business conditions and the financial state of the business. After these, it depends on the value of the worker and the mode of payment used. Appreciation of the factors dependent on the employer's will and the value of employees demands good knowledge of business, judgement and fairness.

Scientific precision in wage setting is impossible but this does not mean that there are no guidelines. Nor does the presence of scientific methods, like job evaluation, mean that executives can dodge responsibility for wage decisions. The main advantage of modern techniques of wage and salary administration is that the wage-setting process is brought more fully under management scrutiny and control. Businesses using casual methods cannot ensure consistent, economical wage administration.

If a grade or other type of scale has upper and lower limits, the actual amount paid may be discretionary but should be assessed on merit at least yearly, based on performance over the period since the last review.

Salaries should be fixed by systematic comparison of the similarities and differences in jobs; not in people. These comparisons can be made by job evaluation which usually considers skills, qualifications, experience, responsibility and seniority. Job evaluation is a way of examining and analysing jobs and fixing their relationship to each other. It follows that, once the relationship has been fixed, evaluation can be the basis for a suitable wage or salary structure. Tables 7.2, 7.3 and 7.4 show how job evaluation could be used on farms.

Table 7.2 Points System for Differing Levels of Skill, Effort, Responsibility and Job Conditions

Factors			Levels		
	1	2	3	4	5
			(points)		
Skill					
Education	14	28	42	56	70
Experience	22	44	66	88	110
Initiative and ingenuity	14	28	42	56	70
Effort					
Physical demand	10	20	30	40	50
Mental and visual demand	5	10	15	20	25
Responsibility					
Equipment or process	5	10	15	20	25
materials or product	5	10	15	20	25
Safety of others	5	10	15	20	25
Work of others	5	10	15	20	25
Job conditions					
Working conditions	10	20	30	40	50
Unavoidable dangers	5	10	15	20	25

Levels are fixed as follows:

- First level: ability to follow simple orders. The worker is told exactly what to do.
- Second level: this requires small decisions and some judgement.
- Third level: this involves responsibility for a sequence of activities and needs general decisions on quantity and quality.
- Fourth level: this needs ability to plan operations and to perform unusual and difficult tasks.
- Fifth level: this needs ability to work alone on complex jobs.

Table 7.3 Total Points Ratings and Corresponding Grades for Farm Jobs

Score range	Grade	Score range	Grade
100–271	6	316–337	3
272–293	5	338–359	2
294–315	4	360–500	1

The main potential for raising productivity exists in the Grade 6 group of general fieldworkers.

Grades might be transferred into wage scales, allowing for yearly rises, as in the example:

Grade	Daily wage		
	$		$
6	1.35	to	2.00
5	1.60	to	2.35
4	1.80	to	2.80
3	2.20	to	3.50
2	2.60	to	4.30

Many farmers use tickets, as shown in figure 7.9, for paying wages and believe this method is popular with their workers, but most systems of payment seem acceptable provided they are clearly explained to and understood by the staff.

Labourers are paid on completing a 30-day ticket, that is after 30 working days, and completed tickets are paid at a given time each week. This system probably helps to reduce absenteeism as no payment is made until a ticket is completed. Also, as workers are not all paid at the same time, they tend to help each other financially as required. However, it has several disadvantages and many farmers find it more convenient to pay all their staff at the end of each month however many days worked in that month.

The two main drawbacks of the ticket system are that it does not allow for single monthly paydays and that it encourages keeping poor workers on the farm. The normal ticket system requires completion of 30 days' work before any payment. Allowing for sickness, leave, malingering or absenteeism for any reason, this 30-day period can extend to as much as three months in *extreme* cases so full rations may be issued to staff who work only about one day in three.

Having one communal payday each month simplifies wages administration and reduces the need to give long-term credit which, when recovered, often causes misunderstanding and resentment. Reduction of long-term credit facilitates dismissing a bad worker who is unlikely to work hard anyway if he suspects that he will be fired as soon as he has completed sufficient tickets to repay his debt. It seems best, therefore, to pay a monthly wage based on the days worked during that month. This is exactly how artisans are paid and gives the farm worker a regular wage and some security.

Estate _____

Name_____

Date _____ Pay _____

M	T	W	Th	F	S

Paid Off _____

Figure 7.9 Example of a Labour Payment Ticket

Table 7.4 Example of Job Classification and Grading

Type of worker	Field labourer	Tobacco grader	Master blender/foreman	Workshop labourer	Driver	Clerk
			(points)			
Factors						
Skill						
Education	14	28	28	42	28	56
Experience	66	66	66	88	88	88
Initiative	42	42	42	56	56	56
Effort						
Physical demand	40	40	30	30	40	40
Mental and visual						
demand	15	15	15	15	20	20
Responsibility						
Equipment	10	5	15	20	15	15
Product	20	15	15	15	15	20
Safety of others	5	5	15	5	5	20
Work of others	5	5	15	5	5	20
Job conditions						
Working conditions						
(contact with						
others)	30	40	30	20	30	30
Dangers	5	5	15	10	20	5
Total points	252	266	286	306	322	355
Grade	6	6	5	4	3	2

Financial incentives

Much use is made in industry of monetary incentive schemes, paying above normal rates for satisfactory work done in less than a fixed basic time. It is important, however, to realise that it is easy to arrange incentives for some jobs but hard for others. In practice, financial incentive schemes work better for labourers than for managers but they can be successful for managers if carefully designed.

Financial incentive schemes, often called payment by results, are attempts to formally relate pay to output or effort, so as to motivate people to maintain or exceed a fixed standard of performance. A well-designed incentive scheme, whether financial or not, is a sound performance improvement technique used either to raise physical output, cut costs, to encourage good workers to stay in the job or to encourage and reward extra care, skill or responsibility.

Financial incentives, like piecework and bonus work, were described in chapter 6. Other forms are profit-sharing schemes and even knowledge that promotion to a better-paid job will really be on merit. Many schemes, simple and complex, have been planned for both individual workers and gangs of workers. They should be based on work study findings if possible.

Current thinking in developed countries is that financial incentive schemes are no longer the main motivators. It might be well here to distinguish between motivation and incentive, as they are different. An incentive is something offered to someone else. Motivation is within a person and is often an irrational, inexplicable impulse. Non-monetary factors become more important once staff have satisfied most of their basic needs. However, there is little doubt that money is still a strong basic motivator, able to raise productivity, in less developed countries.

The advantages of payment by results are that it raises workers' output, often by 30 per cent, lowers the cost of each unit of output and enables good workers to earn more than poor ones. Less supervision may be needed to keep output high enough and well-designed and administered work study incentive schemes can greatly improve labour turnover, absenteeism and management — labour relations. Incentives for manual workers tend to produce quicker, more skillful work and probably more effort. Those designed for managers reduce the need to supervise them and enable them to assess their own performance and understand the results of their efforts.

Financial incentive schemes have advantages and disadvantages. There are often conditions that affect the amount of cash received that the worker cannot control, like the weather, livestock and crop diseases, dependence on others or poor support. These conditions can seldom be eliminated and workers often regard the basis for payment as unjust. The resulting grievance can reduce effort and lead to bad labour relations. Quality tends to worsen unless there is a strict system of checking. For example, land might not be cleared properly under piecework rates and rubber trees might be carelessly tapped by sharecroppers.

When paid by results, workers tend to regard their highest earnings as usual and therefore to press for a higher basic wage. Jealousies may arise between workers because some earn more than others or, in teams, fast workers may be dissatisfied with the slower or older members of the gang. It is hard to set piece- or bonus work rates accurately. If they are too low, workers may feel forced to work hard and become dissatisfied and, possibly, ill; if too high, they may reduce their effort to avoid a revision of rates.

Payment by results has often lowered rather than raised productivity, especially if the payment system is unfair and causes friction. Many firms now prefer to pay a good flat rate of pay, relying on other ways of raising output. Some managers now claim that to introduce an incentive scheme is to admit bad management, as a good manager, they say, can get his men to work hard by providing less crude, non-financial incentives. The solution is good design of any incentive scheme.

Financial incentive schemes for managers often estrange them from their employers. They may argue over the exact way of calculating the bonus, which may be delayed unduly until the year's financial results are known, and they may feel that some of the farmer's decisions reduce the bonus. They tend to earn a good bonus in easy years and a poor one when the work load is highest, so they may leave when needed most.

The main requirements of good financial incentive schemes are as follows:

- They should benefit both employer and staff; be fair to both and produce harmony rather than argument.
- They should be easy to calculate and understand.
- They should cover the maximum number of workers on the farm and not upset the natural balance of wage rates.
- They shoud be paid as soon as possible after the work is done.
- Reward should, if possible, depend upon the results of a person or *small* team and be based on both effort and results.

- They should not promote poor quality work or waste of resources.
- There should be sufficient spread between the guaranteed basic rate and the normal bonus rate to provide sufficient incentive for extra effort. That is, the incentive should be generous. However, the bonus should not usually exceed 20 per cent of regular salary and provision should be made for salary rises besides bonuses.
- The figures on which a bonus is calculated should be within the control of the employee.
- Workers must have all necessary tools and materials and there should be fair adjustment for failure to meet the task if the cause of failure was beyond the worker's control.

Fringe benefits have a financial value beyond the wage or salary paid because they reduce the expenses of the recipients. These include many of the items described under 'welfare' like free housing, farm produce or other rations, medical care, sick leave and paid holidays.

Non-financial incentives

It is almost impossible to assess the relative effect of financial and non-financial incentives in raising labour productivity. Non-financial incentives should induce staff to stay on the farm and make them feel part of the business. Genuine motivators are usually connected with achievements and a feeling of greater skill. There is real reward in knowing that one's work is important. Therefore, non-financial incentives should make workers more skilled and thus benefit the farmer.

What makes people feel 'bad' are background factors, like poor working conditions, job security or supervision, together called 'hygiene' factors.

Most non-financial incentives have been mentioned already. They include good working conditions, sports amenities, quick payment of wages and good communications and human relations.

Opportunity for promotion is both a financial and a prestige incentive — called 'authority pay'.

OFFICE ORGANISATION AND METHODS

The office is an integral part and information centre of the farm. The larger the farm, the more important the office becomes and the need for routines to deal with office work arises. Sadly, many farmers and managers are less familiar with office procedures than they are with other techniques.

The main function of a farm office is processing information needed for informed decision-making; to translate, co-ordinate, record, digest, correlate, display and reproduce information. The farm secretary or clerk, sometimes the farmer or his wife, should control the processing of information to decide what is needed for making informed decisions.

Other functions of a farm office are to enable the business to comply with legal and statutory requirements, especially those concerning taxes and wages; also the functions of any office, like typing and filing. A tidy mind and a tidy office are useful if the neatness is systematic; access is just as important as disposal.

Effective office work is important and the rising cost of office and administrative work is a growing concern on many farms. The aim is to improve office systems, reduce paperwork and introduce suitable mechanisation to both cut costs and provide better management information. Although the volume of output of a production department is a useful guide to efficiency, this is not so in office work. On the contrary, effectiveness is likely to be inversely related to the amount of paperwork. It is easy for systems and procedures, irrelevant to current needs, to develop and therefore there is scope on most farms for critical review of what the office produces. Paperwork should be the servant of a business — it is worthless unless it helps to promote rapid control of its main activities.

Management may not know where the *real* inefficiency in the office lies. The following basic questions should be asked of any office activity:

- Can it be eliminated?
- Does it do more than necessary for its basic purpose?
- Is the cost justified?
- Is there a simpler, cheaper alternative?
- Is there any duplication of effort?

Ideally, a farm secretary or clerk needs some knowledge of both husbandry (production) and general business management to know the sort of information that helps in making informed decisions. This knowledge is desirable in a farm secretary if initiative is to be used within the office. It is also important to know which telephone callers and visitors should be welcomed and which should be told that the boss is out.

Office layout and equipment

Much has been written about office equipment and organisation and those running large estate offices are advised to refer to this. This section, however, refers to the practical situation on most farms where the 'office', if there is one, may be in the farmhouse itself. Even in this case, there are still certain basic essentials — a working surface, a storage system for papers that cannot be dealt with at once and a filing system for storing documents that, though finished with, must be kept. A wastepaper basket is equally needed.

The best siting and layout of an office is aided by making a string diagram. Besides an efficient layout, it is important to have adequate lighting and control of temperature, ventilation and noise.

The ideal working surface is large, clear and on two levels. The lower level is correct only for typewriter use and may therefore be small. The larger main desk is slightly higher (about 72 cm) and so more comfortable for writing and using other equipment. The two parts are best fixed together at right angles with a swivel chair in the middle. Each clerk's own main desk should be about 120–150 cm long and 75 cm deep. From the one chair, all routine matters can be dealt with. The larger desk should be fitted with at least one drawer for writing materials, paper and clips.

Conditions vary so much between farms that one cannot be rigid about the relative merits of machines and equipment needed but the basic equipment needed for financial accounting has already been described. However, it is certain that use of suitable office machines can raise efficiency on any farm. The two almost essential machines are an electronic calculator and a typewriter, both of which should be carefully chosen for the work they are needed to do and for the local servicing facilities.

Small portable typewriters are probably the best value for money if use is slight. However, with greater use, full-size models, preferably electric, are more suitable. Any typewriter should have a tabulator so that the user can stop the carriage at present positions to set out work quickly and accurately. Before purchase, test the typewriter by ensuring that six layers of thin paper will come round the roller without catching, folding or wrinkling. Then type a fullstop at the top of the page, turn to the bottom and back again and make another fullstop on top of the first one to show whether or not the paper slips much during work. Typewriters used to have keys that, suitably hit, directed symbols through a ribbon onto the passing page. With electricity only a mere touch was needed and disposable carbon always guaranteed a crisp image. Now, golf balls or daisy wheels race across stationary stationery with built-in control and memory chips that determine line length and layout. It is not too great a leap to turn to a small on-farm office micro-computer. Ubiquitous silicon chips exist but they are programmed permanently to accept working instructions, in the form of software; to turn them into memory banks, accounting machines or word processors.

Electronic calculators have recently become cheaper and more versatile, therefore no farm office now should be without one. An ideal machine to meet most needs will have memories and be a print-out model, printing both the calculation and result. This enables checking and print-out slips, for example, to be put in pay packets. Accountancy machines may be worthwhile on large estates. Programmable calculators will 'remember' not only individual numbers but also a sequence of arithmetical operations that together form a mathematical formula or 'problem' to, for example, calculate internal rates of return.

On-farm microcomputers should be considered for word-processing. A word processor is a computer programme that lets its user type documents on a computer, eases editing, (corrections, additions, deletions and justification) and allows printing in a chosen style. The editing ability of a word processor allows non-typists to type and skilled typists to work faster without concern for errors. Business letters, agreements and reports can be produced attractively and professionally and stored easily and the ability to run-off extra copies equals that of a costly photocopier.

Computers are both very clever and very stupid. They are good at obeying instructions quickly and accurately. However, they need complete data and precise instructions.

Computers can be used not only for calculating but also for organising and filing data for payroll preparation, invoicing, ledger maintenance, stock control and production control. Use for these applications has led to the phrase 'electronic office!'

Dictating machines, on large farms, enable correspondence and instructions to be recorded as speech and later be transcribed by an audio-typist. Small pocket models are useful for recording work to be done during field inspections.

Machines are not the only aids to calculation. Slide rules help with small figures. Ready reckoners may be useful and in some countries clerks are fast and accurate with an abacus — a calculating frame with balls sliding on wires.

Office procedures

Have you tried to improve office procedures? Recording procedures are much easier if there is order rather than chaos in the office. On farms big enough to have an office apart from the house, it is important that it should be obvious to visitors and have a letter box for papers, like delivery notes, when it is shut.

Much paperwork may be unavoidable but, unless firm control is kept, there is a tendency for the papers circulating internally to grow each year. As many forms are used, even small improvements in design, aimed at simplifying them or reducing their cost, can raise administrative efficiency. It is easy to start a new form but most managers are reluctant to scrap an existing one.

Things are often unnecessarily written on paper. Unless it is necessary to record what has been said and the reply, it is better to use a telephone, if available. Telephone messages received should be noted down in some detail, including date and time.

Incoming papers and post should be sorted and grouped methodically in the office so that they can be found easily and quickly. When papers arrive by post or with goods on the farm, they must sometimes be put aside, after checking, until it is time for further action. Rather than let them grow into a muddled heap, it is better to sort them at once into wire trays. The need for permanency and ease of retrieval are the two main criteria for designing a record-keeping system.

A farm diary is a useful, multi-purpose record book. If it is large enough, it can be used to note important points of telephone calls, like prices quoted. It can be used as a first record of many other items before they are made more permanent; for example, rainfall, deliveries, dispatches, prices and key future dates. Two alternatives to the usual type of desk diary can be very useful. One is a wall diary, with a space for each day, covering three, six or twelve months. The other, for busy offices, is a loose-leaf diary with a page for each day. The advantage of this is that, as well as reminders, letters and papers can be inserted in the appropriate days in the future, thus avoiding forgetting them or having to search for them later. This sort of diary can be continuous; it can cover more or less than 12 months and pages can be added and withdrawn every month or so.

If an invoice may be left unpaid for some time without loss of discount, and is therefore a source of credit, it should be filed so that it is paid as late as possible to get maximum credit without losing discount. Provided that the diary is big enough, it is best 'filed' there as it can be fixed by paper clip to the space for the week before the one when it must be paid.

Goods sold should be invoiced as soon as they leave the farm. The date of dispatch, type and number of goods, price agreed and name of the buyer should be noted to ensure that income is not lost by forgetting a dispatch. A triplicate invoice book helps in this control. The top copy is sent to the consignee, the second is kept in the office until payment has been made and the third stays as a permanent record in the invoice book. The book copy is marked 'paid' when payment is received. The pile of outstanding invoices filed in a 'desk tidy' or in a bulldog clip hanging from the wall is a quick check on debtors at any time.

Written communications

It is well-known that some managers seem to spend all their time writing whereas others seem to get by without ever putting pen to paper. Probably the best course is to use the telephone, within reason, but to follow up the call with a written memo, if a decision has been made that must not be overlooked or misunderstood.

The more formal the organisation, the more communications will be written rather than spoken. Written communications provide a record that can be referred to later. The written word is likely to be more carefully considered than the spoken, so it encourages people to think before they write. Written communications can take less time than spoken ones, depending both on what must be communicated and to whom.

Sadly, many managers do not express themselves clearly in writing and misunderstandings are harder to correct than when speaking. Many people use too much jargon and seem to write to impress rather than to inform readers. Many are too verbose. Why write 'facilitate', 'utilise', 'necessitate' and 'located in the vicinity of' instead of 'ease', 'use', 'need' and 'near'? The main principle of writing is that the words used should convey to the reader the meaning of the writer. Have something to say and say it as clearly and briefly as you can. That is the only secret of style. The language should be basically that of normal speech.

Correspondence

There should be two or three trays for incoming correspondence. All unopened letters (or letters opened by a secretary but not yet seen by the manager), unchecked delivery notes and miscellaneous papers should be placed in the 'in' or no. 1 tray. Once read and considered any letters or other material of no further use may be destroyed; those needing later attention should be placed in the 'pending' or no. 2 tray and others may go straight into the 'filing' or no. 3 tray and be filed immediately. The contents of the 'pending' tray, when dealt with, are similarly either placed in the filing tray or filed at once.

Letters remain the main way of conducting business. Improving letter quality means producing plain letters — written clearly, well set out and easy to read.

Those without a typist and whose own typing skill is restricted to a two-finger exercise must write letters by hand. This needs a large (A5 or A4) duplicate or triplicate book, rubber-stamped, embossed or printed with the farm name and address. Use of carbon paper ensures that a copy is kept. Most letters, however, are typed.

Whether typed or handwritten, copies of outgoing letters should always be kept. Make two copies of each letter when typing; one preferably on coloured paper. The first copy on, say, white paper is immediately filed or put in the filing tray. The second, on coloured paper, can be kept in the pending tray or in a desk tidy until a reply is received or the matter otherwise dealt with. This system ensures that all outstanding correspondence is kept in mind until settled. If a copy is found on the floor or amongst other papers, it is easy to see if it is a file or a pending copy.

Finished letters should either be left for the manager's signature or signed 'p.p.' (per pro — on behalf of) the manager by his secretary. The alternative of signing 'Nkuku Estates Ltd' is too impersonal. Whatever system is used, whoever signs the original initials the copies.

Report writing

A report is a document in which a given subject is examined to convey information, report findings, put forward ideas and, sometimes, make recommendations. It is of a scientific nature, either investigatory or explanatory, and should exactly describe the subject to its intended readers.

In business, routine and special reports are prepared for many reasons. The task of writing reports is probably secondary to the work that demands them. Nevertheless, reports are an essential part of communications in business because they contain both records of the facts of that work and the thinking behind it.

Although information may be gathered by office clerks, the job of writing the report, or at least of editing the final draft, usually falls on the manager.

Preparing reports is an art that can be studied and developed. Information and recommendations placed before top management have little value unless they help in forming judgements and lead to decisions. A report that does not stimulate thought and lead to action may be of passing interest but it serves no useful purpose. It is unlikely to warrant the cost of preparing it.

A report should contain the following: title, table of contents, summary, introduction, main text, conclusions and recommendations, references and appendices. Conclusions should be supported by the evidence presented; otherwise they are merely opinions. The four stages of report writing are preparation, arrangement, writing and revision.

Preparation means deciding why and to whom you are writing, what your subject is and how best to present it. Note down all your facts and ideas. If any main divisions of your subject are obvious, you can allot a separate page or index card to each.

The arrangement should help readers to understand the main points as quickly and easily as possible. Consider your collected facts and ideas. Decide your main divisions and the order in which to present them. Within each division, arrange your material in an easily followed order. Also consider using illustrations to supplement or replace words and whether any factual detail can be removed to appendices.

When writing, simplicity of style and language, short sentences and a logical sequence of thought are essential. The style should vary according to the needs of the readers and the nature of the material.

- Use a high proportion of active verbs, e.g. 'Elias weeded the millet', rather than 'The millet was weeded by Elias'.
- Use concrete rather than abstract nouns where possible.
- Use adjectives sparingly and with care; 'few', 'many' or 'six' buying points is less vague than 'a number of' buying points.
- Use short words if possible and avoid verbosity. If your writing contains more than ten per cent of three or more syllable words you are using too many long words. Why say 'in the vicinity of' instead of 'near' or 'substantiate' instead of 'confirm'?

- Add together your average words/sentences and the percentage of three or more syllable words. If the sum exceeds 30, edit to simplify and thus clarify.

The first and sometimes the second sentence of each paragraph should indicate the idea to be described in it and act as linking sentences between topics. If you know your readers personally, you should know their tastes well enough to avoid irritating them and to judge the right level at which to write — how technical, how informal you can be. However, always write to express, not to impress.

After finishing your rough draft, lay it aside for a few days if time allows. Few people can write a good report at the first attempt. Later re-read it objectively as if it were someone else's work. It is surprising how many needless words can be removed by careful editing. As it is easy to overlook mistakes and clumsy expressions written by oneself, it is worth asking a qualified person to criticise the draft after revision. Your letters and reports are ambassadors both for yourself and your organisation.

Information Filing and Retrieval

This is the process of arranging and storing documents in an orderly way so that they are easily found when needed. Although time-consuming, efficient filing saves time in the long run. Filing is basically a data-retrieval system and the really important part is to be able to find what you want quickly. Consequently, it is wrong to let typists and clerks work out their own system, which only they understand.

An efficient filing system can often improve administrative efficiency and also reduce office costs, time and space. This is one in which the

- records needed always can be found quickly,
- records kept for reference are kept safely,
- cost of installing and maintaining the system is reasonable for the service needed.

Ideally, file classification should relate as closely as possible to the headings in the cash analysis book. Cross-indexing is needed for documents relating to more than one file heading.

Initially, documents are placed into a tray for short-term filing. Longer-term filing needs plain manilla folders. On small farms, box files or concertina files, with pockets lettered in alphabetical order, may be adequate. The old-fashioned way of impaling papers on a spike has little to commend it; the papers collect dust and soon tear.

Various other systems are available from lever-arch files to vertical or horizontal filing cabinets. Vertical cabinets have drawers with suspension rails and labelled suspension units into which the manilla folders are placed.

The disadvantages of drawers are that they need twice the space that they occupy as the drawers must be pulled out, the contents of only one drawer can be seen at a time and, if too many drawers are opened, a multi-drawer vertical cabinet can fall over. All these disadvantages are avoided with a lateral filing cabinet. This slightly less familiar item is a tall cupboard with no front. Either rails or shelves are fitted from side to side at various heights. Special suspension units clip into or onto the rails. Thus each rail carries a complete 'drawer-full' of files, permanently displayed laterally.

Specialist filing systems may be needed for some farm physical records. Card index boxes are familiar enough but a visit to any office equipment supplier will reveal a wide selection of storage methods, like circular carriers and special drawer units in which each card is held in a flip-back holder to help find the needed ones and make entries without changing the order of the cards.

Receipts should be kept for several years in case any dispute over payment arises. Invoices, receipts and payments should be kept in separate, neatly labelled files.

An essential part of an efficient filing system is a suitable coding or indexing system so that records are easily found. The index may be kept apart from the records to which it refers or the records themselves may be so arranged as to be self-indexing.

Simple numerical filing can cope with any need but becomes rather unwieldy if the farm is complex as there is probably no obvious logic in a series of file numbers. This worsens over time as some files die and their numbers are reallocated to new subjects. It is better to use a key consisting of a code of letters and numbers and issue this to all staff who will use the system, for example:

A1 Agricultural returns	B1 Beef — purchases
A2 Mortgage	B2 Beef — records
A3 Bank loan	B3 Beef — diseases
	B4 Beef — sales
	etc.

C1 Crops — seeds
C2 Crops — pesticides
C3 Crops — field machinery
C4 Crops — insurances

The category 'Miscellaneous' should be avoided. There are three alternatives for a document that might otherwise be regarded as a miscellaneous item. One is to throw it away. The second is to start a new file for it. The other is to insert it in a file on a closely related topic.

When you must open a new file, it can be put into the suitable group rather than at the end of the list. Filing is then simple and understood by even the most junior office staff. All that is needed is to mark each incoming paper to be filed with a pencil note, for example, C4. Outgoing letters should be referenced with the relevant filing code. In this way the reply will have its filing code already on it.

No filing system has infinite capacity. Nor is it desirable that individual files should get so fat that they finally break under the excessive pressure of squashing them into a box, drawer or cabinet. Once a year, soon after valuation day, out-dated papers should be removed and either destroyed or stored elsewhere in neatly tied and labelled archives.

GLOSSARY

Executive once a high status word for top managers, it now tends to include anyone who supervises anyone else's work. You no longer have to be an executive to drive an executive car or even to read this — an executive book.

Fringe benefits rewards for employment *beyond* the wage or salary paid. They may include pension arrangements, subsidised meals and provision of housing or goods either free or at a discount.

Labour turnover the rate at which workers leave a firm.

Overtime time worked in excess of a standard time laid down in conditions of employment.

Payment in kind payment in goods or services *instead* of money.

RECOMMENDED READING

Books

Argyle, M. (ed.), *Social Skills and Works*, (Methuen, 1981).

Barnard, C.S. and Nix, J.S., *Farm Planning and Control*, 2nd ed. (C. V. P., 1979).

Bartlett, A., *Effective Team Buildings, Notes for Managers* No. 25 (Industrial Society, 1974).

Berry, T.E., *The Craft of Writing* (McGraw-Hill, 1974).

Blake, R.R. and Mouton J.S., *The Managerial Grid III* (Gulf Publishing Co., 1985).

Bradford, L.P., *Making Meetings Work*, (University Associates, 1976).

Brech, E.F.1. (ed.), *The Principles and Practice of Management*, 3rd ed. (Longman, 1975).

Castle, E.N., Becker, M.H. nd Smith, F.J., *Farm Business Management*, 2nd ed. (Macmillan, 1972).

Deverell, C.S., *Business Administration and Mangement*, 2nd ed. (Gee, 1972).

Forrest, a., *Delegation, Notes for Managers* No. 19 (Industrial Society, 1972).

Foxen, T. & Peck, T., *Supervisors: their Selection, Training, Development. Notes for Managers* No. 28 (Industrial Society, 1976).

Foxen, T., and Smith, E., *The Manager as a Leader. Notes for Managers* No. 14 (Industrial Society, 1976).

Herzberg, F., *Work and the Nature of Man* (Staples Press, 1968).

Hosken, M., *The Farm Office*, 3rd. ed. (Farming Press, 1982).

Humphrey, P., *How to be your own Personnel Manager* (Institute of Personnel Management, 1981).

Kempener, T., (ed.) *A Handbook of Management* (Penguin, 1976).

Kenny, J., Donnelly, E., and Reid, M., *Manpower Training and Development* (Institute of Personnel Planning, 1979).

Koontz, H., and O'Donnell, C., *Principles of Management*, 5th ed. (McGraw-Hill, 1972).

Leeds, C.A., and Stainton, R.S., *Management and Business Studies* (Macdonald & Evans, 1974).

McFarland, D.E., *Management Principles & Practices*, 4th ed. (Collier-Macmillan, 1974).

McGregor D., *The Human Side of Enterprise* (McGraw-Hill, 1961).

Maslow, A.H., *Motivation and Personality* (Harper & Row, 1970).

Prior, P.J., *Leadership is not a Bowler Hat* (David & Charles, 1977).

Pryse, B.E., *Successful Communication in Business* (Blackwell, 1981).

Sargent, A., *Decision-taking. Notes for Managers* No. 29 (Industrial Society, 1976).

——, *Effective Supervision in a Factory. Notes for Managers* No. 5 (Industrial Society, 1976).

Standingford, O., *Simplifying Office Work,* 2nd ed. (Pitman, 1974).

Turrell, M., *Training Analysis* (Macdonald & Evans, 1980.

Vroom, V.H., Deci, E.L., (eds), *Management and Motivation* (Penguin, 1970).

Wilson, B., and Macpherson, G., *Computers in Farm Management* (Northwood Publications, 1982).

——, *Micro-computers on the Farm,* (M.A.F.F.(U.K.), 1981).

Periodicals

Armstrong, J., 'The Role of Authority in Enlightened Man Management', *Farm Management* (1975), 2, 12.

Errington, A., The Delegation of Decision-Taking', *Farm Management* (1985), 5, 10.

——, 'Farmers' Attitudes to Delegation' *Farm Management* (1985), 5, 12.

Giles, A.K., 'Delegation', *Farm Management* (1984) 5,6.

Harrison, S.R. and Course, C.P., 'Word-processing in the Farm Office, *Farm Management* (1985) 5,9.

Hebden, J., & Shaw, G., 'Pitfalls of Participation' *Management Today* (December, 1975).

Johns, E., 'How to Delegate', *Management Today* (June, 1975)

Meyler, S., 'Assistant Managers Need Training Too', *Farm Management* (1981), 4,6.

du Toit, F.P., 'The Accomodation of Permanent Farm Labourers', *Rhod. Agric. J. Tech. Bull. 17* (1977)

EXERCISE 7.1

Managerial styles

This exercise is designed to help you find the type of managerial style typical of yourself.

Answer all the questions according to your opinions and your ways of handling the situations described. If some of them do not apply to you directly, answer them on the basis of past experience or how you think you would behave.

Answer the questions as carefully and honestly as you can. There are no 'right' or 'wrong' answers. Different managerial styles can be equally effective.

For each of the statements below, please tick:

SA   if you STRONGLY AGREE — that is, if the statement definitely expresses your opinion.

TA   if you TEND TO AGREE — that is, if you are not quite sure but think that the statement expresses your opinion.

TD   if you TEND TO DISAGREE — that is, if you are not quite sure but think the statement does *not* express your opinion.

SD   if you STRONGLY DISAGREE — that is, if the statement definitely does *not* express your opinion.

1.   I have succeeded in creating a feeling of need to continually improve personal and group performance.   SA TA TD SD

2.   I do not bother much about organisation and authority but concentrate instead on getting the right people together to do the job.   SA TA TD SD

3.   A lot of my managing is through informal, unplanned friendly talks.   SA TA TD SD

4.   At least once a week I choose someone doing a good job and praise him personally.   SA TA TD SD

5.   I think it is very important to have a free and easy atmosphere and not push people too hard.   SA TA TD SD

6. To be an effective manager you must keep somewhat apart from your junior staff and avoid becoming too friendly with them.   SA TA TD SD
_____

7. I dislike rules and procedures and remove them whenever possible.   SA TA TD SD
_____

8. I expect people to check with me before making decisions.   SA TA TD SD
_____

9. I believe a manager should make efforts to form friendly relationships with his staff.   SA TA TD SD
_____

10. In making promotions or salary rises, I think it is important to consider the person's length of service and financial state.   SA TA TD SD
_____

11. I strongly encourage my staff to try to solve problems by themselves even if they make a few mistakes.   SA TA TD SD
_____

12. I like to maintain a business-like atmosphere so there must not be much social meeting or gossip on the job.   SA TA TD SD
_____

13. I set very high standards for performance   SA TA TD SD
_____

14. I like to ensure that all tasks are clearly defined and reasonably combined.   SA TA TD SD
_____

15. I resent staff checking everything with me. If they think they have got the right approach, they should go ahead.   SA TA TD SD
_____

16. If one of my staff makes a mistake, I will definitely criticise him.   SA TA TD SD
_____

17. I have had to set up some standard practices and procedures to keep the organisation orderly and effective.   SA TA TD SD
_____

18. I reward people according to the standard of their work and ignore their seniority and whether I personally like them or not.   SA TA TD SD
_____

19. I do not rely too much on others' estimates and opinions and check almost everything myself.   SA TA TD SD
_____

20. There is little I can do with people who do not have pride in producing good results.   SA TA TD SD
_____

8 Production planning

LAND USE

Current technical progress in farming is increasing the need for more efficient production management. This includes the whole process of manufacture and means producing the right goods at the right time for the right markets. To achieve this perfection, optimum use of machines, men and materials must be made.

Various techniques exist for production planning and control, ranging from calculations on the backs of envelopes and the use of graphs, charts and tables to network analysis, scheduling routines and linear programming. Each is useful in certain situations, according to the size of the business and the period for which plans are made.

As described before, intensity of land use depends on the combined total of capital, labour and managerial skill applied to each unit of land. In areas where animal draught prevails over motor vehicles for transport, intensity of land use in commercial farming tends to fall with growing distance from the market.

Diversified or specialised land use

There has been much debate for years over whether farmers should diversify or specialise. Peasants tend to be highly self-sufficient and diversified whereas commercial farmers have moved more towards specialisation and in many areas tend to concentrate mainly on one product.

Diversification occurs when a firm produces a new product without ceasing producing of any existing one. It may be either horizontal or vertical. But the word usually refers to horizontal diversification or production of several articles.

Vertical diversification exists if many steps in producing a given product take place on the farm. A vegetable grower with a retail farm store has much vertical diversification. He may not only grow vegetables but also clean and pack them and deliver them to consumers under contract.

The main advantage of diversification is that it spreads risk by providing a hedge against variations in output, cash income, consumption and savings. It also spreads income and the use of labour and machinery over the year and can reduce working capital needs.

Farmers the world over favour diversification in the face of uncertainly. In subsistence farming such diversification extends to mixtures and combinations of crops as well as of livestock enterprises. Many progressive small farmers have non-farming businesses like groceries. Although diversification seems to reduce average annual profit, it raises the stability of that profits so subsistence farmers tend to select enterprises that they think are most stable. Those with a stable main enterprise can afford high-risk ones, raising their profits compared with farmers with a higher-risk staple who must play safe.

If the yield or price of a product falls much for any reason, it is likely to be disastrous if there are several enterprises on the farm rather than one. Some years are bad for some products but good for others.

Beside combating uncertainty, diversification enables spare management and other resources to be used productively and provides more scope for managers to use their skill and knowledge. Different crops and livestock are often complementary in their labour and specialised machinery needs over time and by-products of one enterpise may be useful inputs for another. Complementary and supplementary relationships tend to favour diversification and, with it, crop rotation. It may also result from some specific opportunity that, though not necessarily planned for, may seem too good to miss. Also, many farms have differences in soils and topography and distances from buildings and water supplies that encourage their having several enterprises.

Diversification of production is hard if few enterprises are suited to local ecological conditions. It also has some disadvantages. There is a point of complexity beyond which management is spread too thinly and nothing is done well. The less diverse a business, the more manageable it is and the fewer things can go wrong. Over-diversification, sometimes called 'mixed farming and muddled thinking', may cause the manager to be a 'jack of all trades and master of none'. It is seldom easy to manage three or four enterprises, like cotton, tobacco, dairying and pig production, that all need a high standard of management and are fairly labour-intensive. When farm profits fall, farmers often add another enterprise. It is usually more effective either to improve the existing ones or to take the more drastic step of removing those that give little or no profit. More than two or three enterprises to little to reduce risk.

Specialisation means intensifying effort on one or two enterprises and has advanatages. With a wide range of products, most profit probably comes from only a few of them and, among the rest, several will either be only slightly profitable or be losing money. A product reduction programme may be worthwhile.

Simple farming systems usually earn the most profit. Few farmers or managers are skilled in more than a few enterprises so they should concentrate on those in which they have most skill. Concentrating managerial skill enables managers to be more alert to new technical developments and there is less waste of effort and materials as specialised men and equipment can be fully used. The advantages of specialisation are rising as skilled labour and specialised machinery and equipment become more costly and need a large output to justify their cost by obtaining economies of size. By specialisation, the special comparative advantages of both the farm and the farmer or manager can be fully exploited and it should be possible to benefit by bulk buying and selling.

With livestock, labour use is usually fairly even over the year. When this is so, one livestock enterprise can use the stockmen's time efficiently without any need for diversfication and they become more expert. For this reason, single-

enterprise ranches are more common than single-enterprise crop farms.

The main disadvantage of specialisation is greater uncertainty owing to market changes, weather, pests and diseases. Some risks can be insured against but others can be crippling.

The trend on commercial farms now is to increase specialisation at the expense of both vertical and horizontal diversification. Scientific advances, like availability of fertilisers, pesticides and herbicides, have reduced the need for crop rotations so that, even if there is much diversity within a district, most farms are tending towards specialisation. Farmers are freer to choose enterprises for their profitability than they used to be.

Present thinking is that commercial farmers with limited capital should equip adequately to produce a few products efficiently rather than inadequately for many products. It is usually best to specialise in one or two main enterprises each of a reasonable size, with one or two supplementary ones to use surpluses without additing much to common costs. This is called *simplification*, variety reduction or rationalisation. Concentration of effort on one or two main enterprises, with the typical efficiency that follows this, is especially important to a small commercial farmer doing manual work himself. By limiting his field, he is more likely to be able keep up with technical advances.

In summary, the arguments for and against diversification or specialisation basically revolve around the farmer himself and his ability to control the feasible enterprises on his farm to produce a sufficient gross margin to cover his common costs and leave a reasonable profit. The questions to ask are: 'What is the *least* diversification the business needs to attain its objectives and stay viable and prosperous?' Also, 'What is the *most* diversification we could manage; the most complexity we could bear?' Some diversity is justified, especially in fruit and vegetable production which is characterised by sharp seasonal price and yield changes. The aim is to remove wasteful unprofitable variety; not to remove it completely. However, if you put all your eggs into one basket, look after that basket.

SELECTION OF PRODUCTION METHODS

The most useful techniques for choosing production methods, as against farming systems, are partial budgeting and breakeven analysis. Partial budgets were described in detail in chapter 4 and breakeven analysis was also briefly described there with regard to its use in helping to decide when to replace machines. A fuller description of breakeven analysis is now given in relation to its use in choosing production methods.

Breakeven analysis in production planning

Some farmers are now facing the problems raised by growth through greater sales or extra products. Others are facing problems of contraction owing to the introduction of synthetic materials like textile fibres. In each case it is vital that management is ready to cope with these changing scales of activity. Breakeven analysis is a specific application of marginal costing and requires costs to be classified as either fixed or variable. Information must be available on the forecasted sales, fixed and variable costs for the period. Graphical representation of these variables on a breakeven chart helps to show clearly the relationship between a firm's costs, scale of activity and profit.

A breakeven chart is an elementary form of profit graph that is easy to prepare and understand. It enables the important aspects of a situation to be grasped quickly and is thus a useful planning tool, giving valuable information for management decisions. This technique requires only basic knowledge of costs and income over a range of scales of activity but it helps management to see the whole picture of cost: income: profit relationships without looking too closely at all the components.

In practice, most managers suffer from a wish to see all the available facts and call for all the detailed costs behind breakeven charts until they become confused by detail. Such a habit should be firmly resisted; only the *relevant* facts are needed for a rational decision and it is rarely necessary to know all cost details.

The conventional layout for a breakeven chart is shown in figure 8.1. A graph is drawn, to a suitable scale, recording of activity (in, say, hectares or tonnes) on the horizontal axis. Costs and income are given, at a suitable scale, on the vertical axis that represents fixed cost. Income arising from each scale of activity appears as a diagonal line from the point of origin (zero). It is conventionally assumed that all variables are linear because common costs, variable cost of each unit of scale of activity and the price received for each unit of output are constant.

In figure 8.1, it is assumed that fixed costs are $5,000, variable cost are $100/tonne sold,

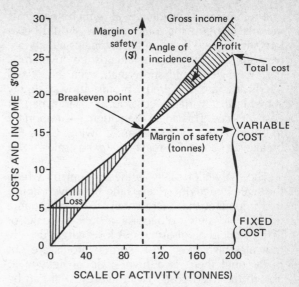

Figure 8.1   A Conventional Breakeven Chart

maximum possible production is 200 tonnes and price/tonne is $150.

The breakeven point is that scale of activity where income equals total cost, so no profit or loss is made. It occurs where the total cost and gross income lines intersect. Below the breakeven point, the shaded area represents loss as total costs exceed total income. Beyond the breakeven point, the shaded area represents profit as total income exceeds total costs. Profit or loss at any given scale of activity is shown by the vertical distance between the gross income and total cost lines.

An alternative to finding the breakeven point graphically is simple arithmetic. In figure 8.1 the breakeven point is 100 tonnes. At this scale of activity, the gross margin (contribution) is $30,000 less $25,000 that is $5000. This is exactly equal to the fixed costs. A characteristic of the breakeven point is that gross margin is just enough to meet all the fixed costs for the period under review and there is no profit or loss.

If gross margin = fixed cost
We know, gross margin = gross income − variable costs
Then,
gross income − variable costs = fixed costs
So, gross income = fixed costs + variable costs
= total costs

The breakeven level of activity can be calculated by formulae, for example,

Breakeven sales revenue = fixed costs/1 − (variable cost/unit/ selling price/unit)
= 5000/1 − (100/150) = $15,000
Breakeven sales volume = fixed costs/gross margin per unit
= 5000/50 = 100 tonnes

The angle of incidence at which the gross income line cuts the total cost line should be as large as possible because this indicates a high rate of profit once the fixed overheads are covered. A narrow angle would show that, even after fixed overheads are covered, profits grows slowly showing that variable costs form a large part of costs of sales. This angle of incidence is important in boom times when sales are growing. Once the breakeven point is reached, extra sales raise profit.

The margin of safety is the amount by which the scale of activity exceeds the breakeven point. It is important that there should be a reasonable margin of safety, otherwise a fall in scale of activity could be disastrous. A low margin usually indicates high fixed costs so that profits are not made until there is a high scale of activity to absorb the fixed costs. From figure 8.1 it can be seen that the margin of safety is 100 (200 − 100) tonnes of sales; therefore the difference is 100 tonnes at $150/tonne, that is, $15,000. The margin of safety is important in times of depression when sales are falling. The more the margin of safety, the more sales can fall before the breakeven point is reached. Once this point is passed, however, a loss will result. It can be expressed by the following equation indicating the percentage by which the scale of activity could be reduced before a loss is incurred. This equation is:

1 − (breakeven level/planned level) × 100%
or, from figure 8.1, 1 − (100/200) × 100%
= 50%

Let us look at a practical example of the use of breakeven analysis. Table 8.1 has been compiled over the years by a pig breeder who sells weaners and wants to calculate the breakeven number of piglets that must be weaned from each sow each year.

Table 8.1 shows that it is necessary to sell between 15 and 16 weaners each year to cover the fixed cost of keeping each sow. This however, could have been calculated more easily and precisely by finding the breakeven number of weaners whose total gross margin equals the fixed cost of a sow, as follows:

Breakeven point in units = fixed costs/(gross income per unit − variable costs per unit)
= 113/(17.22 − 9.72)
= 15.1 weaners/sow/year.

So far we have assumed that all the functions in a breakeven chart are straight lines and are linearly dependent upon scale of activity. This is seldom true and is a gross over-simplification of real life. In practice, the fixed costs are rarely constant through the whole range of activity levels. Moreover, variable costs per unit may fall at certain levels. Moreover, variable costs per unit may fall at certain levels of activity through quantity discounts on bulk purchases of inputs. Again, prices may have to be reduced to obtain large sales. In reality there are curves, steps and different gradients at different levels of activity if one studies a wide enough range of levels. Consequently, there can be more than one breakeven point; as shown in figure 8.2. However, the model is seldom used in this unwieldly form. Instead, managers have found that they can derive acceptable day-to-day decisions from the simple form. It is known not to behave exactly like real life but its accuracy is adequate usually. It is often 'good enough'.

Table 8.1   Sow Breeding Herd — Output and Profits

	Number of weaners/sow/year					
	12	13	14	15	16	17
Fixed costs/sow/year:	$113	$113	$113	$113	$113	$113
Per weaner:						
Fixed costs of sow	9.45	8.72	8.10	7.56	7.09	6.67
Variable costs	9.72	9.72	9.72	9.72	9.72	9.72
Total cost	19.17	18.44	17.82	17.28	16.81	16.39
Income	17.22	17.22	17.22	17.22	17.22	17.22
Profit	(1.95)	(1.22)	(0.60)	(0.06)	0.41	0.83
Profit/sow	(23.40)	(15.86)	(8.40)	(0.90)	6.56	14.11

This situation is typical when a small enterprise is expanded by installing more capital equipment. Profits disappear and can be regained only by much expansion of output. Straight line relationships are valid only within a limited scale of activity.

Figure 8.3 shows figure 8.1 redrawn to show more clearly how the gross margin (contribution) goes first towards covering fixed costs and then, at the breakeven point, the gross margin becomes all profit. This better reveals the effect of fixed overheads on the breakeven scale of activity. For any scale of activity, the gross margin (contribution) can be seen.

Breakeven charts help in relating costs and profits to scale of activity and they are widely used to get a quick idea of the level of production needed for a firm to break even. They help managers to understand the effect of volume on profit.

Breakeven analysis can provide answers to such questions as below:

- What are the likely effects of a change in price?
- What are those of changing the product-mix?
- What are the likely effects on profit if either fixed or variable costs change?
- At which scale of activity does the firm begin to lose?
- By how much can the scale fall before the firm starts to lose?

It shows management clearly what the quantitative effects of decisions could be. For example, breakeven analysis can be used to estimate what level of ouptut is needed to meet known or esti-

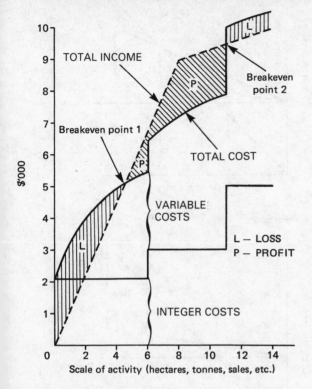

Figure 8.2   A Realistic Breakeven Chart for a Wide Range of Scale of Activity

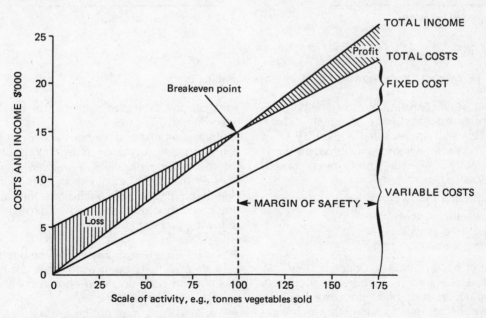

Figure 8.3   A Contribution Breakeven Chart

1. Raise selling price

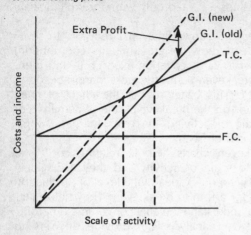

2. Decrease variable costs

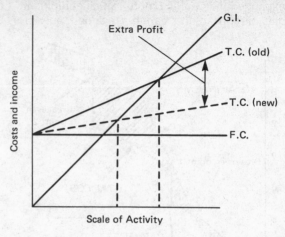

2. Decrease fixed costs

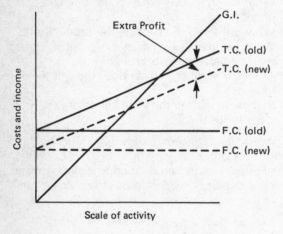

4. Raise output

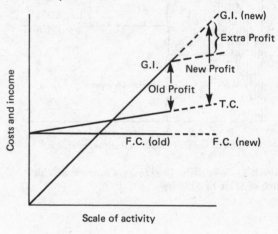

G.I. Gross Income   T.C. Total Cost   F.C. Fixed Cost

Figure 8.4   Breakeven Charts Showing the Main Ways of Raising Profit

mated costs or what level of costs can be borne in the light of an estimated scale of activity.

Breakeven analysis provides a useful way of studying the profit determinants of a business. There are four ways in which profit can be raised, or loss decreased, namely:

- raise the selling price of product(s);
- decrease variable costs;
- decrease fixed costs;
- raise the volume of output.

The effect of each of these changes is shown in figure 8.4. As some farms have heavy overheads and sell products with high gross margins and others have the reverse situation, this technique helps to determine which type of situation pre-

vails on any farm and strategy can be directed accordingly.

Breakeven charts have been shown to provide a useful guide to managers in making decisions on cost-volume problems. However, it must be stressed that there are limits to their usefulness. Before jumping to any conclusions, the validity of any chart should be assessed in the light of the following points:

- Charts are relevant only within the limits of the scale of activity upon which they are based.
- Common costs are fixed only in the short run and within a given range.
- Variable costs may not be linear.
- The income curve may bend a little.

- The relevant time span affects the chart. It is a static picture of a dynamic system and therefore should be revised regularly. Breakeven charts can normally be used for short periods only and usually to check plans rather than to guide large decisions.
- The importance of capital employed and of cash flows is not ignored.
- If the farm sells more than one product, whole farm breakeven charts will show costs and income that do not represent any one product. It may be necessary to draw a breakeven chart for each product or group of products.

Budgets are no more reliable than the data used in making them. Unfortunately, we are often unsure about an important budget item — usually a yield or a price — so that an ordinary partial budget may be hard to produce. Breakeven budgeting is especially useful when a single item in a partial budget is uncertain. The method is applied by representing the uncertain figure with a symbol, say Y, and completing the rest of the partial budget in the usual way. The rise or fall in profit is assumed to be nil; that is, it is assumed that the change just breaks even so that the two sides of the budget equate and the breakeven value for the uncertain figure can be calculated. A manager can then judge his chance of achieving this figure of, say, yield or price in considering the introduction of a new method or enterprise. He may know enough to be sure that the level likely to be achieved is well above or below the breakeven amount and thus be able to decide whether or not the proposed change is worthwhile. Let us look at a simple example.

Table 4.20 in chapter 4 was a partial budget to estimate the effect of substituting a hectare of tobacco for a hectare of maize. Let us assume that the farmer wants to find the breakeven yield of tobacco and the breakeven cost of tobacco fertiliser. These are done separately in tables 8.2 and 8.3 as only one variable can be calculated at a time.

FARM PLANNING

Farm planning involves allocating limited resources among alternative opportunities in order to attain the bounded objective of profit maximisation. It follows therefore that any planning procedure must contain three elements:

- one or more objectives;

- the limitations and possibilities of the available resources;
- alternative ways of combining the available resources to reach the objective(s).

Traditional technical farm planning stressed output and so tended to ignore the level of resource inputs, apart from land, available. Whilst land can be limiting, seasonal labour availability is so more often in many subsistence farming systems and the ratio of labour supply to tillage area can alter optimum farm plans greatly.

Farm planning is a specialised aspect of land-use planning in which economics is highly important. It is the most complex aspect of the work of farmers and their advisers as it must combine all the various financial and technical facts together.

Table 8.2   Breakeven Budget to Estimate Breakeven Yield of Tobacco Needed

Income lost	$	New income	$
Maize	187.20	Y kg tobacco at 50¢	0.5Y
New costs		*Costs saved*	
Fertiliser	73.50	Seed	4.50
Labour	54.00	Fertiliser	37.50
Barn deprec.	24.50	Fumigant	3.30
	= 339.20		45.30+0.5Y

Now, at the breakeven point there will be no profit or loss from the proposed change.

So 45.3 + 0.5Y = 339.2
Therefore   0.5Y = 293.9
And   Y = 588 kg tobacco/ha

Table 8.3   Breakeven Budget to Estimate the Breakeven Cost of Tobacco Fertiliser

Income lost	$	New income	$
Maize	187.20	Tobacco	450.00
New costs		*Costs saved*	
Fertiliser —		Seed	4.50
7 bags at P	7P	Fertiliser	37.50
Labour	54.00	Fumigant	3.30
Barn deprec.	24.50		
	26.570+7P		= 495.30

So 265.7 + 7P = 495.3
Therefore   7P = 229.6
And   P = $32.80/bag of tobacco
                      fertiliser

Owing to the rising commercialisation of farming and the growing number of progressive farmers, individual farm planning is both relevant and necessary, especially if credit is used and in countries heavily dependent upon farming. Although farm planning is often rejected by technical agriculturists, who do not understand how relatively simple, sensible and essential it is, progressive farmers spend much time comparing possible alternative plans and practices.

The object of farm planning is to deliberately select and develop the best steps to achieve a predetermined result. The result is usually to enable the farmer to obtain the greatest return, in the form of cash and food, for his efforts, both in the short and the long term. Plans must be long term and therefore need technically good methods that will maintain or, better, raise soil fertility but they should allow some flexibility. The plan should be more than a mere guideline, but not so rigid that business initiative is stifled. The search is not just for a profitable plan but for *the* one that will attain the farmer's objectives from his available resources, combined with technically good methods. However, objectives are sometimes constrained by a need to conform to social forces.

When replanning is needed, simple techniques — such as partial budgeting — usually suffice. These are readily understood by both farmers and advisers. More intricate techniques should be used when relevant for commercial farmers or for indicative regional or district planning models for target groups of peasant farmers.

Farm plans should be guided by the goals of the family, be geared to the peculiarities of the farm, the time, capital and skill that is available for farm work and the risk the farmer and his family are prepared to a bear. They are most effective when combined with a package of inputs; like water supplies, roads, extension, fertiliser, seeds and biocides. A farm business advisory service, capable of helping both to diagnose the business and to construct plans, is urgently needed in most tropical countries — especially for the smallholder farming sector.

If a farm plan is prepared for a farmer by an outside adviser, it is important that their personal relationship is good. The adviser should hold initial discussions with the farmer and, if possible, walk around the farm with him to get as much information as possible about both the farm and the farmer. The success of the plan will depend upon the rapport established between planner and farmer. The farmer should trust the planner and have confidence in the advice given. On this will depend the farmer's co-operation and his willingness to tell the planner what he really thinks of his ideas. Farmers should be involved as much as possible in the planning process. They should feel committed to the final plan; otherwise it will be simply a 'paper plan' and not be executed. They should be visited often during the early stages of execution because many small problems can be solved by careful advice.

Farm planning means the whole farm approach rather than the single enterprise, technical approach. Each enterprise has a different need for resources and these needs may vary at different times of the year. Attention often has been drawn to the need to regard a farm as a unit for management purposes. Changing farm circumstances do not make this concept any less valid. Farmers are ultimately interested in the profit from their entire farms as they live on it and only from profit can they assess the returns they are getting from their investment of capital, skill and labour and from the risks they take.

A farm plan can help a farmer and his family to do the following:

- choose activities best suited to their conditions;
- look farther ahead than they otherwise would;
- choose an enterprise combination (product-mix) that results in better use of resources;
- time the main jobs so that they do not clash with each other;
- avoid the waste that occurs when resources are poorly used;
- provide a guide for checking progress;
- allocate resources between production for sale, production for consumption and savings.

The gross margin (contribution) concept, as described in chapter 3, is very useful in farm planning — especially for planning small peasant farms where common costs are very low. Few common costs change when a farm plan is varied unless the change is large or a long time-period elapses. Therefore, for small changes in farming systems, only total gross margin will change. Gross margins should be calculated for each existing or proposed enterprise on a per hectare or per head basis.

There are four basic ways of raising profit:

- Raise the income from existing enterprises. This will increase gross margin provided the variable costs are not raised to the same extent.
- Reduce variable costs provided the income is not reduced to the same extent.

These are both methods of raising the gross margins of existing enterprises by better techniques. If a farmer obtains an extra $18 of grain per hectare by applying $9 worth of fertiliser, gross margin is raised but, if he needs $27 worth, it will fall. Similarly if by better rationing a farmer can obtain a litre of milk from 0.4 kg of concentrates instead of 0.5 kg, gross margin will be raised.

- Change common costs to advantage. That is, reduce them if possible by more than any fall in whole farm gross margin or raise them by less than the rise in whole farm gross margin.
- Alter the combination of enterprises by including high gross margin enterprises to the maximum extent within the limits of the constraints of resource availability and good husbandry.

Increase or bring in enterprises with high gross margins and reduce or remove those with low ones. If tobacco has a gross margin of $930/ha and maize only $250/ha then, other things being equal, the more tobacco and the less maize grown the better.

Not only must the farm run with technical efficiency, by means of good yields, good feed conversion, high labour productivity and so on, but also the best *pattern* of farming must be followed in relation to the resources and markets available. It is essential therefore for each farmer to know all about his farm organisation, how well each part works; and how well it *could* work. A farmer may be a skilled husbandman and each enterprise may be technically efficient but, if the *combination* of enterprises is poor, results may be disappointing. It is essential therefore to look both at the efficiency of the individual parts and at the whole farming system; to answer both questions, 'How well does he farm?' and, 'What should be the system?' What is needed is a way of taking the farm system apart. We must look first at the parts to see if they are well run, then at the pattern of enterprises to see if a change in the balance between them, or in a production technique, would make better use of the common services of the farm. We must also check whether new enterprises could be introduced.

Farm planning is simplified and its value to a farmer is raised if the following requirements are met:

Flexibility

Plans should never be regarded as fixed for ever because conditions, like labour supply, prices and costs, change with time. Consequently, any plan should be checked regularly to keep it up-to-date. A farm plan is merely a guide and it should help to prepare farmers for unexpected changes.

Simplicity

Complex plans rarely warrant the time spent making them. Besides, they are seldom followed.

Balance

Plans should take into account only the most important farm enterprises. Essentially, we want to find where the strengths and failings of the business lie. It is a common mistake to record small details and to ignore main points.

Standardisation

The potential for simplifying planning procedures, by using standard methods and forms, is high.

Good decisions must be based on facts, therefore farm records, like labour use and costings, provide important information for planning. Lack of reliable data is a great problem in less developed countries. Either they do not exist or, if they do, they often are not adequate, accurate or well-developed. The sort of information needed includes climate and water supplies, availability of markets, a farm map and land classification map, the interests, aims, likes and dislikes of the farmer, family diet on 'subsistence' farms and the supply of labour and capital over time. Gross margins should preferably be normalised or averaged from the last three years' records with adjustments for any expected future changes in yields, prices or costs.

There are three main sources of data for diagnosing and planning improvements to a farm business:

- Those from the particular farm, supplied by farm records, surveys or interviews. This can be both difficult and time-consuming and the data may be of doubtful quality. However, if adequate records for the particular farm are available, these are undoubtedly the best data to use.
- Those from neighbouring farms run under similar conditions.
- Those from scientific experiments and studies conducted to analyse special problems and often used as standards.

Standards derived from other farms or research data help if data from the particular farm being planned are unavailable, unreliable or unobtainable, like when considering introduction of a new enterprise. Research station responses, in particular, should be viewed with caution as they are usually obtained by more intensive use of labour, capital and supervision than would be practicable on farms and they usually ignore economic realities by trying to maximise rather than to optimise yields.

A permanent recording service to carry out regular farm surveys, enterprise costings and to provide a locally relevant farm management or planning handbook is most useful in any country.

Use of group average input-output standards in planning individual farms may be inevitable if farmers cannot supply adequate and accurate data readily. However, although standards are rough approximations of the situations on farms run under similar conditions, even farms assumed to be similar are never homogeneous. Variations in variable costs, levels of management, soils and other ecological factors all cause differences between farms.

To summarise, the questions to ask when replanning a farm are as follows:

- How efficiently is each enterprise being run? Are the gross margins from each existing enterprise high enough? This is a measure of good farming.
- Which enterprises produce the highest gross margins and could they be expanded profitably to give a more profitable farm system? This is a measure of wise choice of enterprises or product-mix.
- Given a change in conditions, what new farming system is indicated?
- Could the common costs of the existing farming system be reduced? This is a measure of good organisation.

Farm planning is equally important on small subsistence or partly commercial holdings as it is on large commercial farms. Differences in size, economic or social organisation, market orientation, ownership, goals or other attributes of farms and farmers do not cause planning techniques, in general, to be less appropriate in one case than in another. They have general application, whether to a privately owned estate in Malawi, a state farm in Iraq, a rented farm in Britain or a subsistence holding in India. However, modern planning techniques need some modification in detail, but not in concept, when applied to peasant smallholdings because of certain inherent features of them. These include their large numbers and small individual size, communal grazing, fragmentation, mixed cropping and, often, a low level of farmer education and literacy. On many of these holdings, output could be raised by 30 per cent or more simply by improving husbandry methods without changing the combination of enterprises.

There are several ways of farm planning. The older ones include comparison and budgeting. These have the advantage of arithmetical simplicity and are easily used by those with little knowledge of economics or mathematics. Also, the results of budgeting are easily understood and explained to non-professional people. If the change is small; for example, replacing one enterprise by another or changing a production method, then partial budgeting will usually suffice. Complete budgeting may suffice for large changes in organisation.

The newer ways of farm planning are increasingly associated with use of powerful electronic calculators or even computers. These are mathematical optimising techniques, like programme planning or linear programming. Although the basic data used for planning are usually the same whatever the method used, the possible complexity of a new farming system, together with the number of alternatives for consideration and the skill of the planner, should determine the farm planning method used.

Complete budgeting

Complete budgeting, as described in chapter 4, is really an *aid* to more than a *method* of farm planning. It is a method of planning only in the sense that it helps the choice between two or more already made proposed plans.

Basically, it consists of only two steps:

- Preparing a plan that includes the area of each crop, the numbers of each class of livestock and the production methods. The proposed plan is based on subjective judgement, experience and intuition coupled with technical considerations. For example, an agronomist may have suggested a new crop or rotation or a livestock specialist the introduction of a goat enterprise. Thus, several alternative plans may be prepared.
- Budgeting the expected costs, including common costs, and returns to financially evaluate each plan and find which is best in terms of expected net farm income.

As complete budgeting relates to the whole

farm, much physical information is needed. This and future cost and price assumptions should be shown.

Complete budgeting therefore is a crude way of evaluating the effect of large changes in system or plans for a new farm which have been made subjectively. Owing to the many possible interactions of large changes, it is often easier than partial budgeting. It is quick and needs little mathematical skill and is thus a useful and flexible aid to whole farm planning, especially suited to fast-changing conditions.

Programme planning

So far we have seen that profit can be raised by raising the technical efficiency and gross margins of enterprises within the system and that budgeting is not a formal method of finding the plan that will maximise profit. The second way to raise profit is to recombine enterprises to produce a higher total gross margin for a given input of common costs. The problem, therefore, is how to design the best plan to suit a given situation.

Programme planning is a formal planning method, based on economic concepts like fixed and variable costs, gross margin per unit of scarce resource, marginal substitution, optimal allocation of scarce resources and opportunity costs. This methodical approach lies somewhere between simpler 'hand' planning techniques, like budgeting, and more sophisticated ones like linear programming. It achieves the main objective of the latter without need for a computer.

Programme planning is a form of mathematical programming for maximising net farm income by selecting the type and size of enterprises that make the best use of scarce resources within the limits of any other constraints that apply. It involves progressive selection of enterprises using normalised gross margins per unit of the most limiting resources. These constraints can affect the objective. Programme planning has the advantage over budgeting that a near optimal plan is reached without needing a computer. The procedure has the advantage of flexibility in that obviously severe constraints that appear during the calculations can be examined and possibly eased and enterprises that are too small can be avoided deliberately.

The logic of programme planning is easily understood by numerate farmers. If a farmer is happy about the accuracy of the data used, either by himself or an adviser, the resulting programme should be meaningful to him and, in broad terms,

it is likely to be adopted. This method has helped many farmers and managers to reorganise their production programme to raise profits. All managers who help to decide policy should understand and be able to use the technique.

Five sets of information are needed for programme planning:

1. The objective to be reached
2. Resource availability and constraints
3. Feasible activities
4. The gross margins of feasible activities
5. The resources needed for feasible activities

Let us consider each of these in turn:

1. THE OBJECTIVE

From the viewpoint of a farmer, his objective is to maximise whole farm profit within the limits imposed by all costraints. These constraints may include non-budgetable items like his personal wishes and skills or a desire for leisure of status, for example, to produce show cattle even if unprofitable.

The objectives of governments may vary from those of farmers and could include items like providing jobs, settlement, import substitution or promoting exports.

Peasants, growing little more than their subsistence needs, may have the objective of maximising food output in the worst season possible.

2. RESOURCE AVAILABILITY AND   CONSTRAINTS

This limits the further development or modification of enterprises. Finding the constraints can be harder than the later arithmetic calculations. The main constraints are as follows:

Land

This means both the quality and the net amount of arable and grazing land. It also means other natural resources like irrigation water supplies and climate. The quality of land affects erosion hazard and the maximum safe intensity of cropping and stocking. Soil texture affects its suitability for various crops.

Locational

Thse constraints mean, for example, the distance from water for stock watering or cotton spraying, from a barn for curing tobacco, from a milking parlour for a dairy herd or from a market or processing plant for products like sugar cane or cassava. They may prevent a given enterprise on a farm or part of a farm.

Labour

Either the quality of labour or its supply in any month can prevent growth of a given enterprise.

Self-sufficiency

Peasants usually insist on producing their own needs for staple foods for consumption. This can prevent them from adopting the potentially most profitable system with the resources available to them.

Capital

Limited availability of fixed capital, like fencing, buildings or equipment, or of seasonal working capital can prevent further growth of an enterprise.

Sequential

Some enterprises may need as an input the output of another enterprise. For example, beef fattening may depend upon the supply of weaners from the breeding enterprise or of maize silage. Technically sound rotations for pest and disease control or for soil conservation may fix the sequence in which crops should be planted. These ensure that long-term productivity is not harmed in pursuit of short-term profit.

Legal, administrative and political

These constraints include government regulations like banning potato growing on tobacco farms. Constraints owing to tenancy agreements, quotas and contracts are also relevant in this context. Also, banning the inflow of foreign workers could enforce mechanisation and reduce profits.

Managerial

These are difficult constraints to build into a farm planning model as many of them cannot be measured. Examples include the age and skills of the farmer and his wish for leisure for non-farm pursuits. These all affect the complexity of plan of number of enterprises that he will follow willingly. The subjective desire for unprofitable

'status' enterprises has been mentioned already. Other important constraints are risk aversion and some farmers' refusal to consider certain enterprises, however profitable. For example, Muslims will not produce pigmeat or Apostles tobacco.

3. FEASIBLE ACTIVITIES

A farm activity is a specified way of producing a crop or operating a livestock enterprise. Feasible activities are the ways by which various resources could be used to attain the objective. They are conventionally all enterprises that are feasible in the given economic and ecological environment. The feasibility of cropping activities depends greatly on the length and quality of the growing season(s), unless irrigation is available. Considering single enterprises as separate activities, as will be shown later, can lead to a mathematically optimal enterprise combination that would be impossible to build into a sound rotation. A modification to overcome this problem is to treat whole, agronomically sound rotations as single activities.

4. THE GROSS MARGINS OF FEASIBLE ACTIVITIES

The gross margin of an activity is usually expressed per hectare in the case of crops and per head or animal unit in the case of livestock. However, gross margins often must be expressed in terms of monthly use of labour or working capital when these are seasonally limiting.

Some managers argue that return on capital is *the* important thing otherwise the funds might be better invested in a building society or in shares. Others reply that capital can be borrowed but not land, therefore, *the* important thing must be return per hectare. However, what is important to a rancher with 5000 ha and a large overdraft may be irrelevant to his neighbour with 50 ha of intensive irrigated crops and a large bank balance. Each case should be judged separately. The important factor for any farmer at any time is the return he expects from each unit of his most limiting resource — be it land, April labour, June working capital or proneness to ulcers.

Enterprise gross margins are best prepared from the farmer's own records over the last three or so years to give 'normalised' figures of his past results in an average season. If this is impossible, other sources, like surveys and case studies, must be used. Both physical and financial data should be compiled. The information should then be analysed to find any technical weaknesses. Analysis methods include inter-farm comparison to compare results with standards and partial budgeting to assess the likely effect of possible technical improvements. New enterprise gross margins are then prepared for planning use, using expected future methods, costs, yields and prices.

5. THE RESOURCES NEEDED FOR FEASIBLE ACTIVITIES

These are the resource needs for each unit of each activity, for example, per hectare or per head. The data are best obtained from records kept on the farm being planned but, failing this, data from the fifth or so most 'successful' farms in a survey of other holdings using present technology may be used.

Programming procedure

It is hard to explain the method briefly and precisely in the form of steps to be taken. It is best demonstrated by example and learned by experience. However, the following attempts to summarise the process of programme planning:

1. List all resources that could be limiting, for example, arable land, livestock housing and working capital and labour in critical months. Judgement is needed to restrict the number of potentially limiting resources without omitting any critical ones.
2. List all feasible activities, their resource needs and gross margins. Again, judgement is needed to keep this list within manageable proportions. If, for example, an activity has a lower gross margin and needs more of *all* potentially limiting resources than another activity, it should be omitted.
3. Calculate the gross margin of each activity in terms of each potentially scarce, shared resource.
4. Tabulate and grade from first to last the gross margins of each activity to each limiting resource.
5. Decide which resource seems most limiting. This may be obvious but is often not. The final result will not change if you make a wrong decision; there will merely be more arithmetic.
6. Start the plan by building in the maximum

amount of the activity that gives the best return to the most limiting resource, excluding any activity that does not use that resource.
7. If the most limiting resource is not fully used, include other activities, in decreasing order of their return to it, until it or another resource runs out and thus becomes the new limiting one.
8. When a new limiting resource appears, substitute the activity not yet in the plan (or not yet at its maximum) that gives the best return to the new limiting resource for the activity already in the plan that gives the least return to it.
9. Continue these substitutions and modifications to the plan until total gross margin can no longer be raised.
10. Add any activities that need no fully used resources.

This all seems difficult but, in practice, with some intuition, experience and flair it is not. Most users of this technique soon learn shortcuts. For example, the 'rules' do not have to be strictly adhered to and some latitude can be allowed; if regular labour is limiting in a certain month, extra overtime could overcome this.

An example of programme planning should make the method clearer. It will be calculated first in the conventional, trial and error way and, later, by using whole rotations as activities.

Example

A smallholding of 3.2 ha of arable land in Malawi.

OBJECTIVE

To provide the staple grain need for the farmer, his family and labour and then to maximise sustained profit while following a technically sound rotation.

Resource availability and contraints

Land	3.2 ha
Working capital (cash and credit)	$100
Grain needed	13.5 bags/year
Labour available	90 man-days/month (peak months are December, April, May)

Rotational constraints: Cotton and tobacco are poor cover crops so, for soil conservation, should not occupy more than a third of the area.

For pest and disease control reasons, tobacco should not be grown more than once in four years; cotton and groundnuts should not be grown more than once in three years.

Tobacco and cotton may not both be grown on the same farm because of possible DDT contamination of tobacco leaf.

Feasible activities

Maize
Sorghum
Groundnuts
Cotton
Tobacco

Normalised Gross margins

	GM/ha	Yield/ha
Maize	$ 75	30 bags
Sorghum	85	18 bags
Groundnuts	110	
Cotton	100	
Tobacco	160	

Resources needed for a hectare

		Requirements per hectare				
Constraints	Availability	Maize	Sorghum	Groundnuts	Cotton	Tobacco
Land	3.2	1	1	1	1	1
Working capital[a]	100	24	12	15	60	30
Groundnut land	1.07	—	—	1	—	—
Cotton land	1.07	—	—	—	1	—
Tobacco land	0.80	—	—	—	—	1
Dec. labour	90	25	30	25	20	34
Apr. labour	90	15	12	37	40	70
May labour	90	12	10	30	47	44

[a] Variable Costs

CONVENTIONAL PROCEDURE

As the objective is to maximise the return to potentially scarce resources, the first step is to calculate what return each activity will give to each scarce resource. For example, the return given by maize to each unit of working capital is $3.12 ($75 (GM/ha) ÷ 24).

Returns to potentially scarce resources

Resource:	Land	Capital	Dec. lab.	Apr. lab.	May lab.
	(GM/ha)	(GM/VC)		(GM/man-day)	
Activity					
	$	$	$	$	$
Maize	75(5)	3.12(4)	3.00(4)	5.00(2)	6.25(2)
Sorghum	85(4)	7.08(2)	2.83(5)	7.08(1)	8.50(1)
Groundnuts	110(2)	7.33(1)	4.40(3)	2.97(3)	3.67(3)
Cotton	100(3)	1.67(5)	5.00(1)	2.50(4)	2.13(5)
Tobacco	160(1)	5.33(3)	4.71(2)	2.29(5)	3.64(4)

The figures in brackets show the order in which activities (enterprises) maximise return to the various scarce resources.

We can now combine activities so as to maximise return to any given resource. As it is not readily clear which resource is most limiting in this case, we will start by selecting according to return to land, for example, in the order of tobacco, groundnuts, cotton, sorghum and maize. However, the self-sufficiency restriction makes it necessary for us to supply 13.5 bags of grain. This could be achieved with the output of 0.45 ha maize (13.5/30) or 0.75 ha sorghum (13.5/18). Sorghum has the higher return to land so we will incorporate 0.75 ha of it to supply food. After incorporating the maximum tobacco area of 0.80 ha, we turn to groundnuts. However, only 0.67 ha of groundnuts can be incorporated before we run out of April labour. This is set out below as Plan 1.

PLAN 1   SELECT ENTERPRISES ACCORDING TO RETURN TO LAND

Constraint	Availability	Sorghum	Tobacco	Groundnuts	Used	Balance
Land	3.2	0.75	0.80	0.67	2.22	0.98
Capital	100	9	24	10	43	57
Groundnut land	1.07	—	—	0.67	0.67	0.40
Cotton land	1.07	—	—	—	0	1.07
Tobacco land	0.80	—	0.80	—	0.80	0
Dec. labour	90	22.50	27.20	16.75	66.45	23.55
April labour	90	9.00	56.00	24.79	89.79	0
May labour	90	7.50	35.20	20.10	62.80	27.20

April labour and tobacco land are now limiting. Whole farm gross margin is $265.45.

Nothing can be done to increase tobacco land so in Plan 2 we see if we can raise whole farm gross margin by using the balance of resources in such a way as to increase the activity with the greatest return to April labour (sorghum) at the expense of that one with the least return to it (tobacco).

The maximum sorghum we can introduce, because December labour is limiting, is 0.785 ha as shown below.

	Balance	Sorghum need/ha	Maximum ha
Land	0.98	1	0.98
Capital	57	12	4.75
Dec. labour	23.55	30	0.785
April labour	0	12	—
May labour	27.20	10	2.72

We now incorporate this extra sorghum and reduce the tobacco to the maximum amount that remaining resources will allow to be grown.

PLAN 2

Constraint	Availability	Sorghum	Tobacco	Groundnuts	Used	Balance
Land	3.2	1.53	0.67	0.67	2.87	0.33
Capital	100	18.42	20.05	10	48.47	51.33
Groundnut land	1.07	—	—	0.67	0.67	0.4
Cotton land	1.07	—	—	—	0	1.07
Tobacco land	0.80	—	0.67	—	0.67	0.13
Dec. labour	90	46.45	22.73	16.75	85.53	4.47
April labour	90	18.42	46.79	24.79	90	0
May labour	90	15.35	29.41	20.1	64.86	25.14

Gross margin = $311–12

April labour is still limiting so the substitution of sorghum for tobacco is continued with 0.15 ha of sorghum.

PLAN 3

Constraint	Availability	Sorghum	Tobacco	Groundnuts	Used	Balance
Land	3.2	1.68	0.64	0.67	3.00	0.20
Capital	100	20.20	19.29	10	49.49	50.51
Groundnut land	1.07	—	—	0.67	0.67	0.4
Cotton land	1.07	—	—	—	0	1.07
Tobacco land	0.08	—	0.64	—	0.64	0.16
Dec. labour	90	50.52	21.86	16.75	89.13	0.86
April labour	90	20.20	45.00	24.75	90	0
May labour	90	16.84	28.29	20.1	65.23	23.77

Gross margin = $319–71

April labour is still most limiting but December labour is also rapidly becoming limiting. To save time, it is decided to increase sorghum to 1.75 ha at the expense of tobacco.

PLAN 4

Constraint	Availability	Sorghum	Tobacco	Groundnuts	Used	Balance
Land	3.2	1.75	0.61	0.67	3.03	0.17
Capital	100	21	18.31	10	49.31	50.69
Groundnut land	1.07	—	—	0.67	0.67	0.4
Cotton land	1.07	—	—	—	0	1.07
Tobacco land	0.80	—	0.61	—	0.61	0.19
Dec. labour	90	52.5	20.75	16.75	90	0
April labour	90	21	42.71	24.79	88.50	1.50
May labour	90	17.5	26.85	20.1	64.45	25.55

Gross margin = $320.08

Now December labour has become most limiting. Cotton gives the highest return to December labour but cannot be introduced if tobacco is

grown on the farm. Tobacco gives the second highest return and sorghum gives the least return to December labour. Therefore we should substitute some tobacco for sorghum. The maximum tobacco that could be re-introduced with the the remaining resources is 0.02 ha as shown below:

	Balance	Tobacco need/ha	Maximum ha
Land	0.17	1	0.17
Capital	50.69	30	1.69
Tobacco land	0.19	1	0.19
Dec. labour	0	34	—
April labour	1.50	70	0.02
May labour	25.55	44	0.58

PLAN 5

Constraint	Availability	Sorghum	Tobacco	Groundnuts	Used	Balance
Land	3.2	1.75	0.63	0.67	3.05	0.15
Capital	100	21	18.94	10	49.95	50.05
Groundnut land	1.07	—	—	0.67	0.67	0.4
Cotton land	1.07	—	—	—	0	1.07
Tobacco land	0.80	—	0.63	—	0.63	0.17
Dec. labour	90	52.51	21.47	16.75	90.73	(0.73)
April labour	90	21	44.20	24.79	20.1	0
May labour	90	17.5	27.79	20.1	65.39	34.9

Gross margin = $323.52

April labour has now run out again and we have exceeded the labour supply slightly in December. It is clear that the optimum, theoretical solution lies between Plans 4 and 5, namely:

Sorghum	1.75 ha	
Groundnuts	0.67	(Gross margin − $322)
Tobacco	0.62	

The conventional method in the example gave us the following optimum enterprise combination:

1.75 ha	sorghum (SG)
0.62 ha	tobacco (TB)
0.67 ha	groundnuts (GN)
0.16 ha	fallow
—	maize (M)
—	cotton (CT)

Whilst this crop combination would be possible in the first year of a new cropping programme, it is impossible to continue it as a sound rotation. A modification of the conventional method, to give a feasible result, is to treat each technically sound rotation as a single activity with an average gross margin per hectare. These are listed below.

Rotation	GM/ha $	Rotation	GM/ha $
1. All maize	75.00	10. M-M-SG-TB	98.75
2. All sorghum	85.00	11. M-SG-SG-TB	101.25
3. M-M-GN	86.67	12. SG-SF-SG-TB	103.75
4. M-SG-GN	90.00	13. M-GN-CT	95.00

5. SG-SG-GN	93.33	14. SG-GN-CT	98.33
6. M-M-CT	83.33	15. M-M-GN-TB	105.00
7. M-SG-CT	86.67	16. M-M-GN-TB	107.50
8. SG-SG-CT	90.00	17. SG-SG-GN-TB	110.00
9. M-M-M-TB			

Each of these feasible rotations may now be tested for its resource needs over 3.2 ha. If resource needs exceed availability, maximum plantable area can be calculated. This can be multiplied by gross margin/ha to give the whole farm gross margin as shown in Table 8.4.

As shown in the example, the potentially most profitable farming system will often not use all the resources available. This is no cause for alarm. Too many farmers use a resource, like a pig fattening house, merely because it is available. It might be more profitable to sell weaners, eliminate pigs altogether or use the building for something else.

Programme planning maximises whole farm gross margin but we wish to maximise net farm income. Consequently, the optimum plan must be checked to ensure that the increase in gross margin expected more than offsets any likely rise in common costs such as, for example, the cost of any extra buildings, equipment or interest charges.

Programme planning is clearly a time-consuming and sophisticated procedure. It would hardly be practicable to use it for planning individual smallholdings. However, a feature of smallholder agriculture in most tropical countries is that the farms are fairly homogeneous in the enterprises followed, methods and assets used and in market opportunities. The main two variations between individual holdings is in the land, labour ratio and in the farmer's wishes. It is therefore possible to use programme planning on models of representative farms and to provide flexibility by allowing for selection of plan according to an individual farmer's land:labour ratio and wishes.

Linear programming

A common feature of many management problems is the need to use a limited set of total resources, basically land, men and money, to achieve a certain end most effectively by the optimum solution. To perform most operations, many activities are necessary. These all lay claim to the limited resources available and they are all inter-related. The operational research techniques that try to deal with this sort of problem are called allocation models. Operational research is the application of mathematics and other sciences to management problems. It is concerned with current operations and results and ways to improve these.

Linear programming is a mathematical tech-

Table 8.4   Resources Needed for Each Rotation on 3.2 Hectares

Rotation	Resource availability:	Capital 10	Dec. lab. 90	Apr. lab. 90	May lab. 90	Max. hectares	GM/ha	Whole farm GM
							$	$
1		76.8	80.0	48.0	38.4	3.20	75.00	240.00
2		38.4	96.0	38.4	32.0	3.00[a]	85.00	255.00
3		67.2	80.0	71.5	57.6	3.20	86.67	277.34
4		54.4	85.3	68.3	55.5	3.20	90.0	288.00
5		41.6	90.7	54.4	53.3	3.17	93.33	295.86
6		115.2	74.7	74.7	75.7	2.77	83.33	230.82
7		102.4	80.0	71.5	73.6	3.12	86.67	270.41
8		89.6	85.3	68.3	71.5	3.20	90.00	288.00
9		81.6	87.2	92.0	64.0	3.13	96.25	301.26
10		72.0	91.2	89.6	62.4	3.15	98.75	311.06
11		62.4	95.2	87.2	62.4	3.02	101.25	305.78
12		52.8	99.2	84.8	59.2	2.90	103.75	300.88
13		105.6	74.7	98.1	94.9	2.93	95.00	278.35
14		92.8	80.0	94.9	92.8	3.03	98.33	297.94
15		74.4	87.2	109.6	78.4	2.62	105.00	275.10
16		64.8	91.2	107.2	76.8	2.68	107.50	288.10
17		55.2	95.2	104.8	75.2	2.74	110.00	301.40

a $\dfrac{3.20 \times 90}{96}$

Clearly, the optimum rotation is no. 10, that is, to plant 3.15 ha to a rotation of M- M- SG-TB. This would need four rotational blocks of about 0.79 ha each.

nique similar to programme planning but using algebra. Its scope, however, is greater than that of programme planning because it is usually done with a computer. It is a complex technique, much used in industry and government, the details of which can be found in specialist textbooks. The aim her is merely to outline the technique, its possible uses, advantages and disadvantages in farming.

Initially, linear programming found many applications in farm planning (mainly in research institutes and universities) because it offered an elegant, theoretical solution to certain farm problems, given certain assumptions. When these assumptions, and therefore the value of the optimal plans, were questioned, sophisticated variations and extensions were found or developed to evade the invalidating assumptions.

Linear programming is an operational research technique based on matrix algebra whereby a stated objective is either maximised or minimised while satisfying various linear constraints. The result is mathematically exact. The production possibilities for the farm, their resource needs and gross margins (or, more precisely, their net revenues) and the resource constraints are brought together in a tableau, called a matrix, from which the optimum plan is automatically produced.

The technique depends upon obtaining a set of equations describing the resources to be studied. Thus, if a farmer knows that one cow or six goats need 4 ha of grazing land and he has only 120 ha, then:

$$4C + 0.6667 G \leq 120$$

If he employs only one stockman who will work no more than 2300 hours a year and cows and goats each need 115 and 23 hours annual work, then:

$$115 C + 23 G \leq 2300$$

We are starting to collect a set of equations linking output (cows and goats) to the resources and the equations also show limits to the resources, as the symbol $\leq$ means 'must not exceed'. Now we can insert some costs: $150C + 15G - 0.2 L$ means that the farmer can sell his cows for $150 each, his goats for $15 and he pays 20¢ an hour to the stockman. Knowing the cost of feed and other inputs, a complete mathematical model (that is, a set of equations) is developed.

The next step is to find the optimum mix of cows and goats to maximise profit from the known costs and constraints. Linear programming does this by organised search. If the solution says

that 24 cows and 150 goats is best, then this solution cannot be bettered. It is possible to use this technique to test any changes to the original data. The farmer, for example, could ask, 'What would be the best mix of cows and goats if I could persuade the stockman to work an extra 150 hours for 30¢ an hour? or, 'What rent could I afford to pay for an extra 30 ha and what would be the best stock to have then?' A computer can usually be programmed to point out the marginal costs at the same time as the solution.

The main advantage of linear programming over other planning methods is that it can quickly solve complex problems where there may be thousands of alternative choices. A computer makes it possible for far more data to be handled and for fuller solutions to be obtained.

If is often stated that linear programming must replace programme planning by hand after the number of activities and constraints reaches a certain level, such as six or seven of each, because the task then becomes too great to tackle without a computer. I dispute this. After collecting all relevant input: output data for peasant farming conditions in the dry zone of Sri Lanka — the hardest task — I had about a week to produce optimum modal farming systems for both irrigated and rainfed farming. There were over 20 feasible activities and as many constraints, including monthly labour supply, four significantly different soil types and differences between areas where sugar cane was feasible and not. I found it quicker to solve the problem by programme planning than to use a computer. In any case, there was too great a risk, in the short time available, of problems developing in a computer programme and having to be righted. The only facilities I used were a quiet lockable room, very large sheets of squared paper, a soft pencil, an eraser and a mind unbiased by others' opinions or the results they wanted.

Disadvantages of linear programming are as follows:

1. It needs the presentation of planning data in a way that needs special skill.
2. Its advantages are often removed by the inadequacy of data available.
3. Any assumption of linearity, that is, that how ever many units of an activity are included in the plan, unit costs and returns are constant, is often wrong.

Any problem that can be formulated as the maximisation or minimisation of a linear objective with several linear constraints can be solved by linear programming. Farm management uses include planning production functions to maximise profit, finding least cost feed mixes to meet stated nutritional standards and planning the best mix of finance — equity, loan, preference, etc. Besides its use for planning large farms, a more important use is probably to indicate plans for model or representative farms where there is much homogeneity between farms and the solution can guide the planning of individual farms. Linear programming can also help governments to find the optimum settlement holding size for a family and to indicate the changes in farming system that should follow changes in prices, costs, labour supply or technical improvements.

GLOSSARY

Budget　an estimate of income and expenditure for a future period as opposed to an account which records past financial transactions.

Diversification　this occurs when a firm undertakes production of a new product without ceasing production of any existing products.

Operational research　an interdisciplinary field of activity that tries to develop mathematical procedures for finding optimum solutions to management problems, for example, linear programming, discounted cash flow and network analysis.

Production　the transformation of raw materials (inputs) into goods (outputs) for sale or use.

RECOMMENDED READING
Books

Barnard, C.S. and Nix, J.S., *Farm Planning and Control*, 2nd ed. (C.U.P., 1979).

Brown, J.L. and Howard, L.R., *Principles and Practice of Management Accountancy*, 3rd ed. (Macdonald & Evans, 1975).

Castle, E.N., Becker, M.H. and Smith, E.J., *Farm Business Management*, 2nd ed. (Macmillan, 1972).

Harrison, E.F., *The Managerial Decision-Making Process* (Houghton Mifflin, 1975).

Johnson, D.T., *Uncertainty in Subsistence Farming*, Unpublished M. Sc. of dissertation, University of Reading, 1974.

——, *Farm Planning in Malawi* (Ministry of Agriculture, Malawi 1972).

Moore, P. G., *Basic Operational Research* (Pitman, 1979).

Norman, L. and Coote, R.B., *The Farm Business* (Longman, 1971).

Stainton, R.S., *Operational Research and its Management Implications* (Macdonald & Evans, 1977).

Sturrock, F., *Farm Accounting and Management*, 7th ed. (Pitman, 1982).

Upton, M., *Farm Management in Africa* (O.U.P., 1973).

Watts, B.K., *Business and Financial Management*, 2nd ed. (Macdonald & Evans, 1975).

Periodicals

Boyd, D.A., 'Interpreting Published Information', *Farm Management* (1979) 4, 1.

Hosken, A., 'Introducing a Computer into the Farm Office', *Farm Management* (1983) 5, 3.

EXERCISE 8.1

A peasant is thinking of opening 3 ha of fallow land for cotton production. This would require hiring 80 man-days of casual labour per ha at 40¢/day or hiring a tractor to do the same work at $40 per ha. If he hires the tractor he expects to be able to plant earlier and thus increase his usual yield of 850 kg/ha seed cotton. Seed, fertiliser and chemical costs will amount to $38/ha and harvesting, packing and transport add up to 2.75¢/kg.

1. What is the breakeven price per kg cotton to justify cultivation by hand?
2. What is the minimum yield he must get at this price if he hires the tractor?
3. Would you recommend tractor hire?
4. What advice would you give the farmer if the expected price of cotton next year is 18¢/kg?

EXERCISE 8.2

Your managing director has suggested that instead of buying maize for labour rations, you should plant 20 ha, now used for cotton production, to maize. You now buy in maize at $6.90 per 91 kg bag.

Given the following data, calculate the breakeven yield of maize per ha:

		Cotton (Per hectare)	Maize
Variable costs	$	185	140
Yield	kg	1,760	—
Price/kg	¢	18.5	—

EXERCISE 8.3

You are investigating starting a banana plantation that would have an estimated total annual 'fixed' cost of $250,000. The current banana price is $37.50/tonne and a feasibility study reveals that a yield of 70 tonnes/ha/year is expected in the area. At this yield level your expected variable cost of production is $625/ha/year. Find, either arithmetically or graphically:

1. the breakeven area
2. the breakeven area if 'fixed' costs are reduced by a third
3. the breakeven area if variable costs are reduced by a third
4. the breakeven area if both 'fixed' and variable costs are reduced by a third
5. the breakeven yield per hectare if 140 ha is planted

} other factors staying constant.

Dairy manager's salary: $2500
Crop manager's salary: $3125
New capital equipment needed:

Item	Capital cost	Straight line depreciation	Annual repairs and maintenance
	$	%	$
Forage chopper	4500	30	430
Trailer	2000	15	200
Silo	625	5	—

Calculate:

1. The financial feasibility of introducing maize silage into the dairy ration.
2. The breakeven number of cows.
3. The breakeven silage yield per hectare if herd size remains 75.

EXERCISE 8.4

As a dairy section manager, you wish to raise the mean annual milk yield of each cow from 2460 litres to 3000 litres. The present feeding system is based on natural grazing with bought concentrate feeds. You believe that by supplementing this with high-quality maize silage, average annual yield will rise to 3000 litres.

Given the following information:

Herd size:	75 cows
Mean annual yield:	2460 litres/cow
Milk price:	15¢/litre basic
Quality premium:	0.033¢/litre for each per cent rise in SNF over 8%
Present solids-not-fat:	8.5%
Expected solids-not-fat with silage:	10.5%
Silage needed:	18 kg/cow/day for 300 days each year
Silage yield:	16.8 tonnes/ha
Silage variable costs:	$170/ha
Concentrates fed	5.4 kg/cow/day at $103/tonne
Labour needed to feed silage:	2 workers at $348/year

EXERCISE 8.5

As the manager of a market garden selling fresh produce, you are invited to tender for the contract to supply a hotel with its annual vegetable needs. In your tender you must specify a single price/kg for the entire range of produce to be supplied. Using the following data, calculate the breakeven tender price than will include a return on extra capital, including working capital, not less than your existing return on capital.

Present performance

Crop	Area (ha)	Yield (kg/ha)	Market price (c/kg)	Variable
Tomatoes	6	13500	11	770
Cabbage	4	8900	11	150
Carrots	4	5350	8	275
Beans	2	5350	22	250
Cauliflower	2	6700	28	540
Lettuce	2	5350	55	1400
Peas	2	2675	22	375

Annual quantity of total contract for tender

Tomatoes	14.00 tonnes
Cabbage	9.00
Beans	4.50
Cauliflower	3.50
Lettuce	2.75
Peas	2.75
Carrots	2.25

Extra capital equipment needed to fulfil contract

Item	Capital cost	Straight-line depreciation	Driver	Extra electricity charge
	$	%	$	
Small tractor	3125	20	600	—
Irrigation Equipment	1875	15	—	$25/month

Present capital employed is $95,000 and casual overhead costs are $6250.

EXERCISE 8.6

Exercise in simple programme planning

1. Given the following information, use the conventional programme planning technique to calculate the theoretical optimum crop combination.

    (a) Objective — to maximise whole farm gross margin

    (b) Resource availability and constraints —

Annual cropping area	12 ha
Tobacco quota	7500 kg
Expected tobacco yield	1500 kg/ha
Maximum variable costs (VC)	$2250

    (c) Feasible crops and gross margins —

	VC/ha	GM/ha
	$	$
Tobacco	325	325
Groundnuts	80	40
Maize	125	50

2. What would the optimum theoretical crop combination be if the limit of variable cost were raised to $2750?

9   Production control

Production control, or progressing, is concerned with the execution of a present production plan and control of all aspects of it. It means checking that work is proceeding according to the overall plan and the orders given, and that a manager knows the extent to which the plan is being achieved or not.

Planning and control cannot be divorced; neither can stand alone. Unless plans and, by definition, the setting of standards are monitored, the objectives will probably not be achieved. Control, through taking remedial action, in turn, can be meaningful only in the presence of a plan.

Relevant aspects of production control have already been described — stock control, budgetary control and variance analysis in chapter 5 and the control of time use in chapter 7. Network analysis for controlling projects will be described in chapter 11.

The purpose of production control ultimately, is to contribute to efficient production that, in the long run, is measured in terms of profit. The need for control has existed ever since people have been concerned about the results of their efforts. Change is either planned or fortuitous. If a manager wishes to control his own fate, planned action is better than unplanned action. This makes control desirable and necessary; otherwise plans become little more than statements of hope.

Once a production plan is in operation, it is necessary to monitor actual events to see how closely they follow the plan. Standards of performance that have been built into the plan are unlikely to be wholly accurate and failures and mistakes are sure to occur. This is part of any business. To enable managers to take remedial action, variances from plan must be soon detected. More important, though, a control system should be designed to prevent errors from occurring. The purpose of production control is to make the process go smoothly, properly and according to high standards.

A good control system has the following features:

- It maintains the process within a permissible variance with the least effort. To spend a dollar to protect 99 cents is not control; it is waste.
- Decisions have been made on desired performance standards and permissible variances. Control must be basically by 'exception' and only significant variance from the norm triggers the control. Provided a process runs within the preset standards, it is under control and needs no action.
- Control must be by feedback from the work done. The work itself has to provide the information. If it must be checked all the time, there is no control.

There are three essential steps in control. These are setting of standards, comparing actual performance against the prescribed standards and finally, if results are unacceptable, taking remedial action. The latter may mean changing a standard in the plan.

To be useful, control information must be timely. It is no help to find in the calving season that the bull was sterile nine months before. With a complex farming system, a manager may need a weekly or even a daily report on each enterprise, summarising any points in the plan that have not been achieved.

Management without control would be futile. However, a manager who tries to control every detail would be overloaded. Intelligent control should focus only on the performance needed to achieve results in key areas. To try to control the irrelevant always misdirects a manager's energies.

To keep informed without being drowned in detail, he should decide the strategic control points for all activities.

Quantity control, or finding yield levels and total physical output, is perhaps the aspect of control that has received most attention in farming. However, the input side is often equally if not more important. For example, quantity control of inputs allows a manager to check the rate of fertiliser application no his fields without leaving his office. A simple daily record of fertiliser drawn from the store and tractor operating hours enable him to estimate daily fertiliser application. Though this is a rough check, the information is available quickly before much harm can be done.

Quality control is as important as quantity control. Product quality cannot be controlled adequately without preset standards such as the proportion of kernels damaged by a groundnut sheller and the population of cotton stainers that warrants spraying the crop. Input quality may be harder to control but should concern a manager no less. The quality of labour, seed and stockfeed are obvious examples.

Management consultants have long noted the weakness of scanty information for farm management decision-making. As a result, a vast institution has been built around the central theme of record keeping for control purposes. Farmers have been persuaded that without many records they have no chance of success. To some extent, this has been overdone.

CONTROL CHARTS AND DIAGRAMS

Factual information is often clearer and easier to understand if shown graphically. However, if the figures are not given, care is needed to avoid distortion. Tables are best for management control purposes only if the data are unsuitable for graphical display or a manager needs numerical values for calculations. The ability to draw and use suitable charts greatly aids control of all management functions. In the daily control of a business, there is little time to study detailed records. Decision-makers therefore need the information upon which to base decisions in a form that provides an almost instant grasp of the situation. Managers mainly rely, therefore, on summarised statements, ratios and charts.

The value of a control chart depends upon regular recording of actual performance against that planned and plotted. This enables a manager to check progress quickly so that action can be taken before a crisis arises.

Production planning charts, such as those shown for planning seasonal tractor and labour uses in chapters 4 and 6 respectively, also aid control as they enable actual performance to be compared with that planned. These charts should be kept up to date and should be amended on a transparent overlay, if necessary.

Graphs

A simple graph containing two curves or lines suffices for most practical purposes, like comparisons with past results, planned results or with related figures. It shows fluctuations and rates of change. Owing to the frequent fluctuations in output, purchase or sales figures, comparisons are often best shown by cumulative graphs, as in figure 9.1.

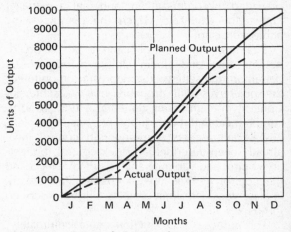

Figure 9.1   A Cumulative Graph

Graphs, and similar ways of displaying facts or estimates, suffer from the problem that (a) they are inexactly defined or (b) they cannot be read to the level of precision to which they were drawn. Nevertheless, they can save much time and effort and, if the following points are remembered when drawing them, graphs can be most useful:

- You must decide which is the independent variable against which the other is measured. Thus, if the variables are poultry-feed intake and time, the latter is the independent one and it should therefore be plotted on the horizontal axis.
- If the graph depicts absolute changes, the zero point(s) should usually be shown; otherwise a wrong impression can be given.

- Avoid too many lines on one graph or they will confuse.
- A clear heading is needed, with further explanation by means of sub-titles or footnotes.
- Wording should all be horizontal.
- The scale usually should be a convenient multiple of 10 and, if it helps, the left-hand one may be repeated on the right-hand side.
- If several curves are needed, distinguish them by different colours or types of line (solid, long dashes, short dashes, a line of dots and alternating dots and dashes, in that order).
- Limit the number of scale division marks for background grid lines to the few needed for rough readings.
- It is good practice to leave the data points on the graph. If they cause crowding or confusion, increase the scale until they do not.
- Plot only those points that are relevant to your purpose. Do not extend the curves simply because the figures are available.

'Z' charts

A 'Z' chart is simply a graph that extends over a single year and consists of three curves that together look like the letter Z. The three curves show:

(a) the current figure for the period concerned
(b) the cumulative amount to the latest date
(c) the moving annual total or 'trend'

A 'Z' chart can avoid the need for lengthy scrutiny of at least three sets of figures that might otherwise be hard to relate. It records three figures that are vital to efficient managing — the result of the current period, the cumulative results of the year or season to date and the moving annual total showing the result of the last 13 periods (or 12 months). The last figure is obtained at each period by summing the figures for each of the past 13 periods (or 12 months) up to and including the current one. This indicates the trend of the variable plotted by smoothing out purely seasonal or unusual variations. The chart may be used to depict any type of data — sales, costs, wages, water charges, etc.

Figure 9.2 shows a typical 'Z' chart.

This useful chart has one drawback. As the trend curve begins at the cumulative figure for the previous period, it is widely separated from the current monthly or weekly figure. A double scale may be used to show current changes more clearly.

Figure 9.2   A 'Z' Chart for 1987 Milk Sales

Bar charts

Bar charts, histograms or stack diagrams are diagrams that can be even more striking than graphs and are convenient for inserting actual figures. They should not be used if the differences in quantities are small and it is misleading to use a broken scale, or one that does not start at zero, to exaggerate differences.

The chart consists of a series of bars on columns whose lengths or heights are proportional to the quantity. In designing a bar chart, remember:

- each bar should be the same width
- this width should be about a fifth to a tenth of the length (or height) of the longest bar
- the length is proportional to the magnitude
- label and scale each bar clearly

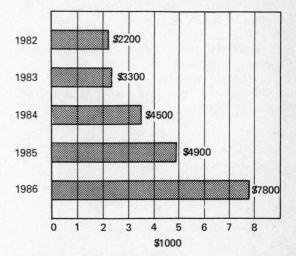

Figure 9.3   A Typical Bar Chart

- give the exact values if they are needed to interpret the results but are not clear from the scale
- allow half-bar width between bars and at each end.

The simplest bar chart shown in figure 9.3 has several variations of which the following help most in production control:

- bringing together two sets of bars for direct comparison of, say, planned and actual results (figure 9.4)
- showing an added set of totals for comparing with an original set (figure 9.5)
- a deviation chart designed to show differences instead of totals (figure 9.6)

A histogram is a figure composed of rectangles drawn next to each other such that the area of each rectangle is proportional to the frequency of observations in the class interval respresented by the width of the rectangle.

Gantt charts

The Gantt chart is a highly flexible development of the bar chart that is often used in progressing output. It has the merit of showing current and cumulative results, both compared with a standard. The chart is a visual display based on time rather than quantity and gives a continuous picture of progress, thus enabling managers to more closely control production. It shows how effectively the plan is being executed and what remedial action, if any, is needed.

The limitations of Gantt charts are that they are hard to modify — rubbing-out and redrawing is time-consuming — and they cannot show the interdependence of activities.

The quantity scale is replaced by a time scale and to achieve uniformity the performance bars represent percentages.

For example, suppose it is desired to show that on Monday and Tuesday 2 and 3 ha are to be harvested. This can be shown as follows:

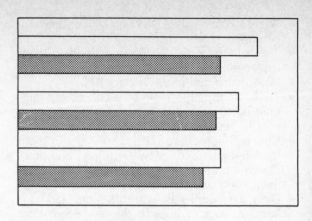

Figure 9.4   A Doubled Bar Chart

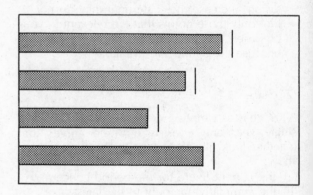

Figure 9.5   A Bar Chart with Totals

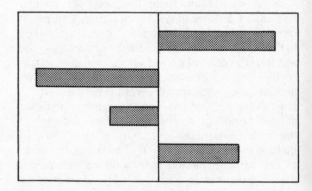

Figure 9.6 . A Deviation Bar Chart

Although the two bars represent different quantities they are drawn to equal length, representing 100 per cent of target. The number at the start of each line represents the individual target figures (two for Monday, three for Tuesday); that at their ends is the cumulative figure.

Assume that actual performance is 1.0 ha on Monday and 3.5 ha on Tuesday.

First a performance line 1.0/2.0 of the target is drawn on Monday and a performance is in excess of plan, the line is drawn in two parts, one full length and the other one-sixth the length of the plan line.

The combined effect of both Monday and uesday is shown by drawing a cumulative bar below the target lines. The position of the end of this cumulative bar shows quite clearly how work is progressing in terms of time — that is, that by the end of Tuesday work is one-sixth of Tuesday behind.

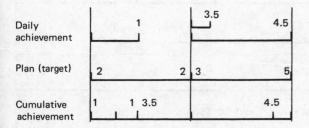

A Gantt chart helps most when several activities are being considered. Consider the target output of six sections of a large estate and their achievements as shown below.

Plotted on a Gantt chart, figure 9.7, the effect is to clearly show which sections are lagging behind and by how much.

PRODUCTION RECORDS

Many physical and financial records that aid production control have already been described. These include cash analysis, labour and machinery use and stock control records. Let us now look at specific input and output records for crop and livestock production to illustrate their use in production control.

Efficient production control is not guaranteed by well-kept records but it will certainly not be achieved with poor ones. Forms and documents are the main means through which control is exercised and they should:

- be clear and accurate
- need the least possible clerical work by managers and operators
- be designed to trigger action besides recording events
- minimise any duplication and repetition in records
- create an integrated system

Crop production records

The physical yield per hectare of each crop grown usually has the most effect on each crop's gross margin. However, the effect of yield per hectare on profitability is modified by both the market price and the cost of each unit of output sold.

Low yields more often cause low profits than excessive inputs. Usually, within limits, the higher the yield per hectare, the lower will be the cost per unit. One has to take care, however, with some crops like tobacco, that high yields are not achieved at the expense of quality. Raising yields depends both upon using more physical inputs and working capital per hectare and also upon inherent soil fertility and a manager's technical skill to do he right job in the best way at the best time.

Plan (Hectares Sugar Cane Cut)

	Week 1	Week 2	Week 3	Week 4	Week 5
Section 1	10 / 10	15 / 25	20 / 45	25 / 70	30 / 100
2	0 / 0	10 / 10	15 / 25	20 / 45	25 / 70
3	0 / 0	0 / 0	10 / 10	15 / 25	20 / 45
4	70 / 70	0 / 70	0 / 70	10 / 80	15 / 95
5	70 / 70	70 / 140	0 / 140	0 / 140	10 / 150
6	60 / 60	70 / 130	70 / 200	0 / 200	0 / 200

Actual (Hectares Sugar Cane Cut)

	Week 1	Week 2	Week 3	Week 4	Week 5
Section 1	5 / 5	10 / 15	25 / 40	20 / 60	30 / 90
2	0 / 0	5 / 5	10 / 15	23 / 30	22 / 60
3	0 / 0	0 / 0	5 / 5	10 / 15	23 / 38
4	60 / 60	10 / 70	0 / 70	5 / 75	8 / 83
5	60 / 60	80 / 140	0 / 140	0 / 140	5 / 145
6	60 / 60	60 / 120	70 / 190	30 / 220	0 / 220

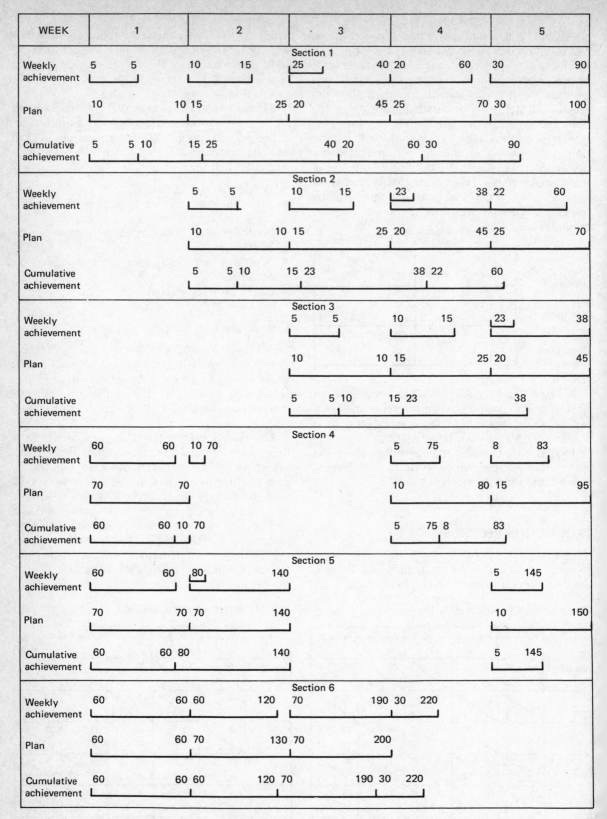

Figure 9.7 A Typical Gantt Chart

The market price per unit of output can be much affected by careful grading of most products. The bottom grade of, say, cotton or sorghum may fetch less than half the price of the top one. Time of sale can also affect the price and it often pays to store crops after harvest for sale later when the price rises. Widely varying prices are characterisic of seasonal crops of perishable fruit and vegetables. Out of season production can turn this feature to advantage.

The main costs are usually labour, fertiliser and machinery. Higher than average costs of these items per hectare is usually associated with reduced costs per unit of output. This relation, naturally, does not apply to high transport costs if the farm is far from the market. High yields, selective mechanisation and a farming system that uses labour productively throughout the year all help to lower unit costs.

A manager's need for field data varies between farms but, however small the farm, a separate record should be kept for each crop on each field. Apart from the obvious need to record inputs and outputs, the other data recorded depends upon the manager's needs. Figure 9.8 shows a suitable land use record sheet for most farms, assuming that separate records are kept of irrigation water use and that details of cultivations are available from labour and tractor use records.

Besides helping in calculating gross margins, field records kept for several years highlight differences between fields so that they can be treated individually in future. They also provide the basis for planning a calendar of operations and can reveal any build-up of pests, diseases or weeds that may affect future use or practices. Although rigid rotations are no longer necessary, rational cropping sequences are and records help to ensure this.

The value of field records lies mainly in amassing readily available data on the success or failure of husbandry practices so that action is taken to maximise the likelihood of future success. It is important to record what really occurred not what should have occurred. If the fertiliser drill was wrongly calibrated or the wrong type of fertiliser was applied, either another dressing can be given or any residual nutrients can be allowed for in fertilising the next crop.

A bound crop record book is less convenient than a loose-leaf file with a separate form, or card storage box with a separate card, for each crop in each field each year. It is handy to pin cards to an outline farm map on the office wall as it is easier to walk to the wall and make an entry on return-ing from the fields than to leaf through a file. At the end of the cropping season, the cards are filed and replaced by new ones. The value of loose-leaf files or cards is that successive records for each field can be assembled into a continuous record and data for a particular crop can be collated quickly.

It is likely to be the manager's job to ensure that records, whether entered by a clerk or not, are correct. This is in his interest because the quality of so many later decisions will depend upon the accuracy of past records.

Livestock production records

Intensive livestock enterprises need good records if the danger of financial loss is to be avoided. The types of records needed depends on the particular enterprises but the most important ones are breeding performance of individual animals, the amount and cost of feed and the amount and value of output. The records to be mentioned are those connected with the factors that affect profit most. It remains true, however, that the eye of the master fattens the beast. No paperwork will isolate a sick pig, spot unhygienic conditions or ensure that a sow really was served.

To assess the efficiency of a livestock enterprise, it is clearly necessary to know how many livestock of each class you have. This is less easy than it seems because numbers vary throughout the year owing to births, purchases, deaths, sales and transfers of stock from one category to another, for example, gilts to sows. A monthly reconciliation of livestock numbers in each category, as shown in figure 9.9, is needed.

The average livestock numbers over the year is found by averaging the 12 month-end totals. This is needed for annual calculations like milk yield per cow, weaners per sow and average stocking rates.

Cattle numbers on ranches are usually checked when dipping or spraying them.

The most important general physical record for livestock enterprises is the use of both bought and home-grown feeds. Feed costs are usually the only ones where great savings are possible, especially with pigs, poultry and dairying where most other costs are fixed. It is essential to be able to allocate feeds to the various consuming enterprises.

A small daily error in feeding concentrates can add up to a large amount over the year and you must be able to calculate feed conversion ratios which are critical to the profitability of most stock

FURA FARMS (PVT) LTD

LAND USE RECORD

Season: 198 –198 .   Field: _____

                                                                Net area: _____ ha

Crop grown _____   Cultivar _____

Planting date(s) _____   Date 50% Emergence _____

Spacing _____   Population ha _____

....

Inputs:

Seed rate (kg/ha) _____   _____

Fertiliser ('Basal') _____   Top Dressing _____

Lime (t/ha) _____

Other: _____

....

                                                        Type   Rate

Pests _____   Control _____

_____   _____

_____   _____

Disease _____   Control _____

_____   _____

Weeds _____   Control _____

_____   _____

....

Output:

Yield/ha _____   Use _____

Harvest date(s) _____

....

Remarks: _____

_____

_____

Soil analysis:

  Texture _____

Mineral N (before and after incubation)   p.p.m. _____   _____

Available P_2O_5 p.p.m.   _____

Exchangeable K   _____

                                    pH ($CaCl_2$)   _____

Figure 9.8   Example of a Land Use Record Sheet

enterprises. Few farmers keep a strict enough eye on concentrate use and much may be wasted. Stockmen often give their favourite animals a bit extra and the difficult ones enough to keep them quiet!

	Oct.	Nov.	Dec.	Jan.	etc.
Start of month	30	26	21	18	15
Births	—	—	—	—	
Purchases	3	4	4	3	
Transfers in	4	4	3	3	
Total	37	34	28	24	
Deaths	1	—	2	—	
Sales	3	2	3	2	
Transfers out	7	11	5	7	
End of month	26	21	18	15	
Total (=Check)	37	34	28	24	

Figure 9.9   Monthly Reconciliation of Livestock Numbers — Heifers, 1–2 years

For precise feed recording, it should be the task of one employee. If he is paid a bonus based on the feed conversion ratio, he will have an incentive to do the job properly. It is as important to record the use of home-grown feeds as that of bought ones. A poultry farmer may think his feed bill satisfactory for the eggs produced until he remembers and costs the home-grown sorghum fed. Proper stock control records are needed.

RECORDS FOR BEEF AND DRAUGHT ANIMALS

Many farmers like to have a few cattle on the farm and, because of a belief that cattle 'grow into money', they do not bother to cost them. The profitability of beef cattle is low in most countries. It is thus essential to keep basic records of the performance of each breeding cow and of the feed used by fattening stock.

Checking the breeding performance of each animal is vital to profitability and is a valuable aid to culling and selecting new breeding stock on their performance and their progeny's performance. National average weaning rates are often little more than 50 per cent. Raising this to 80 per cent by supplementing feeding of grazing cattle in dry seasons would greatly raise productivity. Culling cows that pregnancy tests show not to be in calf for the second year running would also raise calving and weaning percentages. A simple card index system or a bulling and calving register will suffice to record details on each breeding female. The detail needed for the cows should usually include its sire and dam, 205-day weaning weight, 18-month weight, date of birth and date of first mating. Each cow's card should also include details of all her progency, such as bull used, date of service, date of birth and the calf's number, sex, weaning weight and 205-day weight. The year of birth of each calf can be recorded on it by notching the last figure of that year in the right ear and the last figure of the year of birth of each of a cow's calves can be branded on her neck or cheek. These records are the only basis on which to base a sound breeding policy.

Date**	**Stock changes: births, deaths, purchases, sales, farm slaughterings, reclassification	Bulls**	Cows			Heifers					Steers					Calves			Total on farm **	Total dipped **
			Breeding	Cull	Total**	Not bulled			Bulled	Total**	1–2 yrs	2–3 yrs	3–4 yrs	4+ yrs	Total**	0–6 mths	7–12 mths	Total **		
						1–2 yrs	2–3 yrs	3+ yrs												
1.12.87	Stock in hand	8	280		280	45	45		40	130	100	100	75	40	315	220	210	430	1163	
2.12.87	Births															+5		+5		
3.12.87	Sale to CSC													−40	−40					
4.12.87	3 calves killed															−3		−3		
5.12.87	Purchases Kano sale											+50			+50					
7.12.87	Dip day	8	280		280	45	45		40	130	100	150	75		−325	222	210	432	1175	117
12.12.87	Calving finished					−45	−45		−40		−100	−150	−75				102			
	Reclassification		+40			+102	+45		+45		+108	+100	+150	+75			−108			
	Stock in hand	8	320		320	102	45		45	192	108	100	150	75	433	222		222	1175	

Figure 9.10   Livestock Records

The columns asterisked (**) must be completed on dip days and all changes must be recorded within a fortnight of their happening. As in this example, it is recommended that the form is used like a cash account. The first entry should be in the form of a statement of stock in hand by classes. Thereafter, any changes are entered under the correct headings as additions or subtractions. Regularly, or on dip days, the new balance is calculated. At least once a year it will be necessary to reclassify stock under the various headings. Calves will become either one-year steers or heifers; some heifers will have calved and therefore become cows; two-year olds will become three-year olds, and so on.

Naturally, cattle being fattened on concentrates must be weighed so that the feed cost per kg live-weight gain and the amount eaten is known. Feeders should be weighed as soon as they have settled onto their ration and then about every three weeks. Knowing the feed used and the live-weight gain aids detection of feed wastage and culling of poor performers. It also helps you decide when to sell each batch of feeders.

RECORDS FOR DAIRYING

Milk production is relatively intensive for a livestock enterprise and it needs much husbandry skill. Dairying usually consists of two separate enterprises — milk production, producing milk, calves and cull cows, and a unit of followers supplying the milking herd with replacements. Yield per cow and the cost of concentrates affect profit most. The most profitable producers usually have higher than average annual milk yields and lower than average feed costs per litre of milk.

Milk production records are needed for each cow to enable them to be fed correctly and to know whether to breed from them or cull them. The annual milk yield of a cow depends on its lactation yield and calving interval. Calving interval affects the annual gross income from a cow and should be as close to 365 days as possible. Table 9.1 shows that an extended 4500 litre lactation is worth only $51 more, on an

annual basis, than one 1000 litres less and it will almost surely have cost more than $51 in extra concentrates to produce the extra 1000 litres.

Up to at least the 4000-litre lactation yield level, raising yield will raise gross margin per cow, of a suitable dairy type, by spreading the fixed cost of the maintenance ration and thus reducing total feed cost per litre. Fixed cost per litre fall and variable costs per litre rise with greater yields. Low yielders should gradually be replaced by better milking strains found from the individual production records.

The amount of milk produced by a cow at a milking is measured either by weight or volume.

If too much milk is used on the farm, gross margin per cow can be reduced by a quarter. The cost of rearing a calf on whole milk after the first three or four weeks can be almost twice the cost of using milk substitute. However, most farmers still milk-feed for seven or eight weeks instead of starting to use substitute at 10 days, using all substitute by three weeks and weaning onto solids at about 45 days.

Service or insemination dates should be recorded with a note of possible returns. When each date is passed without a return occurring, the likely calving date should be noted. By each return, note whether first, second, third, etc., to ensure that breeding faults are soon detected. A pegboard is useful for recording services and expected calving dates. Important heats and inseminations are often missed through lack of clear accessible calving, service and pregnancy diagnosis information.

Each cow's breeding record is written on the back of its lactation record card. Yields are set out in the form of a pedigree with, if possible, yields of daughters. This gives information about families that aids informed decisions on which heifer calves to keep as replacements. Choosing the best bull is a specialist job and advice is worth seeking. Not only milk yield but also milk quality and longevity should be considered.

A high annual replacement cost can greatly reduce profit. Cows are often culled before

Table 9.1   The Influence of Calving Interval and Lactation Yield on Gross Income/Cow/Year from Milk Sales — ($)

Lactation yield at 11¢/litre	Calving interval (days)			
	350	365	380	400
2500	287	275	264	251
3000	344	330	317	301
3500	401	385	369	352
4000	459	440	422	402
4500	516	495	475	452

Table 9.2   The Relation between Daily Milk Yield, Feed and Margin over Feed Costs per Cow

Daily yield (litres)	Concentrate (kg/litre)	Feed cost at 8.2 ¢/kg (¢/litre)	Receipts (¢/litre)	Margin over feed costs (¢/litre)	Total margin/ cow/day (¢)
5	0.35	2.87	12.4	9.53	47.65
10	0.45	3.69	12.4	8.71	91.20
15	0.60	4.92	12.4	7.48	128.60
20	0.85	6.97	12.4	5.43	155.75
25	1.15	9.43	12.4	2.97	170.60

reaching their peak yields because, 'There is a wonderful bunch of heifers coming along.' These heifers are often little or no better than the cows they replace. The most profitable dairy herds have an annual replacement rate of about 16–18 per cent. It is usually important to reduce the number of heifers being reared in relation to the number of milking cows, especially if land is scarce. Age at first calving and the herd life of cows is more important. It is vital to study milk records to decide which cows to cull and which heifer calves to rear so as to raise herd performance. The time to cull poor yielders is arguable but it is is usually after the second lactation. However, replacement heifers can safely be selected on first lactation performance. A bonus could be based on calving index.

The amount and cost of concentrates fed is usually the greatest determinant of profit. It is important therefore to feed the cheapest possible ration that can maintain body weight and give optimum milk yield and quality. Each cow should be fed to give her full economic potential yield. A 'natural' 3500-litre cow may well be profitable if fed accordingly; excessive feeding may turn it into a 3900-litre cow but also lower its profitability. Strict rationing is essential at all times.

Much substitution is possible in mixing rations so there is great scope for cost reduction. Feed cost per litre rises at a growing rate with greater yield because the feed conversion ratio falls as a cow nears her inherent yield potential so that more feed is needed for each successive litre and, secondly, because the ration has to be more concentrated if she is to eat enough to sustain a high yield. This is shown in table 9.2.

A sound rationing system for concentrates depends on the supply of bulk fodder and the daily yield of individual cows. Feed records sheets are needed for both bought and home-grown concentrates. Each cow's allowance is chalked on its name plate in the parlour and accurate feed measures or scoops must be used. Data from daily production and feed records should be transferred to a weekly or monthly summary output, including home use and rearing, and concentrate use. This is a check on the concentrates fed per litre. An error of overfeeding by 0.05 kg/litre can greatly lower profitability. It is equally important to record and supervise the feeding of followers. Surprisingly, farmers are often unaware of how much feed their cows are eating.

Cow health records, used together with a veterinary surgeon, are most helpful in maintaining long-term health and in culling. Mastitis can

lower both the quantity and quality of milk. Subclinical mastitis is thought to reduce yield by 10 per cent in many dairy herds.

Labour costs are only about 10 per cent of total variable costs in tropical dairying but there is a wide range. However, cost alone does not reveal the true value of labour as the quality of work affects both yields and other costs. With the same resources, good trained stockmen produce more milk than poor ones. The herd should be big enough to keep one or more men fully employed and make it worth while for an able supervisor to give ample time to its management. Labour use per cow falls as herd size rises.

RECORDS FOR PIG FARMING

Like poultry farming, pig farming is a useful supplementary enterprise needing little land. They enable business size to grow by profitably using spare labour and buildings without ousting high value, arable crops. These 'factory' enterprises bring in regular income and thus lower the average working capital in use.

Quality of output varies more than in dairying but it is mainly under management control. The production process is more precise than it is with grazing livestock. Most production comes from costly concentrates and the profit margin on each pig is low, so attention to detail is essential. An economic unit is usually 100 sows or more; less than 20 is uneconomic as the cost of the boar is shared by too few sows.

Breeding and fattening usually occur on the same farm. However, they should be costed as separate enterprises. The critical factors affecting profit are pigs marketed from each litter, pigs sold from each sow in a year, the amount and cost of feed used and the quality of pigs sold.

Table 9.3   Effect on Number of Piglets Weaned/Sow/Year on Food Cost per Weaner

Piglets weaned/sow/year	10	14	18	22
Basic/feed/sow/year + share for boar (kg)	765	765	765	765
Extra for sow while lactating (kg)	142	198	255	312
Creep feed for piglets (kg)	169	236	304	371
Cost of sow feed at 7.9/kg ($)	71.65	76.08	80.58	85.08
Cost of creep feed at 10.2/kg ($)	17.24	24.07	31.01	37.84
Total feed cost ($)	88.89	100.15	111.59	122.92
Total feed cost/piglet weaned ($)	8.89	7.15	6.60	5.59

Individual sow cards or a sow performance book should be kept. It should record the sow's number, dates of birth and service, returns to service, the boars used, the expected and actual farrowing date, the number of piglets born and weaned and the date and total weaning weight of each litter. This enables sows that fail to conceive regularly, to rear large litters or that develop weak legs to be culled. The number of piglets weaned in a litter and the farrowing interval are critical. These are controlled more by good management than by genetics. An acceptable standard is 15–16 piglets weaned/sow/year. Performance is improved by weaning at five to six weeks instead of eight and by serving the sow twice with a performance-tested boar within a few days of weaning. It might pay to have a bonus system for the stockman of, say, 50¢ for each weaner over 14 a year for each sow.

A specially designed sow calendar, with coloured pin coding, shows a manager at a glance when any sow is due to farrow, be weaned or come on heat. He can easily see if individual sows are cycling normally or not without checking paper records.

Dry and lactating sows should be rationed according to their individual needs and the method of recording feed issues should separate issues to breeding and fattening pigs. Feed absorbs such a high share of the value of output that waste must be avoided. This is aided by floor feeding twice a day to appetite instead of feeding *ad lib* from hoppers. If this is done and suitable rations are used, there is little scope for lowering feed costs except to provide an optimum environment and hygiene to allow pigs selected for improved feed efficiency to express their potential.

It is hard to use recorded data for pig fattening unless the pigs are weighed regularly. Not only is it desirable to know the feed conversion ratio for the whole feeding period of a batch of pigs, it is best to have this information every week or two. A feed conversion ratio 3.4 kg feed per kg liveweight gain is suggested as a suitable standard for porkers and 3.6:1 for baconers. Pen records are kept for fattening stock to enable the batch to be rationed on weight and age. With correct feeding and management, and tested breeding stock, top-grade carcases should be obtained.

RECORDS FOR POULTRY PRODUCTION

Farmyard, pocket money enterprises are not worth recording but specialist units must be recorded fully to relate output to controllable inputs, especially feed. A poultryman cannot afford not to spend a few minutes each day on record keeping as it is otherwise hard to find the reasons for poor performance.

As with pigs, rearing and production should be costed separately. You must know the cost of rearing a pullet to be able to cost egg production or to decide whether it would pay to buy point-of-lay pullets instead of day-old chicks.

Rearing records are usually kept over 20 weeks. They record the daily feed consumption and number of chickens in the house or unit. Thus, at the end of the rearing period, you have an exact record of the feed eaten and of any mortality, which should not exceed 5 per cent. Total rearing feed consumption should be about 9.5 kg. If it is much over this, feed has been wasted; if much less, pullets will be undersized and lay too few and small eggs.

The cost of a bag of feed is usually fixed as balanced feeds are usually bought, although it is possible to mix a balancer with home-grown grain. Waste of 1 kg/100 hens a day is common. This must be avoided by not filling hoppers more than a third full and by raising them off the floor so that the troughs are level with the hens' backs. It is hard to overfeed a hen provided it is of good genotype, produces well and the feed really goes into her. Reasonable standards for kg feed/dozen eggs are:

Battery	—	2.2
Deep litter	—	2.4
Free range	—	2.5

In laying trials 1.8 kg/dozen eggs has been achieved. Heavy breeds should not exceed 142 grams/feed/bird/day and light ones 128 grams.

If feed cannot be weighed each day, it is easy to calibrate automatic feed hoppers from the top down to show how *empty* they are when topped up each morning. A quick fall in feed or water intake can indicate disease, for example, coccidiosis.

Daily egg output naturally should be recorded and any marked fall should be investigated at once. The number of different sized and of cracked eggs should be noted. Output varies with breed, production system and culling policy. After all pullets are in lay, output should rise rapidly to a peak of 80–85 per cent at about 29–32 weeks. It should then fall by no more than 2–3 per cent a month but temperatures over 27°C lower yield.

During the second six months of lay all battery

birds whose output is below 50–60 per cent should be culled. These will include sick, broody, thin or overfat birds. Up to 30 per cent of the pullet flock may lay for 15 months before moulting. These birds should be kept for a second year to lower replacement costs. Many large producers, however, do not cull but sell the whole batch when average production drops to 60 per cent or less after 10–13 months. This is also necessary with deep litter or free-range birds. Reasonable standards for eggs/hen/year are: battery — 208; deep litter — 196 and free range — 164.

Mortality records indicate the severity of any disease outbreak. it should not exceed 0.25 per cent a week or 0.75 per cent a month and can be kept low by isolating rearing birds from adults and fumigating buildings between batches of birds.

RECORDS FOR GOATS AND SHEEP

Although some flocks are well managed, most get little attention. Individual breeding performance should be recorded so that future breeding stock are selected from those ewe lambs likely to have good breeding and mothering abilities and so that unprolific ewes can be culled. Each ewe should be identifiable by ear notching or by tags, the colour of which is changed each year. At mating time the rams' chests should be colour-marked; either with a proprietary colouring or with floor polish — green for the first 18 days, red for the next and black for the last heat period. This shows which ewes have not been served and should be culled and enables the lambing pattern to be forecast.

At weaning, 16-week weight gains for lambs are calculated by subtracting birth weight (or 3.6 kg) from final weaning weight, dividing by the days to weaning and multiplying by 112. This converted weight gain reflects the milking and mothering ability of the dam. Ewes whose lambs have inferior converted weight gains and those with slated mouths can be culled.

Lambing percentage is a familiar average that is supposed to measure the success of the lambing season but it can also measure stupidity. The true figure is

Total lambs weaned × 100
Total ewes put to ram

but it is easy to sell suspected barren ewes and to count the lambs too soon.

Total lambs born × 100
Total ewes giving birth

sounds better in the hotel bar but it does not bring in any more cash. Weaning percentage is usually well below 100 although this figure should be attained easily.

Adequate attention and supervision during the six-to-eight-week lambing period is essential. A bonus of, say, 17¢ a ewe with a lamb at foot at lambing or 68¢ a lamb weaned is usually worthwhile as an incentive to conscientious work.

INTER-FARM COMPARISONS

Sometimes called comparative account analysis, inter–farm comparison, on the basis of uniform costing, is merely scrutiny of a farm business to produce a statement of its costs and returns for comparing with one for similar farms. Comparing the accounts with some efficiency standards helps to show if the business is well-organised and run. Inter-farm comparisons are based upon past results but it is hoped that their analysis will give useful guidelines for future action.

Inter-farm comparison is a diagnostic technique requiring the calculation of 'efficiency factors' or indices for comparing with standards obtained from groups of similar farms. The main source of material is the profit and loss account but other financial or physical records may also be used. This method of business analysis was popular in Britain in the 1950s as a means of finding the strengths and weaknesses of a farm business so that attempts can be made to remedy the weaknesses.

Like more sophisticated measures, such as liquidity ratio, inter-firm comparison began in industrial accounting where it is still widely used. The British Institute of Management set up a centre for Inter-firm Comparison in 1959 that is still very active. The Centre has found huge differences between the profitability of otherwise similar firms in any one industry. Each company submits information in confidence and can compare this with the grouped data for unidentified similar firms. The differences in industry have not only shown that there is much scope for improvement; the comparisons that revealed these differences also showed each participating firm why its profits differed from those of others, where it was weak and which of its departments or activities could be improved. Inter-farm comparison can be equally helpful if interpreted carefully.

Clearly the data for a given farm and those with which they are compared must be produced in the same way, otherwise comparison would be invalid. The profit shown on financial statements is affected by matters other than farming efficiency. These include whether the farmer pays rent or is an owner-occupier, whether he uses unpaid family labour or has to pay for it all and whether he pays interest on borrowed capital or uses his own. Uniform costing is therefore necessary to standardise whole farm results before comparing them with standards. This does not mean that farms must be the same size, produce the same products or have the same plant and machinery but that they must be sufficiently comparable for rational conclusions to be drawn from reports based on comparison.

The following adjustments are needed to standardise the profit and loss accounts of all farms taking part in a standard costing scheme or comparing their own results with the results from such a scheme:

Add   If the farm is owned, a notional rent comparable to that paid for farms of a similar size and quality in the district.
The value of any unpaid farmily manual labour.
the value of farm produce or stores used by the family or labour force.

Remove   Mortgage payments and other capital repayments.
All landlord-type expenses, for example, owner-occupier's repairs.
Interest charges (but not bank handling charges).
Manager's salary.
All private or non-farm income and expenses.

All farms are treated as if they were tennanted, pay for all their manual labour and have no charges for interest or management. Standard values for livestock may need to be changed to realistic ones. Output is revenue plus closing valuation less opening valuation, livestock purchases and any other produce bought for resale. Costs are expenditure plus opening valuation less closing valuation.

The first step in inter-farm comparison is to select suitable comparative farm data. These should be for a farming system as like as possible to that being followed. The main sources are government departments, universities and consultants. Next compare your management and investment income to that of the group of farms

Table 9.4   Inter-farm Comparison of Cotton Results per Hectare

	Own farm	Standard (Top 20% of farms)
Yield (kg)	1760	2150
price/kg (¢)	14.8	14.2
Gross income ($)	260.5	305.3
Labour	46.0	45.7
Tractors/machinery	21.7	22.0
Seed	3.0	3.1
Fertiliser/lime	24.2	28.3
Chemicals	33.1	29.2
Packing	4.0	4.9
Transport	6.2	7.6
Insurance	0.5	0.6
Sundries	9.6	9.5
Variable Costs ($)	148.3	150.9
Gross margin ($)	112.2	154.4

and try to find the reason if it is lower. There is no need to bother about cents; figures can be rounded to the nearest dollar or 10 dollars. If management and investment income per hectare is poor, either output is too low, costs are too high or both. Each in turn can be studied in detail. To trace weaknesses it helps to consider those factors that contribute to profit, as shown in figure 9.11. The dependencies that produce profit are shown in growing detail from left to right of the network. The factors that usually affect profit most are shown towards the top of the diagram and those with least effect towards the bottom.

Records may show that a farm business is becoming more efficient each year — its labour productivity, fuel efficiency, return on capital and so on all improving. But this may not be the key factor; what really matters may not be how much better it is than before but how much better it is than its competitors. Inter-farm comparison can be valuable in indicating weaknesses and provoking action to improve them. Uniform costing enables the exchange or comparison of costs prepared on a similar basis in each business and this provides a means of raising efficiency throughout the industry. By profiting from the experience of the most efficient farmer, others can imrpove their results.

Inter-farm comparison was developed basically to guide farmers with few records apart from annual books of account, crop areas and livestock numbers. It was never meant as more than a rough check and guide on areas likely to repay detailed scrutiny.

Table 9.4 is an example of the use of inter-farm comparison of one enterprise.

Initial scrutiny of these figures shows that gross income, variable costs and gross margin per hectare are 15, 2 and 27 per cent respectively below standard. A closer look shows that the low gross income is due to a low yield and the breakdown of variable costs shows that this may be associated with using too little fertiliser. Chemical costs are 13 per cent above standard for a yield of 18 per cent below standard so it may well be worth looking critically at the type, timing and amount of pesticide used.

Common efficiency factors, or indices, include the following:

Gross output = (total sales + closing valuation) −
 (livestock purchases + opening valuation).
Net output = gross output less purchases of seed
 and stockfeed.
(This is better for efficiency comparisons as it allows for internal transfers on the farm reflecting different policies of self-sufficiency.)
Net income = (total sales + closing valuation) −
 (total costs + opening valuation).
Management and investment income = net
 income less value of the farmer's manual work.
(All of the above can be expressed per arable hectare, per individual crop hectares, per animal unit, per $100 labour costs, per $100 machinery costs or per $100 combined labour and machinery costs.)
Feed cost/$100 livestock sales.
Animal units/feed hectare.
Common costs/cultivated hectare.
Gross margin/cultivated hectare.

Many other efficiency factors have been used and others could be derived to suit specific needs.

Inter-farm comparison is a crude tool and has the following drawbacks:

- The delay between the year's end and the availability of the profit and loss account.
- The information in a profit and loss account, even after standardisation, is too scanty to help much in taking specific action and improve the business without further information.
- The diversity of farming systems in some areas makes it hard to obtain ecological and economic comparability and a reasonably representative group standard. If farms are grouped by main enterprise, they may come from different areas with different soils and climate and each farm has different objectives and resources.

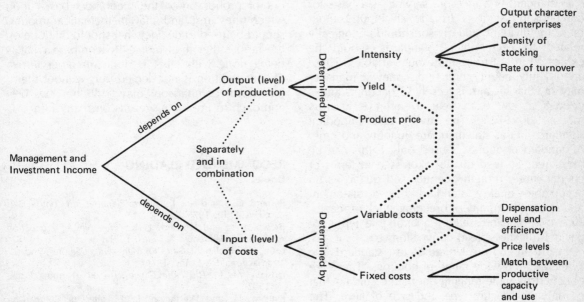

Strong links (broken lines) exist between level of yield and level of variable costs, and intensity and levels of variable and fixed costs.

Figure 9.11   A Profit Derivation Network

- Comparison is usually made with data derived from the results of the most profitable group of farms. Nevertheless, mistakes may have been made on these farms and are thus likely to be repeated.
- The whole farm approach is too general and more attention should be given to specific enterprises.
- It is concerned with average relationships whereas it is what happens at the margin that is relevant for good decision-making.

The moral seems to be that inter-farm comparison should be used only to pinpoint areas for closer scrutiny. It is little value comparing only one ratio and taking action to improve it, and the only true criterion is overall efficiency for which no single, universally reliable ratio so far exists.

Standards or targets

The value of any technique depends ultimately on what that technique achieves and this implies either setting a standard where none existed before or evaluation against acceptable standards. The establishment of standards where none existed was the original idea behind the scientific management movement pioneered by F.W. Taylor.

A standard is a measure serving as a basis for comparison. As it is usually an input: output ratio, the dividend and divisor should be logically related to each other. Standards should be expressed in not more than four variables. These are quantity, quality, time and money. A manager may say that his aim is to sell 13 baconers and 3 porkers a sow each year to achieve a gross margin of $125. The two most important standards in this statement are quantity and time. A standard of quality is used only if this can be measured. So feed quality expressed in terms of energy, fibre, protein quantity and quality is an acceptable quality standard but a statement relating to the quality of breeding stock is acceptable only if expressed in quantifiable terms — backfat thickness, feed conversion, etc.

In Britain and the United States, standards are available for a wide range of holdings. These are obtained by wide-ranging economic surveys that cover each specific type and size of farm. The results obtained from a representative sample of farms in a distinct group are used as standards by which the comparative efficiency of other farms in the same group can be roughly judged. Standards are also essential for variance analysis.

Given that standards must be quantifiable so that actual performance can be measured against planned performance, it need hardly be stated that the standards set must be realistic. Too high a standard results in disillusionment and abandonment; too low a standard is no standard at all. It should be attainable with effort.

In comparing gross margins for a given farm with suitable standards, it is usual to look at output before variable costs, in cases where the farm's gross margin is below average. Physical data relating to output should be known besides the value of output. If these are available, it should be easy to discern whether low yields or low prices cause the lower than average output per hectare. If output is normal then study of the individual items of variable costs is needed to find where these costs vary from the standard. Usually, low gross margins for cash crops result from low output, rather than high variable costs. For livestock, other than goats and sheep, it is impossible to generalise as a low gross margin is just as likely to be the result of high variable costs as of low output. Having decided what has caused the poor gross margin, the next step naturally is to try to remedy the fault.

It has been said of the use of farm standards and inter-farm comparisons that the effort made to collect and publish farm standards has been out of proportion to the theoretical basis upon which they rest and that mathematical models should be used to decide what should be done on the farm rather than indices to compare what is being done with what is done on other farms. This contention is not denied but, without them, inter-farm comparison may still be the best approach to improving results on many farms.

RECOMMENDED READING
Books

Barnard, C. S. and Nix, J. S., *Farm Planning and Control*, 2nd ed. (C.U.P., 1979).

Brown, J. L. and Howard, L. R., *Principles of Production Control* (Macdonald & Evans, 1971).

Croft, D., *Applied Statistics for Management Studies*, 2nd ed. (Macdonald & Evans, 1976).

Hosken, M. J., *The Farm Office*, 3rd ed. (Farming Press, 1982).

Ingham, H. and Harrington, L. T., *Interfirm Comparison* (Heinemann, 1980).

Lockyer, K. G., *Production Control in Practice* (Pitman, 1975).

Norman, L. and Coote, R. B., *The Farm Business* (Longman, 1971).

Taylor, A. H. and Shearing, H., *Financial and Cost Accounting for Management*, 6th ed. (Macdonald & Evans, 1974).

10 Marketing management

Much research and study of production methods has enabled greater production, often at lower cost. Mechanisation of farming, efficent ways of transporting goods, refrigeration, higher levels of consumer awareness and purchasing power now enable farmers to serve national and international markets. However, marketing has had less attention than production. We may ask of the marketing of any farm product:

- How does the market function?
- Are all the present market activities needed?
- Are they efficient?
- Do farmers receive a fair share of the price paid by consumers?

Markets and marketing

A market exists when buyers wishing to exchange money for a good or service are in contact with sellers wishing to exchange the good or service for money. It is a system of inter-communication or a network of dealings that exists to enable buyers and sellers of a good, service or resource to make contact. In marketing terms, a market is made up of people who use, need or want a product and who have the money to buy it. It does not simply mean a physical marketplace. There are many different markets for a product — each with different characteristics and a different purchasing power.

The term 'marketing' can be defined in many ways and several definitions by well-known authorities follow:

The management function that organises and directs all business activities involved in assessing and converting consumer purchasing power into effective demand for a specific product or service and in moving it to the final consumer or user so as to achieve the profit target or other objectives set by the company. (Institute of Marketing)

Marketing is not just selling; selling is a part of marketing. Advertizing and promotion are not marketing; they are part of it. Marketing is not market research; research is also part of marketing, and indeed I believe that manufacture is also a part of marketing.

Marketing is a perspective, a way of looking at the operations of a business — a way of doing business. A marketing perspective puts the customer at the centre of the firm's activities and orientates the firm towards its markets and customers rather than towards its factory and production lines. (L.W. Rodger)

Marketing would be better understood if one definition had universal acceptance. However, we should not be too worried. The term is open to varying definitions as each firm applies marketing in the way most suited to its business.

The common theme is that marketing is more than selling; it is the whole process that occurs between the production of any surplus goods or services and their consumption or use and it is consumer orientated. The most important need is not an exact definition but to acquire a sound understanding of what marketing means.

Marketing is the function that assesses consumer needs and then satisfies them by creating an effective demand for, and providing, the goods and services at a profit. Its main aspects are market research, product planning, merchandising, distribution, selling, sales promotion and public relations. That is, those activities that directly involve contact with the consumer, assessment of his needs and translation of this information into outputs for sale consistent with the firm's objectives. Marketing adds value to goods by changing the form, time and place in which they are available to meet an effective demand.

The essential principle of marketing is that businesses thrive by producing what can profitably be sold rather than easily be produced. For a business to exist, or to continue, it must identify and satisfy consumer needs profitably. Unless the consumer thinks he is getting value for money, he probably will not buy the product. Successful firms are market- or customer-orientated rather than product-orientated. A market-orientated firm finds out what people need or want, arranges all its resources so that the end-product will meet this need or want, uses suitable marketing techniques to sell the product and ensures that both buyer and seller profit from the transaction. It also seeks new market opportunities and adapts to the needs of potential as well as existing customers. By contrast, production-orientated firms either launch new products without market research or keep offering existing products year after year. They seem blind to competition and to ignore changing market conditions.

The importance of marketing

For economic development it is important to raise farming output but equally so to develop marketing so that the extra production reaches consumers efficiently. In a competitive economy, greater marketing efficiency will not only give farmers higher prices but also give consumers lower ones and thus expand their buying power.

It is often hard for people trained in production to see that the marketing function dictates production policy, which must be firmly based on market potential. Market conditions basically influence production so much that production and marketing are often best regarded as an integrated whole. Finance, production and marketing form a trinity of interests. Neither can be truly effective without a right balance of the other two.

Good marketing facilities are especially important for peasants with small family holdings and only small surpluses over their subsistence needs for sale. The small scale of their invididual activities tends to raise input prices and lower product prices. Economies of size seem to be more common in market-related activities than in production-related ones. A market structure exists in which, as both a seller and input buyer, an individual farmer is a price taker with no power, by himself, to affect the market.

It is often time-consuming for small farmers to take their surpluses to a marketplace, so often a middleman, a storekeeper or itinerant trader, buys from many farmers and sells to a marketing organisation or directly to consumers. For export products, there are often several 'middlemen'. Farmers often complain that the distributive margin between the prices they receive and those consumers pay is too high.

Vertical integration exists if one firm owns or fully controls several stages of production of a commodity and if a commodity is produced at each stage that could be sold. In farming, it occurs mainly in enterprises where there is much contract selling, as with the broiler industry. Hatching, rearing, fattening and even consumer selling outlets may all be under the control of a single enterprise. In these cases, farmers are little more than paid staff, though they are paid for their output rather than their time.

Perfect and imperfect competition

It is often said that farming occurs in conditions of perfect competition. These are that:

- each market has many independent producers none of whom produces enough in total to be able to affect market price by offering or withholding his product;
- there are also many similarly independent buyers, none of whom can individually affect price;
- the produce in any one market is homogeneous;
- all buyers and sellers always have full knowledge of the current market price.

With modern communications, sales can occur almost simultaneously, thus fulfilling another condition for perfect competition. A Sri Lankan buyer, for example, can compare offers of rice from Thailand, Burma, China and the USA in a few minutes.

In practice, however, imperfect competition is common. This is partly because of government or marketing board intervention to reduce the effect on farmers of changes in demand or supply by fixing prices or issuing production quotas. Also, some buyers and sellers are large enough to affect market prices by their own action. For example, a peasant growing chillies for sale at a small village market can produce a high enough share of total sales there to affect the local price. If there is only one producer of a commodity for a given market, he is said to have a monopoly in the sale of his product. This is more likely with a producers' marketing co-operative, a state farm or a plantation than with a small farmer. With imperfect competition, a farmer can sell more only if consumers will buy more and that usually means a fall in price.

Imperfect competition also occurs if there are few buyers of a product or input. This is called monopsony. This is common at small local auctions when buying is controlled by a few local dealers and butchers who can combine to keep prices down, a small rise in supply can lead to a drastic price fall.

Marketing functions

Describing the functions of marketing means describing what happens or can happen to products between the time of production and the time of purchase by the consumer. This means how marketing can help in presenting goods to consumers so that both producers and consumers gain. One or more of the following marketing functions are essential: assembling, preparation, grading, processing, packing, storing, transport, distribution, publicity and selling.

Assembling

Collecting the product so that it is available in large enough amounts to attract buyers. This function may be fulfilled by itinerant buyers who visit farms or by local marketplaces.

Preparation

Also called merchandising, this is the act of presenting goods to potential customers in the most appealing and convenient way. Not all products are suitable for consumption in the state in which they are sold from the farm. For example, a farmer sells a steer but a consumer does not want to buy or cook a whole steer. Farmers initially sort tobacco leaf into a few grades for sale to a marketing body but international buyers require a finer grading before they will bid for the leaf at auctions.

Middlemen play a large part in preparing many farm products into forms that the consumer wants and will buy. This raises the value of the product to the consumer. Naturally, middlemen charge for this service to cover their costs and leave a profit (unless the middleman is a state-subsidised organisation). Preparation includes grading, processing and packing.

Grading

Grading and inspection are needed to maintain high quality, enable good pricing policies and to promote exports. It may be done by farmers, by marketing organisations or both. Farmers soon learn that good grading of, say, seed cotton will give them a better price as each grade has its own consumer market. Marketing organisation graders must be honest and skilled, otherwise farmers soon become dissatisfied.

For good grading, exact, standard specifications must be known by all concerned. Some factors affecting quality and grade include evenness of size, shape and quality, condition, purity, flavour and freedom from pests and diseases.

Processing

Many farm products are processed before consumption. For example, steers are slaughtered, dressed and cut into joints of meat, some meat is mixed and made into sausages, fruit and vegetables may be canned, seed cotton is ginned and spun into yarn and maize is pounded or milled before being used for human food or stock-feed. Preparation increases the utility of most farm products and, therefore, also their value and price. Thus the prepared product costs more than the raw one.

Packing

This makes some products easier to handle; for example, eggs and sugar. It helps to compact others, like tobacco and cotton, by pressing them into bales. This saves storage space and transport costs. Packing also helps to keep produce clean, protects if from damage and adds value; for example, tea bags instead of bulk tea in chests. Food must be wrapped hygienically and the package should offer the consumer a recognisable quantity and standard of quality. The outward appearance, the design and presentation of consumer goods should be constantly reviewed.

Packing was once regarded as a production function but now it takes its correct place as part of the 'marketing mix'. Packing is especially important to farmers who sell to wholesalers or retailers or direct to consumers. Potatoes, for example, could be packed in sacks or in small, transparent plastic bags with the name of the farm, or the farmers' co-operative, printed on them to help establish a brand image. Every package should help to sell the product.

The growth of self-service shops has raised the importance of packaging. For example, the package must not be too tall for the average shelf and, if it gets turned round, it should be printed on all sides.

Bulking of goods for transport requires large containers, crates, casses or boxes. There is a growing trend towards using lighter forms of non-returnable packing material to save transport costs and the cost of recording and handling returns.

Storage

Storage of goods, sometimes called warehousing, is another charge that can widen the gap between the farmer price and the consumer price if it is done by a middleman. Consumers' demands for farm produce vary little but most farm production is seasonal, so storage becomes necessary either by the farmer or a marketing body. Costs are incurred for protective chemicals and interest so consumers usually pay more for out-of-season produce. This is especially true of perishable products that need controlled air temperature and humidity.

Transport

This is a crucial part of marketing. All surplus farm produce needs some transport, provided either by farmers or middlemen, to the marketplace. Transport costs for export products is especially high, particularly where roads are poor. These affect both consumer and producer prices. It is important to try to reduce cost by avoiding the waste of half-empty vehicles or empty return trips.

The main points affecting the type of transport to use are the time factor and the weight and bulk of the goods. The fastest means of transport, like aircraft, cost most but are sometimes justified for exporting high value, non-bulky goods, like day-old chicks, fresh vegetables and flowers from Kenya to Europe or frozen beef from Botswana.

Unit loads are eased by using pallets and fork lift trucks.

Distribution

Traders, wholesalers and retailers help in the distribution network. Owing to long distances and high transport costs, distribution is costly and adds much to the consumer price.

Publicity

Before a product is distributed, a demand should be created or stimulated for it. This is the task of advertising and other promotional activities and it is useful to farmers selling uncontrolled products on the free market; like certified seeds, pedigree breeding stock, hay or machinery contract services. The newspapers, farming press, sales literature, agricultural show and trade fairs can all be used.

Packaging publicity is cheap and effective because the container stays with the consumer until its contents are all used.

SPECIAL PROBLEMS IN AGRICULTURAL MARKETING

Marketing farm produce is affected by certain features of farming that together are unique to the industry. There are:

- the seasonal pattern of production
- inelastic demand for many farm products
- the large number of small producers
- the length of the production cycle
- the dangers of bad weather, pests or diseases
- the range of product quality
- perishability of many products
- low bulk value of many products, and often
- long distances from farms to marketplaces

The seasonal nature of the climate causes seasonal production of most products, especially crop products. The surpluses at harvest must be either stored or sold.

The demand for much agricultural production is inelastic. That is, it does not respond quickly to price changes. This is explained, for example, by a person who has sufficient food of the type he likes. He is not then likely to buy more of it, even if he has spare cash. However, with luxury goods, like radios and bicycles, the more spare cash a person has, the more he is likely to buy. If the price drops, he is more likely to buy *extra* luxury goods. Thus it is said that the demand for luxury goods is elastic. As a result of the seasonal supply and the inelastic demand for farm products, their market price can vary widely.

Where both buyers and sellers operate as small units, none can individually affect price. The smaller a farmer's scale of business and the less sound his financial position, the more he is at the mercy of the market.

The longer the production time, the more uncertain farmers are of the prices they will receive. Coffee, for example, therefore involves more uncertainty than maize. Besides price uncertainty, there is also much yield uncertainty in farming because output is not fully under farmers' control owing to weather, pests and diseases.

Because of bad weather, poor handling or failure to control pests or diseases, farm products are often of mixed quality. This leads to a marketing problem in fixing fair prices and necessitates grading.

As most farm produce is perishable, storage problems arise. Processing can solve this problem but it is costly and hard to organise because most of a given crop is harvested at about the same time. Consequently, the production of perishable products is often limited to areas near to their markets. Also, much farm produce has a low value per tonne or cubic metre compared with industrial products. Cassava is a typical example. This raises transport costs as a proportion of the product's value. This problem can often be relieved by initial, local processing. For example, peeling, chipping and sun-drying cassava before dispatch instead of transporting whole, heavy and perishable tubers.

The distance of many farms from the market for one or more of their products can greatly affect the on-farm price if the farmer has to deliver the product to a marketplace before receiving a fixed market price. In many countries, however, marketing boards have many local buying points and a uniform national price for both farm products and farm inputs for purposes of social equity and regional balance. Clearly, this distorts the free working of the principle of comparative advantage and implies subsidisation of producers in remote areas by those more favourably located. Without this subsidisation, there would be more geographical concentration of production of bulky, low value products for sale and this would lead to economies of scale and savings in both product assembly and distribution and in input distribution costs. Nevertheless, the location of farms still strongly affects farming systems.

Remote areas have a comparative advantage in cattle production, because they can be trekked to market, whereas areas near towns have a comparative advantage in milk, vegetables and any other products without a controlled price and a widespread network of buying points. Besides distance from towns, distance from good roads or the line of rail, rivers or processing plants greatly affects transport costs and thus the supply zone for bulky farm products.

MARKETING CHANNELS

The channels through which farm produce is marketed have adapted themselves over the years to the special needs of given products so that many different systems now exist and new developments are frequent.

Marketing channels vary from individual farmers or companies acting alone, farmers contracting to supply produce to processors and other bodies, and farmers acting together in co-operative societies to statutory marketing bodies. The marketing of a given product need not be confined to only one of these channels. Several may exist side by side although, with some products, there may be good reasons for using only one channel.

Choosing suitable distribution outlets, wholesale or retail for example, is usually of basic importance to successful selling. Wholesalers, retailers and consumers all buy some farm produce direct from farmers but increasingly the first buyer is either a co-operative, a food processor or a statutory marketing board.

In the simplest marketing channel, producers are in direct contact with consumers and this is still important in peasant farming. An unknown amount of produce is traded locally between individuals and villages and much of this trade is conducted by barter. Some of it occurs illegally across national frontiers if official prices are below the open market ones.

Products sold privately, without the aid of formal marketing organisations often include poultry, eggs, fruit and vegetables and estate-grown tea. Merchants sometimes buy produce on the farm by private treaty for resale to processors or shippers. Much private selling of cattle is done 'on the hoof' on farms.

Auction selling is common, especially for export crops like tea, coffee, cocoa, tobacco and also for cattle. Produce from many scattered farms is assembled at one place, either by the growers or by a marketing board, and buyers come from a distance to bid for what is on offer.

Local produce markets, usually run by urban or district councils, are widely used for perishable fruit and vegetables and the local staple food if it is not a fully controlled product that must be offered for sale only to a marketing board.

It has been described already how farmers can help themselves by forming buying groups and machinery syndicates. It is hard for small individual buyers and sellers to establish contact but the position can be much improved by co-operative buying and selling. A co-operative is a group working together towards a common aim for mutual benefit. There can be production marketing, supply and services co-operatives. Producers' co-operatives can not only obtain economies of scale in marketing but also in buying inputs which can be brought back in lorries after delivering produce to the best market.

Co-operative societies may market single products, such as ghee, or a range of products. Besides buying activities, they may engage in the whole range of marketing functions — assembling, grading, processing, packing, storage, transport and selling. They also often provide credit to members. These are multi-purpose co-operatives.

Trading co-operatives often buy the produce of small producers and process it quickly, cheaply and efficiently. The society can afford better processing equipment than an individual farmer, for example, coffee processing and tobacco curing. A natural development from co-operative selling is co-operative buying of seeds, machinery, sacks, fertilisers and biocides. Quantity discounts are obtained by the ability of the society to buy in bulk and the common problem of farmers having to buy retail and sell wholesale is reduced. Members increase their bargaining powers in relation to large-scale commercial firms.

Societies can usually obtain loans and provide seasonal credit to their members whose produce is marketed through the co-operative. They thus provide machinery through which credit can be covered. This is particularly important for cultivators on communal land who have no title or acceptable security to offer.

Starting a co-operative

Members must be prepared to help themselves and each other without relying *wholly* on outside help although, when it is proposed to start a society, advice should be obtained from the representative of the relevant government department. Leaders whom the members can trust and follow are essential. An able and honest secretary is also necessary to run the day-to-day affairs of the society. The co-operative officer should avoid giving directions rather than advice or members are likely to regard the society as an imposition rather than as an aid to orderly buying and selling.

IMPORTANT CO-OPERATIVE PRINCIPLES

1. Unrestricted, voluntary membership should be open to anyone of good character.

2. Roughly equal shares in money contributed to stop the richer members trying to control the affairs of the society (limited share-holding). Share capital is returned if a member leaves.
3. A low, fixed rate of interest on capital as a first charge on profits.
4. Profits not used to buy new equipment for the society are distributed to members as a dividend in proportion to the amount of trade each member has conducted through the society.
5. Equal voice in control. 'One man, one vote.'
6. Trade is limited to members who undertake to deliver all their sales of certain products to the society. If members use co-operatives as support buyers and desert them when a short-term price advantage offers elsewhere, the movement will not be viable. Members cannot decide to sell the best part of their produce themselves and the poorest part to the co-op.

Some enterprising farmers sell produce on contract directly to institutions, like hospitals, schools or prisons. Others sell, often on contract, to processors. For example, smallholders, called outgrowers, may sell sugar cane or tea in green leaf form to estates that are large enough to have the necessary factory for processing. Flower and vegetable seeds may be grown on contract for seed merchants and fruit or vegetables for food processing firms. Farmers can also sometimes secure contracts with wholesalers or retailers.

The value of contract selling to farmers is that they know beforehand what price they will obtain and how much they can sell at that price. They may also obtain cheap credit and specialist technical advice from the processor to whom they can pass all marketing problems. However, farmers lose the benefit of a price rise if there is a shortage of the product.

There is some similarity between the functions of marketing boards and of co-operatives. However, a board whose activities are under government control may be more effective in support buying and price stabilisation for producers. A marketing board is a producer-controlled, compulsory, horizontal organisation, authorised by government to perform specific marketing functions in the interests of the producers of a product(s). They may or may not trade and, if they do, they may or may not have a monopoly.

A criticism of monopolistic marketing boards is that, because they must buy produce offered by all the members they can keep in being inefficient production patterns. If there is a truly competitive demand, co-operatives without compulsory powers may work just as well as marketing boards. However, in conditions that tend to over-supply and if buyers are few, marketing boards, answerable to the government, seem to be better marketing channels.

If a farmer sells to a producers' co-operative or marketing board, he may still be unprotected from changes in demand — nor indeed should he be, except in the short term, as the continued production of unwanted surpluses of goods is an uneconomic use of resources. However, he is more likely to be able to plan his production for any one year without bothering what demand will be when his produce is mature, as producers' marketing agencies usually try to run price stabilising policies.

Government intervention and its influence on marketing

Many governments have recently tried to build marketing policies into their economic development plans. There has been a growing realisation that progress in smallholder credit, land reform and extension programmes is slowed by ineffective marketing systems.

The reasons that governments often intervene in marketing more in agriculture than in other industries are as follows:

- The strategic value of agriculture to the nation.
- To help the farmer improve his competitive position, to provide confidence for long-term production, to obtain economies of scale, to even out seasonal supplies, to stabilise prices, to make the best of differing markets and to raise both demand and quality.
- To help consumers by ensuring and expanding supplies and by stabilsing prices.

I pointed out before that prices of farm products tend to vary widely because of erratic supply and inelastic demand. It is hard to plan production rationally with unstable prices. This is bad for the whole farming industry and can be disastrous for individual farmers. Therefore, to give some stability to the industry and to encourage growth, most governments intervene in the marketing of many of the main farm products, both economically and physically. Governments can intervene economically in the agricultural market by guaranteeing produce prices, controlling inputs or controlling output.

Provision by government of physical infrastructure, to aid efficient marketing, includes roads, railways, telephones, market places, storage and handling depots and processing plants.

Producer price and consumer price stabilisation are important aspects of government intervention. This is usually achieved by holding either a buffer fund or buffer stocks to protect farmers from wildly changing prices on world commodity markets. Government price stabilisation and support programmes, especially announcement of minimum guaranteed prices before planting time, improves peasant farmers' decisions by reducing price uncertainty.

Few economists have distinguished between yield variations and price variations but have concentrated on their combined effect — income variability. The price of export crops may be unrelated to national yield so price stabilisation greatly helps growers. However, there is a negative correlation between the price and market supply of the local staple food crop and price stabilisation for these crops would increase income variability.

Governments sometimes intervene by controlling the supply or price of farm inputs like seed, fertiliser or biocides. The supply of a product, or its quality, can often be raised by supplying inputs at a subsidised price in suitably sized packs at the right time and within easy access for farmers.

Output can be controlled not only by controlling input supply and prices but also by a system of producer registration and the issue of production or area quotas. Quota systems are often arranged so that a good grower, who produces his full quota of high quality produce, can obtain a larger quota in the next year. Sometimes, quotas are bought and sold between farmers.

PRICE FIXING WHEN PRICE IS NOT CONTROLLED

The selling price of a product is vital in marketing. It is important, therefore, to understand how selling prices are fixed and what research should be done beforehand. As sales finally depend on consumers' attitudes to the selling price, their reactions are important for price fixing.

The price of a product in an open market is basically controlled by supply and demand. If the supply of a product is small and many people want to buy it, the price will be high. When it is in plentiful supply but there are few buyers, its price will fall. This is common straight after harvest with seasonal crops for which the price is uncontrolled.

The complexity and importance of deciding the 'right price' varies with the size of the business and the product to be sold. This difficult decision is easier for those, like stall-holders, who are continually facing their customers and can quickly adjust their prices. Two basic factors, however, affect the price of a product — the *cost* and the *market*. The most common pricing methods are as follows:

The cost-plus method

For example,

Sales forecast:	6000 gunny bags of yams
Estimated costs:	$4000
Unit cost:	$0.67
Add profit margin:	20 % = $0.13
Selling price:	$0.80

Marginal pricing

In the last method the estimated costs of $4000 covered both variable costs and a share of common costs, say, $1000. The sale of 6000 bags will cover the common costs to be met by the contribution from the yam enterprise. Now, if a sudden order arrives for a further 1000 bags of yams, the common costs have already been met and any price over the variable cost of production will add to profit. For example,

Variable cost/unit: $(4000−1000/6000) = $0.50
Add profit margin: 20% = $0.10
Selling price: $0.60

Competitive pricing

Price is fixed after considering the price range between competitors offering the same or substitute products and after allowing for any difference in quality.

'What the market will bear' method

The product is priced according to assessments of how much consumers would be willing to pay. A balance must be found between too high a price, which might attract competition, and too low a one to yield an adequate return.

Farmers can often raise their product prices, a part from sheer bargaining power and marketing skill, by improving quality or the timing of sales. In each case, extra costs are usually incurred so a partial budget should be prepared to ensure that extra income will exceed these.

Many products, like tobacco, groundnuts, cotton and eggs, are priced on quality and to some extent this can be raised merely by more careful management of production and product preparation before sale.

Price improvement by timely selling often entails storage and other costs. For example, the costs of grain storage may include depreciation and interest on the plant and equipment, interest on extra working capital needed, loss of weight through drying and extra handling and drying costs. Nevertheless, it often pays farmers to store products unit prices rise or to plan for livestock to be ready for sale during a main festival. Another way of producing out of season, for sale at times of higher prices, is to use full or supplementary irrigation.

GLOSSARY

Barter direct exchange of goods or service without use of money.

Competition the environment in which production and distribution are carried on. The two extremes are perfect competition and monopoly, both being extreme conditions that do not exist in real life.

Contract a legal agreement between two parties.

Demand the willingness and ability to pay a sum of money for some amount of a given good or service. Sometimes called effective demand to distinguish it from need.

Elasticity the degree of responsiveness of demand or supply to a change in price.

Elasticity of demand the responsiveness of demand to a change in price. If a small price change results in a large demand change then demand is elastic.

Market forces the forces of supply and demand that together fix the price at which a product is sold and the amount that will be traded.

Monopoly strictly, this occurs when there is only one producer of a commodity for which there is no substitute. Widely used to mean near-monopoly or very imperfect competition.

Monopsony the situation in which there is only one buyer in a market.

Open market a market in which there are no restrictions on buyers and sellers and where prices are fixed by supply and demand.

Price the cost of one unit of a good or service. It is usually expressed in money but not always — the price of a bride may be 25 cows.

Private treaty a method of sale where the price of the commodity is decided by bargaining between the buyer and seller.

Quota in the sense of this chapter, a quantitative limit on the production of a specified commodity.

Subsidy a payment by the state to an industry to prevent its decline or rise in its prices.

Utility the satisfaction derived from some quantity of a good or service.

RECOMMENDED READING
Books

Abbot, J. C., and Makeham, J. P., *Agricultural Economics and Marketing in the Tropics*, (Longman, 1979).

Barker, J. W., *Agricultural Marketing*, (O.U.P., 1981).

Bateman, D. I. (ed.), *Marketing Management in Agriculture* (Department of Agricultural Economics, University College Wales, 1972).

Brech, E. L. (ed.), *The Principles and Practice of Management*, 3rd ed. (Longman, 1975).

Dean, E., *The Supply Response of African Farmers* (Amsterdam, North Holland Publishing Co., 1966).

Jones, W. O., *Marketing Staple Food Crops in Tropical Africa* (Cornell University Press, 1972).

Upton, M., *Farm Management in Africa* (O.U.P., 1973).

Whetham, E. H., *Agricultural Marketing in Africa* (Nairobi, O.U.P., 1972).

Wilmshurst, J., *The Fundamentals and Practice of Marketing* (Heinemann, 1978).

11 Project management

The subjects of this chapter are capital investment appraisal — mainly a planning tool — and network analysis — both a planning and a control tool. Neither the techniques of capital investment appraisal nor those of network analysis are as complex as they may seem at first. Sadly, they are used far too little by farm managers, although they are both useful in business and private life.

Project management is described in this chapter from the viewpoint of a private farmer or company and not from that of the state. Consequently, the stress is on financial costs and returns and no mention will be made of economic or social costs and benefits or benefit: cost analysis. As technology advances, new projects become ever more complex and competition rises. This makes it harder to make good investment decisions.

A separate chapter is given to project management because of the unique one-off nature of a project which is defined as any task that has a beginning and an end and during which each job occurs only once. It usually involves fresh injections of capital. Projects should have identifiable goals to be achieved within a set time limit. Unfortunately these often remain unspecified. Managers have two overlapping functions to perform: to plan and control — to plan the work and to work the plan. The planning task is to:

- forecast the relationships between all the different activities,
- estimate the duration of the activities,
- channel materials and other resources to the right place when needed.

Specialised topics are best covered by specialists in specialised texts. This chapter is intended only to introduce investment appraisal and network analysis. Those needing more detail should read Sturrock and Mulvaney respectively.

CAPITAL INVESTMENT APPRAISAL

It was once said that agricultural projects in their early stages proceed as follows: (a) get it to grow; (b) stop it being eaten by pests and (c) see if anyone will buy it. Great strides have been made in improving yields and biocides; less so in marketing. Simultaneously, scrutiny of the likely results of potential projects is now much more thorough than it was immediately after World War II. Much literature on investment appraisal has dwelt on mathematical techniques but the Tanganyika groundnut scheme did not fail because someone forgot to discount the cash flows. One should not be diverted from the basic exercise which is to check the assumptions on which the technical analysis is based. These include capital cost, development time, markets and prices, yields, running costs, availability of raw materials, water, transport, storage, management and labour. Investment appraisal techniques do not replace managerial judgement but aid it by showing how profitable a given project will be if expectations are realised.

No investment in a farm business should be undertaken unless the benefits exceed the costs sufficiently to repay the initial capital cost over the economic life of the project and also give a yield equal or better than what could be obtained elsewhere by investing at compound interest. There is always a cost in acquiring capital, even if only an opportunity cost, so a firm must invest only in projects that yield a return greater than the cost of finance.

Investment appraisal now features more strongly than before in farm management literature. It means choosing between alternative investments; otherwise there is no decision to make. Some factors to be assumed can be quantified; others not. The most important factors in most projects are the net cash flows, the life of the project, the rate of interest or cost of capital, taxation, risk and uncertainty. All assumptions should be strictly checked.

The farm capital that should be examined closest is the small share of it that is being newly invested and is uncommitted until the investment occurs. In theory, even past investment is always being newly invested as, if it is decided to leave it alone, it is really being reinvested in its current use.

Investment appraisal means analysing the prospective costs and benefits of possible new investments and evaluating the desirability of committing resources to them. It is about the future management of capital, whether fixed or working capital and whatever the source of finance.

Careful investment appraisal is important because it is rarely possible to remedy mistakes and these can affect business results for years ahead. It will not guarantee wise investment decisions but it provides a sound basis for making them. A methodical approach to the planning and control of projects is essential for the best use to be made of resources. Few farmers have capital reserves to finance large projects and must borrow at least some of the funds needed. To borrow funds, the lender must believe in the profit ability of the proposed project. The appraisal should assure the farmer that he can repay interest on any loan, obtain sufficient reward for his risk and recover his capital investment during the life of the project.

All investment decisions refer to the future so the factors determining those decisions can never be certain. The further into the future a forecast is made, the less certainty there is about possible changes in demand, supply, prices, costs or methods. Managers, therefore, should try to build safety measures into their appraisals. Estimates of a project's future cash flows, and often its cost, are inevitably uncertain. One theory of profit is that it is the reward needed to induce firms to take risks and accept uncertainty; so the more the uncertainty, the more profit required.

Possible ways of coping with the risk and uncertainty underlying investment decisions are to require a higher return before accepting a project regarded as risky, weighted returns and sensitivity analysis. The latter will be described later. A weighted return requires an estimate of the most likely outcome, a fairly optimistic estimate and a moderately pessimistic one. The usual weights used are 50, 25 and 25 per cent respectively.

The rather crude but common methods of appraising possible projects by pay-back periods or simple average rate of return on capital were described in chapter 4. Tables 4.12 and 4.13 showed that neither of these methods are satisfactory on their own for choosing between alternative projects and can lead to bad decisions. Traditional return on capital and payback are useful screening devices to reject projects that are 'non-starters' but the final decision should be based on the results of discounting the net cash flow over the life of the project to allow for the time value of money.

THE TIME VALUE OF MONEY

This is a simple notion that is often dressed up to make it seem complex. It is based on time preference — a theory of interest based on the idea that some people prefer to have money to spend at present, and are willing to pay for this privilege, whilst others, if paid interest for doing so, are willing to delay their spending. Money in one's pocket now is worth more than the same sum in a year's time, even in the absence of inflation, because during that year it could have been invested at interest. It is also worth correspondingly more than the same sum several years ahead because the interest could have been compounded. Thus the future earnings of a capital investment project become less valuable the further ahead they accrue and the sooner the profits come, the sooner they can be reinvested to earn more income.

To allow for this simple fact, which has nothing to do with the fact that the value of money is already often falling over time through inflation, although one can allow for this also, future earnings are discounted back to the present. Long-term investment is thus more complex than simply ensuring that total benefit exceeds total cost as time itself is a constraint. People have limited lives and companies do not last for ever. Time preference is why people expect interest on their savings and to pay interest on mortgages. It is better, even if future cash flows were certain, which they never are, to receive money sooner rather than later and to pay it later rather than sooner. This is because of time preference, inflation and uncertainty.

Clearly, the interest rate a person will accept to forgo present consumption varies with his financial position and expectations for the future. That needed by an ascetic monk may be very low but an alcoholic father of 10 starving children, paying alimony, would need a high interest rate.

The time value of money is built into the principle of compounding and discounting. Most people are familiar with the idea of compound interest. This is the addition of interest to cumulative debt at the end of each time period. It is expressed by the following equation:

$$F = P(1 + r)^n$$

where   F = future value, r = rate of interest for the time period in decimals,
P = present value, n = number of time periods.

For example, the future value of $5000 invested now at 25 per cent compound interest paid annually for three years is $5000 $(1 + 0.25)^3$

= $5000 × 1.25^3
= $5000 × 1.953125 = $9765.62

Future value factors are given in Appendix A.1.

If $5000 now, at an interest rate of 25 per cent paid at the end of each year, is worth $9766 in three years' time, then, conversely, the present value of $9766 to be received in three years' time is $5000. This can be shown by reversing the equation to become:

$P = F/(1 + r)^n$ or $P = F[1/(1 + r)^n]$

The expression $[1/(1 + r)^n]$ is called the discount factor or present value factor and is given in Appendix A.2. In this case, it is 1/1.953125 or 0.5120.

$9765.62 × 0.5120 = $5000.00

The reverse of compounding, that is finding the present value of sums to be received or paid in the future, is called discounting. This is a way of assessing the present value of a cash flow at various future dates by allowing for the time value of money. It enables projects to be compared over the whole of their productive lives, even though the returns from the projects vary each year. Failure to discount future cash flows can easily cause poor decisions, especially if two projects are being compared.

———————— COMPOUNDING ————————→

P   F

←———————— DISCOUNTING ————————

Discounting helps managers more than compounding as they are concerned with *present* values in making current decisions. It enables the present value of cash flows that vary in time pattern to be directly compared.

DISCOUNTED CASH FLOWS

Many companies now accept discounting methods of screening capital investment proposals. However, it should be stressed that DCF techniques are not the beginning and end of investment appraisal. Comparison is made of the present values of the flows of cash that are expected from each project during its life. It is possible to assess precisely the return obtained on capital even if net cash flows vary over the project's life. These techniques are likely to become used more in farming, especially if it is capital-intensive.

DCF methods should ensure that investment decisions are soundly based and often indicate decisions different from those arrived at without discounting and show that a project deemed excellent by traditional methods is marginal or worse. However, provided the estimated net cash flows are fairly accurate, decisions made after discounting will be more valid than those based on actual net cash flows. This is especially so when governments use the tax system, or direct incentives like subsidies or grants, to hasten development. It should be clearly understood however that no technique can foretell the future.

Cash-flow budgeting, without discounting, and the concept of the net cash flow for a given period was described in chapter 4. Cash outflows are subtracted from cash inflows to give the net cash flows. No distinction is made between capital and current operating items; tax payments and subsidies are included and depreciation is excluded. For discounted cash flows, however, we also exclude interest payments and repayments of principal. Cash flow prediction is essential before investment to ensure that ends will meet. As the flow are available, it is little more trouble to discount them. There is usually an initial cash outflow followed by later cash flows that can be any combination of inflows and outflows. The cash flows in question are those that are incremental if the project is accepted. In other words, they are measured by the difference between the cash flow that would exist if the project were accepted and those that would exist if it were not.

It is important to estimate as accurately as possible the time during which an investment project will yield returns — its economic life. Some enterprises have predictable lives. Tea can last over 50 years; some vegetable crops only three months. It is harder to set a limit for livestock enterprises. It is seldom worth budgeting for more than 15 years as estimates beyond this time are likely to be unreliable and the remoter the period the less is the present value of the incremental net cash flow. Machinery and equipment usually should be costed as having a life of less than 10 years. The effects of obsolescence, wear and tear and residual value should all be estimated. Projects that take a long time to provide any return, like forestry, are rarely competitive with alternative and uses in DCF calculations.

Although the receipt and payment of cash usually occur throughout the year, cash flows are normally assumed to occur at the end of each year as lump sums. This assumption is merely for ease of calculation as it would be impracticable to discount cash flows for each day of the year. In any case, most discount tables that are widely used are based on end of period cash flows.

If the assets used in a project still have value at the end of the economic life assumed, this terminal or salvage value, and any recoverable working capital, is added to the last year's cash flow. As mentioned, tax payments and savings should be included in the cash flows. Taxation and investment incentives are too important to be omitted. They should be inserted in the time periods during which payment is likely to be made or received.

Working capital, cash needed to complete a single production cycle, is often high in relation to fixed capital needs in farming projects. In this case, it is misleading to regard all cash flows as arising at the end of each year. It is probably wise to use quarterly or monthly cash flows and discount these with the relevant discount factors for the shorter time periods. In any case, it is often important to estimate short-period cash flows for

other reasons in assessing a capital investment project.

However refined the calculations may be, the validity of the conclusions depends on the accuracy of the data used. There are many variable factors in DCF calculations so, unless the input figures are realistic, the output information will have little use. These variables include future selling prices, future sales volume, future production costs and the estimated life and residual value of the asset.

The main DCF methods are net present value and internal rate of return.

Net present value

Sometimes called excess present value, this is a single figure for a series of net cash flows. It is the present value of future net cash flows less the initial capital cost. In other words, it is the present value of the surplus profits expected after repayment of principal and interest. Exchanging a stream of future receipts for a single, discounted present value is called capitalising. This makes it easy to compare alternative projects.

Let us look at the example illustrated in table 4.13, where pay-back showed projects D and E to be equally desirable and both better than project F. This example is re-worked to find the net present values in table 11.1

When the timing of the net cash flows is accounted for, it is clear that none of the projects can repay the capital cost with interest at 20 per cent over eight years. However, project E is less unfavourable than project D. If we invested $8500 in it now, we would be worse off by only $774 than if we invested it at 20 per cent fixed compound interest for eight years.

Table 11.1   An Illustration of the Calculation of Net Present Values in Three Projects

Year	Discount factor (20%)	Project D		Project E		Project F	
		NCF	PV	NCF	PV	NCF	PV
		$	$	$	$	$	$
0	1.0000	(8500)	(8500)	(8500)	(8500)	(8500)	(8500)
1	0.8333	2000	1667	4000	3333	1000	833
2	0.6944	2000	1389	3000	2083	1000	694
3	0.5787	2000	1157	1000	579	1000	579
4	0.4823	4000	1929	1000	482	3000	1447
5	0.4019	1000	402	1000	402	4000	1608
6	0.3349	—	—	1000	335	1000	335
7	0.2791	—	—	1000	279	1000	279
8	0.2326	—	—	1000	233	1000	233
Cumulative NCF		2500	—	4500	—	4500	—
Net present value			(1956)		(774)		(2492)

Table 11.2   An Illustration of the Effect of the Timing of Cash Flows on Net Present Value

Year	Discount factor (16%)	Project D		Project E		Project F	
		NCF	PV	NCF	PV	NCF	PV
		$	$	$	$	$	$
0	1.0000	(200)	(200)	(200)	(200)	(200)	(200)
1	0.8621	70	60.35	100	86.21	40	34.48
2	0.7482	70	52.37	80	59.86	60	44.89
3	0.6407	70	44.85	60	38.44	80	51.26
4	0.5523	70	38.66	40	22.09	100	55.23
Cumulative NCF		80	—	80	—	80	—
Net present value		—	(3.77)	—	6.60	—	(14.14)

In table 4.12, simple average rate of return on capital was 10 per cent whether cash flows stayed constant, rose or fell over time. Table 11.2 looks at this example again using a discount factor of 16 per cent and shows that only project B is worthwhile at this discount rate. It clearly shows the advantage of high returns in the early years.

To calculate the net present value a discount rate must be assumed. This rate is important as different discount rates can give different rankings between projects. The rate used is usually the minimum acceptable return to the business after allowing for risk. This could be the greatest of the firm's marginal cost of borrowing fresh capital or the opportunity cost of capital — what it would earn in its most profitable alternative use either on or off the farm. It is thus necessary to assume what source of finance will be used and to set a cut-off rate below which any project giving a negative net present value will be rejected. In periods of high inflation, one could use the real interest rate (see p. 97) or the 'net interest' rate — the difference between the rate of interest and that of general inflation.

If the cost of capital is expected to vary, it is possible to use different discount rates over the life of the project and the cut-off rates chosen may be varied for different classes of fixed assets.

In using net present value to decide whether a project is acceptable, if the figure is negative at the discount rate chosen, the project is considered not worthwhile. If the present value is positive, the project is viable as it gives the required return and a bonus of the excess of net present value over zero. You should go ahead if this excess is great enough to compensate for the uncertainty and responsibility involved.

Net present value has the limitation that it fails as an index of profitability for different-sized projects. The lump-sum answers can be hard to evaluate and this complicates the problem of comparing different projects costing different amounts.

Internal rate of return

The measure of project performance most often used now is the internal rate of return (IRR). This is that discount rate that, when applied to the net cash flows results in a net present value of nil. There is an advantage in stating the outcomes in terms of percentage rates of return as farmers can compare these rates directly with other projects or investments. The net cash flows are considered as a recovery of the initial capital plus a rate of return on it after tax.

IRR is also called the DCF rate of return, the overdraft rate of return, the breakeven rate of return, the discounted yield and the yield. At the IRR the project would just break even, if funds

Table 11.3   Calculation of Internal Rate of Return

	Net cash flows	Discount factor 12%	Present value	Discount factor 13%	Present value
Y 0	(200)	1.000	(200.00)	1.000	(200.00)
Y 1	40	0.893	37.72	0.885	35.40
Y 2	60	0.797	47.82	0.753	46.98
Y 3	80	0.712	56.96	0.692	55.44
Y 4	100	0.636	63.60	0.613	61.30
	N.P.V.	—	+4.10	—	(0.88)

were borrowed at this rate, in its final year with no profit or loss. After paying interest, the rest of last year's income would just repay the remaining principal. It is the rate of compound interest earned by the investment during its life.

The actual value of internal rate of return is found by interpolation between those discount rates that give the smallest positive and negative net present values. Let us return to the table 11.2: projects A, B and C all had a simple rate of return of 10 per cent. However, the internal rates of return are 15.0 per cent, 17.8 per cent and 12.8 per cent respectively. These are found by trial and error, as shown in table 11.3 for project C.

Now if the NPV at 12 per cent is $4.10 and at 13 per cent is $(0.88), the internal rate of return (with a NPV of zero) must lie between 12 per cent and 13 per cent. It is found by interpolation, for example,

Discount rate	Net present value
12%	+4.10
13%	−0.88
Differences 1%	4.98

Internal rate of return $= 12 + (4.10/4.98 \times 1)\%$
$= 12.82\%$

This may be checked as shown in table 11.4.

Internal rate of return is a percentage figure that managers seem to find easier to understand than NPV. It needs no assumptions about uncertain future interest rates. Any risk-free project is likely to be worthwhile if its internal rate of return is higher than the objective rate of return which may be:

- the interest rate,
- the firm's cost of capital, or
- a minimum cut-off rate set by top management.

Internal rate of return has some limitations and anomalies. Its trial and error calculation is no problem as a good modern hand calculator can handle this. However, some patterns of cash flows have impossible or multiple solutions. It is also an over-simplification to assume a uniform return that is constant over time and a constant borrowing rate. If there is a negative cash flow in any year after the first, adjustment is necessary. However, few problems are likely to arise in practical farming situations.

PROJECT CHOICE

No method of project appraisal can relieve management of the need and responsibility for judgement and final choice, taking objectives and fads into account. Furthermore, before investing in new enterprises, it is always wise to check if the same income could be obtained by raising the technical performance of existing ones.

Investments can be mutually exclusive but, even if not, IRR and NPV can rank them differently if they differ much in size. In general, it seems safest to rely mainly on IRR, together with regard for both uncertainty and the pay-back period.

SENSITIVITY ANALYSIS

The main variables in an investment decision are usually the capital cost, the costs of production and the level of sales and prices. Sensitivity checks show the effect on the expected return of varying any one of these factors independently. While providing no overall measure of the level of uncertainty, sensitivity analysis shows the most critical areas of an investment project.

Sensitivity analysis or simulation is defined as the analysis of how errors in one or more estimates could affect the conclusions drawn from these estimates. It measures the responsiveness of

Table 11.4   Check on Internal Rate of Return for Project C

	Net cash flow	Interest at 12.82%	Principal repayment	Balance owed at year's end
Y 0	(200)	—	—	(200.00)
Y 1	40	25.64	14.36	(185.64)
Y 2	60	23.80	36.20	(149.44)
Y 3	80	19.16	60.84	(88.60)
Y 4	100	11.36	88.64	(+ 0.04)
	280.00	79.96	200.04	

a margin to changes in one or more of the elements comprising that margin. It can be applied to net cash flows, initial investment cost, cost of capital and so on. It is a process that features changing a planning coefficient within reasonable bounds of the original estimate to determine if the original ranking of alternatives is affected. It provides a way of answering 'What would happen if ...?' type of questions. When chancy (stochastic) processes are involved, simulation is sometimes referred to as the Monte Carlo method.

Simulation is an extremely versatile management tool. Indeed, it has been said that *any* process that can be fully described in words can also be simulated. The most valuable perception that a manager gains is a recognition of the most sensitive influences on the result. Once the system is understood well enough for the important factors to be revealed, an adequate solution is usually possible.

Sensitivity analysis, using a spreadsheet, demonstrates computer effectiveness most clearly. Having input the data, a suitable programme can get the computer to print out the whole year's cash-flow budget, including the monthly interest charges that it has calculated. But *what if* a machine replacement were to be delayed and a cash crop sale brought forward? All that is needed is to input the two changes and the computation is re-done to show the effect, printing out the revised budget if required.

So far I have expressed cash flows on the basis of a single set of the most likely figures and then calculated the desirability of the project using this single set of figures. However, each annual cash flow is a forecast and it is unlikely to be fully accurate in practice because we can never be sure of future costs or prices or of the length of life of a project.

By quantifying uncertainties to identify those key factors that, if they vary, will have a marked effect on profit, those key factors can then be given extra attention either to improve the accuracy of the estimate or to reduce its uncertainty.

Let us look at an example of sensitivity analysis. Suppose we are considering setting up a banana plantation and we have the following data with which to work:

- Initial capital investment — $300,000 plus $500/ha planted.
- Annual fixed costs — $50,000.
- Variable costs/hectare/year —

	$
Labour	97.20
tractors/ machinery	111.10
fertilisers	41.65
sprays	83.30
packing	27.75
transport	55.55
water	69.40
sundries	13.87
Total	500.00

- Price — $30/tonne.
- Yields — 40 tonnes/ha in the first year, 60 tonnes/ha in the second year and 70 tonnes/ha afterwards.
- Expected life of the project — five years.
- Area to be planted — 180 hectares.

Table 11.5 shows the net present value of this project, discounted at 10 per cent in the conventional form using expected values.

However, we are uncertain of some of the data. The planted area could vary by 10 per cent according to the method of soil conservation adopted, poor rains could increase water use by about 10 per cent, a further rise in the price of oil could raise the cost of both fertiliser and fuel (one-third of tractor and machinery costs) and government intervention could raise labour wages. We are fairly sure of the yield levels but price depends on the world market. Furthermore, there is a possibility of nematode infestation reducing the effective life of the crop to four years. The results of sensitivity analysis are shown in table 11.6.

Table 11.6 clearly shows that the three key factors are area planted, life of the plantation and (because we are fairly sure of the yield level) price. This draws our attention to the need to maximise area planted, to increase the economic life of the crop by, for example, use of a nematicide and, if we think the world price may fall, to try to arrange a contract in advance.

Table 11.5 Net Present Value of Banana Project

	NCF
	$
Y 0	(390000)
Y 1	76000
Y 2	184000
Y 3	238000
Y 4	238000
Y 5	238000
NPV (10%)	320158

If a very sensitive element is identified, one should take steps to try to safeguard it, for example, by using insurance, by forward contracting, by various technical safeguards such as spraying, irrigation, seed dressing, by diversifying, by having slightly excess or multi-purpose resources. One should if necessary go back and modify the budget or cash flow accordingly. Notice that, like all forms of insurance the measures cost money. But they do tend to safeguard the outcome.

Mathematical probability is sometimes useful. This is possible results multiplied by the probabilities, usually subjective, of achieving them to produce weighted averages.

There are two ways of setting about the calculation. In the first, having ascribed probabilities to values, one merely totals their products to arrive at the best estimate. Table 11.7 demonstrates such a calculation for hypothetical sugar-cane yields for the first five crops of a planting. Crop yields are especially suited to probability analysis as they are variable by nature. Occasionally, past data are available, which will give an indication of the values to be expected and of their frequency distributions. If best estimates of all the doubtful variables are produced in this way and projected, one then proceeds to a calculation of the best estimate of IRR.

Three problems must be overcome. The first is persuading the technical experts that their quantitative opinions may be expressed in the form of probabilities rather than best estimates. Secondly, if using past results, one must ensure that the reasons for variations do not themselves invalidate the data. For example, sugar-cane yields depend, among other things, on weather, on the incidence of disease and on cultivars. On the face of it, all three factors seem to justify, even to demand, probability analysis. However, cane cultivars are selected mainly to raise yields and resistance to disease. The success of these aims may be measured statistically. Distributions of yields are particular to each cultivar; therefore the analyst must ensure that his data are for the one in which he is interested.

PROJECT CONTROL

In this chapter so far, I have concentrated on project planning. The detailed plan, however, should also serve as a cost control tool. The amount and cost of goods and services ordered should be monitored in the light of the planned amounts and costs and any variances should be investigated. When a project is under way and

Table 11.6   Sensitivity Analysis of the Banana Project (NPV (10%) − $)

| | | | | Cost/price/life changes | | | |
	Base case	10% yield/ price drop	10% more water	10% fertiliser price rise	10% labour cost rise	10% fuel cost rise	Life of four years
Area							
162 ha	238687	128798	234437	236138	232727	236445	108783
180 ha	320158	195719	315424	317316	313526	317657	172360
198 ha	401124	264640	395916	397498	393828	398372	235441

Table 11.7   Best Estimate of Sugar-cane Yields according to Mathematical Probability — an Example

| | Assumptions Probabilities | | | | | Calculation Products — tonnes/ha/crop | | | | |
Yield (t/ha/crop)	Plant cane	1	Ratoons 2	3	4	Plant cane	1	Ratoons 2	3	4
90										
95	0.1		0.1	0.2	0.2	9.5		9.5	19.0	19.0
100	0.2	0.1	0.2	0.3	0.3	20.0	10.0	20.0	30.0	30.0
105	0.4	0.2	0.3	0.2	0.3	42.0	21.0	31.5	21.0	31.5
110	0.2	0.3	0.3	0.2	0.2	22.0	33.0	33.0	22.0	22.0
115	0.1	0.3	0.1	0.1		1.5	34.5	11.5	11.5	
120		0.1					12.0			
	1.0	1.0	1.0	1.0	1.0	105.0	110.5	105.5	103.5	102.5

the inevitable deviations from plan occur, network techniques help managers to know the importance of these deviations and to take remedial action.

Network analysis

This consists of a group of similar techniques used both for planning and controlling projects. The planning aspect is sometimes called production scheduling.

To complete any large project, many activities must be separately completed. If each activity must be completed in turn before the next one can start, then total completion time is clearly the sum of the times of the component activities. Usually, however, several activities can be undertaken simultaneously or 'in parallel'. For example, a car can be refuelled while the windscreen is being wiped and the oil and water levels are checked. The complete service time, therefore, is less than that of all the component activities added together. Indeed, it is only the time of the longest activity.

Managers today face a great rise in the complexity of their work and because they are usually dealing with the future they also face uncertainty. Network techniques were designed specifically to deal with these two factors. They have survived and prospered because they all bring the same powerful discipline to bear on the task of planning and controlling projects. Their power lies in this basic discipline and not in any of the elaborate adornments some of them exhibit. They have been devised to enable projects to be finished sooner with the same or less resources. The technique of network analysis, once mastered, is remarkably simple.

Originally, the main use of network analysis was in planning and controlling construction programmes for large projects, like power stations or missile building, but it can be used whenever there are several dependent, individual tasks that must be ordered logically if the project is to be finished in the least time. It is often suggested that network analysis is worth using only for complex projects with many activities. This is false. It is also valuable for small projects. An auditor, for example, might cut the time to audit a client's books from eight weeks to three by drawing the network of activities on the back of an envelope.

There are several project network techniques; like critical path analysis, project evaluation and review technique (PERT) and analysis bar charting (ABC). They all require drawing a diagrammatic network showing the activities, the sequence in which they must be performed and the expected time each one will take. The network is made up of some combination of lines, arrows and circles or boxes produced by systematic reasoning to find the 'critical path', ways of reducing it and to use the critical path for control purposes.

The critical path is crucial because it consists of those activities that form the longest route through the whole network. If any activity on the critical path is delayed, then final completion time of the project will be delayed. Other paths that are not critical can be 'fitted in' around the critical one so as to reduce costs but the main attention of managers should be directed at activities on the critical path to reduce project time. It is possible for there to be more than one critical path through the network.

The procedures and depiction of the network vary with the technique used. The simplest and easiest method is analysis bar charting. This is fully described in Mulvaney's excellent short, simple book, the basics of which any manager should readily master in an evening.

The advantages of network analysis are that it:

- forces the application of logic and concentration to the project,
- is a visual, readily understood record of the logical ordering of activities,
- enables a realistic completion time to be estimated,
- identifies the decisive jobs on the critical path and therefore the priorities for control,
- schedules the earliest and latest possible dates for all activities,
- allows production control by exception. Differences between actual and forecast achievements become spurs for action.

Monitoring and evaluation

The phrase 'monitoring and evaluation' is now popular in most public sector agricultural and rural development projects. Despite the aura of modernity surrounding the activity and all the fashionable jargon it has generated, there is nothing new about it. Unfortunately, however, there seem to be as many definitions of M and E as there are workers in this 'new' discipline.

In my view, 'monitoring' means measuring what is happening and 'evaluation' means comparing what is happening with what it was intended should have happened. It follows therefore, that you can have monitoring without evaluation but the latter is impossible without monitoring.

The provision of information by monitoring based on feedback, and the use of that information by project managers plays a key role in keeping implementation on schedule in both physical and budgetary terms. Feedback is obtained from farmers and others to assess the results of past project activities.

On-going and ex-*post* evaluation is aimed to assess whether or not the project is or has produced its expected benefits and which factors have contributed to failure or success. It may lead to a revision of objectives.

RECOMMENDED READING
Books

Anderson, R.G., *Corporate Planning and Control* (Macdonald & Evans, 1975).

Anon — *Handbook on Monitory and Formulation of Agric. and Rural Development Projects* (Ind. Bank for Reconst. & Dev., 1981).

Barnard, C.S. and Nix, J.S., *Farm Planning and Control*, 2nd ed. (C.U.P., 1979).

Boland, R.G.A. and Oxtoby, R.M., *D.C.F. for Capital Investment Analysis* (Hodder & Stoughton, 1975).

Brech, E.F. (ed.), *The Principles and Practice of Management*, 3rd ed. (Longman, 1975).

Brown, J.L., and Howard, L.R., *Principles and Practice of Management Accounting*, 3rd ed. (Macdonald & Evans, 1975).

Harper, W.M., *Operational Research* (Macdonald & Evans, 1976).

Kharbanda, O.P., Stallworthy, E.A., and Williams, L.F., *Project Cost Control in Action* (Gower Publishing Co., 1980).

Lock, D., *Project Management* (Gower, 1970).

Lockyer, K.G., *Introduction to Critical Path Analysis*, 3rd ed. (Pitman 1969).

Merrett, A.J. and Sykes, A., *The Finance and Analysis of Capital Projects*, 2nd ed. (Longman, 1973).

Mulvaney, J., *Analysis Bar Charting*, 3rd ed. (John Mulvaney & Associates, 1978).

Murphy, J. and Sprey, L.H., *Monitoring and Evaluation of Agricultural Changes*. (int, Inst. of Land Recl & Improvement, Wargeningen, 1982).

Smith, P.J., *Agricultural Project Management; Monitoring and Control of Implementation* (Elsevier, 1984).

Stafford, C., (ed), *Project Cost Control using Networks* (Heinemann, 1980).

Sturrock, F., *Farm Accounting and Management*, 7th ed. (Pitman, 1982).

Taylor, W.J. and Watling, T.F., *Successful Project Management* (Business Books, 1970).

Villemain, C., *Use of Network Analysis in Project Implementation In FAO Agricultural Planning Studies 20: Near East Readings on Agric. Inv. Projects* (FAO, Rome, 1979).

Woodgate, H.S., *Planning by Network* (Business Books, 1977).

Periodicals

Elrick, J. and Bright, G., 'Inflation Proof Budgeting', *Farm Management* (1982), 4, 12.

Mitchell, W., 'Choosing Between Alternative Farm Commitments', *Farm Management* (1975) 2, 11.

Smith, P.J., 'Monitoring & Evaluation of Agricultural Development Projects: Definition & Methodology', *Agricultural Administration* (1985) 18, 107–20.

Sutherland, R.M., 'Costing of Capital in Partial Budgeting', *Farm Management* (1980) 4, 2.

Temple M., 'Spread sheet Programs for Farm Management Advisory Work, *Farm Management* (1985), 5, 12.

Williams, N.T., 'Appraising the Profitability and Feasibility of an Agricultural Investment under Inflation', *Occasional Paper No. 5, Wye College* (1981)

EXERCISE 11.1

Wasantha expects to receive the following amounts of income at the end of each of the next four years:

	$
1	400
2	260
3	540
4	635

How much is all this worth in terms of present value if his rate of time preference is 9 per cent, and these amounts are discounted at that rate?

EXERCISE 11.2

At a capital cost of 10 per cent calculate which you would prefer to receive in payment for a house:
(a) $5000 now.
(b) Annual rent of $500 for 15 years starting now, but with no payment at the end of the fifteenth year, the house being fit for demolition then and having no residual value.
(c) $7000 in three years' time.

EXERCISE 11.3

In the eight remaining years of his tenancy agreement, a farmer wishes to invest in and operate a mushroom production and canning enterprise. Existing buildings could be modified and equipped for $14,000. Annual output could be sold on a contract basis for $10,000 each year. Annual production costs would be $45,000. The initial capital investment would be financed with the farmer's own capital, which has an opportunity cost of 15 per cent, and the terminal value of the buildings and equipment would be $2500. Calculate the net present value of the project.

Appendix A   Mathematical tables

APPENDIX A.1   FUTURE VALUE OF $1 AT COMPOUND INTEREST

No. periods	4	5	6	7	8	9	10
			% Interest Per Period				
1	1.0400	1.0500	1.0600	1.0700	1.0800	1.0900	1.1000
2	1.0816	1.1025	1.1236	1.1149	1.1664	1.1881	1.2100
3	1.1249	1.1576	1.1910	1.2250	1.2597	1.2950	1.3310
4	1.1699	1.2155	1.2625	1.3108	1.3605	1.4112	1.4641
5	1.2167	1.2763	1.3382	1.4026	1.4693	1.5386	1.6105
6	1.2653	1.3401	1.4185	1.5007	1.5869	1.6771	1.7716
7	1.3159	1.4071	1.5036	1.6058	1.7138	1.8280	1.9487
8	1.3686	1.4775	1.5938	1.7182	1.8509	1.9926	2.1436
9	1.4233	1.5513	1.6895	1.8385	1.9990	2.1719	2.3579
10	1.4802	1.6289	1.7908	1.9672	2.1589	2.3674	2.5937
11	1.5394	1.7103	1.8983	2.1049	2.3316	2.5804	2.8531
12	1.6010	1.7059	2.0122	2.2522	2.5182	2.8127	3.1384
13	1.6651	1.8856	2.1329	2.4098	2.7196	3.0658	3.4523
14	1.7317	1.9799	2.2609	2.5785	2.9372	3.3417	3.7975
15	1.8009	2.0789	2.3966	2.7590	3.1722	3.6425	4.1772
16	1.8730	2.1829	2.544	2.9522	3.4259	3.9703	4.5950
17	1.9479	2.2920	2.6928	3.1588	3.7000	4.3276	5.0545
18	2.0258	2.4 66	2.8543	3.3799	3.9960	4.7171	5.5599
19	2.1068	2.5270	3.0256	3.6165	4.3157	5.1417	6.1159
20	2.1911	2.6533	3.2071	3.8697	4.6610	5.6044	6.7275
21	2.2788	2.7860	3.3996	4.1406	5.0338	6.1088	7.4002
22	2.3699	2.9253	3.6035	4.4304	5.4365	6.6586	8.1403
23	2.4647	3.0715	3.8197	4.7405	5.8715	7.2579	8.9543
24	2.5633	3.2251	4.0489	5.0724	6.3412	7.9111	9.8497
25	2.6658	3.3864	4.2919	5.4274	6.8485	8.6231	10.835
26	2.7725	3.5557	4.5494	5.8074	7.3964	9.3992	11.918
27	2.8834	3.7335	4.8223	6.2139	7.9881	10.245	13.110
28	2.9987	3.9201	5.1117	6.6488	8.6271	11.167	14.421
29	3.1186	4.1161	5.4184	7.1143	9.3173	12.172	15.863
30	3.2434	4.3219	5.7435	7.6123	10.063	13.267	17.449
35	3.9461	5.5160	7.6861	10.677	14.785	20.414	28.102
40	4.8010	7.0400	10.286	14.974	21.724	31.409	45.259
45	5.8412	8.9850	13.765	21.002	31.620	48.327	71.890
50	7.1067	11.467	18.420	29.457	46.902	74.358	117.39

APPENDIX A.1 *(continued)*

No. periods	12	14	16	18	20	25	30
				% Interest Per Period			
1	1.1200	1.1400	1.1600	1.1800	1.2000	1.2500	1.3000
2	1.2544	1.2996	1.3456	1.3924	1.4400	1.5625	1.6900
3	1.4049	1.4815	1.5609	1.6430	1.7280	1.9531	2.1970
4	1.5735	1.6890	1.8106	1.9388	2.0736	2.4414	2.8561
5	1.7623	1.9254	2.1003	2.2878	2.4883	3.0518	3.7129
6	1.9738	2.1950	2.4364	2.6996	2.9860	3.8147	4.8268
7	2.2107	2.5023	2.8262	3.1855	3.5832	4.7684	6.2749
8	2.4760	2.8526	3.2784	3.7589	4.2998	5.9605	8.1573
9	2.7731	3.2519	3.8030	4.4354	5.1598	7.4506	10.604
10	3.1058	3.7072	4.4114	5.2338	6.1917	9.3123	13.786
11	3.4785	4.2262	5.1173	6.1759	7.4301	11.642	17.922
12	3.8960	4.8179	5.9360	7.2876	8.9161	14.552	23.298
13	4.3635	5.4924	6.8858	8.5994	10.699	18.190	30.288
14	4.8871	6.2613	7.9875	10.147	12.839	22.737	39.374
15	5.4736	7.1379	9.2655	11.974	15.407	28.422	51.186
16	6.1304	8.1372	10.748	14.128	18.488	35.527	66.542
17	6.8660	9.2765	12.468	16.672	22.186	44.409	86.505
18	7.6900	10.575	14.462	19.673	26.623	55.511	112.46
19	8.6128	12.056	16.776	23.214	31.948	69.389	146.19
20	9.6463	13.744	19.461	27.393	38.338	86.726	190.05
21	10.804	15.668	22.574	32.324	46.005	108.42	247.07
22	12.100	17.861	26.186	38.142	55.206	135.53	321.18
23	13.552	20.362	30.376	45.008	66.247	169.41	417.54
24	15.179	23.212	35.236	53.109	79.497	211.76	542.80
25	17.000	26.462	40.874	62.669	95.396	264.70	705.64
26	19.040	30.167	47.414	73.949	114.48	330.87	917.33
27	21.325	34.390	55.000	87.260	137.37	413.59	1192.5
28	23.884	39.204	63.800	102.97	164.84	516.99	1550.3
29	26.750	44.693	74.008	121.50	197.81	646.23	2015.4
30	29.960	50.950	85.850	143.37	237.38	807.79	2620.0
35	52.800	98.100	180.31	328.00	590.67	2465.2	9729.9
40	93.051	188.88	378.72	750.38	1469.8	7523.2	36119
45	163.99	363.68	795.44	1716.7	3657.3	22959	134110
50	289.00	700.23	1670.7	3927.4	9100.4	70065	497930

Example: If a sum of $1000 is placed on deposit for 4 years at a rate of compound interest of 12%, it will amount to: $1000 × 1.5735 = $1573.50

APPENDIX A.2   PRESENT VALUE OF $1 RECEIVED IN ONE PAYMENT AT THE END OF INDIVIDUAL YEARS AFTER n YEARS (ANNUAL COMPOUNDING)

Year(s)	4	5	6	7	8	9	10
			% Discount Rate				
1	0.9615	0.9524	0.9434	0.9346	0.9259	0.9174	0.9091
2	0.9246	0.9070	0.8900	0.8734	0.8573	0.8417	0.8264
3	0.8890	0.8638	0.8396	0.8163	0.7938	0.7722	0.7513
4	0.8548	0.8227	0.7921	0.7629	0.7350	0.7084	0.6830
5	0.8219	0.7835	0.7473	0.7130	0.6806	0.6499	0.6209
6	0.7903	0.7462	0.7050	0.6663	0.6302	0.5963	0.5645
7	0.7599	0.7107	0.6651	0.6227	0.5835	0.5470	0.5132
8	0.7307	0.6768	0.6274	0.5820	0.5403	0.5019	0.4665
9	0.7026	0.6446	0.5919	0.5439	0.5002	0.4604	0.4241
10	0.6756	0.6139	0.5584	0.5083	0.4632	0.4224	0.3855
11	0.6496	0.5847	0.5268	0.4751	0.4289	0.3875	0.3505
12	0.6246	0.5568	0.4970	0.4440	0.3971	0.3555	0.3186
13	0.6006	0.5303	0.4688	0.4150	0.3677	0.3262	0.2897
14	0.5775	0.5051	0.4423	0.3878	0.3405	0.2992	0.2633
15	0.5553	0.4810	0.4173	0.3624	0.3152	0.2745	0.2394
16	0.5339	0.4581	0.3936	0.3387	0.2919	0.2519	0.2176
17	0.5134	0.4363	0.3714	0.3166	0.2703	0.2311	0.1978
18	0.4936	0.4155	0.3503	0.2959	0.2502	0.2120	0.1799
19	0.4746	0.3957	0.3305	0.2765	0.2317	0.1945	0.1635
20	0.4564	0.3760	0.3118	0.2584	0.2145	0.1784	0.1486
21	0.4388	0.3589	0.2942	0.2415	0.1978	0.1637	0.1351
22	0.4220	0.3418	0.2775	0.2257	0.1839	0.1502	0.1228
23	0.4057	0.3256	0.2618	0.2109	0.1703	0.1378	0.1117
24	0.3901	0.3101	0.2407	0.1971	0.1577	0.1264	0.1015
25	0.3751	0.2953	0.2330	0.1842	0.1460	0.1160	0.0923
26	0.3607	0.2812	0.2198	0.1722	0.1352	0.1064	0.0839
27	0.3468	0.2678	0.2074	0.1609	0.1252	0.0976	0.0763
28	0.335	0.2551	0.1956	0.1504	0.1159	0.9895	0.0693
29	0.3207	0.2429	0.1846	0.1406	0.1073	0.0822	0.0630
30	0.3083	0.2314	0.1741	0.1314	0.0994	0.0754	0.0573
35	0.2534	0.1813	0.1301	0.0937	0.0676	0.0490	0.0356
40	0.2083	0.1420	0.0972	0.0668	0.0460	0.0318	0.0221
45	0.1712	0.1113	0.0727	0.0476	0.0313	0.0207	0.0137
50	0.1407	0.0872	0.0543	0.0339	0.0213	0.0134	0.0085

Example: If interest is 9 per cent per year, a promise to pay $100 at the end of year 7 is now worth $100 × 0.5470 = $54.70.

APPENDIX A.2 *(continued)*

Year(s)	12	14	16	18	20	25	30
			% Discount Rate				
1	0.8929	0.8772	0.8621	0.8475	0.8333	0.8000	0.7692
2	0.7972	0.7695	0.7482	0.7182	0.6944	0.6400	0.5917
3	0.7118	0.6750	0.6407	0.6086	0.5787	0.5120	0.4552
4	0.6355	0.5921	0.5523	0.5158	0.4823	0.4096	0.3501
5	0.5674	0.5194	0.4761	0.4371	0.4019	0.3277	0.2693
6	0.5066	0.4556	0.4104	0.3704	0.3349	0.2621	0.2072
7	0.4523	0.3996	0.3538	0.3139	0.2791	0.2097	0.1594
8	0.4039	0.3506	0.3050	0.2660	0.2326	0.1678	0.1226
9	0.3606	0.3075	0.2630	0.2255	0.1938	0.1342	0.0943
10	0.3220	0.2696	0.2267	0.1911	0.1615	0.1074	0.0725
11	0.2875	0.2366	0.1954	0.1619	0.1346	0.0859	0.0558
12	0.2567	0.2076	0.1685	0.1372	0.1122	0.0687	0.0429
13	0.2292	0.1821	0.1452	0.1163	0.0935	0.0550	0.0330
14	0.2046	0.1597	0.1252	0.0985	0.0779	0.0440	0.0254
15	0.1827	0.1401	0.1070	0.0835	0.0649	0.0352	0.0195
16	0.1631	0.1229	0.0930	0.0708	0.0541	0.0281	0.0150
17	0.1456	0.1078	0.0802	0.0600	0.0451	0.0225	0.0116
18	0.1300	0.0946	0.0691	0.0508	0.0376	0.0180	0.0089
19	0.1161	0.0829	0.0596	0.0431	0.0313	0.0144	0.0068
20	0.1037	0.0728	0.0514	0.0365	0.0261	0.0115	0.0053
21	0.0926	0.0638	0.0443	0.0309	0.0217	0.0092	0.0040
22	0.0826	0.0560	0.0382	0.0262	0.0181	0.0074	0.0031
23	0.0738	0.0491	0.0329	0.0222	0.0151	0.0059	0.0024
24	0.0659	0.0431	0.0284	0.0188	0.0126	0.0047	0.0018
25	0.0588	0.0378	0.0245	0.0160	0.0105	0.0038	0.0014
26	0.0525	0.0311	0.0211	0.0135	0.0087	0.0030	0.0011
27	0.0469	0.0291	0.0182	0.0115	0.0073	0.0024	0.0008
28	0.0419	0.0255	0.0157	0.0097	0.0061	0.0019	0.0006
29	0.0374	0.0244	0.0135	0.0082	0.0051	0.0015	0.0005
30	0.0334	0.0196	0.0116	0.0070	0.0042	0.0012	0.0004
35	0.0189	0.0102	0.0055	0.0030	0.0017	0.0004	0.0001
40	0.0107	0.0053	0.0026	0.0013	0.0007	0.0001	0.0000
45	0.0061	0.0027	0.0013	0.0006	0.0003	0.0000	
50	0.0035	0.0014	0.0006	0.0003	0.0001		

APPENDIX A.3   PRESENT VALUE OF AN ANNUITY OF $1 PER YEAR FOR n YEARS

Year(s)	4	5	6	7	8	9	10
			% Discount Rate				
1	0.9615	0.9524	0.9434	0.9346	0.9259	0.9174	0.9091
2	1.8861	1.8594	1.8334	1.8080	1.7833	1.7591	1.7355
3	2.7751	2.7232	2.6730	2.6243	2.5771	2.5313	2.4869
4	3.6299	3.5460	3.4561	3.3872	3.3121	3.2397	3.1699
5	4.4518	4.3295	4.2124	4.1002	3.9927	3.8897	3.7908
6	5.2421	5.0757	4.9173	4.7665	4.6229	4.4859	4.3553
7	6.0021	5.7864	5.5824	5.3893	5.2064	5.0330	4.8684
8	6.7327	6.4632	6.2098	5.9713	5.7466	5.5348	5.3349
9	7.4353	7.1078	6.8017	6.5152	6.2469	5.9952	5.7590
10	8.1109	7.7217	7.3601	7.0236	6.7101	6.4177	6.1446
11	8.7605	8.3061	7.8869	7.4987	7.1390	6.8051	6.4951
12	9.3851	8.8632	8.3838	7.9427	7.5361	7.1607	6.8137
13	9.9856	9.3936	8.8527	8.3577	7.9038	7.4869	7.1034
14	10.563	9.8986	9.2950	8.7455	8.2442	7.7862	7.3667
15	11.118	10.380	9.7122	9.1079	8.5595	8.0607	7.6061
16	11.652	10.838	10.106	9.4466	8.8514	8.3126	7.8237
17	12.116	11.274	10.477	9.7632	9.1216	8.5436	8.0216
18	12.659	11.690	10.828	10.059	9.3719	8.7556	8.2014
19	13.134	12.085	11.158	10.336	9.6036	8.9501	8.3649
20	13.590	12.462	11.470	10.594	9.8181	9.1285	8.5136
21	14.029	12.821	11.764	10.836	10.017	9.2922	8.6487
22	14.451	13.163	12.046	11.061	10.201	9.4424	8.7715
23	14.857	13.489	12.305	11.272	10.371	9.5802	8.8832
24	15.247	13.799	12.550	11.469	10.529	9.7066	8.9847
25	15.622	14.094	12.783	11.654	10.675	9.8226	9.0770
26	15.983	14.375	13.003	11.828	10.810	9.9290	9.1609
27	16.330	14.643	13.210	11.987	10.935	10.027	9.2372
28	16.663	14.898	13.406	12.137	11.051	10.116	9.3066
29	16.984	15.141	13.591	12.278	11.158	10.198	9.3696
30	17.292	15.372	13.765	12.409	11.258	10.274	9.4269
35	18.665	16.374	14.498	12.948	11.655	10.567	9.6442
40	19.793	17.159	15.046	13.332	11.925	10.757	9.7791
45	20.720	17.774	15.456	13.606	12.108	10.881	9.8628
50	21.482	18.256	15.762	13.801	12.234	10.962	9.9148

Example: If interest is 9%, a promise to pay $1 each year for 7 years is now worth $5.0330.

APPENDIX A.3 *(continued)*

Year(s)	12	14	16	18	20	25	30
			% Discount Rate				
1	0.8929	0.8772	0.8621	0.8475	0.8333	0.8000	0.7692
2	1.6901	1.6467	1.6052	1.5656	1.5278	1.4400	1.3609
3	2.4018	2.3216	2.2459	2.1743	2.1065	1.9520	1.8161
4	3.0373	2.9137	2.7982	2.6901	2.5887	2.3616	2.1662
5	3.6048	3.4331	3.2743	3.1272	2.9906	2.6893	2.4356
6	4.114	3.887	3.6847	3.4976	3.3255	2.9514	2.6427
7	4.5638	4.2883	4.0386	3.8115	3.6046	3.1611	2.8021
8	4.9676	4.6289	4.3436	4.0776	3.8372	3.3289	2.9247
9	5.3282	4.9464	4.6065	4.3030	4.0310	3.4631	3.0190
10	5.5602	5.2161	4.8332	4.4941	4.1925	3.5705	3.0915
11	5.9377	5.4527	5.0286	4.6560	4.3271	3.6564	3.1473
12	6.1944	5.6603	5.1971	4.7932	4.4392	3.7251	3.1903
13	6.4235	5.8424	5.3423	4.9095	4.5372	3.7801	3.2233
14	6.6282	6.0021	5.4675	5.0081	4.6106	3.8241	3.2487
15	5.8109	6.1422	5.5755	5.0916	4.6755	3.8593	3.2682
16	6.9740	6.2651	5.6685	5.1624	4.7296	3.8874	3.2832
17	7.1196	6.3729	5.7487	5.2223	4.7746	3.9099	3.2948
18	7.2497	6.4674	5.8178	5.2732	4.8122	3.9279	3.3037
19	7.3658	6.5504	5.8775	5.3162	4.8435	3.9424	3.3105
20	7.4694	6.6231	5.9288	5.3527	4.8696	3.9539	3.3158
21	7.5620	6.6870	5.9731	5.3837	4.8914	3.9631	3.3198
22	7.6446	6.7429	6.0113	5.4099	4.9094	3.9705	3.3230
23	7.7184	6.7921	6.0442	5.4321	4.9245	3.9764	3.3253
24	7.7843	6.8351	6.0726	5.4509	4.9371	3.9811	3.3272
25	7.8431	6.8729	6.0971	5.4669	4.9476	3.9849	3.3286
26	7.8957	6.9061	6.1182	5.4804	4.9563	3.9879	3.3297
27	7.9426	6.9352	6.1364	5.4919	4.9636	3.9903	3.3305
28	7.9844	6.9607	6.1520	5.5016	4.9697	3.9923	3.3312
29	8.0218	6.9830	6.1656	5.5098	4.9747	3.9938	3.3316
30	8.0552	7.0027	6.1772	5.5168	4.9789	3.9950	3.3321
35	8.1755	7.0700	6.2153	5.5386	4.9915	3.9984	3.3330
40	8.2438	7.1050	6.2335	5.5482	4.9966	3.9995	3.3332
45	8.2825	7.1232	6.2421	5.5523	4.9986	3.9998	3.3333
50	8.3045	7.1327	6.2463	5.5541	4.9995	3.9999	3.3333

APPENDIX A.4   ANNUAL AMORTISATION CHARGE TO REPAY $1 WITH INTEREST OVER n YEARS (ANNUITY WHOSE PRESENT VALUE IS $1)

Year(s)	4	5	6	7	8	9	10
			% Interest per Year				
1	1.0400	1.0500	1.0600	1.0700	1.0800	1.0900	1.1000
2	0.5302	0.5378	0.5454	0.5531	0.5608	0.5685	0.5762
3	0.3603	0.3672	0.3741	0.3811	0.3880	0.3951	0.4021
4	0.2755	0.2820	0.2886	0.2952	0.3019	0.3087	0.3155
5	0.2246	0.2310	0.2374	0.2439	0.2505	0.2571	0.2638
6	0.1908	0.1970	0.2043	0.2098	0.2163	0.2229	0.2296
7	0.1666	0.1728	0.1791	0.1856	0.1921	0.1987	0.2054
8	0.1485	0.1547	0.1610	0.1675	0.1740	0.1807	0.1874
9	0.1345	0.1407	0.1470	0.1535	0.1601	0.1668	0.1736
10	0.1233	0.1295	0.1359	0.1424	0.1490	0.1558	0.1627
11	0.1141	0.1204	0.1268	0.1334	0.1401	0.1469	0.1540
12	0.1066	0.1128	0.1193	0.1259	0.1327	0.1397	0.1468
13	0.1001	0.1065	0.1130	0.1197	0.1265	0.1336	0.1408
14	0.0947	0.1010	0.1076	0.1143	0.1213	0.1284	0.1357
15	0.0899	0.0963	0.1030	0.1098	0.1168	0.1241	0.1315
16	0.0858	0.0923	0.0990	0.1059	0.1130	0.1203	0.1278
17	0.0822	0.0887	0.0954	0.1024	0.1096	0.1170	0.1267
18	0.0790	0.0855	0.0924	0.0994	0.1067	0.1142	0.1219
19	0.0761	0.0827	0.0896	0.0968	0.1041	0.1117	0.1195
20	0.0736	0.0802	0.0872	0.0944	0.1019	0.1095	0.1175
21	0.0713	0.0780	0.0850	0.0923	0.0998	0.1076	0.1156
22	0.0672	0.0760	0.0830	0.0904	0.0980	0.1059	0.1140
23	0.0673	0.0741	0.0813	0.0887	0.0964	0.1044	0.1126
24	0.0656	0.0725	0.0797	0.0872	0.0950	0.1030	0.1113
25	0.0640	0.0710	0.0782	0.0858	0.0937	0.1018	0.1102
26	0.0626	0.0696	0.0796	0.0846	0.0925	0.1072	0.1092
27	0.0612	0.0683	0.0757	0.0834	0.0914	0.0997	0.1083
28	0.0600	0.0671	0.0746	0.0824	0.0905	0.0989	0.1075
29	0.0589	0.0660	0.0736	0.0814	0.0896	0.0981	0.1067
30	0.0578	0.0651	0.0726	0.0806	0.0888	0.0973	0.1061
35	0.0536	0.0611	0.0690	0.0772	0.0858	0.0946	0.1037
40	0.0505	0.0583	0.0665	0.0750	0.0839	0.0930	0.1023
45	0.0483	0.0563	0.0647	0.0735	0.0826	0.0919	0.1014
50	0.0466	0.0548	0.0634	0.0725	0.0817	0.0912	0.1009

APPENDIX A.4 *(continued)*

Year(s)	12	14	16	18	20	25	30
				% Interest per Year			
1	1.1200	1.1400	1.1600	0.1800	1.2000	1.2500	1.3000
2	0.5917	0.6073	0.6230	0.6387	0.6545	0.6944	0.7348
3	0.4164	0.4307	0.4453	0.4599	0.4747	0.5123	0.5506
4	0.3292	0.3432	0.3574	0.3863	0.3863	0.4234	0.4616
5	0.2774	0.2913	0.3054	0.3198	0.3344	0.3718	0.4106
6	0.2432	0.2572	0.2714	0.2859	0.3007	0.3388	0.3784
7	0.2191	0.2332	0.2476	0.2624	0.2774	0.3163	0.3569
8	0.2013	0.2156	0.2302	0.2452	0.2606	0.3004	0.3419
9	0.1877	0.2022	0.2171	0.2324	0.2481	0.2888	0.3312
10	0.1770	0.1917	0.2069	0.2225	0.2385	0.2801	0.3235
11	0.1684	0.1834	0.1989	0.2148	0.2311	0.2735	0.3177
12	0.1614	0.1767	0.1924	0.2086	0.2253	0.2684	0.3135
13	0.1557	0.1712	0.1872	0.2037	0.2204	0.2645	0.3102
14	0.1509	0.1666	0.1829	0.1997	0.2169	0.2615	0.3078
15	0.1458	0.1628	0.1714	0.1964	0.2139	0.2591	0.3060
16	0.1434	0.1596	0.1764	0.1937	0.2114	0.2572	0.3046
17	0.1405	0.1569	0.1740	0.1915	0.2094	0.2558	0.3035
18	0.1379	0.1546	0.1719	0.1896	0.2078	0.2546	0.3027
19	0.1358	0.1527	0.1701	0.1881	0.2065	0.2537	0.3021
20	0.1339	0.1510	0.1687	0.1868	0.2054	0.2529	0.3016
21	0.1322	0.1495	0.1674	0.1857	0.2044	0.2523	0.3012
22	0.1308	0.1489	0.1664	0.1848	0.2037	0.2519	0.3009
23	0.1296	0.1472	0.1654	0.1841	0.2031	0.2515	0.3007
24	0.1285	0.1463	0.1647	0.1835	0.2025	0.2512	0.3006
25	0.1275	0.1455	0.1640	0.1829	0.2021	0.2509	0.3004
26	0.1266	0.1448	0.1634	0.1825	0.2018	0.2508	0.3003
27	0.1259	0.1442	0.1630	0.1821	0.2015	0.2506	0.3003
28	0.1252	0.1437	0.1625	0.1818	0.2012	0.2505	0.3002
29	0.1247	0.1432	0.1622	0.1815	0.2010	0.2504	0.3002
30	0.1241	0.1428	0.1619	0.1813	0.2008	0.2503	0.3001
35	0.1223	0.1414	0.1609	0.1806	0.2003	0.2501	0.3000
40	0.1213	0.1407	0.1604	0.1802	0.2001	0.2500	0.3000
45	0.1207	0.1404	0.1602	0.1801	0.2001	0.2500	0.3000
50	0.1204	0.1402	0.1601	0.1800	0.2000	0.2500	0.3000

Appendix B　Metric conversions

APPENDIX B　CONVERSION FROM METRIC (S1) TO TRADITIONAL UNITS

	Metric	UK	USA
Length	Kilometre (km)	0.6214 miles	
	Metre (m)	1.094 yard	
	Centimetre (cm)	0.3937 inches	
Area	Square kilometre (km^2) (100 ha)	247.1 acres	
	Hectare (ha) (10^4 m^2)	2.471 acres	
	Square metre (m^2)	1.196 sq. yds	
Mass	Tonne (t) (10^3 kg)	0.9842 long tons (2240 lb)	1.102 short tons (2000 lb)
	Quintal (q) (10^2 kg)	220.5 pounds	
	Kilogram (kg)	2.205 pounds	
	Gram (g)	0.03527 ounces	
Volume	Cubic metre (m^3) (10^3 l)	220.0 gallons	264.2 gallons
Capacity		35.31 cu. ft	
		1.308 cu. yds	
	Hectolitre (hl)	22.00 gallons	26.42 gallons
		2.750 bushels	2.838 bushels
	Litre (l)*	0.2200 gallons	0.2642 gallons
		1.760 pints	2.113 pints
		61.02 cu. in.	
	Mililitre (ml)	0.06102 cu. in.	
		0.03520 fl. oz	
Other	Kilogram/hectare (kg/ha)	0.8922 lb/acre	
	Kilograms/litre	10.02 lb/gallon	
	Cumec (m^3/5)	35.31 cusecs (ft^3/s)	
		13208 galls/min.	
	Litre/second (l/s)	13.21 galls/min.	
	Bar (b)	14.7 lb/in^2	
	Kilowatt (kW)	1.341 h.p.	
	Kilometre/hour (km/h)	0.6214 m.p.h.	

* The symbol for litre can be confused with the figure 'one' so it is common to write it in full.

Appendix C   Worked answers to exercises

Exercise 3.1

Kg fertiliser/ ha	Kg maize	50% response (Kg)	Value of response less its harvesting, etc., costs (a) ($56/tonne)	(b) ($66/tonne)
0	2714	—	—	—
200	3724	505	$28.28	$33.33
400	4556	416	$23.30	$27.45
600	5130	287	$16.07	$18.94

(a) The optimum, theoretical rate of fertilisation is where marginal cost equals marginal revenue. The marginal cost of 200 kg fertiliser is $28.00. The best fertilisation rate, therefore, is about 200 kg/ha unless the funds needed could be used more profitably elsewhere.
(b) The marginal cost of fertiliser is now $36.40/200 kg. The best rate is now less than 200 kg/ha, say, about 100 kg/ha.

Exercise 3.2

Fertiliser for 1.5 ha	Yield from 1.5 ha (bags)	Response (bags)	Marginal revenue at $3.85	Marginal revenue at $5.90	Total fertiliser at $7	Cost at $10
0	22.5	—	—	—	0	0
225	36.0	13.5	51.98	79.65	31.50	45.00
450	45.0	9.0	34.65	53.10	63.00	90.00
675	52.5	7.5	28.88	44.25	94.50	135.00
900	58.5	6.0	23.10	35.40	126.00	180.00

ADVICE
(1) Mr Seushi should try to store his maize for six months to get the higher price.
(2) If he stores it and fertiliser costs $7/bag, he should try to get credit and apply 600 kg/ha (the marginal cost of 225 kg fertiliser is $31.50); if the price goes up to $10/bag (marginal cost of 225 kg fertiliser is $45.00), he should apply for sufficient credit to apply about 400 kg/ha.

If he stores his maize and cannot get credit, he should use up all his $70 by applying 333 kg/ha fertiliser at $7/bag or 233 kg/ha at $10/bag.

(3) If he has to sell maize at harvest to raise cash, and fertiliser costs $7/bag, he should apply for credit to apply 375 kg/ha. Without credit, he should spend all his $70 by applying about 333 kg/ha. If the fertiliser price rises to $10/bag, he should apply about 175 kg/ha.

Exercise 3.3

Fixed costs/year		*Variable costs/km*	
	($)		(¢)
Annual depreciation	1183.33	Fuel	3.11
Licence and insurance	180.00	Tyres	0.95
Hire purchase	2880.00	Servicing	0.60
		Repairs and maintenance	0.95
	4243.33		5.61

You should accept the company's offer of 8¢/km as it covers the variable, running costs and helps to meet fixed costs.

	Without official business	*With official business*
	($)	($)
Fixed costs	4243.33	4243.33
Variable costs	1009.80	1211.76
Total costs	5253.13	5455.09
less income	—	288.00
	5253.13	5167.09
Gain	—	86.04

Exercise 3.4

Tractor no.	Age	Total depreciation	Total annual depreciation	Average hours/ year
		$	$	
1	1	600	600.00	10
2	4	3150	787.50	1000
3	3	3135	1045.00	1500
4	3	3675	1225.00	2000
5	1	2125	2125.00	2500

In the graph total, annual depreciation is plotted against level of activity, measured in average hours worked each year, and the points are connected by the straight line that seems to fit them best.

(a) The yearly obsolescence cost, which occurs whether tractors are used or not, seems to be about $400. If we take this figure, we can then work out the hourly wear and tear cost of each tractor.

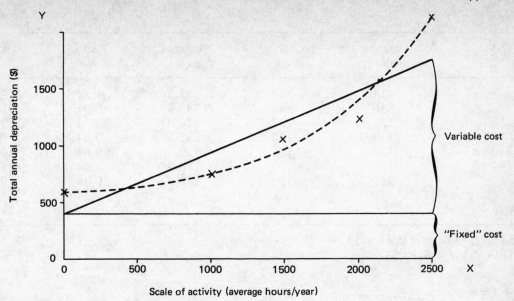

Scale of activity (average hours/year)

(b)

Tractor no.	Total depreciation	Obsolescence	Wear and tear	Total hours	Wear and tear cost/hr
	$	$	$		$
1	600	400	200	10	20.00
2	3150	1600	1550	4000	0.38
3	3135	1200	1935	4500	0.43
4	3675	1200	2475	6000	0.41
5	3125	400	2725	2500	1.09

(c) Clearly, obsolescence is 'fixed' and wear and tear is variable. However, it would seem from the graph that, in this case, it would be more realistic to fix obsolescence at about $600 each year and show wear and tear as rising hourly as annual use rises.

(The statistical method of linear regression gives the relationship as $Y = 387 + 0.549 X$.)

Exercise 3.5

(a)

	$	$
Sales	65550	
Closing valuation	197820	263370
Purchases	7625	
Opening valuation	195125	202750
Gross income		60 620

(b)

Type	Opening stock	Closing stock	Average stock	Factor	A.U.S
Bulls	14	13	13.5	1.27	17.14
Cows	350	345	347.5	0.91	316.22
Heifers 2–3 yrs	125	127	126.0	0.73	91.98
Heifers 1–2 yrs	130	137	133.5	0.54	72.09
Steers 2–3 yrs	120	125	122.5	0.82	100.45
Steers 1–2 yrs	127	110	118.5	0.64	75.84
Calves	255	300	277.5	0.36	99.90
Total AUs					773.62

(c) Total average head of stock = 1139
$$1139 \times 35¢ = \$398.65$$

Exercise 3.6

	Maize	Tobacco	Ground-nuts	Cotton
Gross income	281.05	385.00	241.72	312.00
Variable costs				
Seed	20.00	—	45.00	—
Fertiliser	238.00	110.00	—	70.00
Chemicals	3.50	—	—	40.00
Total variable costs	261.50	110.00	45.00	110.00
Gross margin	19.55	275.00	196.72	202.00

Total farm gross margin	693.27
Common costs	
Annual sprayer depreciation	12.50
Annual barn depreciation	40.00
Annual implement depreciation	12.50
Total common costs	65.00
Net farm income	628.27

Exercise 3.7

(a) *Farm results per hectare*

	Maize	Groundnuts	Cotton	Vegetables
Yield	20 bags	10 bags	1500 kg	2000 kg
	$	$	$	$
Price	6.30	21.00	0.42	0.28
Gross income	126.00	210.00	630.00	560.00
Seed	10.50	56.70	6.60	10.50
Fertiliser	14.00	—	7.00	21.00
Chemicals	—	—	14.00	42.00
Casual labour	2.80	44.80	52.50	—
Mech. unit	17.50	26.25	17.50	35.00
Transport	2.25	—	3.75	—
Total variable costs	47.05	127.75	101.35	108.50
Gross margin	78.95	82.25	528.65	451.50
Hectares	2.0	0.4	0.4	0.2
Gross margin, total	157.90	32.90	211.46	90.30

(Whole farm gross margin therefore, is $492.56)

(b) Technical efficiency of groundnut and cotton production seems high. With maize and vegetables, gross margin per hectare is low. This is mainly owing to low physical yields and, in the case of vegetables, a low price per kg. It seems worth examining whether it would pay to use more fertiliser and better seed of more productive cultivars in these enterprises.

With regard to enterprise combination, any increase in the area of cotton or vegetables at the expense of maize or groundnuts would raise whole farm gross margin.

(c) Substituting 0.35 ha of tobacco for 0.38 ha groundnuts, leaving 0.03 ha fallow, would raise whole farm gross margin by $143.75 without increasing working capital needs.

Owing to the limited capital available for barn construction, the tobacco area is limited to 0.35 ha. This would need $49 working capital and give a gross margin of $175. To release $49 of working capital (total variable costs), the farmer would have to forgo one of the following crop areas:

	Area needing $49 working capital	Gross margin of that area
Maize	1.04 ha	$ 82.10
Groundnuts	0.38	31.25
Cotton	0.48	253.75
Vegetables	0.45	203.17

Clearly, it would not pay to substitute tobacco for either cotton or vegetables. Substitution of tobacco for maize would raise whole farm gross margin by $92.90 ($175 − $82.10) and leave 0.69 ha (1.04 − 0.35 ha) fallow.

Exercise 4.1

See Appendix A.3.

(a) $2000 × 8.0607 = $16,121.40
(b) $1600 × 7.0236 = $11,237.76
(c) $3000 × 3.6048 = $10,814.40

Exercise 4.2

See Appendix A.1.

$1 set aside at the end of year 1 for five years will grow to $1.5386
$1 set aside at the end of year 2 for four years will grow to $1.4112
$1 set aside at the end of year 3 for three years will grow to $1.2950
$1 set aside at the end of year 4 for two years will grow to $1.1881
$1 set aside at the end of year 5 for one year will grow to $1.0900
  Therefore $1 set aside at the end of each year will grow to $6.5229 by the end of year 6.
  To accumulate $5500 by this time, you need to set aside $843.18 each year ($5500/6.5229).

Exercise 4.3

	Plant A $	Plant B $
Cost of plant	50 000	80 000
Estimated life (years)	6	8
Annual gains	35 000	46 000
Annual extra costs	7 000	10 000
Extra profit (before tax)	28 000	36 000
Tax	8 400	10 800
Extra profit (after tax)	19 600	25 200
Pay-back period (years):		
Before tax	1.8	2.2
After tax	2.6	3.2

The pay-back period for plant A is less than that for B so the former would be preferred.
  However, the simple pay-back period considers only the years during which the plant is paying for itself in savings, not the whole savings that accrue over its working life. This is often regarded as too incomplete a picture so a refinement can be introduced to show the extra savings; as follows:

	Plant A		Plant B	
	Before tax $	After tax $	Before tax $	After t $
Extra annual profit	28 000	19 600	36 000	25 20
Working life after pay-back (years)	4.2	3.4	5.8	4.
Total extra profit after pay-back	117 600	66 640	208 800	120 96

In this example, if the pay-back period is used, the firm would buy plant A; but if profitability after the pay-back is considered, plant B would be better.

Exercise 4.4

Partial Budget for Introduction of Cotton Fertilisation — (1 hectare)

Losses		Gains	
New costs		*Net income*	
Fertiliser	$ 30.00	Extra cotton sales	$ 54.90
Boron	4.50		
Extra reaping cost	6.00		
Net gain	14.40		
	$ 54.90		$ 54.90

Fertilising his cotton is likely to raise Mr Ndhlovu's profit by $14.40 a hectare.

Exercise 4.5

Partial Budget for Substitution of Machine for Hand-shelling

Losses		Gains	
New costs		*Net income*	
Annual sheller repayments	$ 30.00	2340 kg whole nuts @ 20¢/kg	$ 468.00
Average annual interest	3.00	1260 kg damaged nuts @ 8.5 ¢/kg	107.10
Income lost		*Costs saved*	
3600 kg whole nuts @ 20¢/kg	720.00	Hand-shelling labour	75.00
		Net loss	102.90
	$ 753.00		$ 753.00

Use of a mechanical sheller is likely to reduce Mr Kanyemba's yearly profit by $102.90.

Exercise 4.6

Partial Budget for Getting Equipment on Loan

Losses		Gains	
New costs	$	*New income*	$
Annual depreciation	14.00	Extra maize income	47.50
Annual repairs	3.00	Extra groundnut income	18.00
Av. annual interest	3.73		
Extra feed	3.00		
Income lost		*Costs saved*	
Loss from contract carting	6.25	Labour for hand weeding	10.00
Net gain	45.52		
	75.50		75.50

(a) Getting the loan should raise Chief Ode's yearly profit by $45.52.
(b) Maximum loan payment in the first year is $28.93 and easily covered.

Exercise 4.7

Partial Budget to Estimate the Effect of Storing instead of Selling Maize at Harvest

Losses		Gains	
Income lost	$	New income	$
10 tonnes maize @ $3.85/91 kg bag	423.08	9.5 tonnes maize @ 7¢/kg	
New costs		Costs saved	
Annual bin depreciation	35.00	110 bags at 45¢	49.50
Annual bin cleaning	7.00	Transport	30.00
Annual fumigation	50.00		
Net gain	229.42		
	744.50		744.50

Exercise 4.8

Partial Budget for Introduction of Cotton Fertilisation — (1 hectare)

Losses		Gains	
Income lost		Net income	
6 ha maize @ $84	504.00	6 ha cotton @ $198	1188.00
New costs		Costs saved	
Annual machine depreciation	157.50	8 ha casual weeding labour @ $28	224.00
Annual running costs	42.00		
14 ha weeding @ $7	98.00		
14 ha herbicide @ $11.20	156.80		
Net gain	453.70		
	1412.00		1412.00

In this example, the gross margin of cotton before use of herbicide is $170/ha. After the innovation, it is expected to be $179.80 (170 + 28 − 7 − 11.20). This partial budget could therefore be set out, more simply, as follows:

Losses		Gains	
Income lost	$	New income	$
6 ha maize @ $84	504.00	14 ha cotton @ $179.80	2517.20
6 ha cotton @ $170	1360.00		
Net costs		Costs saved	
Annual machine depreciation	157.50		
Annual running costs	42.00		
Net gain	453.70		
	2517.20		2517.20

Exercise 4.9

Partial Budget for Substitution of Cane for Cotton, 100 ha/year

Losses		$	Gains		$
Income lost			*New income*		
Gross margin 100 ha cotton		62500	Gross margin 100 ha cane		75 000
Net costs			*Costs saved*		
Extra machine depreciation (1)		5000			
Extra repairs (2)		4310			
Extra fuel/oil (3)		1565			
Net gain		1625			
		75000			75000

		Tractor	Two trailers	Crane	Total
Capital	($)	5625	6250	16250	28125 (4)
Life (years)		5	10	5	—
Annual depreciation	($)	1125	625	3250	5000 (1)
Repairs		810	1000	2500	4310 (2)
Fuel/oil		625	—	940	1565 (3)

The expected change in trading profit is a rise of $1625.

The likely return on the extra investment is (1625 × 100)/28125 (4) = 5.78 %/year.

Exercise 4.10

Partial Budget for Substitution of a Mist Blower for a Sprayer — 40 ha

Losses		$	Gains		$
Income lost		—	*New income*		—
New costs			*Costs saved*		
Annual machine depreciation (1)		1145	Reduced chemical cost (2)		5214
			Reduced labour cost (3)		1512
			Reduced tractor running costs (4)		1452
Net gain		7003			
		8178			8178

(1) $(6000 − 275)/5 = $1145
(2) $(1.76 − 1.43) × 395 × 40 = $5214
(3) *Annual mist blower labour*   *Annual sprayer labour*
     $(40 × 5 × 2.40)/4 = $120 $(40 × 5 × 10.20)/1.25 = $1632
(4) *Mist blower days*   *Sprayer days*
     (40 × 5)/4 = 50   (40 × 5)/1.25 = 160
     110 days saved at $13.20/day = $142

Exercise 4.11

Partial Budget for the Thinning of a 40 ha Citrus Grove — Five Years

	Losses		Gains	
		$		$
Income lost			*New income*	
Gross income over five years without thinning (1)		1158916	Gross income over five years after thinning (2)	832312
New costs			*Costs forgone*	
Chemical spray cost after thinning (3)		21330	Chemical spray cost without thinning (4)	109020
Labour cost after thinning (5)		1450	Labour cost without thinning (6)	19033
Fertiliser cost after thinning (6)		11246	Fertiliser cost without thinning (8)	18281
Irrigation cost after thinning (9)		7505	Irrigation cost without thinning (10)	15010
Pruning cost after thinning (11)		4938	Pruning cost without thinning (12)	9857
Harvest cost after thinning (13)		27460	Harvest cost without thinning (14)	38236
Tree removal cost (15)		15010	*Net loss*	206088
		1247855		1247855

(1) Net realisation/tonne fruit =
$[0.65 \times (1.50 \times 1000)/15] + [0.20 \times (0.30 \times 1000/10] + [0.15 \times 20.63]$
= $65.00 + 6.00 + 3.09 = $74.09/tonne
Lugs/tree/five years before thinning =
$(55 \times 18 \times 395 \times 40 \times 74.09/1000 = $1158916

(2) Gross income over five years after thinning =
$(79 \times 18 \times 197.5 \times 40 \times 74.09)/1000 = $832312

(3) $(5 \times 90 \times 197.5 \times 40 \times 5 \times 0.0012) = $21330

(4) $(5 \times 230 \times 395 \times 40 \times 5 \times 0.0012) = $109020

(5) $(40 \times 5 \times 5 \times 5.80/4) = $1450

(6) $(40 \times 5 \times 5 \times 28.55/1.5) = $19033

(7–8)	*After thinning*	¢	*Without thinning*	¢
Year 1	12.5 + 9.4 + 4.96	26.86	12.5 + 9.4 + 2.48	24.38
Year 2	12.5 + 9.4 + 6.2 28.10		12.5 + 9.4 + 1.86	23.76
Year 3	12.5 + 9.4 + 6.2 + 0.62	28.72	12.5 + 9.4 + 1.24	23.14
Year 4	12.5 + 9.4 + 6.2 + 1.24	29.34	12.5 + 9.4 + 0.62	22.52
Year 5	12.5 + 9.4 + 6.2 + 1.24	29.34	12.5 + 9.4	21.90

Total cost/tree/five years	$1.4236		$1.1570

Total cost: $1.4236 \times 197.5 \times 40 = $11246 $1.157 \times 395 \times 40 = $18281

(9–10)   *After thinning*   *Without thinning*
$0.19 \times 19 \times 197.5 \times 40 \times 5 = $7505 $0.19 \times 395 \times 40 \times 5 = $15010

(11–12)   *After thinning*   *Without thinning*
$0.125 \times 197.5 \times 40 \times 5 = $4938 $0.125 \times 395 \times 40 \times 5 = $9875

(13–14)   *After thinning*   *Without thinning*
Total lugs/five years: 79   55
Cost ($)   79 \times 0.044 \times 197.5 \times 40 = $27460   55 \times 0.044 \times 395 \times 40 = $38236

(15) $1.9 \times 197.5 \times 40 = $15.010

Financially, it is best not to thin as the 44 per cent rise in yield of each tree does not compensate for halving the number of trees; even though all annual costs are reduced.

Exercise 4.12

Breakeven use = (3014 − 2189)/1.76 − 1.15) = 1352 hours/year.
   It does not seem to be worthwhile financially to get the new tractor unless use can be increased by, for example, working on contract for neighbours.

Exercise 4.13

Breakeven area = (7200 − 3600)/55.50 − 40.30) = 237 ha/year.
   It would not pay to replace the old machine with a new one but non-financial considerations may persuade Mr Kamanga to do so.

Exercise 4.14

(a) The cumulative depreciation is clearly the new price less the resale price for each year.

Year	1	2	3	4	5	6
Cumulative operating costs	1250	2675	4250	6150	8350	11100
Cumulative depreciation	1600	2950	4000	4800	5325	5600
Cumulative total cost	2850	5625	8250	10950	13675	16700
Average annual cost	2850	2812	2750	2738	2735	2783

Hence the lorry should be replaced between the fifth and sixth years.

(b) Using the holding cost concept, the situation is as follows:

Year	Purchase price	Cumulative operating costs	Total cost	Average annual cost
	$	$	$	$
1	8200	1250	9450	9450
2	8200	2675	10875	5438
3	8200	4250	12450	4150
4	8200	6150	14350	3588
5	8200	8350	16550	3310
6	8200	11100	19300	3217

Exercise 5.3

Profit and Loss Account for the Year Ended 30.9.86
Penga-Penga Estates Ltd

Income	$	$
Tobacco	27000	
Maize	13000	
Beef	7000	47000
Value of home consumption		630
Valuation appreciation:		
Plant and machinery	4313	
Cattle	2000	6313
Closing debtors	1400	
less Opening debtors	600	800
Expenses		
Seed and fertiliser	9000	
Pesticides	1500	
Wages	5500	
Repairs and maintenance	2300	
Overheads	2700	
Tractor	7500	
Feedstuffs	4500	33000
Valuation depreciation:		
Buildings	640	
Fencing	30	
Vehicles	1100	
Office equipment	50	
Stocks in hand	250	2070
Closing creditors	1400	
less Opening creditors	1200	200
Net profit before tax		19473

Exercise 5.6

	$
(a) Net worth at 30.9.86	26835
Net worth at 30.9.85	23042
Increase in net worth	3793

The increase in net worth during 1985—6 is attributed to the following factors:

	Rise in value of $	Fall in value of $
Fixed assets	6519	—
Stock	—	941
Debtors	1166	—
Bank	—	4974
Creditors	2023	—
Increase in net worth	—	3793
	9708	9708

(b)   Working capital at 30.9.85 (13488 − 5058) = $8430
      Working capital at 30.9.86 (11465 − 5761) = $5704

Reconciliation	$		$
Working capital at 30.9.85			8430
Add Extra cash introduced	3513		
Net profit	9554		
Annual depreciation	2192		15259
			23689
Less Cost of new fixed assets	8711		
Drawings	9274		17985
			5704

Exercise 5.7

Modified Trading Account to Include Creditors and Debtors

Income	$	$
Sales	123.00	
Valuation appreciation —		
livestock	24.60	
Home consumption	14.76	
Closing debtors	4.00	
less Opening debtors	21.00	145.36
Expenses		
Purchases	12.30	
Valuation depreciation —		
deadstock	45.92	
Closing creditors	6.15	
less Opening creditors	32.15	32.22
Net profit before tax		113.14

Exercise 5.8

(a)   *Summarised cash account for the year*

	$	$
Opening cash		165
Plus Rebate on rotten vegetables returned	88	
Sales	2475	2563
Less Loan interest	40	
Payments to suppliers	1815	
Vehicle running expenses	572	
Sundry expenses	312	
(Private drawings)	(187)	2552
Closing cash		176

N.B. The figure for private drawings is a residual one. In this case, as it is negative, Mr Yesufu must have placed $187 of personal cash at the disposal of his business.

(b)   *Profit and Loss account for the year*

		$	$
Income			
Sales		2475	
Rotten vegetables		88	
Valuation appreciation —			
vegetable stocks		88	2651
Expenses			
Payments to suppliers		1815	
Vehicle running expenses		572	
Sundry expenses		312	
Valuation depreciation —			
pick-up truck		114	
Interest		40	
Closing creditors	792		
Less Opening creditors	715	77	2930
Net profit before tax			(279)

He made a trading loss of $279.

(c)   *Balance sheet as at 31.12.86. Mr Yesufu*

	1985	1986
Current assets		
Cash in hand	165	176
Stocks	270	358
Total current assets	435	534
Current liabilities		
Creditors	385	462
Net current assets	50	72
Fixed assets		
Pick-up truck	572	458
Cash register	55	55
Total fixed assets	627	513
Total net assets	677	585
These net assets are financed by:		
Loan from Mr Edje	330	330
Net worth	347	255
	677	585

Total funds employed	
Opening net worth	347
Plus profit	(279)
Private funds injected	187
Closing net worth	$ 255

Exercuse 7.1

Managerial styles

To score the questionnaire:

* Find the item number in one of the two columns below.
* Score each item as follows:
     Positive items: SA–4; TA–3; TD–2; SD–1
     Negative items: SA–1; TA–2; TD–3; SD–4
* Total the scores for each area by adding all four item scores
* Plot the total score for each area on the managerial-style profile.

Area	Positive items	Negative items	Total score
Conformity: the extent to which a manager demands that people comply with rules, policies and procedures rather than do their job as they wish	14 17	2 7	_____
Responsibility: the extent to which a manager delegates responsibility for the job and the results to his staff	11 15	8 19	_____
Standards: the stress a manager puts on setting high performance standards	1 13	5 20	_____
Rewards: the extent to which a manager rewards excellent performance rather than seniority, friendship, etc., and instead of criticising errors	4 18	10 16	_____
Warmth and support: the extent to which a manager builds warm, trusting relationships with his staff instead of keeping his distance	3 9	6 12	_____

Managerial Style Profile

Conformity	Responsibility	Standards	Rewards	Warmth and support
—— 16	—— 16	—— 16	—— 16	—— 16
—— 15				
	—— 15	—— 15	—— 15	
				—— 15
—— 14		—— 14	—— 14	
				—— 14

High

Conformity	Responsibility	Standards	Rewards	Warmth and support
	—— 14	—— 13	—— 13	
—— 13				
—— 12	—— 13	—— 12	—— 12	—— 13
			—— 11	
—— 11	—— 12	—— 11		
—— 10	—— 11		—— 10	—— 12

Average

Conformity	Responsibility	Standards	Rewards	Warmth and support
—— 9	—— 10	—— 10		
			—— 9	—— 11
—— 8	—— 9	—— 9	—— 8	—— 10
—— 7	—— 8	—— 8	—— 7	—— 9
—— 6				—— 8

Low

Conformity	Responsibility	Standards	Rewards	Warmth and support
	—— 7	—— 7	—— 6	—— 7
—— 5				—— 6
		—— 6	—— 5	
	—— 6			—— 5
—— 4	—— 5	—— 5	—— 4	
	—— 4	—— 4		—— 4

Exercise 8.1

1. Breakeven Budget for the Introudction of 3 ha Hand-cultivated Cotton

Losses		Gains	
	$		$
Income lost		*Net income*	
	—	3 ha cotton	2550 × P
New cost		*Cost saved*	
Casual labour	96.00		
Seed, fertiliser and chemicals	114.00		
Harvesting, packing and transport	70.12		
	280.12		2.550P

Where P is the breakeven price of seed cotton in DOLLARS.
Therefore, the breakeven price in cents/kg = (280.12 × 100)/2550 = 10.99¢

2. Breakeven Budget for the Introduction of 3 ha Tractor-planted Cotton

Losses		Gains	
	$		$
Income lost		*New income*	
		3 ha cotton	3 × 0.1099 Y
New costs		*Cost saved*	
Tractor	120.00		
Seed, fertiliser and chemicals	114.00		
Harvesting, packing and			
transport	3 × Y × 0.0275		
	234 + 0.825 Y		0.3297 Y

Where Y is the total yield in kg/ha
  Now $0.3297 Y = 234 + 0.0825 Y$
Therefore, $0.2472 Y = 234$
  So $Y = 947$ kg/ha

3. Tractor hire is advisable if it is likely to raise yield/ha by over 97 kg.
4. The farmer should grow cotton if he has adequate resources and no more profitable alternative enterprise. The question is, should he clear the land manually or by tractor. This can be assessed by finding out by how much yield per hectare would have to rise to make clearing by tractor more profitable than clearing by hand, as in the following example:

	Hand	(Per hectare)	Tractor
Yield (kg/ha)	850		Y
Price (¢/kg)	18		18
Gross income ($/ha)	153.00		0.18 Y
Clearing ($/ha)	32.00		40
Seed, fertiliser and chemicals ($/ha)	38.00		38
Harvesting, packing and transport ($/ha)	23.28		0.0275 Y
Total variable cost	93.38		78 + 0.0275 Y
Gross margin/ha	59.62		59.62

To obtain the same gross margin per hectare by tractor clearing instead of manual clearing, $0.18 Y - (78 + 0.275 Y) = 59.62$ and $Y = 902$ kg/ha. In other words, if the price of seed cotton becomes 18¢/kg, it would be better to clear by tractor provided yield is expected to rise from 850 kg/ha to 902 kg/ha or more.

Exercise 8.2

Cotton gross margin/ha is $140.60. To break even, therefore, maize would need to produce a gross income of $280.60. This requires a yield of 40.7 bags/ha.

Exercise 8.3

1. 125 ha $(250000 + 625A = 70A \times 37.5)$
2. 83 ha $(166667 + 625A = 70A \times 37.5)$
3. 113 ha $(250000 + 417A = 70A \times 37.5)$
4. 75 ha $(166667 + 417A = 70A \times 37.5)$
5. 64.3 tonnes/ha $(250000 + (140 \times 625) = 140Y \times 37.5)$

Exercise 8.4

1. Partial Budget for the Addition of Silage to the Dairy Ration

Gains		Losses	
New costs	$	*New income*	
Silage variable costs (1)	4098	Extra milk (5)	6075
Machinery depreciation (2)	1681	Extra quality premium (6)	148
Machinery repair and maintenance (3)	630		
Feeding labour (4)	696	*Loss*	882
	7105		7105

(1) $(18 \times 300 \times 75 \times 170)/(1000 \times 16.8)$
(2) $(4500 \times 0.3) + (2000 \times 0.15) + (625 \times 0.05)$
(3) $430 + 200$
(4) (2×348)
(5) $(540 + 75 \times 0.15)$
(6) $(3000 \times 75 \times 0.066)/100$
   The introduction of silage is not financially worthwhile.
2. The breakdown number of cows (C) needed to make silage supplementation worthwhile is as follows: $(18 \times 300 \times C \times 170/1000 + 16.8) + 1681 + 630 + 696 = (540 \times C \times 0.15) + (3000 \times C \times 0.066/1000)$
   Therefore, $68850/Y + 3007 = 6075 + 148$
   So $28.34 \ C = 3007$
   and $C = 106$ cows.
3. The breakeven silage yield/ha if herd size remains at 75 can be calculated as follows:
   $(18 \times 300 \times 75 \times 170/1000 \ Y) + 3007 = 6075 + 148$
   Therefore, $68850/Y + 3007 = 6075 + 148$
   So $68850 \ Y = 3216$
   and $Y = 21.4$ tonnes/ha

Exercise 8.5

1. Present situation

Crop	Hectares	Yield (kg/ha)	Market price (¢/kg)	Gross income	Variable costs	Gross margin
				$	$	$
Tomatoes	6	13500	11	8910	4620	4290
Cabbage	4	8900	11	3916	600	3316
Carrots	4	5350	8	1712	1100	612
Beans	2	5350	22	2354	500	1854
Cauliflowers	2	6700	28	3752	1080	2672
Lettuce	2	5350	55	5885	2800	3085
Peas	2	2675	22	1177	750	427
Totals				27706	11450	16256
Less annual overhead costs						6250
Net farm income						*10006*

2. *Present return on capital:* $10006/95000 \times 100\% = 10.53\%$
3. *Variable costs/kg:*

Tomatoes	5.7037
Cabbage	1.6854
Carrots	5.1402
Beans	4.6729
Cauliflower	8.0597
Lettuce	26.1682
Peas	14.0187

4. New costs to be incurred

	Tractor	Irrigation	Total
	$	$	$
Capital	3125	1875	—
Depreciation	20%	15%	—
Annual depreciation	625	281	906
Driver	600	—	600
Annual electricity	—	300	300
	1225	581	1806

5. Breakeven Tender Price Calculation

New variable costs	$
14.00 tonnes tomatoes @ 5.7037¢/kg	798.52
9.00 tonnes cabbage @ 1.6854¢/kg	151.69
4.50 tonnes beans @ 4.6729¢/kg	210.28
3.50 tonnes cauliflower @ 8.0597¢/kg	282.06
2.75 tonnes lettuce @ 26.1683¢/kg	719.63
2.25 tonnes carrots @ 5.1402¢/kg	115.65
2.75 tonnes peas @ 14.0187¢/kg	385.51

38.75 t	Total	2663
	Extra annual machinery and equipment costs	1806
	Sub-total	4469
	Return on extra investment at 10.53%	471
		4940

6. Total mass to be tendered for is 38.75 tonnes.
   Therefore, the breakeven price is 494000/38750 = 12.75¢/kg.

Exercise 8.6

A. Returns to Potentially Scarce Resources

Activity	Resource	Land (GM/ha)	Working capital (GM/$ variable cost)
		$	$
Tobacco		325 (1)	1.00 (1)
Groundnuts		40 (3)	0.50 (2)
Maize		50 (2)	0.40 (3)

B.I. Plan 1. Enterprise Selected on Return to Land

Constraint	Availability	Tobacco	Groundnuts	Maize	Used	Balance
Land	12.0	5.0	—	5.0	10.0	2.0
Capital	2250	1625	—	625	2250	0
Tobacco land	5.0	5.0	—	—	5.0	0

Gross margin is $1875 but capital is limiting and both tobacco and groundnuts give a better return on capital than maize. The plan contains the maximum tobacco therefore substitute groundnuts for maize.

Plan 2

Contraints	Availability	Tobacco	Groundnuts	Maize	Used	Balance
Land	12.0	5.0	7.0	—	12.0	0
Capital	2250	1625	560	—	2185	65
Tobacco land	5.0	5.0	—	—	5.0	0

Gross margin is raised to $1905 now land is limiting so some substitution of maize for groundnuts should occur. The area of maize needed is $65/(125 - 8) = 1.44$ ha.

Plan 3. Optimal Plan

Constraint	Availability	Tobacco	Groundnuts	Maize	Used	Balance
Land	12.0	5.0	5.56	1.44	12.0	0
Capital	2250	1625	445	180	2250	0
Tobacco land	5.0	5.0	—	—	5.0	0

Gross margin is $1919.40 and both land and capital are fully used.
B.11. The optimum solution is simply: 5 ha tobacco and 7 ha maize with a gross margin of $1975.

Exercise 11.1

	NCF	Discount factor	PV
	$		$
71	400	0.9174	366.96
71	260	0.8417	218.84
73	540	0.7722	416.99
74	635	0.7084	449.83
	Net present value		1452.62

Exercise 11.2

Present values:
(a) $5000
(b) $500 × 7.6061 (Appendix A.3) = $3803.05
(c) $7000 × 0.7513 (Appendix A.2) = $5259.10
Alternative (c) has the highest present value.

Exercise 11.3

Net present value is $11.497.

Index

SOCIAL SCIENCE LIBRARY

Oxford University Library Services
Manor Road
Oxford OX1 3UQ

WITHDRAWN

Tel: (2)71093 (enquiries and renewals)
http://www.ssl.ox.ac.uk

This is a NORMAL LOAN item.

We will email you a reminder before this item is due.

Please see http://www.ssl.ox.ac.uk/lending.html
for details on:

- loan policies these are also displayed on the
  notice boards and in our library guide.

WITHDRAWN

- how to check when your books are due back.

- how to renew your books, including information
  on the maximum number of renewals.
  Items may be renewed if not reserved by
  another reader. Items must be renewed before
  the library closes on the due date.

- level of fines; fines are charged on overdue books.

Please note that this item may be recalled during Term.

WITHDRAWN